Classroom Companion: Economics

The Classroom Companion series in Economics features fundamental textbooks aimed at introducing students to the core concepts, empirical methods, theories and tools of the subject. The books offer a firm foundation for students preparing to move towards advanced learning. Each book follows a clear didactic structure and presents easy adoption opportunities for lecturers.

More information about this series at http://www.springer.com/series/16375

Peter Zweifel • Roland Eisen •
David L. Eckles

Insurance Economics

Second Edition

 Springer

Peter Zweifel
University of Zurich
Bad Bleiberg, Austria

Roland Eisen
Goethe University Frankfurt
Munich, Germany

David L. Eckles
University of Georgia
Athens, GA, USA

ISSN 2662-2882 ISSN 2662-2890 (electronic)
Classroom Companion: Economics
ISBN 978-3-030-80392-6 ISBN 978-3-030-80390-2 (eBook)
https://doi.org/10.1007/978-3-030-80390-2

Original German edition published with title "Versicherungsökonom"
1st edition: © Springer-Verlag Berlin Heidelberg 2012
2nd edition: © The Editor(s) (if applicable) and The Author(s), under exclusive license to Springer Nature Switzerland AG 2021

This work is subject to copyright. All rights are solely and exclusively licensed by the Publisher, whether the whole or part of the material is concerned, specifically the rights of translation, reprinting, reuse of illustrations, recitation, broadcasting, reproduction on microfilms or in any other physical way, and transmission or information storage and retrieval, electronic adaptation, computer software, or by similar or dissimilar methodology now known or hereafter developed.

The use of general descriptive names, registered names, trademarks, service marks, etc. in this publication does not imply, even in the absence of a specific statement, that such names are exempt from the relevant protective laws and regulations and therefore free for general use.

The publisher, the authors and the editors are safe to assume that the advice and information in this book are believed to be true and accurate at the date of publication. Neither the publisher nor the authors or the editors give a warranty, expressed or implied, with respect to the material contained herein or for any errors or omissions that may have been made. The publisher remains neutral with regard to jurisdictional claims in published maps and institutional affiliations.

This Springer imprint is published by the registered company Springer Nature Switzerland AG
The registered company address is: Gewerbestrasse 11, 6330 Cham, Switzerland

Foreword

While insurance can be traced back over millennia, it is only in the last half century that we have come to a comprehensive and deep understanding of this most vital, yet complex, economic institution. To really understand insurance takes a deep knowledge of the subtleties of risk and probability, of how rational (and not so rational) people behave when faced with risk, of how insurance companies can be structured to cope with risk, and of how governments can effectively intercede when insurance markets fail to deliver. Such a journey will take us to such interesting phenomena as "adverse selection" and "moral hazard"; it will expose us to modern financial theories such as asset pricing theory and option theory and, in doing so, will expose us to such exotic financial instruments as catastrophe bonds. It will take us deep into public policy and the welfare state and into the challenges of operating universal health insurance programs. And it will face us with the challenges of a world where new and unpredicted risks (many of which were revealed in 2007/9 financial crisis) are appearing and for which normal insurance mechanisms may not function. Such is the journey for which Peter Zweifel, Roland Eisen, and now, David Eckles will guide us.

It would be hazardous to try to pinpoint the first attempts to explain an economic theory of insurance. Adam Smith, predictably, had something sensible to say. With commendable parsimony, he captures in two sentences the essential ingredients for such a theory, risk aversion, diversification, and the need for capital.

"The trade of insurance gives great security to the fortunes of private people, and, by dividing among a great many that loss which would ruin an individual, makes it fall light and easy upon the whole society. In order to give this security, however, it is necessary that the insurers should have a very large capital. (Wealth of Nations, page 619)".

But Smith does not have too much more to offer on insurance.

While the actuarial processes for insurance have been in continuous development since Adam Smith's time, it really took till the second half of the twentieth century for a modern theory of insurance economics to emerge. I would suggest that the catalyst was Kenneth Arrow's 1963 paper in the American Economic Review which laid out a model of an optimal insurance contract between risk-averse consumers and an insurance company capable of diversification. From this seminal paper has sprung an ever-growing field of enquiry in which rational consumers and rational insurers come together in a mutually beneficial trade of risk. While this line of enquiry deepened our understanding of how people come to share risk in an insurance market, and the natural frictions that occur (particularly the conflicting incentives of the policyholders and insurers), there was growing dissatisfaction with a theory that ignored the quirkiness of actual behavior; in real life, people might not be quite so rational. How would insurance market work in a world of limited rationality?

At about the same time that Kenneth Arrow was describing how insurance might be explained from rational consumer behavior, the Norwegian actuary and economist, Karl Borch, was to produce another idea that was to have profound implications, not only for our understanding of insurance but also for the manner in which capital markets functioned. In a paper nominally about reinsurance, Borch laid out a complete theory of asset market pricing, which appeared shortly thereafter as the Capital Asset Pricing Model (and for which Bill Sharpe was awarded the Nobel Prize). This had fascinating implications for insurance, for it implied that publicly traded firms should not need insurance—their shareholders could diversify risk just as well as insurance companies. Thus, we needed a new theory for corporate insurance which would explain not only why firms bought insurance but also how insurance companies manage risk.

But modern financial theory has another set of fascinating implications for those interested in insurance. An insurance policy is simply a financial instrument—strictly speaking, it is of the class of instruments known as options. This insight itself may help us to understand, and to price, insurance policies in different ways. But it also reveals why the function of insurance (to transfer risk from one person to another) can also be achieved with other financial instruments. These include options and forward and future contracts in many simple and complex forms. Such instruments not only are now used routinely as alternatives to insurance by sophisticated risk managers but are also used by insurance companies to offload their own risk so that they can enhance the security they provide to their own policyholders. But such strategies are not for the faint of heart. Derivatives also have their dark side and have been at the heart of several financial crises, including the recession of 2007/9 (in this case, in the form of credit derivatives such as default credit swaps). Thus, rather like a surgeon's scalpel, derivatives can be used for good or bad and their treatment will demand some care and subtlety.

These historical illustrations reveal what a rich and complex phenomenon modern insurance is. And the delight of this book is that it addresses insurance in all its subtleties and richness. Any foundational book on modern insurance will need to prepare students well in the basic disciplines of probability, economics, and finance, and this is achieved admirably by Roland Eisen, Peter Zweifel, and David Eckles. After preparing the reader with a thorough grounding in risk and diversification, they introduce the theory of decision-making under risk which leads seamlessly to a model of insurance in which all can benefit by a pooling of risk. They guide us through the frictions that can hinder insurance markets when information is not available or shared. With appropriately dismal titles, moral hazard (the lack of care people often exercise when they are protected by insurance) and adverse selection (the tendency for insurance to be bought by those most at risk) are examined along with the clever strategies for their resolution. The financial theory of risk and insurance is dealt with in similar exquisite detail including, not only a rationale for corporate risk management but also detailed explanation of the financial and operational management of insurance companies, and of the use of reinsurance options and other financial instruments for hedging risk.

Given their intellectual background, it is not surprising that the authors go much further than simply explaining the economic and financial foundations of insurance. Insurance markets are, not surprisingly, quite heavily regulated and such regulation presumably should improve equity and efficiency. Having examined theories of regulation, and described existing and new regulatory initiatives (such as Solvency II), they review the evidence and show when regulation contributes to the common good and when it does not. In a similar vein, their treatment of Social Insurance goes far beyond a simple description of programs in place, to address the political–economic foundations for state insurance programs and to embark on a critical examination of such programs and the challenges these face. Throughout all, Peter Zweifel, Roland Eisen, and David Eckles are careful to blend the basic theoretical concepts, with real-life illustration and a dispassionate review of the evidence.

But the world is changing. And as it does, new risks are appearing which will create a demand for their management and will present new challenges to insurers. Certainly, cyber risk has emerged as a mega concern. Climate change is creating new risks of quite unknown magnitude and who will ever be quite so complacent about systemic risk after the recent financial crisis. No treatise can accurately anticipate these new risks. But by careful preparation, we can be prepared, not only to respond to these risks as they arise but also to structure our affairs so that we can be more robust in the face of unknowable shocks. As Louis Pasteur famously said, "chance favors only the prepared mind". This is a book of surprising substance and, in the changing world of risk, it will prepare us well.

Philadelphia
January 2021

Neil A. Doherty

Preface to the Second Edition

This second edition improves the text in several ways. Most importantly, several changes have been included to represent the evolution of risk-related research. A new chapter, Insurance Demand II: Nontraditional Approaches, provides a timely addition in view of recent developments in risk theory and insurance. Previous discussions of enterprise risk management, long-term care, insurance, adverse selection, and moral hazard have all been updated. In an effort to expand the global reach of the text, evidence and research from the U.S. and China have been added. Finally, ths second edition of *Insurance Economics* now includes a third author, David L. Eckles (University of Georgia). An accomplished researcher and author on his own, David contributed importantly to this text. In particular, he suggested and facilitated the new behavioral-based chapter. Throughout the book, he increased its familiarity to readers from the United States notably by discussing U. S. best accounting practice and solvency regulation. Finally, he was willing to deal with the nitty-gritty of updating statistical information and getting rid of (fortunately minor) errors.

We hope that *Insurance Economics* will receive the same favorable reception as its first edition almost a decade ago!

Bad Bleiberg, Austria	Peter Zweifel
Munich, Germany	Roland Eisen
Athens, GA, USA	David L. Eckles
April 2021	

Preface to the First Edition

This book is dedicated to the memory of Wolfgang Müller († 1993), who made the two authors join him for performing a critical review of insurance regulation as envisaged by the European Union at the time. The three of us soon noticed that there did not seem to exist a textbook on insurance economics we could refer to. It would have to be at the crossroads of insurance as one way to cope with risk and insurance as an industry with its own peculiarities. We therefore planned to write a textbook that would be quantitative enough to permit its readers to understand a concept such as "ex-ante moral hazard" while being accessible to (future) practitioners who also want to know "how insurance works". The two survivors finally realized this plan in 2000, when a first edition of the textbook was published in German. Its success suggested that it indeed filled a gap.

Since then, important works have appeared, notably Risk Management and Insurance by Harold W. Skipper and W. Jean Kwon (2007) and Economic and Financial Decisions under Risk by Louis Eeckhoudt, Christan Gollier, and Harris Schlesinger (2005). The first covers a very wide range of topics but in turn avoids all those mathematical formulations and graphical illustrations that are so heavily used in the pertinent scientific literature. The other goes to the other extreme by providing a great deal of advanced theory without ever saying anything about the insurance industry. There still seems to be a gap to be closed.

This volume is unique in at least three ways. First, it clearly distinguishes between the demand for insurance by individuals who lack other alternatives of risk diversification and by those who also have diversification possibilities through the capital market. Accordingly, the reader is made familiar not only with the conventional theory pioneered by Arrow and Borch but with the Capital Asset Pricing and the Option Pricing models developed by Doherty as well. Second, the supply of insurance is given due attention. Analysis of the decision-making problems facing the management of an insurance company are of importance not only to students of Business Administration but also to policy makers inside and outside government who are confronted with initiatives of deregulation and re-regulation and wish to predict the likely consequences of these initiatives. And third, the book devotes an entire chapter to social insurance, whose importance exceeds that of private insurance by far, at least in industrial countries. To this topic, economic analysis is brought to bear and tested against empirical evidence in precisely the same way as in the remainder of the book.

This text is designed for advanced MBA, MSc Fin, and MSc Econ and beginning PhD students. It can be taught as a two-semester course, with the first term devoted to Chaps. 1–5. Chapters 6–10 are recommended for the second term. Chapter 11 (on future challenges confronting insurance) provides not only "food for thought" but also many linkages with the body of the book.

The transition from the original to the present English version called for many modifications, hopefully also resulting in improvements. It required the continuous and concentrated effort of Philippe Widmer, our project coordinator. Without him, the book would not be in existence! We also owe a great deal to Susan Danuser, who transformed hours of voice recording into a raw text. At that point, our two formatting specialists took over. For weeks and months, Christian Elsasser and Alexander Ziegenbein homogenized text, drew graphs, and perfected the layout of tables. Finally, George Elias, Maurus Rischatsch, Johannes Schoder, Michèle Sennhauser, Maria Trottmann, Philippe Widmer, and Alexander Ziegenbein read parts of the manuscript. They went far beyond pointing out typos, suggesting many clarifications in exposition. To all of them, we would like to express our gratitude.

For the first edition, we are also grateful for the support of the Verein zur Förderung der Versicherungswissenschaft an der Johann-Wolfgang-Goethe Universität (Association for the Advancement of the Science of Insurance at Frankfurt University) and the American Risk and Insurance Association in the awarding of the Kulp-Wright prize.

Last but not least, there are several generations of students who were exposed to the original German and precursors of the present English text. We are very thankful for their queries and comments. And now we hope that a few more generations of readers will be stimulated by the arguments and insights of the pages that follow. At a minimum, they should not be bored; better still, they may at times even find pleasure reading!

Zurich and Munich

Peter Zweifel
Roland Eisen

Contents

1	**Introduction: Insurance and Its Economic Role**		1
	1.1	Basics and Definitions	1
	1.2	Risks and Their Development Over Time	4
	1.3	Macroeconomic Importance of Insurance	4
	1.4	Functions of Insurance	11
	1.5	Major Determinants of the Demand for Insurance	15
		1.5.1 The Effects of Wealth and Income	16
		1.5.2 The Effect of Price	18
	1.6	System Analysis and Organization of the Book	22
	Exercises		24
	References		24
2	**Risk: Measurement, Perception, and Management**		27
	2.1	Definition and Measurement of Risk	28
		2.1.1 Definition of Risk	28
		2.1.2 Measurement of Risk	29
	2.2	Subjective Perception of Risk, Risk Aversion, and the Risk Utility Function	34
		2.2.1 Risk Perception as a Subjective and Cultural Phenomenon	34
		2.2.2 Risk Aversion and the Risk Utility Function	38
	2.3	Willingness to Pay for Certainty, Risk Aversion, and Prudence	46
		2.3.1 Willingness to Pay for Certainty, Certainty Equivalent, and Risk Premium	46
		2.3.2 Risk Premium and Coefficients of Risk Aversion	49
		2.3.3 Prudence and Higher Order Derivatives of the Risk Utility Function	54
	2.4	Estimates of Risk Aversion	56
		2.4.1 Microeconomic Evidence	57
		2.4.2 Macroeconomic Evidence	58
	2.5	Instruments of Risk Management	61

	2.6	Effectiveness of Risk Management and Risk Policy Measures	63
	2.A	Appendix: Stochastic Dominance	69
		2.A.1 First-Degree Stochastic Dominance	69
		2.A.2 Second-Degree Stochastic Dominance	71
	Exercises	75
	References	75
3	**Insurance Demand I: Decisions Under Risk Without Diversification Possibilities**	79
	3.1	The Expected Utility Maximization Hypothesis	80
	3.2	Theory of Insurance Demand	84
		3.2.1 The Basic Model	84
		3.2.2 Insurance Demand in the Presence of Irreplaceable Assets	89
	3.3	Demand for Insurance Without Fair Premiums	93
		3.3.1 Optimal Degree of Coverage Without Fair Premiums	93
		3.3.2 Risk Aversion as a Determinant of Insurance Demand	97
		3.3.3 Premium Rate and Wealth as Determinants of Insurance Demand	98
		3.3.4 Pareto-Optimal Insurance Contracts	103
	3.4	Demand for Insurance with Multiple Risks	104
	3.5	Relation Between Insurance Demand and Prevention	109
	Exercises	113
	References	113
4	**Insurance Demand II: Nontraditional Approaches to Decisions Under Risk**	115
	4.1	Expected Utility Theory Revisited	116
		4.1.1 The Axiomatic Basis of Expected Utility Theory	116
		4.1.2 Other Weaknesses of Expected Utility Theory	120
	4.2	Prospect Theory	121
		4.2.1 The Editing Phase	122
		4.2.2 The Evaluation Phase	123
	4.3	Other Non-Traditional Approaches	125
		4.3.1 Rank Dependent Expected Utility	125
		4.3.2 Ambiguity Aversion	126
		4.3.3 Linear Partial Information	128
		4.3.4 Regret Theory	130
		4.3.5 Representative and Availability Heuristics	131
		4.3.6 Other Biases	132

	4.4	Empirical Evidence	133
		4.4.1 Prospect Theory and Rank Dependent Expected Utility	133
		4.4.2 Ambiguity Aversion and Regret	134
		4.4.3 Representative Heuristic	136
	Exercises		137
	References		138
5	**Insurance Demand III: Decisions Under Risk with Diversification Possibilities**		**141**
	5.1	Risk Management and Diversification	142
		5.1.1 Risk Management and Internal Diversification	142
		5.1.2 Risk Diversification Through the Capital Market	148
		5.1.3 The Capital Asset Pricing Model (CAPM)	156
		5.1.4 Empirical Asset Pricing Model (EAPM) and Arbitrage Pricing Theory (APT)	162
	5.2	Risk Management, Forward Contracts, Futures, and Options	164
		5.2.1 Hedging Through Forward Contracts and Options	164
		5.2.2 Hedging Through Stock Options	167
	5.3	Corporate Demand for Insurance	173
		5.3.1 Demand for Insurance in the Light of Capital Market Theory	173
		5.3.2 Reasons for Corporate Demand for Insurance Not Related to the Capital Market	177
		5.3.3 Empirical Studies of Corporate Demand for Insurance	179
	Exercises		182
	References		183
6	**The Insurance Company and Its Insurance Technology**		**185**
	6.1	Financial Statements of an Insurance Company	186
		6.1.1 The Balance Sheet	186
		6.1.2 The Income Statement	191
	6.2	Objectives of the IC	196
		6.2.1 Theoretical Considerations	196
		6.2.2 A Descriptive Study Concerning the Importance of IC Objectives	200
	6.3	Survey of Insurance Technology of an IC	203
		6.3.1 What Is the Output of an IC?	204
		6.3.2 Instruments of Insurance Technology	205
	6.4	Choice of Distribution Channel	206
		6.4.1 Main Distribution Channels for Insurance Products	206
		6.4.2 The Principal-Agent Relationship as the Underlying Problem	208

		6.4.3	A Comparison of the Cost of Distribution Channels Using U.S. Data	209
		6.4.4	A Study Relating Performance to Incentives	212
	6.5	Underwriting Policy		214
		6.5.1	Instruments of Underwriting Policy	214
		6.5.2	A Simple Model of Risk Selection	215
	6.6	Controlling Moral Hazard Effects		218
	6.7	Reinsurance		221
		6.7.1	Functions of Reinsurance	222
		6.7.2	Types of Reinsurance	223
		6.7.3	A Model of Demand for Reinsurance Based on Option Pricing Theory	226
		6.7.4	Empirical Testing of the Model	232
	6.8	Capital Investment Policy		235
	6.9	New Elements of Insurance Technology		241
		6.9.1	The Value at Risk Concept	241
		6.9.2	Copulas for Dealing with Tail Dependence	243
		6.9.3	Alternative Risk Transfer (ART) Through Capital Markets	245
	Exercises			249
	References			250
7	**The Supply of Insurance**			253
	7.1	Traditional Premium Calculation		254
		7.1.1	Claims Process and Loss Distribution	254
		7.1.2	Basics of Probability Theory and Insurer's Risk	263
		7.1.3	Premium Principles	267
	7.2	Financial Models of Insurance Pricing		270
		7.2.1	Portfolio Optimization by the IC	271
		7.2.2	Pricing According to the Insurance CAPM	272
		7.2.3	Pricing According to Option Pricing Theory	277
		7.2.4	Evidence on the Actual Behavior of the IC	282
	7.3	Economies of Scope		289
		7.3.1	Economies of Scope and Properties of the Cost Function	290
		7.3.2	Empirical Relevance of Economies of Scope	291
		7.3.3	Stochastic Economies of Scope	294
	7.4	Economies of Scale		295
		7.4.1	Definitional Issues	295
		7.4.2	Empirical Relevance of Economies of Scale in Life Insurance	298
		7.4.3	Empirical Relevance of Economies of Scale in Non-life Insurance	302

		7.4.4	Alternatives and Extensions	303
		7.4.5	Scale Economies and Size of Market	307
	Exercises			308
	References			312

8 Insurance Markets and Asymmetric Information 315
- 8.1 Asymmetric Information and Its Consequences 315
- 8.2 Moral Hazard 318
  - 8.2.1 Definition and Importance of Moral Hazard 318
  - 8.2.2 *Ex-Ante* Moral Hazard 320
  - 8.2.3 Market Equilibrium with *Ex-Ante* Moral Hazard 327
  - 8.2.4 Empirical Evidence on *Ex-Ante* Moral Hazard 331
  - 8.2.5 *Ex-Post* Moral Hazard in Short-Term Disability Insurance 334
  - 8.2.6 Relational Moral Hazard in Long-Term Care Insurance 341
- 8.3 Adverse Selection 352
  - 8.3.1 Adverse Selection in a Single-Period Framework 352
  - 8.3.2 Empirical Relevance of Adverse Selection 361
  - 8.3.3 Adverse Selection in a Multi-period Context 366
  - 8.3.4 Empirical Evidence Regarding the Experience-Rating Model 371
- 8.4 Adverse Selection and Moral Hazard in Combination 373
- Exercises 378
- References 379

9 Regulation of Insurance 383
- 9.1 Objectives and Types of Insurance Regulation 383
  - 9.1.1 Objectives of Insurance Regulation 383
  - 9.1.2 Avoidance of Negative Externalities 384
  - 9.1.3 Material Regulation 384
  - 9.1.4 Regulation Limited to Formal Requirements 388
  - 9.1.5 Historical Differences in Insurance Regulation Between Countries 390
- 9.2 Three Competing Theories of Regulation 391
  - 9.2.1 Public Interest Theory 391
  - 9.2.2 Capture Theory 392
  - 9.2.3 Market for Regulation Theory 392
  - 9.2.4 Empirically Testable Implications for Insurance 395
- 9.3 Effects of Insurance Regulation 396
  - 9.3.1 Evidence from the United States 397
  - 9.3.2 Risk-Based Capital as the U.S. Regulatory Response 403
  - 9.3.3 Evidence from Europe 404

	9.4	Recent Trends in Insurance Regulation	409
		9.4.1 The Financial Crisis of 2007–2009	409
		9.4.2 The Convergence of Banking and Insurance and Their Regulation	411
		9.4.3 Is there Systemic Risk in Insurance Markets?	412
		9.4.4 Characterization of Recent Regulatory Initiatives	413
	Exercises		416
	References		416
10	**Social Insurance**		419
	10.1	Importance of Social Insurance	420
	10.2	Why Social Insurance?	422
		10.2.1 Social Insurance as an Efficiency-Enhancing Institution	422
		10.2.2 Social Insurance as an Instrument Wielded by Political Decision-Makers	429
	10.3	Overview of the Branches of Social Insurance	433
	10.4	Requirements for Efficient Social Insurance	435
		10.4.1 Comparing the Efficiency of Provision for Old Age	435
		10.4.2 Efficiency Assessment from a Portfolio Theory Perspective	438
	10.5	Impacts of Social Insurance Beyond Insurance Markets	443
		10.5.1 Impacts of Provision for Old Age	444
		10.5.2 Impacts of Social Health Insurance	453
		10.5.3 Impacts of Unemployment Insurance	457
		10.5.4 Optimal Amount of Social Insurance	459
	Exercises		464
	References		464
11	**Challenges Confronting Insurance**		467
	11.1	Globalization of International Economic Relations	468
		11.1.1 Globalization and Corporate Insurance	468
		11.1.2 Globalization and Individual Insurance	469
	11.2	Changes in Science and Technology	470
		11.2.1 Genetic Information	471
		11.2.2 Advances in Information Technology	476
	11.3	Changes in Legal Norms	479
		11.3.1 Principal Elements of Insurance Contract Law	480
		11.3.2 Consequences of Deregulation	481
	11.4	Demographic Change	482
		11.4.1 Aging of Population	482
		11.4.2 Increasing Share of One-Person Households	484
	11.5	Final Remarks	485
	References		488

References 489
Author Index 513
Subject Index 519

Introduction: Insurance and Its Economic Role

1.1 Basics and Definitions

Uncertainty is at the heart of insurance. This is already manifested in our limited knowledge about (observable) past events. In the "real" world, all our activities depend on uncertain and unknown circumstances beyond the control of a single individual. Unambiguous, deterministic cause–effect relationships are replaced by ambiguity in the perception of the economic environment. With respect to the future, uncertainty looms even larger. However, it is possible to make forecasts about future events even with incomplete knowledge of past ones.

There are different *degrees of uncertainty*. (1) Uncertainty where the structure of the system and the cause–effect relationships are known. (2) Uncertainty with known probability distributions (uncertainty of the first degree, usually called risk). Individuals have a probabilistic but exact notion of the world. Probabilities can be based on objective evidence or on subjective judgment. (3) Uncertainty without the knowledge of probabilities. (4) Uncertainty arising out of game situations (uncertainty of the second degree), where individuals play against each other and choose from a set of feasible strategies. Note that the adversary also can be "Nature" rather than a human being, in which case the situation belongs to category No. 2. (5) Complete ignorance with regard to the set of feasible strategies and hence their probabilities of occurrence.

In formal decision theory, these five cases are commonly classified according to their information structure:

- Risk situation [case No. 2];
- Uncertainty in the narrow sense [case No. 3];
- Game situation [case No. 4].

This classification is based on the terminology introduced by Knight (1921). "Risk" is measurable and thus insurable uncertainty while "true uncertainty" is non-measurable and therefore cannot be insured [Knight (1921), 233]: "The practical difference between the two categories, risk and uncertainty, is that in the former the distribution of the outcome in a group of instances is known (either through calculation *a priori* or from statistics of past experience), while in the case of uncertainty this is not true, the reason being in general that it is impossible to form a group of instances, because the situation dealt with is in a high degree unique."

Admittedly, there are major reservations with respect to this terminology as well as to the above classification. For example, neither ignorance nor complete knowledge adequately describe the economic environment. If one possesses knowledge however vague, partial, and incomplete it may be, it would be inefficient not to use it.[1] Inasmuch partial knowledge is the rule, the degree of confidence in probabilistic statements becomes important. Since past observations allow only for relative frequencies as estimates of the unknown probabilities, it remains a subjective judgment whether such estimates are sufficiently solid. The probabilities entering individual decision processes are therefore subjective, blurring the distinction between uncertainty and risk. Basically, only two cases remain, viz., decisions under *uncertainty* and *game situations*.

Irrespective of this classification, a decision under uncertainty depends on the probabilities of possible events and on the evaluation of each event. In economic theory, this is summarized by the concept of expected utility (see Sect. 2.1).

Insurance is occasionally called the "business of uncertainty". On the one hand, insurance is only possible in the presence of uncertainty; on the other hand, insurance is supplied by firms who seek to make a profit out of this. Insurance is, however, not the only way to cope with risk or uncertainty; rather, there exists a variety of other measures, methods, and institutions individuals use to create, influence, transfer, and finally bear risks. These measures, methods, and institutions are usually summarized under the heading of *risk policy* or *risk management* (see Sect. 2.3).

Insurance is, however, of particular importance for risks with negative consequences (which corresponds to the colloquial usage of "risk"). Usually, risk is understood as the danger of incurring a loss. This danger can materialize in different ways, ranging from complete loss, impairment or reduction of value of an asset, or even to the loss of a limb or loss of life. Consider the example of a car accident. The driver is injured to a greater or lesser extent, possibly impairing his or her ability to work, which may also be true of a victim, and the car is possibly damaged. The driver thus suffers a multiple loss of assets [health, wealth, and wisdom (meaning skills and knowledge)]. Depending on the design of criminal and liability law, claims against the (guilty) car driver may also arise motivating the purchase of liability insurance.

---

[1] One approach to the use of partial knowledge is the theory of fuzzy sets [see, e.g., Negoita (1981)]. A more practical alternative is Linear Partial Information theory [see, e.g., Kofler and Zweifel (1988); see Sect. 4.3.3 for details].

But what is meant by "insurance"? The pertinent literature gives various *definitions* of insurance. Problems arise because the term originates from business practice. Therefore, there is the scientific task of coming up with a definition which comprises to the greatest extent possible what is meant by the colloquial use of "insurance" while being precise. Yet, a definition should impart a notion of insurance that will guide readers through this book. On the other hand, a vague and ambiguous concept cannot be characterized accurately and unequivocally. There is thus a fundamental dilemma. One can, however, turn the tables by introducing a nominal definition, i.e., one that does not describe any specific phenomenon in the real world [see Hempel (1965)]. In this sense, insurance can be said to be a means or a procedure that *reduces uncertainty* with respect to the future. The following definitions from the literature take this aspect of risk mitigation into account.

- Insurance is the exchange of an uncertain loss of unknown magnitude for a small and known loss (the premium) [translated from Hax (1964)].
- Insurance is the exchange of money now for money payable contingent on the occurrence of certain events [see Arrow (1965, 45)].

Admittedly, the first definition is on the one hand too broad since it encompasses all kinds of loss prevention activities while on the other hand, it is too narrow since it excludes mutual insurance. At least in some countries, the law obliges members of mutual insurance associations to pay additional contributions in case of a major loss, which contradicts the clause "small and known". The second definition is also problematic because it fails to distinguish insurance from games and lotteries. However, one could counter this criticism by adding the differentiation that games create uncertainty (through a bet, say), while insurance aims at reducing an already existing uncertainty.

A third definition pointing to the informational aspect of insurance was developed by Müller (1995). Here, the insurance product is defined to be "information guaranteeing the specific states of wealth of the insurance buyer" (Eisen et al. 1993, p. 15).[2] This guarantee improves the IB's imperfect information with respect to outcomes of their actions, "and thus improves the decision situation... of the insurance buyer enabling him/her to attain a less risky position. It is this risk-reducing information which constitutes the economic benefit of insurance" (Eisen et al. 1993, p. 14).

---

[2]For a critical assessment of this definition, see the discussion about how to define the output of an insurance company in Sect. 6.3.1.

1.2 Risks and Their Development Over Time

Individuals seek to protect themselves against irregular but probabilistic shocks impinging on their assets "health", "wealth", and "wisdom" by employing one or several tools of risk management such as saving or in particular, purchasing insurance. Therefore, the importance of insurance presumably increases with *growth in the value of these assets*. During the middle ages, merchant ships became increasingly valuable assets, and indeed the first insurance contracts were struck by Italian merchants who sought to transfer the risks faced by their ships sailing in the Mediterranean. The oldest still existing non-life insurer is the Hamburger Feuerkasse in Germany. It has been predominantly insuring residential buildings, business premises, and inventories against fire.

In step with the growth in general wealth, the concentration of assets has also increased, leading to so-called *catastrophic risks*. Examples are oil-tanker disasters, plane crashes, and the loss of a satellite, each involving several hundred million US Dollars.

Such catastrophes are either triggered by human failure (man-made disasters) or by nature (natural disasters). As shown in Table 1.1, disasters with the largest financial consequences fall into the category of natural disasters in 19 out of 20 cases. The remaining disaster is the horrible attack on the World Trade Center in New York on September 11, 2001. Note that the more recent disasters are also the more severe ones. This gives rise to the conjecture that increasingly, natural disasters are potentially man-made (perhaps caused by environmental pollution through CO_2 and global warming).

A breakdown of catastrophic losses by category is provided by Table 1.2. In the year 2017, natural disasters accounted for 74.3% of lives lost and 95.7% of insured losses. Conversely, man-made disasters are responsible for 25.7% of all victims but only 4.3% of insured losses.

▶ **Conclusion 1.1** Insurance is understood as a means of economic agents to reduce their uncertainty concerning the future. As an industry, it is confronted also with catastrophic risks whose severity tends to increase with time.

1.3 Macroeconomic Importance of Insurance

Insurance is a secondary branch of economic activity. Its effect is essentially indirect because it deals with consequences of economic activity that would occur if insurance did not exist. Insurance serves production and consumption, international and interpersonal trade, payment and credit transactions, as well as the conservation of existing and creation of new wealth. However, the insurance industry has developed differently across industrialized countries due to differences in regulatory regimes (see Chap. 9) and in the role and size of social insurance (see Chap. 10).

1.3 Macroeconomic Importance of Insurance

Table 1.1 The 20 largest insurance losses between 1970 and 2017

Insured losses[a]	in %[b]	Victims[c]	Date	Event[d]	Country
82,394		1,836	August 25, 2005	Hurricane Katrina	USA
38,128		18,451	March 3, 2011	Tohoku earthquake, tsunami	Japan
32,000		136	September 19, 2017	Hurricane Maria	USA, Puerto Rico, Caribbean
30,774		237	October 24, 2012	Hurricane Sandy	USA, Caribbean, Canada
30,000		126	September 6, 2017	Hurricane Irma	USA, Puerto Rico, Caribbean
30,000		89	August 25, 2017	Hurricane Harvey	USA
27,943		65	August 23, 1992	Hurricane Andrew	USA, Bahamas
25,991	4.1	2,982	September 9, 2001	*Terrorist attacks on WTC, and Pentagon*	USA
25,293	4.0	61	January 17, 1994	Northridge earthquake in Southern California	USA
23,051		193	September 6, 2008	Hurricane Ike	USA, Caribbean, Gulf of Mexico
19,070		185	February 22, 2011	Christchurch earthquake	New Zealand
11,740		123	July 15, 2012	Corn Belt drought	USA
10,244		36	August 11, 2004	Hurricane Charley	USA, Caribbean, Gulf of Mexico
10,159	6.5	51	September 27, 1991	Typhoon Mireille	Japan
9,038		78	September 15, 1989	Hurricane Hugo	USA, Caribbean
8,989		562	February 27, 2010	2010 Chile earthquake, tsunami	Chile

[a] Excluding liability damages (in mn. US$, 2017 prices)
[b] Share of non-life premiums of that year, 2009 prices
[c] Persons killed and missing
[d] Man-made disasters in italics
Source Swiss Re (2018a), 48

The insurance industry is part of the services sector. As such, it is influenced by two recent trends that are changing the role of services including insurance [see Giarini and Stahel (2000)]. First, the importance of production cost in total value added diminishes while the cost of activities that ensure the functioning of the productive system (transportation, information) keeps increasing. Second, with modern technology, the *vulnerability* of these systems increases. Vulnerability is the result of a paradoxical evolution: The more sophisticated a technology is, the narrower the range of tolerable error is because accidents and managerial failures have more severe consequences. This is, however, the domain of the insurer who deals with low prob-

Table 1.2 Summary of catastrophic losses in 2017

	Number	in %[a]	Victims	in %[a]	Insured loss[b]	in %[a]
A. All natural disasters	183	60.8	8,470	74.3	138,057	95.7
Flooding	55		3,515		2,144	
Storms	82		1,642		111,475	
Earthquakes	12		1,184		1,615	
Drought, bush fire	14		435		14,237	
Cold, frost	5		153		1,038	
Hail	8		0		7,549	
Other	7		1,541		0	
B. All man-made disasters	118	39.2	2,934	25.7	6,246	4.3
Fire, explosions	45	15.0	477	4.2	5,439	3.8
Industry, stock	14		73		1,845	
Crude oil, natural gas	15		36		3,056	
Hotel, department store, other	16		368		539	
Airborne, space travel	7	2.3	165	1.4	410	0.3
Crash	3		165		131	
Space	2		0		188	
Damage on ground	2		0		90	
Naval disasters	33	11.0	1,163	10.2	197	0.1
Freighter	2		22		75	
Liner	27		1,087		0	
Tankship, drilling platform, other	4		54		122	
Railway disasters	10	3.3	140	0.6	0	0.0
Mine disasters	2	0.6	64	1.2	0	0.0
Other major disasters	21	7.0	925	8.1	200	0.1
Social unrest	1		0		200	
Terrorism	13		731		0	
Others	7		194		0	
Total, all catastrophic losses	301	100	11,404	100	144,303	100.0

[a]Percentage share of category
[b]Persons killed and missing
[c]Excluding liability damage, in US$ mn.
Source Swiss Re (2018a), 27

1.3 Macroeconomic Importance of Insurance

Table 1.3 Number of insurance and reinsurance companies

Company Type	USA		UK		Japan	
	2010	2018	2010	2018	2010	2018
Life	902	715	185	69	47	41
Non-life	3,448	3,444	261	165	50	51
Life and Non-life	0	0	23	20	0	0
Reinsurers	259	235	34	18	n.a.	0
Total	4,609	4,442	503	272	n.a.	92

Source OECD (2019)

abilities but big consequences. Controlling this vulnerability calls for sophisticated insurance products.

Both trends point to an increasing importance of insurance as a means for risk management in a risky environment. Therefore, a well-functioning insurance industry should play a major part in economic development. This conjecture creates an interest in measuring the industry's relevance for an economy. There are several ways to do this.

1. The first indication of its importance is the *number of insurance companies* in a given country. There is a positive correlation between market size and the number of insurance companies (see Sect. 7.4). As illustrated in Table 1.3 for the case of the three most important insurance markets, the United States has by far the largest number of companies in life and non-life businesses compared to the UK and Japan. Part of the difference is caused by regulation. For instance, the combination of life and non-life insurance is only permitted in Austria. This is the likely reason for Austria having four times fewer life insurers than Switzerland, although both countries have a comparable population of roughly 8.5 million. Switzerland's specialty in turn is reinsurance (i.e., insurance of insurers; see Sect. 6.7), which requires great amounts of capital made available by the country's banking industry.
2. The importance of insurance can also be measured by the *number of employees* in the industry (see Table 1.4). Again, the United States dominates the other two countries with a workforce in excess of 1.2 percentage points of total employment.
3. The importance of insurance is even more pronounced if one considers *premium income*. This is the indicator preferred by the industry as it suggests importance (see Table 1.5). The gross premium income of US insurance companies (excluding reinsurance) amounted to US$ 2,632 bn. in 2018, of which life insurance accounted for approximately 35%. Premium income in the UK is seven times lower and with an almost mirror-image split between life and non-life businesses (72% and 28%, respectively). In Japan, approximately three-quarters of premium income is derived from life insurance. An important reason is that employment-related provision for old age is delegated to private insurance rather than provided by public insurance.

Table 1.4 Employment in insurance companies[a]

	USA	% of total employment	UK	% of total employment	Japan	% of total employment
1985	1,112	0.95	234	0.85	557	0.93
1990	1,463	1.20	263	0.97	541	0.87
1995	1,541	1.20	204	0.74	622	0.93
2000	n.a.		225	0.76	484	0.75
2005	n.a.		n.a.		431	0.68

[a] In thousands
Source OECD (2010b)

In taking an international perspective however, the U.S. insurance industry is of medium importance only. A popular indicator for such comparisons is *premium income relative to GDP*. As evidenced in Fig. 1.1, the U.S. value of around 7% is in the middle of the pack among the larger markets, with Asian countries such as Hong Kong, South Korea, and Taiwan leading the way. Consistent with this explanation, the United Kingdom, Ireland, South Africa, and South Korea also show large premium income compared to GDP. Non-life insurance is much less affected by regulation, which is also brought out by the fact that its ratio of premium income to GDP varies less across countries.

4. From an economic point of view, the preferable indicator is the industry's *contribution to GDP*. By way of contrast, the ratio of premium income (a turnover quantity) to GDP (a value-added concept) greatly exaggerates the importance of the insurance industry. Value added is the value of output less the value of intermediate inputs, i.e., the sum of labor income, capital income (from investments), and profits earned in an industry. For the UK, the GDP share is 1.5% as of 2017 (see Table 1.6). In Germany and France, this share is slightly lower.

The comparison with banks, the other important providers of financial services, is instructive. In Germany, their contribution to GDP is 75% higher than that of insurance. In the UK, it is more than double.

The discrepancy between intuition and large premium incomes on the one hand and the small contribution to GDP according to national accounts has prompted a lot of criticism. For instance, the German Insurance Association proposed to add loss payments to the contribution to GDP. This would put the industry's share in GDP to 7%.[3]

5. The theoretically correct measure of the industry's contribution to economic welfare is the *sum of consumer and producer surplus*. Consumer surplus is the difference between buyers' willingness to pay (as indicated by the demand function) and the price actually paid. In Fig. 1.2, it is represented by the area DEF. In the case of insurance, price does not refer to the total premium paid since part of pre-

---

[3] National Accounts include payments which are used to repair damages or are spent on consumption. However, these payments do not enter into the insurance account but into repair services. The proposed modification thus leads to double counting.

1.3 Macroeconomic Importance of Insurance

Table 1.5 Gross premium income of insurers (excluding reinsurance), in US$ mn

	USA		UK		Japan	
	2010	2018	2010	2018	2010	2018
Life	799,066	909,623	239,465	336,122	415,985	302,966
Non-Life	1,236,495	1,492,630	85,677	132,998	102,790	99,808
Total	2,035,562	2,632,284	325,141	469,120	518,775	402,773

Source OECD (2019)

Table 1.6 GDP share of insurance and banks, in percent

	UK		Germany		France	
	Insurance	Banks	Insurance	Banks	Insurance	Banks
2000	1.6	2.7	1.4	2.3	1.1	2.8
2007	1.8	3.4	1.5	2.1	1.3	2.4
2007	1.8	4.9	1.0	4.0	0.8	1.9
2012	1.6	4.3	0.9	2.6	0.5	2.1
2017	1.5	3.9	0.8	2.2	0.3	2.5

Note Values for 2007 changed between the 2010 and 2019 reports. We leave both for transparency.
Source Eurostat (2010, 2019)

mium income is used to pay the losses of the insured. Thus, part of the premium flows back to the consumer, at least in expectation (admittedly not in each individual case because some people never incur a loss while paying premiums). The true price of insurance therefore amounts to the excess of the premium over the expected loss. In the literature, it is common to use the ratio of premium volume over total indemnity payments as the indicator of the price at the company and industry levels.

The demand function EJ in Fig. 1.2 is decreasing in this price while the supply function is increasing in price (which must cover the marginal cost under perfect competition). Indemnities paid serve as an indicator of the quantity of insurance services. Assuming perfect competition, the equilibrium price is at D and the premium volume is given by the area $ODFH$. Since the supply function corresponds to (the increasing part of) the marginal cost curve, the area $COHF$ below the function reflects the total cost of input factors and intermediate goods. The difference between revenue (the area $ODFH$) and these costs corresponds to the *producer surplus DCF* (which is negative at low quantities of services but then turns positive). Adding consumer surplus DEF, one obtains total net gain from insurance transactions, amounting to ECF. Finally, note that value added is given by the area $DBGF$.

Whether the insurance industry appears in a more favorable light using this criterion rather than its share in GDP remains an open question. Other industries generate consumer and producer surplus as well, and their relative contribution to welfare might exceed that of the insurance industry. Furthermore, it is debatable whether the assumption of perfect competition is an acceptable representation of the insurance market. Many important insurance markets were highly regulated

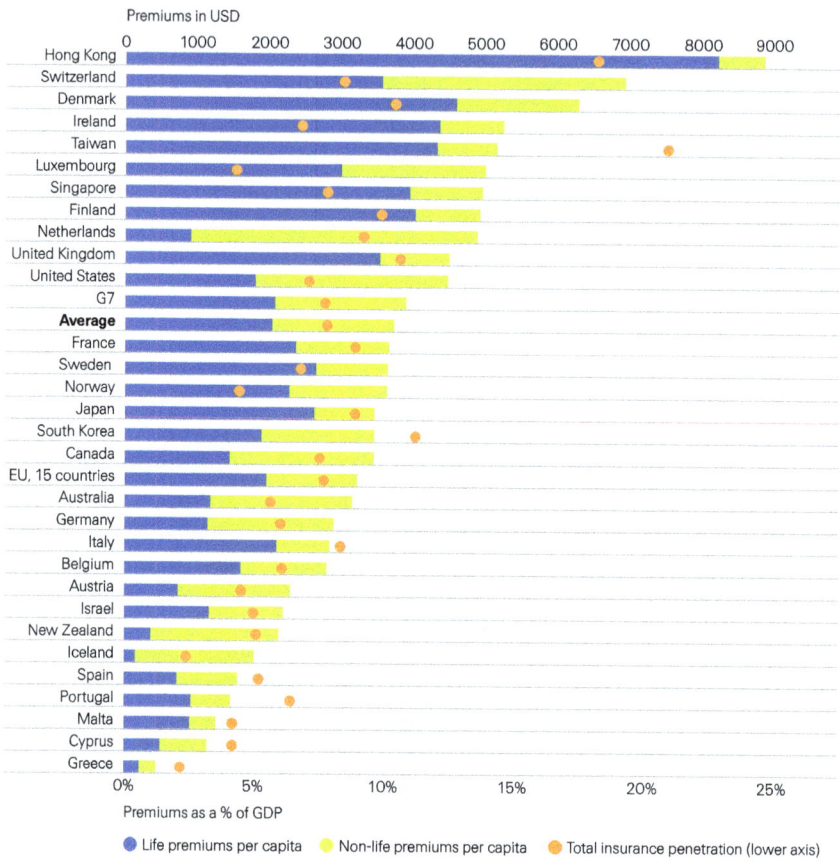

Fig. 1.1 Insurance premiums per capita and relative to GDP in 2018. *Source* Swiss Re (2019)

until recently (not only Germany, France, and Spain, but also Japan). By creating barriers to entry for new competitors, regulation undermines competition. In particular, it encourages companies to form a cartel because firms have a common interest in influencing the regulator. Therefore, suppose that a country's insurance companies form a cartel, i.e., a collective monopoly. Figure 1.3 illustrates a monopoly that seeks to equalize marginal revenue (*MR*) with marginal cost. This causes the quantity of insurance services transacted to decline from H under perfect competition to H' while price increases from D to D'. With the same demand and marginal cost functions as in Fig. 1.2, consumer surplus reduces to $D'EK'$, while producer surplus increases to $D'CI'K'$. However, note that there is a reduction of consumer surplus (area $I'FK'$) caused by monopolization that cannot be appropriated by producers because consumers simply withdraw from the market in response to the increased price. Therefore, if the degree of monopolization of the insurance industry exceeds the economy-wide average, its contribution to welfare falls short of its contribution to GDP. Finally, the effect of monopolization on fac-

1.3 Macroeconomic Importance of Insurance

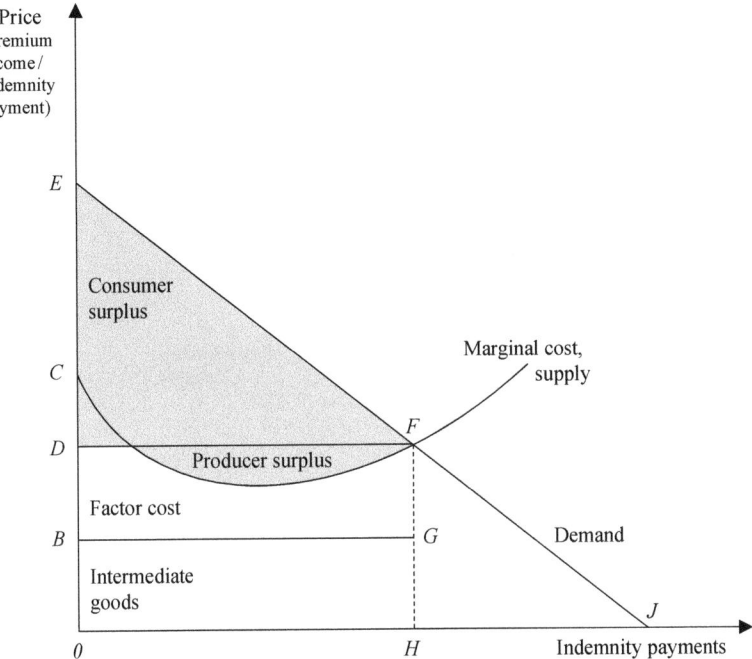

Fig. 1.2 Contribution of the insurance industry to welfare under perfect competition

tor cost is ambiguous because, on the one hand, the quantity of insurance services sold is reduced but, on the other hand, the increased price permits higher wages.

▶ **Conclusion 1.2** There are at least five alternative ways to assess the economic importance of the insurance industry, ranging from the number of insurance companies to their contribution to overall welfare. Using its contribution to GDP as a measure of importance, insurers account for at most 1.5% of GDP, substantially less than banks.

1.4 Functions of Insurance

The importance of insurance can also be inferred indirectly from its economic functions. In a modern economy, insurance fosters efficiency in at least six different ways, by

1. Improving risk allocation;
2. Protecting existing wealth;
3. Accumulating capital;
4. Mobilizing financial resources;
5. Providing governance;

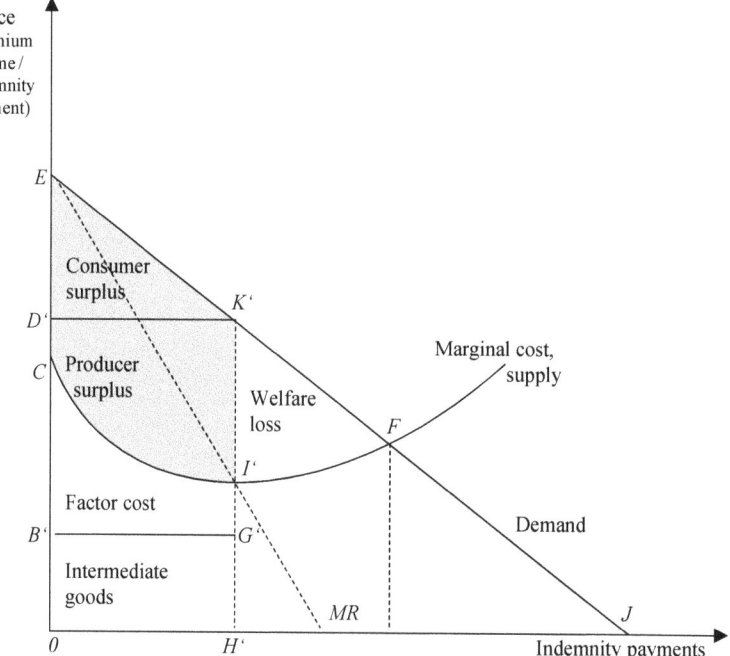

Fig. 1.3 Contribution of the insurance industry to welfare under monopoly conditions

6. Relieving the public purse.

These functions are explained in detail below.

1. *An efficient allocation of risk* minimizes transaction costs. Insurers limit losses by technical inspections and speedy settlement. Both measures increase the efficiency of the economy and contribute to its stability and growth.

   From an individual perspective, insurance reduces or even eliminates losses lurking in life's uncertainties. However, economically the loss persists since assets have been destroyed. Nonetheless, insurers alleviate losses in the following ways:

   - They reduce potential losses by inspection, auditing, and consulting. Here, insurers play the role of technical supervisors. They also undertake research into the causes of losses with the goal of limiting or even eliminating them.
   - They immediately provide the financial means for making up losses incurred, thus preventing further damage (including losses to a third party). For instance, when a storage building burns down, goods to be delivered may be damaged by bad weather unless the insurer covers the cost of building a makeshift replacement.

One might argue that due to moral hazard effects (see Chap. 8), risks might actually increase due to insurance. However, such effects are recognized by insurers, who only bear them in return for a higher premium. If purchasers continue to take out insurance, evidently they still must expect a net benefit from it. Therefore, the possibility of insuring risks increases expected net returns, incomes, and utility, such that despite an increase in risk, insurance contributes to welfare.

2. *By protecting existing wealth*, insurers provide economic agents with a more stable basis for their planning. Often stability is an essential prerequisite for undertaking risky but profitable ventures. For example, a small innovating firm may shy away from launching a new product unless it knows to be safe from the threat of product liability [see Zweifel (2009)]. According to Sinn (1988), fostering the willingness to undertake risky ventures is the main contribution of insurance to economic welfare. In addition, insurance also serves to level out the individual income stream by shifting income from the productive to the retirement phase of life. One may conjecture that this also results in a stabilization of macroeconomic variables such as consumption.

3. Insurance serves *capital accumulation*. Capital accrues naturally since premiums are paid at the beginning of the insured period while losses occur with a lag, which may amount to 20 years and more in the case of life (or some types of liability) insurance. Over the life of a contract, indemnities and insurers' administrative expenses are paid out of this fund, which usually is at least half of total premium income.[4]

In addition, there are the reserves for catastrophic events and the provisions for losses incurred but not (yet) reported (IBNR). These reserves are built up by premium loadings, i.e., markups on expected loss (see Sect. 7.1). The so-called fund generating factor (i.e., the ratio of funds to premium income) is often larger than 1 (roughly 1.5–2). These funds can be invested in the capital and money market, generating returns.

Capital funds are of particular importance in *life and health insurance*. So-called universal life insurance (which pays a capital benefit to subscribers who live up to a given age (e.g., 60 years) contains a savings component and can be viewed as a combination of insurance and precautionary savings. The savings markup and investment income are used to build up the insured capital during the life of the contract. Similarly, health insurance often is designed in a way that premiums do not increase with age. This requires an upfront loading, i.e., a savings component again.

The difference between life and non-life insurance is reflected in their respective capital stocks (see Table 1.7). Life insurance (including health insurance) is responsible for at least two-thirds of the capital stock.

Investments by insurers have considerable influence on the supply of capital

---

[4]Insurers sometimes erroneously claim to provide liquidity to the economy by freeing up cash that would otherwise be tied up as individual savings. However, insurers also need to hold part of their capital in liquid form. As Sinn (1988) points out, this is a case of fallacy of aggregation, i.e., on an incorrect inference from an individual to a macroeconomic phenomenon.

to the economy. Table 1.8 demonstrates that insurance companies act to a considerable degree as lenders (with more than one-half of their funds invested in various lending roles).[5] As a consequence, governments have a vested interest in *regulating* the capital flows controlled by insurance companies, frequently by mandating them to hold a minimum share in government bonds.

4. *Mobilization of capital* is another key function of insurance, except in a stationary economy where premium income from life insurance is paid out as benefits during a given year (after deduction of administrative expenses). As long as the capital stock used to provide for old age is being built up, however, saving takes place. Admittedly, individuals could generate these savings themselves rather than through private insurance; nevertheless, insurers mobilize additional capital, at least during the buildup. This is, however, not true of purely redistributive schemes, as are typical of public provision for old age (see Sect. 10.4).
5. Insurance *fosters governance*, i.e., management in the best interest of the stakeholders of the firm. Since premiums reflect risks, they put a cost on risk-taking by management, generating incentives to avoid excessive risks (see Sect. 5.3.1.3). Consequently, and similar to banks and capital markets, insurers monitor management, thereby encouraging the efficient use of an economy's resources.

    This monitoring function is of particular importance when it comes to the environment. Overuse of environmental resources is caused by their nature as a public good. The production of positively valued private goods inevitably generates negatively valued public goods or externalities (such as waste). Since there exist possibilities for input substitution, *internalization taxes*, better defined *property rights* as well as *technical monitoring* can help to mitigate environmental damage. With respect to the latter instrument, insurers provide a valuable service. When affected individuals or communities raise claims against polluters, there is usually a demand for insurance protection against such claims. To the extent that premiums reflect the underlying risk and thus respond to efforts to preserve the environment, insurance provides incentives to prevent or mitigate environmental damage to an optimal degree (see Sects. 6.6 and 8.2).
6. Finally, insurance provides *financial relief to the government* and ultimately, taxpayers. By purchasing insurance, individuals protect themselves against the risks of daily life, the consequences of which would have otherwise often be borne by the community. By the principle of solidarity or due to liability, the community cannot refuse to help its members in adversity. These might be citizens suffering a loss (ill health, accident, and unemployment) or being victims of damage inflicted by negligence (e.g., failure to operate an effective public fire department). Another example is flooding, where the government usually provides relief. One could conjecture that it is more efficient to introduce (mandatory, possibly subsi-

---

[5]In several countries, however, fixed-income instruments (mainly government bonds) take the lion's share.

1.4 Functions of Insurance

Table 1.7 Outstanding investment by line of insurance, in US$ bn

	US		UK		Japan	
	2000	2018	2000	2018	2000	2018
Life	3,800	4,162	1,878	7,287	3,167	3,429
Non-life	2,025	1,442	220	473	296	249
Total	5,466	6,187	2,098	7,760	3,463	3,678

Source OECD (2010b, 2020)

Table 1.8 Outstanding investment (life and non-life) by direct insurance companies, in US$ bn

	US		UK[a]		Japan[b]	
	2008	2015	2008	2015	2008	2015
Real estate	41	37	78	57	76	59
Mortgage loans	407	415	2	42	21[c]	10
Stocks, Equity	569	469	827	191	116[c]	273
Fixed-income (e.g., bonds)	3,733	3,739	715	704	155[c]	2,165
Other loans	146	127	39	16	570	296
Other investments	570	483	436	293	2,584	411
Total	5,466	5,271	2,098	1,304	3,463	3,214

[a] Exchange rate on 31 Dec. 2009: 1 GBP = 1.61 US$ (for 2008 values)
[b] Exchange rate on 31 Dec. 2009: 1 Yen = 0.011 US$ (for 2008 values)
[c] Values as of 2007
Source OECD (2010b, 2017)

dized) insurance against flooding rather than having the community make up for the losses.

▶ **Conclusion 1.3** Insurance contributes to economic efficiency and fosters economic growth in several ways. In particular, it allows individuals to venture into new and profitable businesses by protecting existing wealth.

1.5 Major Determinants of the Demand for Insurance

A systematic analysis of the factors determining a demand for insurance can be found in Chap. 3. For the moment, these determinants are discussed in general terms only. According to functions (1)–(4) of Sect. 1.4, demand for insurance mainly depends on assets in their three forms (health, wealth, and wisdom). Clearly, also the price of insurance (defined here as the ratio of premium paid to the indemnity provided) plays an important role as well (see Fig. 1.2).

1.5.1 The Effects of Wealth and Income

As shown in Table 1.9, the United States, the United Kingdom, and Japan are leading insurance markets. Taking the OECD countries as the reference, the United States accounts for more than 50% of gross premiums, followed by Japan and the UK with 9 and 8%, respectively, as of 2017. All three countries have a large capital stock. Population does not seem to play a major role. For example, comparing Switzerland to Austria (some 8.5 mn. each), one finds substantial variation. Premiums in Switzerland are more than double the premiums in Austria.

Since only a few countries publish reliable data on wealth, GDP is often used as a proxy. Figure 1.4 shows the partial relationship between gross life premiums and GDP for the most important insurance markets. The linear regression shown is in logarithmic values in order to estimate income elasticities.[6]

The estimated elasticity of premium income w.r.t. GDP of 1.22 suggests that the demand for life insurance increases more than proportionally with GDP, *ceteris paribus* (i.e., all other things held constant).

Figure 1.5 establishes a similar conclusion for non-life premiums. The estimated elasticity w.r.t. to GDP amounts to 1.08, which again exceeds 1.00. Therefore, economic growth is accompanied by a roughly proportional growth in the demand for

Table 1.9 World insurance market (total gross premiums, US$ bn.)

	2007		2017	
	US$ mn.	% OECD	US$ mn.	% OECD
Austria	19.7	0.50	19.7	0.47
Canada	112.1	2.83	68.9[a]	1.65
France	293.8	7.42	273.8	6.57
Germany	222.5	5.62	234.0	5.62
Italy	141.8	3.58	147.3	3.54
Japan	338.4	8.54	366.7	8.80
Switzerland	42.0	1.06	58.1	1.40
United Kingdom	611.6	15.44	337.7	8.11
United States	1,564.2	39.49	2,185.3	52.46
EU15	1,639.2	41.38	1,376.7	33.05
NAFTA	1,693.7	42.76	2,211.4[b]	53.09
OECD	3,961.4	100.00	4,165.7[c]	100.00

[a] Canadian values as of 2015
[b] Does not include Canada
[c] Does not include Canada and S. Korea
Source OECD iLibrary; Extracted 28 Mar 2019

---

[6] Note that from $lny = a + blnx$ one has $dlny = b \cdot dlnx$. Since $dlnx\, dx = 1/x$, the percentage changes are given by $dlnx = dx/x$ and $dlny = dy/y$, respectively. Therefore, $b = \frac{dy}{y} / \frac{dx}{x}$, i.e., the regression coefficient b corresponds to the income elasticity of premiums.

1.5 Major Determinants of the Demand for Insurance

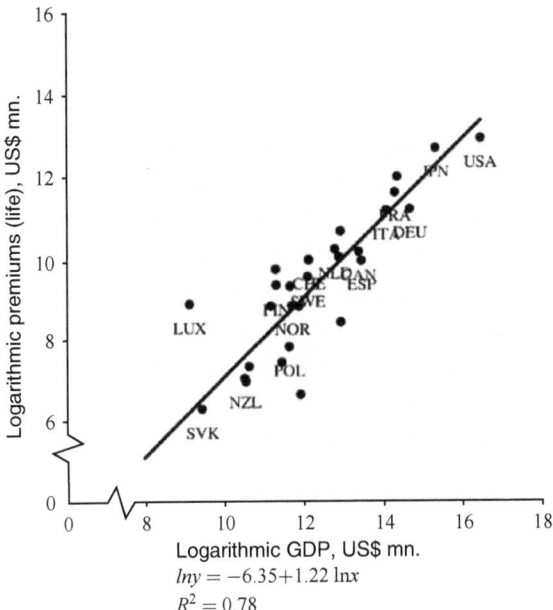

Fig. 1.4 Life insurance premiums and GDP for various countries (2004). *Source* Statistisches Bundesamt (1999) and Swiss Re (1999). CHE: Czechoslovakia; DAN: Denmark; DEU: Germany; ESP: Spain; FIN: Finland; FRA: France; ITA: Italy; JPN: Japan; LUX: Luxembourg; NLD: the Netherlands; NOR: Norway; NZL: New Zealand; POL: Poland; SKV: Slovakia; SWE: Sweden; USA: United States

non-life insurance, *ceteris paribus*. Combining the two estimates, one may infer that the importance of the insurance industry as part of the economy increases over time.

Since insurance is a financial service, one can argue that demand for insurance should primarily depend on financial developments rather than the GDP. However, due to great differences in institutional arrangements (role of banks, insurance, and capital markets; type and intensity of competition), measuring financial development is controversial. A common inverse indicator is the ratio of cash held to money supply (M1). Recall that M1 consists of cash and demand deposits of domestic nonbank agents with banks. Thus, a high ratio points to a little developed financial sector and a low ratio, to a highly developed one.

A second commonly used indicator of financial development is the ratio of money broadly defined (M2 or M3) to GDP. In addition to M1, M2 includes saving deposits and time deposits and M3, still more assets that can be transformed into liquidity at low cost. The higher the ratios M2/GDP and M3/GDP, respectively, the more developed is the financial sector of an economy.

The issue now becomes to assess the role and importance of the insurance industry within the financial sector rather than the economy as a whole. Since the explanatory variable is a ratio (M2/GDP), it makes sense to relate premiums to GDP as well (as in Fig. 1.2 above). Figure 1.6 depicts this relationship for a few industrial countries. Again, a clear tendency shows up. The higher the M2/GDP as an

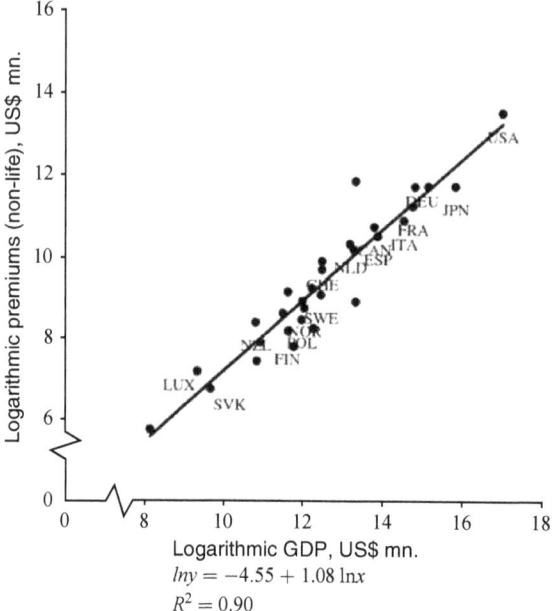

Fig. 1.5 Non-life insurance premiums and GDP for various countries (2004). *Source* Statistisches Bundesamt (1999) and Swiss Re (1999). CHE: Czechoslovakia; DAN: Denmark; DEU: Germany; ESP: Spain; FIN: Finland; FRA: France; ITA: Italy; JPN: Japan; LUX: Luxembourg; NLD: the Netherlands; NOR: Norway; NZL: New Zealand; POL: Poland; SKV: Slovakia; SWE: Sweden; USA: United States

indicator of the monetization of an economy, the higher the premiums relative to GDP (often called "insurance penetration"). An increase of M2/GDP by 10 percentage points (e.g., from 60% to 70%) suggests an increase in insurance penetration by 0.8 percentage points (e.g., from roughly 6% to 6.8% relative to GDP), *ceteris paribus*.

▶ **Conclusion 1.4** Demand for insurance increases more than proportionally with GDP in life insurance and proportionally in non-life insurance. It is positively related to the financial development of industrialized countries.

1.5.2 The Effect of Price

As argued in the text accompanying Fig. 1.2, the premium must not be equated with the price of insurance because the premium includes expected losses, which are distributed back to the insured as an indemnity. If insurers were to charge the fair premium (covering just expected loss), insurance coverage would be costless on average. The literature thus often uses the ratio of premium volume (PV) to indemnity payments (I) as an indicator of price. In insurance practice, a different

1.5 Major Determinants of the Demand for Insurance

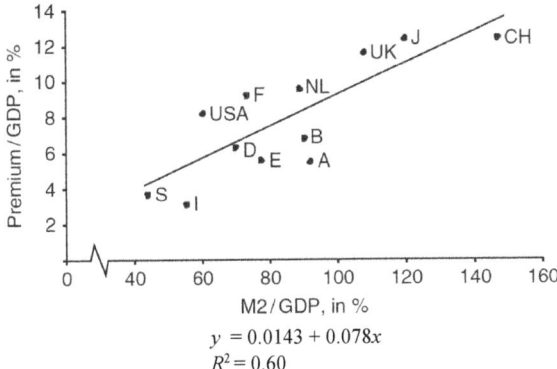

Fig. 1.6 Insurance penetration and M2/GDP. *Source* International Monetary Fund (1998) and Swiss Re (1999). A: Austria; B: Belgium; CH: Switzerland; D: Germany; E: Spain; F: France; I: Italy; J: Japan; NL: the Netherlands; S: Sweden; UK: United Kingdom; USA: United States

indicator is common, viz., the premium rate p. It can be used to decompose the premium volume into a price and quantity component as follows:

$$PV = p \cdot I, \tag{1.1}$$

PV: Premium volume;
p: Premium rate, premium charged per money unit of coverage;
I: Amount of coverage written.

Total differentiation yields

$$dPV = dp \cdot I + p \cdot dI. \tag{1.2}$$

Division by $PV = pI$ results in

$$\frac{dPV}{PV} = \frac{dp}{p} + \frac{dI}{I}. \tag{1.3}$$

Thus, the percentage change in premium volume can be split into the percentage change in the premium rate and the percentage change in the sum insured. Assuming that the sum insured depends on the premium rate p and income Y, one posits

$$I = I(p, Y). \tag{1.4}$$

Therefore, changes in I are determined by changes in p and Y as follows:

$$dI = \frac{\partial I}{\partial p}dp + \frac{\partial I}{\partial Y}dY. \tag{1.5}$$

After division by I and expanding by $1 = p/p$ and $1 = Y/Y$, this becomes

$$\frac{dI}{I} = \left(\frac{\partial I}{\partial p} \cdot \frac{p}{I}\right) \frac{dp}{p} + \left(\frac{\partial I}{\partial Y} \cdot \frac{Y}{I}\right) \frac{dY}{Y}$$

$$= \eta \cdot \frac{dp}{p} + \varepsilon \cdot \frac{dY}{Y}, \qquad (1.6)$$

with $\eta := \frac{\partial I}{\partial p} \cdot \frac{p}{I} < 0$ denoting the price elasticity of the demand for insurance and

$\varepsilon := \frac{\partial I}{\partial Y} \cdot \frac{Y}{I} > 0$ denoting the income elasticity of the demand for insurance.

The (approximate) percentage change in the demand for insurance can thus be split into a percentage change in prices (weighted with the price elasticity of demand) and a percentage change in income (weighted with the income elasticity of demand).

Table 1.10 exhibits some estimates of country-specific elasticities of the demand for insurance w.r.t. price and GDP. These estimates were obtained by a regression using yearly data from 1969 to 1990. For instance, commercial fire insurance in Germany has a relatively low *price elasticity* of demand of −0.2 to −0.3. This could be caused by the regulation at the time which imposed uniform products and premiums (see Chap. 9). Price elasticities for Chile—where insurance markets were liberalized earlier—are much more pronounced.

▶ **Conclusion 1.5** The price elasticity of the demand for insurance is negative but relatively low for regulated markets. In liberalized insurance markets, it tends to be noticeably higher (up to unit elasticity).

Table 1.10 Price and income elasticities of the demand for insurance

	Price Elasticity	GDP Elasticity
Germany		
Industry fire	−0.2 to −0.3	1.5–2
Chile		
Fire	−0.9 to −1.2	3–4
Earthquake	−1	3
Marine	−1	2–2.5
Motor vehicles	−0.8	2.8
Japan		
Fire	−1	1.7
USA		
Life	−0.7	2–2.5

Source Swiss Re (1993)

1.5 Major Determinants of the Demand for Insurance

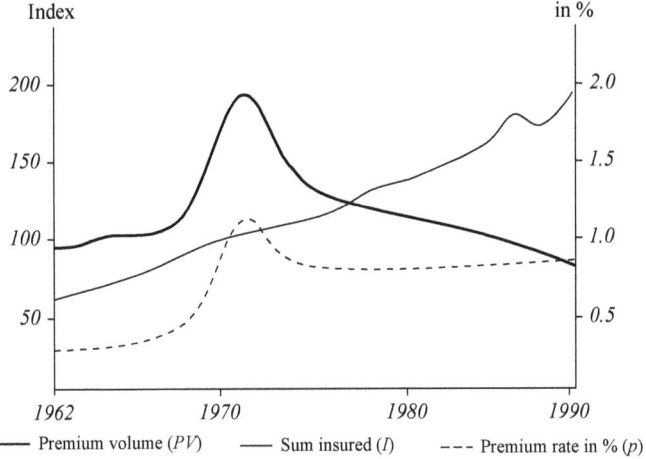

Fig. 1.7 Commercial fire insurance in West Germany (1962–1991). *Source* Swiss Re (1993)

Estimated *income elasticities* are consistently above one. For Germany, however, it lies below 2. This could indicate a relatively slow growth of the insurance industry if compared, e.g., to Chile or the United States, especially if German GDP should continue to grow at a comparatively low rate.

Finally, equation (1.3) can be used to relate the change in premium volume to the price and income elasticity of demand. Substituting (1.6) into equation (1.3) yields

$$\frac{dPV}{PV} = \frac{dp}{p} + \eta \cdot \frac{dI}{I} + \varepsilon \cdot \frac{dY}{Y}. \tag{1.7}$$

Collecting terms, one obtains

$$\frac{dPV}{PV} = (1+\eta)\frac{dp}{p} + \varepsilon \cdot \frac{dY}{Y}. \tag{1.8}$$

If $|\eta| < 1$, as in German commercial fire insurance, an increase in the premium rate increases the premium volume, *ceteris paribus*. Because of $\epsilon > 0$, the same is true for an increase in GDP. Figure 1.7 illustrates the development in German commercial fire insurance over the period 1962–1991. A massive but temporary increase in premium rates took place around 1972. Due to the low price elasticity and fast economic growth at that time, the sum insured increased rather than decreased. As a consequence, premium volume developed parallel to the premium rate. Conversely, the decrease in premium rates in 1990/91 resulted in a decrease in the premium volume. Due to the low price elasticity of demand, the sum insured increased but little, and GDP grew slowly. Thus, the increase in quantity failed to compensate for the decrease in price.

▶ **Conclusion 1.6** The income elasticity of the demand for insurance is positive and above one in relatively more dynamic markets.

1.6 System Analysis and Organization of the Book

Figure 1.8 illustrates the organization of the book. At the center are the three assets health, wealth, and wisdom, with wealth corresponding to "marketed assets" and health and wisdom to "human capital". Individuals have to manage these assets optimally considering their budget constraints as well as other constraints (notably time).

The objective of Chap. 1 was to show why it is worthwhile to acquire knowledge on insurance and the insurance industry. Chapter 2 presents the basics of insurance. In particular, it revolves around the conceptual problems of risk assessment and risk management.

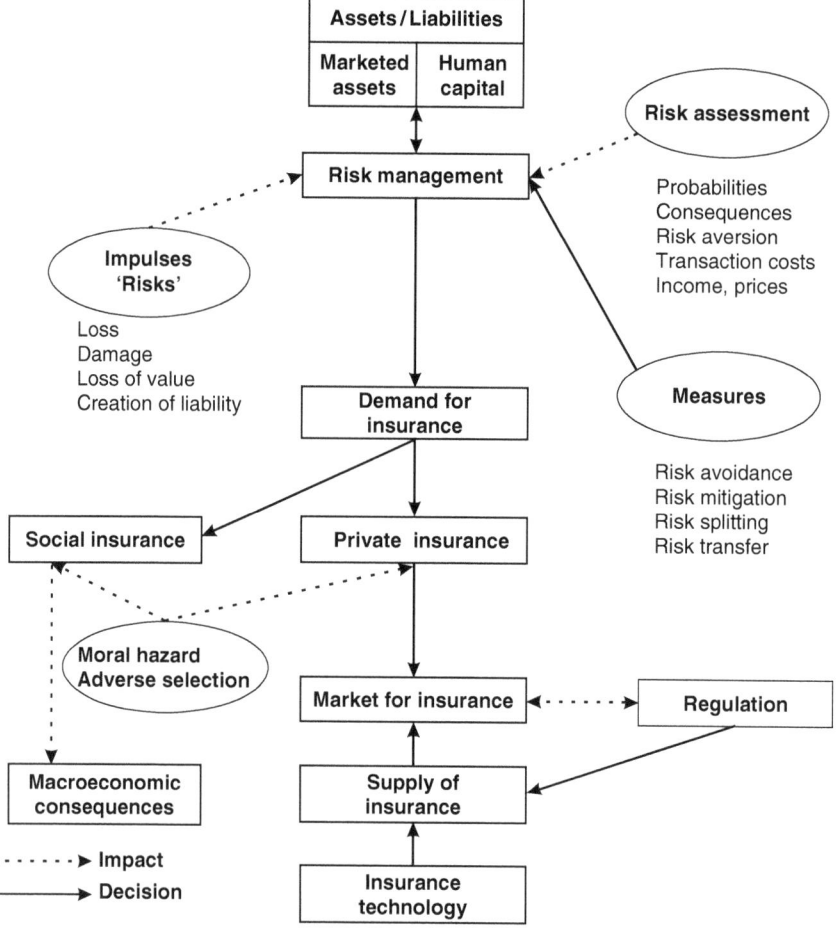

Fig. 1.8 Flowchart and overview

1.6 System Analysis and Organization of the Book

In Chap. 3, a model of the demand for insurance is developed in an expected utility framework. Starting with the simple basics, it addresses increasingly complex questions such as the role of irreplaceable assets and non-insurable risks.

Expected utility theory has its weaknesses, which are examined in Chap. 4. There, alternative approaches are also discussed, including ways to deal with probabilities that are not precisely known.

Contrary to individuals, firms can diversify risks through the capital market. Therefore, modern capital market theory is introduced and combined with insurance theory in Chap. 5. An important question is, why a company—especially an incorporated company—should exhibit a demand for insurance at all. Insurance demand thus becomes an element of comprehensive risk policies of individuals and firms (see Fig. 1.8).

Chapter 6 looks at the supply side of insurance markets. First, the so-called insurance technology of private insurance companies is considered, the foundation being accounting. The elements of this technology studied are acquisition of risks, risk selection, control of moral hazard effects, taking out reinsurance, and investment of disposable assets.

Since pricing has a very prominent role, a separate chapter is devoted to it. Thus, Chap. 7 presents the conventional rules of premium calculation, but its focus lies on the application of the capital asset pricing model (CAPM) to insurance.

Chapter 8 deals with equilibrium on insurance markets. The crucial phenomenon is information asymmetry in the guise of moral hazard and adverse selection effects. Moral hazard refers to the change in the behavior of buyers of insurance induced by the presence of insurance, which weakens buyers' incentive to undertake preventive efforts. Adverse selection refers to another informational problem. Absent detailed knowledge concerning the quality of a buyer, insurance companies have to charge an average risk premium. This attracts "bad" risks while driving away "good" risks.

Chapter 9 addresses the regulation of the insurance industry and its side effects. Competing hypotheses explaining the need for regulation are presented. The analysis suggests that regulation is mainly driven by the interests of politicians and civil servants on the one hand and by the intent of producers to gain protection from competition on the other hand.

Figure 1.8 also shows that citizens have a choice between private and social insurance. While Chap. 10 does not exclude the possibility of social insurance being efficiency-enhancing, it concludes that the fast growth of social insurance cannot be explained on efficiency grounds alone. As with regulation, the expansion of social insurance seems to be driven by the (personal) interests of political decision-makers.

Future challenges are the topic of the concluding Chap. 11. It discusses the use of genetic diagnostics, the fast development of information technologies, and demographic changes. The book concludes with a number of policy recommendations which aim at enhancing the combination of insurance and society's welfare.

Exercises

1.1

(a) State at least two indicators for the macroeconomic significance of the insurance industry.
(b) What are the weaknesses of the indicator, "Premium volume over GDP"? Explain.
(c) Comment on the following statement: "The price of insurance coverage is the premium. The dependent variable is premium income. Independent and dependent variables thus coincide, making it impossible to estimate the price elasticity of the demand for insurance."

1.2

(a) Discuss the meaning of loss prevention and loss reduction.
(b) Why is it not appropriate to talk of a liquidity effect of insurance in spite of the accumulation of capital?
(c) In which respect is the controlling function of insurance especially relevant for environmental problems?

1.3 To laypersons, the premium is the price of insurance.

(a) Explain why this is not true.
(b) An insurance policy is revamped to cover an additional risk. Its premium goes up. Do you predict a negative demand response by policyholders? Why or why not?
(c) The premium of an insurance policy goes up because the insurance company is confronted with an increase in administrative expenses. Do you predict a negative demand response by policyholders? Why or why not?

References

Arrow, K.J. (1965). *Aspects of the Theory of Risk-Bearing*, Yrjö Johnsson Lectures, Säätiö, Helsinki Y.J.
Eisen, R., Müller, W., & Zweifel, P. (1993). Entrepreneurial insurance: a new paradigm for deregulated markets. *Geneva Papers on Risk and Insurance, 66*, 3–56.
Eurostat database, European Commission. Accessed 2010.
Eurostat database, European Commission. Accessed 2019.
Giarini, O., & Stahel, W. (2000). *Die Performance-Gesellschaft (The Performance Society)*. Marburg: Metropolis Verlag.
Hax, K. (1964). *Grundlagen des Versicherungswesens* (Basics of Insurance), Wiesbaden: Gabler.
Hempel, C.G. (1965). *Aspects of Scientific Explanation, and Other Essays in the Philosophy of Science*. New York: Free Press.

References

Knight, F.H. (1921). *Risk, Uncertainty and Profit*. Chicago: Chicago University Press.

Kofler, E., & Zweifel, P. (1988). Exploiting linear partial information for optimal use of forecasts: with an application to U.S. economic policy. *International Journal of Forecasting, 4*(1), 15–32.

Müller, W. (1995). Informationsprodukte (Informational products). *Zeitschrift für Betriebswirtschaft, 65*, 1017–1044.

Negoita, C.V. (1981). The current interest in fuzzy optimization. *Fuzzy Sets and Systems, 6*(3), 261–269.

OECD (2010b). *Insurance Statistics Yearbook 2010*. Paris: OECD Publishing.

OECD (2017). *OECD Insurance Statistics 2016*. Paris: OECD Publishing.

OECD (2019). *OECD Insurance Statistics 2018*. Paris: OECD Publishing.

OECD (2020). *OECD Insurance Statistics 2019*. Paris: OECD Publishing.

Sinn, H.W. (1988). Gedanken zur volkswirtschaftlichen Bedeutung des Versicherungswesens (On the importance of insurance for the economy). *Zeitschrift für die gesamte Versicherungswissenschaft, 77*, 1–27.

Statistisches Bundesamt (1999). *Statistisches Jahrbuch für das Ausland (Statistical Yearbook for Foreign Countries)* Wiesbaden.

Swiss Re (1993). Ökonomische Analyse der Versicherungsnachfrage (Economic analysis of insurance demand). *sigma* 5/1993.

Swiss Re (1999). Assekuranz Global 1997: Stark expandierendes Lebensgeschäft. Stagnierendes Nicht-Lebengeschäft (Global Insurance 1997: Strongly expanding life business. Stagnating non-life business). *sigma* 3/1997.

Swiss Re (2018a). Natural Catastrophes and man-made disasters in 2017: A year of record-breaking losses. *sigma* 1/2018.

Swiss Re (2019). World insurance: the great pivot east continues. *sigma* 3/2019.

Zweifel, P. (2009). Technological change and health insurance. In J. Costa-Font, C. Courbage, & A. McGuire (Eds.), *The Economics of New Health Technologies. Incentives, Organization and Financing* (pp. 93–107). Oxford: Oxford University Press.

Risk: Measurement, Perception, and Management   2

In ordinary language, "risk" is mainly used in conjunction with "chance". In Chinese language, "risk" indeed has two characters, one for risk proper (sometimes called negative consequences) and the other for chance. In insurance economics, however, the word has a specific meaning to be defined in Sect. 2.1. Also, the statistical measurement of risk turns out to be an endeavor fraught with difficulties.

Section 2.2 is devoted to the perception of risk, i.e., the fact that different persons (and perhaps even the same person under differing circumstances) recognize and evaluate risk differently. It seems that human behavior may be determined by a general biological law—characterized by aversion against risk. In insurance economics this fact is taken into account by the concept of a *risk utility function*, which comprises both objective properties of a risk and its subjective valuation.

The risk utility function is used in Sect. 2.3 to derive willingness to pay for certainty. Furthermore, different types of risk aversion and their implication for the willingness-to-pay value are discussed.

Since the degree of risk aversion is a crucial determinant of the demand for security, in general, and for insurance, in particular, attempts at measuring risk aversion are presented in Sect. 2.4.

In Sect. 2.5, instruments and actions designed to influence, mitigate, or prevent risks are discussed, giving rise to the concept of risk management.

In Sect. 2.6, the instruments of risk management are assessed with respect to their appropriateness and effectiveness. The importance of the subjective valuation of risk mentioned in Sect. 2.2 will show up again.

Finally, an appendix is devoted to stochastic dominance, a concept that permits the construction of a preference ordering over arbitrary risks.

2.1 Definition and Measurement of Risk

2.1.1 Definition of Risk

Individuals own three assets that they must manage over their life cycle, viz., health, wealth (financial capital), and wisdom (skills). These three assets enable them to buy consumption goods, to enjoy them, and to earn a labor and capital income. Later (in particular during retirement), wealth accumulation through savings can be used to buy consumption goods.

These three assets are subject to random shocks, causing their value to fluctuate. In insurance practice, these shocks are called *perils*. These perils may emanate from "nature" (e.g., death, illness, disability) or from human activity (such as changes in market prices). They lead to discrepancies between originally *planned* and actually *realized values* of the three assets.

In ordinary language, the possible losses due to these discrepancies in terms of money or utility are called *risk*. This colloquial notion of risk therefore encompasses only the possibility of an unfavorable event occurring. Opposed to this is the *chance* of suffering no loss or even reaping a profit.

For analytical purposes, however, it is preferable to simply consolidate risk and chance into one random variable. Such a random variable X is shown in panel A of Fig. 2.1. Its density function $f(X)$ indicates a prevalence of small values; high values are rare but not impossible. This skewness is typical in the present context since the three assets are frequently subject to shocks resulting in small losses, while shocks resulting in large losses are infrequent (fortunately!). If the asset in question is insured, $f(X)$ is called a loss distribution function (often loss distribution for short) or *pure risk* of a certain line of insurance.

If the random variable takes on positive as well as negative values, the distinction between risk and chance is difficult and even unnecessary. It is often difficult because one has to find or define the zero point. It is unnecessary because it is the distribution over all possible values of the random variable which counts for decision-making. The *risk* of an activity (or of an unintentional process) is therefore best represented by the *density function defined over possible consequences*. In an insurance context, the loss distribution can be combined with the initial value of the asset affected to obtain an asset density function (see panel B of Fig. 2.1). In economic theory, the notion of risk thus has two dimensions, the probability of occurrence and the importance of the associated consequence.

▶ **Conclusion 2.1** In economic terms, the risk of an activity is represented by the probability density defined over possible consequences. In insurance economics, in particular, the probability of occurrence and the severity of the consequence are distinguished in the definition of risk.

In insurance practice, sometimes the buyer of insurance or the insurance contract representing a buyer is also called a risk. The insurance company (insurance enterprise or insurer, IC) receives the premium income, while booking the risk as

2.1 Definition and Measurement of Risk

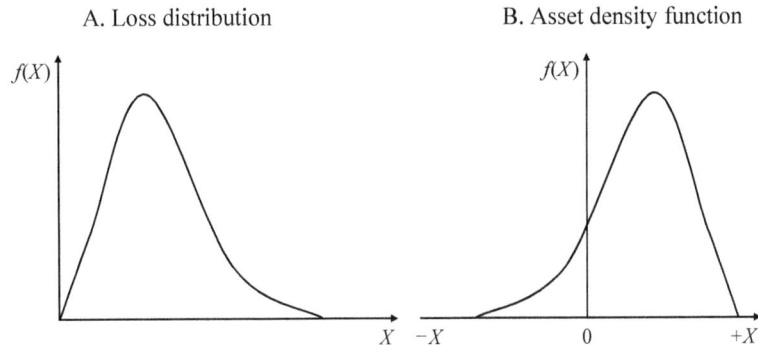

Fig. 2.1 Risk as a random variable

a potential "loss producer". Hence, for the actuarial analysis of a risk, only the *difference* between *premium income* and *loss payments* during a certain time interval is relevant. Since the premium income is often treated as (almost) non-random, this difference is mainly determined by the loss distribution.

2.1.2 Measurement of Risk

In accordance with Conclusion 2.1, the measurement of risk from an insurer's point of view amounts to two tasks:

- assess the probability of occurrence and
- assess the severity of the consequences.

As is evident from Fig. 2.1, these two dimensions of risk should be measured jointly. Still, one can select a certain value of the random variable X and seek to determine the associated probability of occurrence.

2.1.2.1 Measuring the Probability of Occurrence

From an insurer's point of view the probability of occurrence is the probability of having to pay a loss of a certain amount (in insurance also called indemnity). It lies between 0 and 1 (bounds are inclusive because in the case of zero, the event will not happen, while in the case of one, the event happens for sure). Probabilities can be determined either by logic or pure reason, e.g., in the case of an ideal dice where the numbers are equally probable and hence have probability 1/6 (or 16.67%) each. Or they are inferred from an experiment, a series of experiments, or by experience,

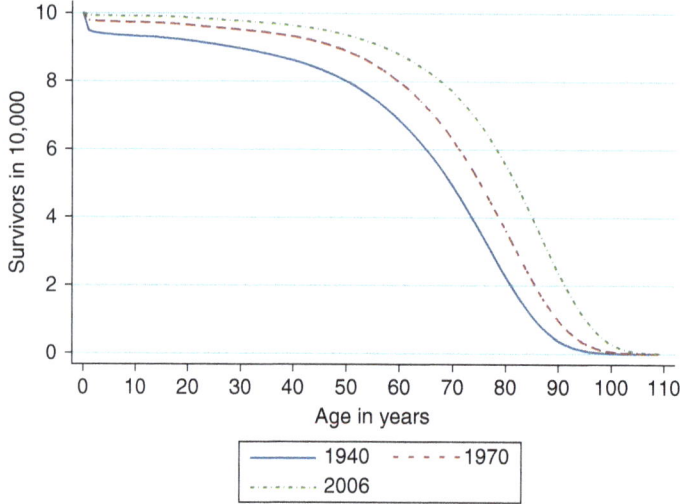

Fig. 2.2 Rectangularization of the survival curve for the U.S. population. *Source* Human Mortality Database (2010), own calculation

with the conditions governing the experiments held as constant as possible. For these cases, the probabilities are called *relative frequencies*.[1]

However, most relative frequencies shift over time because of changes in the environment and, in particular, changes in technical knowledge and its application in society. An interesting example is life expectancy. Since death is a rather homogeneous consequence, it suffices as a first approximation to focus on the probability dimension of risk. Many life insurance policies pay a capital benefit in case the insured dies prematurely and another capital benefit in case he or she lives to a certain age (60, say). When mortality risk decreases, resulting in an increasing life expectancy, the insurer becomes less likely to have to pay the capital benefit due to early death but more likely to have to pay the benefit to surviving policy holders. Now in the United States, for instance, the change of life expectancy has been spectacular. While in 1940 only 58,000 out of 100,000 persons reached the age of 60 years, at present more than 90,000 do. With early death increasingly becoming the exception, the survival curve approaches a rectangle (see Fig. 2.2).[2]

But also the general climatic conditions are changing (regardless of whether due to human or natural causes), resulting in greater frequencies of natural disasters (as shown in Table 1.1 of Sect. 1.2).

---

[1] In probability theory the (weak) law of large numbers is proved, stating that relative frequencies approach the (objective) probabilities when the number of experiments increases (for an application to insurance, see Sect. 7.1.2).

[2] The so-called rectangularization of the survival curve plays an important role in health economics, see, e.g., Zweifel et al. (2009, Chap. 10).

Often, however, estimation of probabilities is merely based on subjective experience or on imperfect experiments. Accordingly, they are called *subjective probabilities* (in contrast to the objective probabilities above). Since most experiments (because of changing side conditions) are imperfect, in insurance practice, subjective rather than objective probabilities prevail. Still, they can be used in the same manner as the objective ones. But insurers (and decision-makers generally) must take into account that this imparts an additional degree of uncertainty to the outcome of their actions. Or put another way, (almost) all probabilities are subjective, yet their degree of credibility varies, i.e., the degree with which they are believed or accepted as correct by a person or a group.[3]

2.1.2.2 Determining the Severity of the Consequences

The second dimension of a risk is the severity (or more generally, importance) of the consequences. It is always possible to enumerate or verbally describe consequences, such as the number of deaths or injured, material and immaterial damages caused, or indemnities paid (see Table 1.2 of Sect. 1.3). This is, however, not enough to compare them. Which event is "worse", hurricane Katrina which destroyed parts of New Orleans on August 28/29, 2005 or the explosion of Deepwater Horizon, the BP oil drilling platform on April 20, 2010 causing the largest oil spill in recorded history?

Comparing events with different consequences is impossible without a value judgment. These valuations may change over the course of time, are based on culture and religion, and therefore are subjective, differing between individuals. One possibility to circumvent these difficulties is to distinguish consequences by type (deaths, injured, damages) rather than aggregating them into one risk measure. For example, by singling out the consequence "number of injuries" in isolation, one can focus on the probability of occurrence as the one relevant dimension of risk. But even then, ambiguities arise. Take, for instance, the simple question, What is the probability of being injured in a traffic accident? First, a period of time has to be specified (a year, say). As columns 1 and 2 of Table 2.1 show for the case of Germany, the choice of time period is important since despite a sharp increase in traffic density, the absolute number of deaths has decreased over time. In addition, for an auto insurer it may be important to know that the most marked decrease occurred in the cities, causing them to be much safer for driving than the countryside.

---

[3] Assuming that all probabilities are subjective makes the famous distinction (introduced by Knight (1921)) between risk and (pure) uncertainty lose its relevance. Risk is obtained when probabilities of occurrence are known, uncertainty, when they are not known. However, the discussion above shows that for the purpose of decision-making, the ability to assign subjective probabilities to events (consequences, respectively) is sufficient (see also Sect. 1.1).

Table 2.1 Deaths in road traffic, Germany (1970–2017)

Year	Total (1)	Index (2)	Freeway (3)	Index (4)	Countryside (5)	Index (6)	City (7)	Index (8)
1970	21,332	100.0	1,093	100.0	10,682	100.0	9,557	100.0
1975	17,011	79.7	1,079	98.7	8,793	82.3	7,139	74.7
1980	15,050	70.6	943	86.3	7,976	74.7	6,131	64.2
1985	10,070	47.2	777	71.7	5,570	52.1	3,723	39.0
1990	11,046	51.8	1,470	134.5	6,215	58.2	3,361	35.2
1995	9,454	44.3	978	89.5	6,041	56.6	2,435	25.5
2000	7,503	35.2	907	83.0	4,767	44.6	1,829	19.1
2005	5,361	25.1	662	60.6	3,228	30.2	1,471	15.4
2006	5,091	23.9	645	59.0	3,062	28.7	1,384	14.5
2007	4,949	23.2	602	55.1	3,012	28.2	1,335	14.0
2010	3,648	17.1	430	39.3	2,207	20.7	1,011	10.7
2015	3,459	16.2	414	37.6	1,997	18.7	1,048	10.9
2017	3,180	14.8	409	37.5	1,795	16.8	976	10.2

Source ADAC (2009, Table 16), Statistisches Bundesamt (2018)

Table 2.2 Deaths on U.S. roads by mode of transportation, 1975–2019

Year	Total	Passenger car	Motorcycle	Trucks (large)	Bus	Pedestrian	Bicycle
1975	38,651	25,929	3,189	961	53	7,516	1,003
1980	42,936	27,449	5,144	1,262	46	8,070	965
1985	36,508	23,212	4,564	977	57	6,808	890
1990	35,414	24,092	3,244	705	32	6,482	859
1995	31,748	22,423	2,227	648	33	5,584	833
2000	29,828	20,699	2,897	754	22	4,763	693
2005	29,628	18,512	4,576	804	58	4,892	786
2009	22,808	13,095	4,462	503	26	4,092	630
2014	32,675	11,926	4,586	657	n.a.	4,884	726
2019	36,096	12,239	5,014	892	n.a.	6,205	846

Source US Department of Transportation, Bureau of Transportation Statistics (2011, 2016, 2020)

In turn both types of traffic continue to be far more risky than driving on freeways (see Col. 3 of Table 2.1), even though on freeways the reduction in the number of injuries has been less marked.[4]

Still, injuries constitute a consequence that arguably are heterogeneous. Therefore, one may only retain those resulting in deaths. Table 2.2 shows the number

---

[4]The spike in 1990 is potentially due to reunification with formerly communist Eastern Germany, with each of the two parts having little experience with the road network of the other.

2.1 Definition and Measurement of Risk

of persons killed on U.S. roads according to mode of transportation. Clearly, most deaths involve passenger cars throughout the observation period, followed by pedestrians and increasingly motorcycles. However, this does not mean that driving a car is the most risky mode of road transportation. A probability or relative frequency always is defined in terms of a reference population. For example, in 2009 there were 260 million passenger cars in circulation. Thus, the 13,095 lives lost amount to a relative frequency of 0.000065475. By way of comparison, the 4,462 deaths involving a motorcycle relate to around 7 million motorcycles, corresponding to a risk of 0.00066597, which is more than nine times larger.[5]

Mathematically, the statement, "1 death per 100, 1,000, 10,000 persons" can be written as $1 \cdot 10^{-2}$, $1 \cdot 10^{-3}$, $1 \cdot 10^{-4}$, etc. Thus, the higher this negative exponent, the less frequent is the event. Urquhart and Heilmann (1983) use these exponents to construct a logarithmic "safety scale",[6] ranging from 1 to 8 but open-ended in principle. With every next higher number on this scale, the probability of occurrence diminishes by the power of ten (see Fig. 2.3).

The scale indicates that (maybe surprisingly) being a regular smoker entails a higher risk than dying from lung cancer. The reason is that other organs are negatively affected by smoking as well. Conversely, bicyclists incur a slightly lower risk of death than pedestrians.

The picture is complicated by the fact that individuals are threatened not only by one risk, but simultaneously by a *multitude of perils*. Car drivers and cyclists are also pedestrians, engage in sports, are exposed to different perils at work and in the household, and may smoke and drink. Since the realization of one of them suffices to cause death, their probabilities must be added up in the case of independence.

▶ **Conclusion 2.2** The risk of a defined event or consequence usually is measured by its relative frequency of occurrence. However, the choice of the correct population of reference can pose problems.

All of this still leaves the question of evaluating the importance of a risk unanswered. For example, is "death like death", as the safety scale of Fig. 2.3 assumes? Is there really no difference between dying in an air crash and dying from cancer? It is obvious that the subjective perception of risk matters. This is discussed in the next section.

---

[5] Alternatively, one may ask how many deadly injured persons occur per 100 million kilometers driven (maybe subdivided by type of road). These figures are called death rates. The death rate for transportation by passenger car is 0.5, and by motorcycle, 6.1, i.e., higher by a factor of 11.4 (instead of only 9 as above).
[6] The logarithm to the base 10 of the number 100 is 2, of 1,000 is 3, etc.

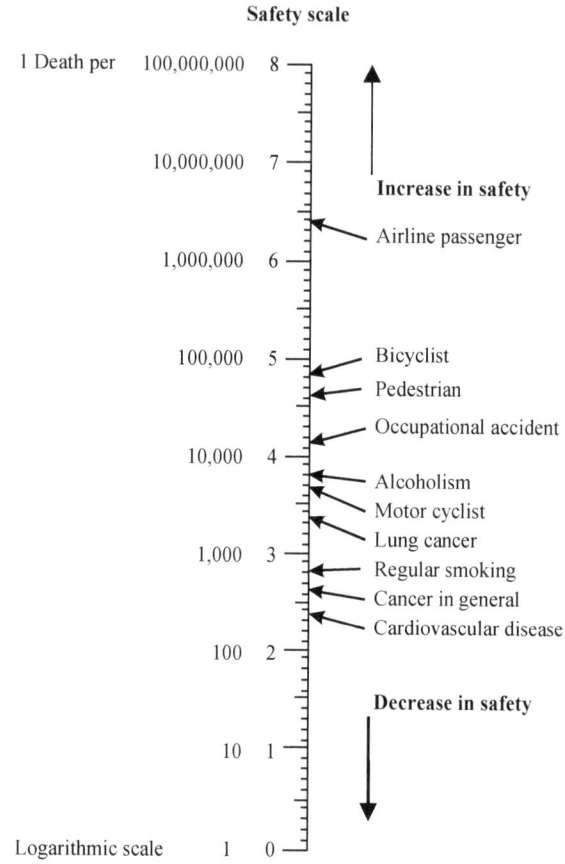

Fig. 2.3 Probabilities of death. *Source* Urquhart and Heilmann (1983)

2.2 Subjective Perception of Risk, Risk Aversion, and the Risk Utility Function

2.2.1 Risk Perception as a Subjective and Cultural Phenomenon

To measure a risk in an objective statistical manner is one thing; to subjectively perceive it, quite another. This discrepancy already makes itself felt in the estimation of the first component of a risk, its probability of occurrence (the second component, the importance of consequences, will be discussed below).

(1) *Perception of probabilities of occurrence*

In Fig. 2.4, the estimated number of deaths per year (with the population of the United States serving as the reference) is plotted against their actual number of deaths. The 45°-line indicates a perfect estimate. Three features are noteworthy. First, while people can estimate some risks quite precisely (e.g., all accidents, car accidents), they misjudge others by a magnitude up to two powers of ten

2.2 Subjective Perception of Risk, Risk Aversion, and the Risk Utility Function

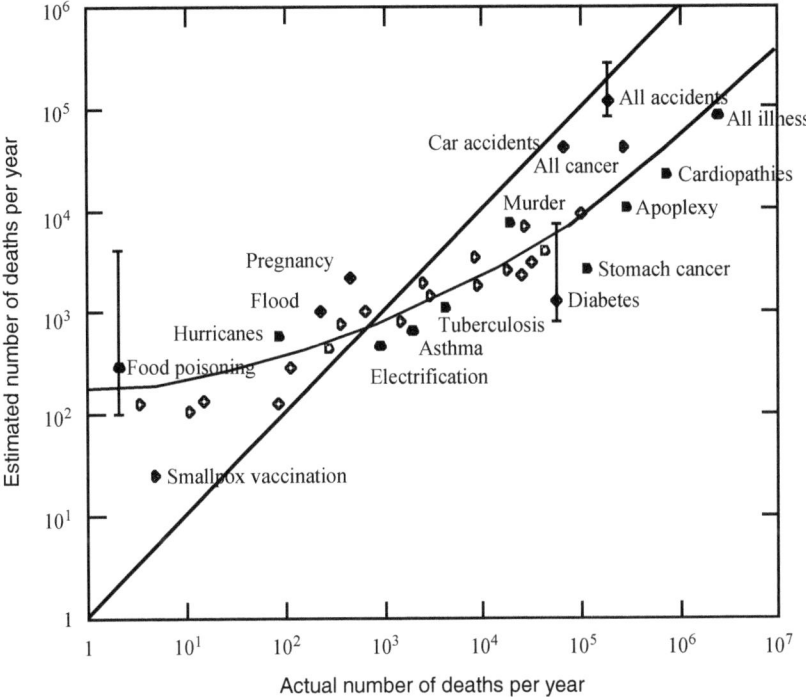

Fig. 2.4 Estimated and actual number of deaths, United States. *Source* Lichtenstein et al. (1978), Slovic et al. (1980, 19)

(e.g., floods, food poisoning, diabetes). Second, low probabilities of occurrence are in general overestimated whereas relatively high ones are underestimated (the regression line runs flatter than the 45°-line). Third, the standard errors of estimated frequencies (indicated by vertical bars) are higher for rare events and smaller for frequent ones. Here, survey participants acted in the manner predicted by mathematical statistics, which relates standard errors inversely to (the square root) of sample size.

One might expect that individuals specializing in probability estimation do better, benefiting from an increased degree of "risk intelligence". This expectation is borne out for U.S. weather forecasters but much less so for physicians who were asked to estimate the probability that real patients had pneumonia (knowing their medical history and the results of a physical examination). For instance, patients who were assigned a 90% probability turned out to have pneumonia in just 15% of cases (see Fig. 2.5).

Intuitively, the closer the points of Figs. 2.4 and 2.5 to the 45° line, the higher the risk intelligence in a population. This can be measured inversely by the area between the line connecting the actual estimates and the 45° line (see Fig. 2.5 again), resulting in the so-called RQ score, which is closely related to the Gini coefficient used for measuring inequality. Thus, the RQ score is lower when

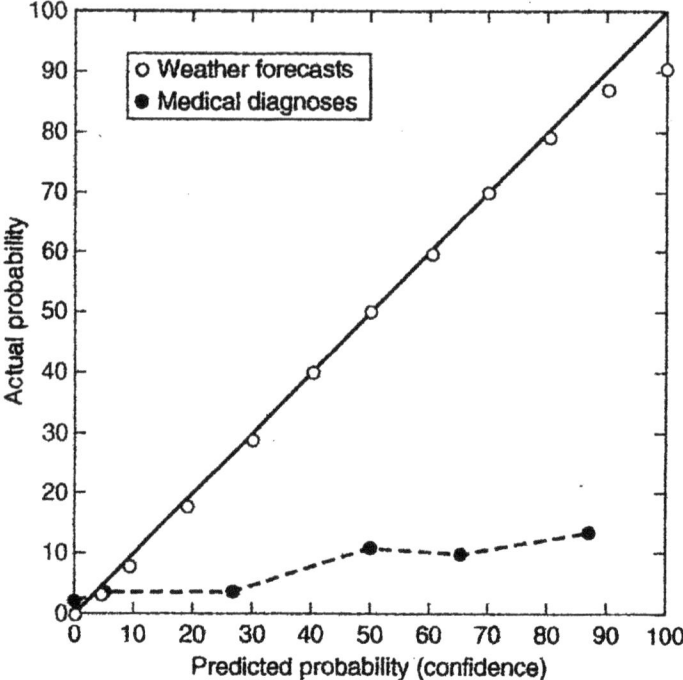

Fig. 2.5 Predicted and actual probabilities in two special cases. *Source* Murphy and Winkler (1977), Christiansen-Szalanski and Bushyhead (1981)

the two probabilities are a better match, lying between 0 (if the predictions are correct) and 1/2 (if there is a total failure). In a study involving 6,700 individuals from 20 nationalities, Evans (2012) finds a decrease in the RQ score with higher education (especially at the bachelor and master levels), indicating a higher degree of risk intelligence.

This type of research is of considerable importance for the economics of insurance, where a reasonably priced estimate of loss probabilities is a standard assumption. Still, there are several reasons for discrepancies between actual and perceived frequencies. Here are a few.

- A first reason is due to the fact that most judgments about probabilities are not based on statistical data but on *personal experience*. Here, a *systematic estimation error* is manifest. Very small probabilities almost always are overestimated, likely because they are unexpected. Conversely, risks associated with known, common conditions and familiar technologies (not shown in Fig. 2.4) are underestimated. Errors of this type are deep-seated and not easily corrected in the light of facts, which often are only accepted if they correspond to preconceptions.

- Risks which can *be influenced* (regardless of whether they are accepted consciously or unconsciously) are estimated quite differently from risks which are beyond the individual's control. For example, the probability of dying from stomach cancer (which can be influenced) is underestimated while that of dying from a hurricane (beyond control) is overestimated.
- Estimates also differ by *age* and *gender*, a fact that will also be relevant for their subjective valuation (see, however, the remarks at the end of Sect. 2.4.1.2).
- The perception of probabilities also depends upon the *social, political, and cultural environment*. This encompasses not only socioeconomic determinants like degree of industrialization or peculiarities of press reporting, but also the culture-specific interpretation of risk characteristics. Common social values lead to common anxieties and selective perceptions of perils. A typical example is the responses in France and Germany to the nuclear accident of Chernobyl (Ukraine) in 1986. While in Germany the risks of this technology were stressed, in France its positive aspects were discussed. And in response to the nuclear accident of Fukushima (Japan) in March 2011, the German government ordered a stop of production at the country's nuclear sites; France continued its production at the same rate.

On a final note, estimation error may in part be caused by an artificial homogeneity of consequences imposed on respondents. One example is the overestimation of the probability of death due to a hurricane. Respondents may in fact be concerned not only about their personal survival but that of their relatives and friends in the area. The likelihood of their simultaneously surviving a hurricane is of course much lower than that of the individual respondent, who has little interest in spending his or her remaining life in isolation. Thus, "death is not death", contrary to what is imposed by the survey. One way to correct for this is to adjust upward the estimated probability of occurrence.

(2) *Subjective valuation of consequences*

Again, there are several reasons why consequences are valued differently by different people and in different situations.

- The *path* leading to a given consequence is part of its valuation. For instance, a smoker considering to quit in view of the high risk of cancer takes into account not only the benefit in the guise of a higher probability of survival. To stop smoking entails opportunity costs, viz., a sacrifice of utility. This investment in a healthier future turns out to be a bad investment if the former smoker dies nevertheless (e.g., because of another cause related to smoking or because of another cause altogether such as an accident) (see Zweifel (2001) for a theoretical treatment).
- Contrary to many experts, *subjectively "death is not like death"*. As argued above, it matters whether oneself only is in danger or whether relatives and friends are at risk as well. The number of possibly affected persons related to a single event influences its importance.

- Here again, the *cultural environment* has an impact on the way consequences are valued. In industrial countries, the death of a cow can be expressed in financial terms, using a market price. In India, this is the loss of a sacred animal that many people would refuse to value in terms of money (at least officially). It may also make a difference whether the cow died due to malnutrition or to an accident caused by a negligent driver.

2.2.2 Risk Aversion and the Risk Utility Function

2.2.2.1 Origin and Prevalence of Risk Aversion

As shown in the preceding section, many different factors determine the evaluation of risk, such as experience, information about consequences of alternate actions, but also temperament, anxiety, economic situation, cultural values, and many more. Hence, risk perception and risk evaluation are topics of psychology, sociology, and anthropology but not of economics, at least traditionally. For risk management—be it at the individual or the societal level—these factors play certainly a role. Indeed, one need to think only of the public debate about the clean-up of the oil platform Brent Spar (1997) or the Creutzfeldt–Jakob ("mad cow") disease caused by BSE.[7]

According to the technical or engineering point of view, risk simply amounts to the expected damage (the so-called expected loss). This point of view certainly is too narrow because it implicitly supposes risk-neutral behavior, thus *neglecting risk aversion*. In economics, risk aversion is seen as a typical characteristic of human beings. It implies that when comparing choices with uncertain (stochastic, random) results, people dislike and avoid dispersion around a given expected value. They accept dispersion (often called volatility) only if it comes with a higher expected value.

The origin of risk aversion can be explained as the result of an evolutionary process (see, e.g., Sinn and Weichenrieder (1993) or Szpiro (1997)). Hunger, thirst, and sexuality cause human beings to adopt certain behaviors which serve a genetic desire to survive. In the course of many generations, different rules of decision under uncertainty were tested in a lot of natural trials. Decision rules that were successful in fostering survival govern behavior still today. More specifically, decision rules have a higher selective quality than others if they produce a bigger population; they show selective dominance.[8] This is the decision rule which results in the highest expected value of the logarithmic ratio, "number of children to size of the parent generation", i.e., the stochastic growth factor of a generation. The stochastic growth factor is maximized if parents value the logarithm of the number of their offspring, implying that they value the possible addition of a child less than the possible loss

---

[7] See, e.g., Setbon et al. (2005).
[8] A preference has selective dominance "if it induces a growth of population so strong as to cause the relative size of populations resulting from other preferences to converge to zero almost with certainty" (Sinn and Weichenrieder (1993, p. 76)).

2.2 Subjective Perception of Risk, Risk Aversion, and the Risk Utility Function

of a child because the logarithmic function runs concave. As will be shown below, concavity of the so-called *risk utility function* is equivalent to risk aversion.

A rule taking into account risk aversion therefore dominates, in terms of selective quality, the simple maximization of the expected value of the growth factor. In economic applications, this rule amounts to maximizing a concave risk utility function (see below). This implies that an increase in dispersion must be compensated by an increase in expected value; otherwise, the individual will not accept the higher risk (second-order stochastic dominance, see appendix to this chapter). Risk aversion also means that downward stochastic deviations from the expected value receive a higher subjective weight than upward deviations.

Risk aversion—or "certainty preference"—characterizes the preferences of most individuals regardless of their income or wealth.

Example 2.1

Saint Petersburg Paradox
In the so-called Saint Petersburg game, a coin is tossed; if "heads" comes up the first time, the player receives two money units (MU) and zero otherwise; if "heads" appears also the second time, the player receives four MU and zero otherwise; the third time, eight MU, etc. How much should a player pay to be able to participate in this game?
The probability of "heads" on the first flip is equal to $1/2$. Moreover, tosses of a coin are independent events; therefore, the probability of having "heads" twice equals $1/2 \cdot 1/2 = 1/4$. The expected value of wealth (EW) from the payout series thus amounts to

$$EW = 2 \cdot \frac{1}{2} + 4 \cdot \frac{1}{2} \cdot \frac{1}{2} + 8 \cdot \frac{1}{2} \cdot \frac{1}{2} \cdot \frac{1}{2} + \ldots = 1 + 1 + 1 + \ldots = +\infty. \quad (2.1)$$

For the player, this game has an expected value that grows beyond limit. Nevertheless, few will pay more than a small amount of money to play it. ∎

2.2.2.2 The Risk Utility Function and Expected Utility

To solve the Saint Petersburg paradox, the decision problem under uncertainty involved was divided into three steps in the literature. In the first step, the individual risk situation is verified. This consists of assigning mutually exclusive *consequences* (usually measured in terms of money) to choices, along with their *probabilities* of occurrence. In the second step, these consequences are *valued*. Daniel Bernoulli (1738)[9] recommended the logarithmic function, a member of the class of concave

---

[9] This was the youngest member of a family of mathematicians from Basel (Switzerland) comprising Jacob (1654–1705), Johann (1667–1748), and Daniel Bernoulli (1700–1782).

Table 2.3 Decision matrix

Set of actions $a_i \in A$ ($i = 1, \ldots, n$)	States of nature $s_j \in S$ ($j = 1, \ldots, m$)				Valuation of consequences
	s_1	s_2	$\ldots$	s_m	C
a_1	c_{11}	c_{12}		c_{1m}	$v[c_{1j}]$
a_2	c_{21}	c_{22}	$\ldots$	c_{2m}	$v[c_{2j}]$
a_3	c_{31}	c_{32}	$\ldots$	c_{3m}	$v[c_{3j}]$
$\ldots$	$\ldots$	$\ldots$	$\ldots$	$\ldots$	$\ldots$
a_n				c_{nm}	$v[c_{nj}]$
Probabilities of occurrence	π_1	π_2	$\ldots$	π_m	

functions.[10] For the third step, he suggested forming the *expected utility* by multiplying (weighting) the utility values with their respective probability of occurrence and then summing up.

The decision rule then becomes to choose the alternative with the *highest* expected utility. Formally, the *Bernoulli principle* is an operator (i.e., a prescription) with which decision situations under uncertainty (or risk) are reduced to the *maximization* of a uniquely defined objective function, the so-called *risk utility function*.

The first two steps can be represented with the help of a matrix. The rows of Table 2.3 show the set of possible actions (a_i, $i = 1, \ldots, n$), i.e., the action space A. Its columns contain the mutually exclusive and exhaustive states of nature (s_j, $j = 1, \ldots, m$), i.e., the state space S. The cells exhibit the consequences or results, i.e., the consequence space, C. The last column contains the valuation of the consequences using the risk utility function $v(\cdot)$. The Bernoulli principle says to calculate the *expected utility* of an action a_i by weighting the utility values by π_j and summing up,

$$EU[a_i] = \sum_j \pi_j v[c_{ij}, a_j], \quad EU\text{: expected utility.} \tag{2.2}$$

The best decision given uncertainty amounts to the choice of the action a_i that has maximum expected utility. From Table 2.3, it is evident that this decision implies the choice of a row of the decision matrix, i.e., a so-called risky prospect.

Example 2.2

Risk utility function and risk aversion

To illustrate the concepts of risk utility function and risk aversion (see Fig. 2.6), let there be just one action a_1, with two consequences measured in terms of wealth,

---

[10] Arrow (1951) demands for the solution of the "Saint Petersburg Paradox" a function which is constrained from above and from below. In fact, the logarithmic function grows without limits as well. But this problem will not be taken up here.

2.2 Subjective Perception of Risk, Risk Aversion, and the Risk Utility Function

$W_1 (= c_{11})$ and $W_2 (= c_{12})$. Accordingly, π_1 and π_2 of Table 2.3 can be simply written as π and $(1 - \pi)$, respectively. Moreover, let $\pi = (1 - \pi) = 1/2$. This means that the expected value of wealth $EW = \pi W_1 + (1 - \pi) W_2$ is half-way in between W_1 and W_2.

An intuitive understanding of risk aversion is that a loss weighs heavier subjectively than a gain of the same size. In Fig. 2.6, let wealth W_0 (associated with utility $v[W_0]$) shown by point D serve as the initial point. Then, the loss $W_0 - W_1$ with probability 1/2 causes a decrease in utility given by the distance between $v[W_0]$ and $v[W_1]$. It must be larger than the increase in utility (the distance between $v[W_0]$ and $v[W_2]$), caused by the same but positive change of wealth with equal probability. This can only be obtained if the risk utility function $v(W)$ is concave. Now let point A show the subjective evaluation of the possible unfavorable consequence W_1. Point B shows the same for the favorable consequence. It must lie higher on the vertical axis by an arbitrary amount because more wealth is better than less wealth. Denote these two subjective values by $v[W_1]$ and $v[W_2]$, respectively. To apply the Bernoulli principle, the expected value of these two utility values, $EU = \pi \cdot v[W_1] + (1 - \pi) \cdot v[W_2]$, needs to be formed. This is a linear combination with weights $\{\pi, (1 - \pi)\}$. Graphically, linear combinations are on the straight line connecting points A and B of Fig. 2.6. If the probabilities of a gain and a loss are equal (1/2, 1/2), then the expected utility of this lottery is shown by point C, with equal distance from A and B. Note that if an alternate action a_2 were available resulting in the same consequences but with $\pi < 1/2$ and hence $(1 - \pi) > 1/2$, the corresponding point (such as A') on AB will lie northeast of C, indicating higher expected utility. The Bernoulli principle would then advise the decision-maker to opt for a_2 rather than a_1.

By assumption, the risk utility function passes through points A and B. Intermediate points can be established using the following argument. Point C of Fig. 2.6 shows the expected utility associated with the risky prospect. Now consider the offer of a *sure alternative*, D, with equal expected value of wealth EW to a risk-averse individual. He or she will value this sure alternative higher than the risky situation. Thus, point D lies above point C, implying $v[EW] > EU[EW_0]$ in Fig. 2.6. However, point D cannot lie higher than point B because this would mean that "more is less" in terms of sure wealth. This establishes the position and concavity of the risk utility function in the case of risk aversion. ■

The following statements can be deduced from the concavity of the risk utility function.

- If a risk-averse individual can choose between two risky prospects (often referred to as "lotteries") with equal gains, he or she will choose the one *with the smaller loss*. In Fig. 2.6, let two lotteries have the same consequence W_2, subjectively valued corresponding to point B. One of them generates a possible loss such that wealth W_1 results, with valuation corresponding to point A. The other results in higher wealth in the unfavorable case, with valuation corresponding to point A'. Since the line $A'B$ (not shown in Fig. 2.6) runs always above the line AB, the

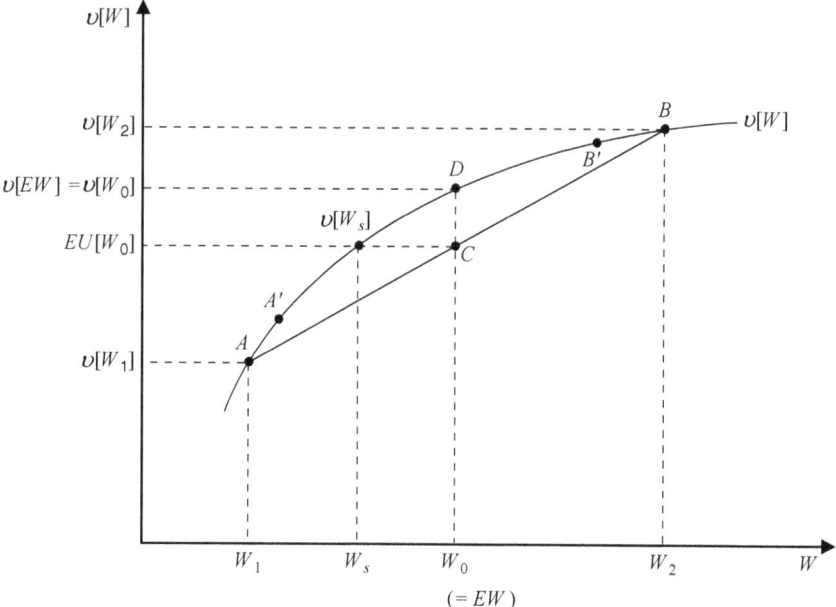

Fig. 2.6 Concave risk utility function

lottery with the smaller loss (and the same gain as the other) has higher expected utility and is therefore preferred for all possible probabilities $\{\pi, (1-\pi)\}$.
- A risk-averse individual prefers the risky prospect which has *smaller fluctuations* around a given expected value. In Fig. 2.6, the lottery characterized by the pair of utilities $\{A', B'\}$ has the same mean but smaller dispersion than the lottery with utilities $\{A, B\}$. The connecting line $A'B'$ (not shown) runs above the connecting line AB for all probabilities $\{\pi, (1-\pi)\}$, indicating higher expected utility. Hence, the lottery with the smaller fluctuations of wealth (around the common expected value EW) is preferred.
- A risk-averse individual is prepared to *pay a price to avoid* a risky prospect in favor of the sure alternative. This also can be shown with the help of Fig. 2.6. Recall that the expected utility associated with risky wealth is given by point C and that the risk utility function shows the valuation of the risk-free alternative. Now a horizontal line through point C indicates indifference. Its intersection with the risk utility function therefore shows the lower (but sure) amount of wealth that is subjectively equivalent to risky wealth with expected value EW. This is utility $v[W_s]$ pertaining to sure wealth $W_s < EW$ such that $v[W_s] = EU[EW]$. The difference between EW and the so-called *certainty equivalent* W_s is the price a risk-averse individual is prepared to pay for sure wealth. Therefore, risk aversion not only results in a (difficult to observe) difference of utilities (the vertical distance between points D and C or $v[W_0]$ and EU, respectively, in

2.2 Subjective Perception of Risk, Risk Aversion, and the Risk Utility Function

Fig. 2.6), but also in the horizontal distance between EW and W_s (which can be observed and measured in money terms). It is this *willingness to pay* for certainty that can cover the cost of insurance (acquisition, administration, risk bearing).

- Both in the case of a sure gain and in the case of a sure loss, there is *no willingness to pay* for avoiding the risky alternative. If the gain occurs with certainty (i.e., $W = W_2$ with probability $1 - \pi = 1$), the risky and the sure prospects coincide in point B of Fig. 2.6. If, to the contrary, the loss is certain ($W = W_1$ with probability $\pi = 1$), then the expected utility EU and the utility of the sure alternative $v[W_1]$ coincide in point A. One might argue that security (to be provided by insurance) is especially valued when a loss has already occurred ($\pi = 1$). However, a payment in this situation amounts to a subsidy. It is the very property of insurance that it is contracted *ex ante*, i.e., when the occurrence of a loss is still uncertain ($\pi < 1$). Hence, in points A and B, the value of the risk utility function $v[W]$ is identical with the expected utility $EU[W]$.

▶ **Conclusion 2.3** It is very plausible and in accordance with socio-biological arguments to assume that human beings behave in a risk-averse manner. This implies that when choosing between two risky prospects with the same expected value, they prefer the one with the smaller dispersion, and that downward deviations from the expected mean ("losses") are valued more than upward deviations ("gains") of equal size.

2.2.2.3 Construction of a Risk Utility Function

Evidently, it is of great interest to know whether the risk utility function $v(\cdot)$ can be constructed. As a fist step, $v(\cdot)$ is shown to be equivalent to the probability $(1 - \pi)$ of a favorable outcome after suitable normalizations.

Let the consequences be known and expressed in money units. Since only one action and two consequences will be considered as before, the set $\{c_{ij}\}$ reduces to $\{W_1, W_2\}$, with W_1 the unfavorable outcome and W_2 the favorable one (a so-called binary prospect). To simplify matters, let W_1 be the lowest level and W_2, the highest level of wealth the individual can think of. Now while cardinal utility permits to choose the zero point and the units arbitrarily, relative differences are still unambiguous.[11] Therefore, one can put $v[W_1] = 0$ and $v[W_2] = 1$. Consider some value of certain wealth between the two extremes, W_s. Assume you are capable of assessing a reference lottery with wealth levels W_1 and W_2 and a probability of occurrence π^* such that you are *indifferent* between this lottery and the certain value W_s. Therefore,

$$v[W_s] = \pi^* v[W_1] + (1 - \pi^*) v[W_2]. \tag{2.3}$$

---

[11] See the analogy with measuring temperature, which can be measured in Celsius, Fahrenheit, Kelvin, and in Réaumur with different zero points and units. However, relative differences in temperature do not depend on the choice of scale.

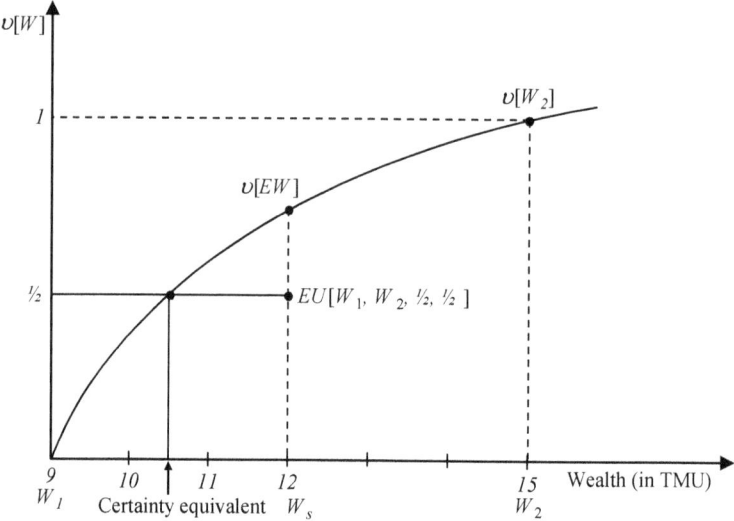

Fig. 2.7 Construction of the risk utility function $v(W)$

Now the utility associated with W_s turns out to be the probability $1 - \pi^*$ of the favorable outcome, which is perfectly measurable. Indeed, if one uses the normalizations introduced above, the right-hand side of (2.3) becomes

$$v[W_s] = v[W_1, W_2; \pi^*, (1-\pi^*)] = \pi^* \cdot 0 + (1-\pi^*) \cdot 1 = 1 - \pi^*. \quad (2.4)$$

In this way, one point on the risk utility function is constructed. By varying the value of W_s, other points can be constructed as well. In particular, by choosing $W_s = W_1$ and $W_s = W_2$, respectively, $v[W_1]$ and $v[W_2]$ are established. Since this procedure can be performed for any extreme values $\{W_1, W_2\}$ associated with an arbitrary action a_i, the $v[c_{ij}]$ values of Table 2.3 can be determined, at least for binary prospects.[12] Evidently, the risk utility function shows the subjective value of having wealth W_s rather than the risky prospect.

Construction of the risk utility function $v(W)$ is illustrated in Figure 2.7. Here, $W_1 = 9$ TMU (thousand money units), $W_2 = 15$ TMU, $W_s = 10.5$ TMU, and $1 - \pi^* = 1/2$. This means that the individual considered is indifferent between certain wealth amounting to 10.5 TMU and participating in a lottery with a 50–50 chance of final wealth of 15 TMU and 9 TMU, respectively. In this case, the certain utility associated with these consequences amounts to exactly 1/2, i.e., $v[10.5] = 1/2$.

The example also shows that $v(\cdot)$ is indeed cardinally defined and has the properties of a probability. This insight also permits constructing the risk utility function using Method No. 2.

---

[12] For the preference ordering of more general risky prospects, see Sect. 2.4.

2.2 Subjective Perception of Risk, Risk Aversion, and the Risk Utility Function

Method No. 1 for the construction of the risk utility function
The respondent can choose between the two alternatives below.

Alternative A	Alternative B
"You consider investing your wealth of 10 TMU in shares, with the probability of 50% of your wealth being 15 TMU and 50% of being only 9 TMU after one year".	"Alternatively, you consider an investment of your wealth of 10 TMU in a savings account that guarantees a wealth of W TMU after one year".
How large must W be so that A and B are *equivalent* to you?	

The construction of the risk utility function proceeds as follows:

- Alternative A is entered as expected utility $EU[W_1, W_2; 1/2, 1/2]$ in Fig. 2.7.
- For indifference, alternative B must be on the same height on the $v(W)$ axis as A. In view of the normalization introduced above, $v[W_1] = 0$, $v[W_2] = 1$. Therefore, this height is 1/2.
- Assume that the respondent indicates $W_s = 10.5$ TMU. This value determines the value on the horizontal axis. The intersection with the horizontal line at height 1/2 corresponds to a point on the $v(W)$ function.
- Because alternative B entails certain wealth, the wealth level $W = W_s$ is called the *certainty equivalent* of the risky alternative A; it is 10.5 TMU in this example.

Method No. 2 for the construction of the risk utility function
Again, the respondent can choose between two alternatives:

Alternative A	Alternative B
"You have assets amounting of 10 TMU in your savings account. At the end of the year, it is guaranteed to have grown to 10.5 TMU".	"You have shares amounting to 10 TMU, the value of which at the end of the year will amount to 15 TMU (dividends included) with probability of $(1 - \pi)$. However, with probability π, it will drop to 9 TMU".
How high must the *probability* $1 - \pi^*$ be so that you consider alternatives A and B as *equivalent*?	

- In this case, alternative A determines the location on the W-axis of the point to be found on the $v(W)$ function of Fig. 2.7.
- The horizontal coordinate is given by the answer to B (recall the equivalence between the $v(W)$ function and the probability measure).
- In the present example, the respondent indicates $(1 - \pi^*) = 1/2$.

2.3 Willingness to Pay for Certainty, Risk Aversion, and Prudence

2.3.1 Willingness to Pay for Certainty, Certainty Equivalent, and Risk Premium

For the remainder of this chapter, let the consequences listed in Table 2.3 be defined in terms of resulting (final) wealth, using the symbol W. Therefore, W_1 symbolizes final wealth in state 1 (of low wealth) and W_2 in state 2 (high wealth). The risk utility function $v(W)$ of Fig. 2.8[13] shows decreasing marginal utility of *risky* wealth, $\partial v(W)/\partial W > 0$ and $\partial^2 v/\partial W^2 < 0$. Thus, the function runs concave from the origin, indicating that a possible increase in wealth is valued less than an equally likely decrease in wealth of the same amount. An individual characterized by such a risk utility function is called *risk-averse*. In accordance with the definitions introduced above, a risk-averse individual prefers the certain wealth corresponding to the expected value of any risky prospect to that risky prospect itself.

Formally, the (certain) utility associated with the expected value of a binary prospect is given by

$$v[EW] := v[\pi W_1 + (1 - \pi) W_2]. \qquad (2.5)$$

This utility is to be compared with the expected utility of the risky prospect,

$$EU[W] := \pi \cdot v[W_1] + (1 - \pi) v[W_2]. \qquad (2.6)$$

In the case of risk aversion, it is true that $v[EW] > EU[W]$ for all $\pi \in (0, 1)$ and two arbitrary wealth levels $W_1 < W_2$. This is an application of *Jensen's inequality* for concave functions.

For deriving the willingness to pay (WTP) for certainty, one evidently needs to focus on deviations of wealth from its expected value. Let such deviations be a random variable $\tilde{X}$ with expected value $E\tilde{X}$. Then, risk aversion prevails if

$$v[EW + E\tilde{X}] > EU[EW + \tilde{X}], \qquad (2.7)$$

---

[13] The argument is written without the tilde for simplicity, i.e., W rather than $\tilde{W}$.

2.3 Willingness to Pay for Certainty, Risk Aversion, and Prudence

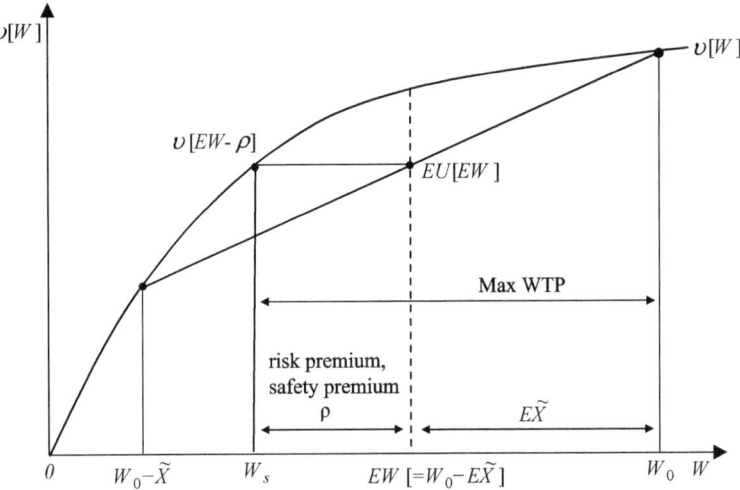

Fig. 2.8 Willingness to pay for certainty, risk premium, and certainty equivalent

which again implies that $v(\cdot)$ is concave to the origin throughout. This can be seen from Fig. 2.8, for the case $E\tilde{X} < 0$ which is typical of insurance.

Given risk aversion, a natural question to ask is what individuals are willing to pay for getting rid of a risky prospect in favor of a riskless alternative. The answer to this question lies at the very heart of the demand for insurance (see Chap. 3). Willingness to pay for certainty can be expressed in several ways:

- as the risk premium (also called safety premium) in absolute terms;
- as the maximum insurance premium the consumer would accept; or
- as the relative risk premium, expressed as a share of wealth.

Thus, willingness to pay for certainty will be used as the generic term comprising these concepts.

The first is the *risk premium* ρ. It generally depends on the individual's wealth level, the random shock $\tilde{X}$ (the amount of variation if the probability of occurrence is held constant), and the shape of the risk utility function. The risk premium has the property of making the individual indifferent between an alternative with certain wealth and one with risky wealth,

$$v[EW - \rho] = EU[EW + \tilde{X}]. \tag{2.8}$$

On the left-hand side, one has the certain utility evaluated at the expected value of wealth EW less the risk premium ρ; on the right-hand side, this is the expected utility associated with risky wealth subject to a variation $\tilde{X}$ around its expected value, EW. Since $v(W)$ is monotonically increasing in W, there must exist an inverted function

$v^{-1}(W)$ which is also monotonically increasing in W. Therefore, one can solve for ρ in the following way:

$$\rho = EW - v^{-1}[EU[EW + \tilde{X}]]. \tag{2.9}$$

The risk premium is the monetary payment that makes the individual indifferent between the risky prospect itself and the expected utility associated with the risky prospect. It indicates the maximum one is prepared to *pay for certainty*, i.e., equality of wealth in both states. One way to achieve this is to buy insurance coverage. The maximum willingness to pay for insurance (Max WTP in Fig. 2.8) can be deducted from Fig. 2.8 as well. Simply redefine expected wealth as

$$EW = W_0 - E\tilde{X}, \tag{2.10}$$

with $E\tilde{X}$: expected value of the loss.

In Fig. 2.8, the case $\tilde{X} = \{0, -X\}$ typical of insurance is illustrated. The equality of wealth across states can be achieved by buying insurance (ideally at a fair premium equal to the expected value of the loss $E\tilde{X}$). The individual then has the choice between a certain final wealth $(W_0 - E\tilde{X})$ net of premium and the risky prospect $\{W_0, W_0 - \tilde{X}; (1-\pi), \pi\}$. Given risk aversion, maximum willingness to pay (WTP) for insurance exceeds the expected value of the loss.

Returning to the risk premium ρ, one sees that it evidently depends

- on the curvature of the *risk utility function* $v(W)$ and hence the degree of risk aversion ("subjective component");
- on the *probability (density) function* of the risky prospect ("objective component"); and
- on the amount of *initial wealth* unless one makes the assumption that the curvature of the risk utility function does not depend on wealth (for more detail, see Sect. 2.4).

Panel A of Fig. 2.9 illustrates the connection between the curvature of the risk utility function and the risk premium. The more marked the curvature of the risk utility function, the greater is risk aversion, and the greater is the risk premium. Accordingly, the small risk premium ρ_1 belongs to the slightly curved risk utility function $v_1(W)$, while the larger risk premium ρ_2 is derived from the strongly curved risk utility function $v_2(W)$.

Panel B of Fig. 2.9 illustrates the dependency of the two premiums on the *probability (density) function* (the loss distribution function in an insurance context). A random variable that can assume only two values (binary prospect) follows the binomial distribution. The variance of such a distribution is given by two factors. It can be large because the two possible values of final wealth are *far apart* (i.e., the possible loss is large). In this case, the prospect with loss $\tilde{X}_2$ is associated with a greater risk premium ρ than the one associated with $\tilde{X}_1 < \tilde{X}_2$ (not shown in the figure). Second, the variance of a binary distribution can be large because the *probability of occurrence* and the counter-probability of non-occurrence are almost equal. This follows

2.3 Willingness to Pay for Certainty, Risk Aversion, and Prudence

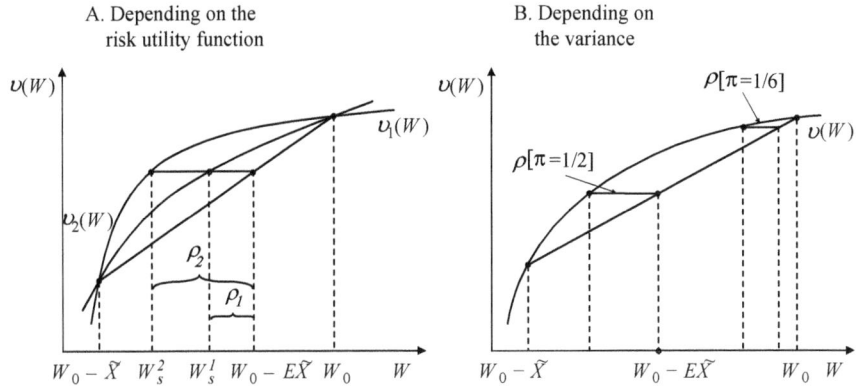

Fig. 2.9 Risk premium $\rho(W, \tilde{X})$

from the fact that the variance of a binomial random variable is $\sigma^2 = \pi(1-\pi)$, which reaches a maximum at $\pi = 1/2$. In Panel B of Fig. 2.9, the risk premium is shown for $\pi = 1/6$ and $\pi = 1/2$, respectively; it is clearly greater in the latter case.

2.3.2 Risk Premium and Coefficients of Risk Aversion

The fact that the willingness to pay for certainty depends on the shape of the risk utility function can be used to measure risk aversion (see Pratt (1964)). The more marked risk aversion, the higher is maximum willingness to pay for certainty.

The mathematical definition of the risk premium ρ can be derived from Fig. 2.8. Evidently, the individual is *indifferent* between the certain wealth W_s after deduction of the risk premium ρ and the risky prospect. Therefore, one has the following equality:

$$v[W_0 - \rho] = EU[W_0 + \tilde{X}]. \quad (2.11)$$

Let $\tilde{X}$ symbolize a variation of wealth that can be positive or negative such that $E\tilde{X} = 0$. Since this variation is typically small, ρ must be small as well. Accordingly, a Taylor approximation of the left-hand side of (2.11) can be limited to the first order,

$$v[W_0 - \rho] = v[W_0] - \rho v'[W_0] + \text{terms in } \rho^2 \text{ and higher order.} \quad (2.12)$$

For the right-hand side of (2.11), one has to acknowledge that $\tilde{X}$, while small, definitely is greater than ρ. Therefore, the Taylor approximation needs to be extended to the second order,

$$EU[W_0 + \tilde{X}] = v[W_0] + v'[W_0] \cdot E\tilde{X} + \frac{1}{2!}v''[W_0] \cdot E(\tilde{X})^2$$
$$+ E(\text{remainder}), \quad (2.13)$$

with the remainder containing terms in $\tilde{X}^3$ and higher order. Note that the value of the risk utility function and its derivatives are non-stochastic so can be factored out of the expectation operator. Moreover, one has $E(\tilde{X})^2 = E(\tilde{X} - 0)^2 = E(\tilde{X} - E\tilde{X})^2$ since $E\tilde{X} = 0$ and therefore $E(\tilde{X})^2 = E(\tilde{X} - E\tilde{X})^2 = \sigma_X^2$, denoting variance. Substituting this into (2.13) and neglecting higher order terms, one obtains[14]

$$EU[W_0 + \tilde{X}] \cong v[W_0] + \frac{1}{2}\sigma_X^2 \cdot v''[W_0]. \tag{2.14}$$

Substituting (2.12) and (2.14) into equality (2.11) yields

$$v[W_0] - \rho v'[W_0] = v[W_0] + \frac{1}{2}\sigma_X^2 \cdot v''[W_0] \tag{2.15}$$

and after division by $v'[W_0]$,

$$\rho = -\frac{1}{2}\sigma_X^2 \cdot \frac{v''[W_0]}{v'[W_0]}. \tag{2.16}$$

This can be rewritten to read

$$\rho = \frac{1}{2}\sigma_X^2 \cdot R_A, \text{ with} \tag{2.17}$$

$$R_A := -\frac{v''[W_0]}{v'[W_0]}, \qquad R_A : \textit{coefficient of absolute risk aversion}. \tag{2.18}$$

Note that while the development leading to expression (2.17) was motivated by the binary prospect shown in Fig. 2.8, it holds for any distribution function. The risk premium, i.e., the maximum *willingness to pay for certainty* of an individual, therefore is given by the product of

- (one half of) the *variance of wealth* (objective component) and
- the *coefficient of absolute risk aversion* (subjective component, generally depending on initial wealth).

The risk premium is zero if one of these two quantities is zero. Therefore, it takes a variation of wealth as well as risk aversion for a positive willingness to pay for certainty to obtain.

---

[14]The approximation (2.13) holds as an equality if wealth (more generally, the random variable considered) is normally distributed and the risk utility function is of the exponential form $[v(W) = 1 - \exp(-\rho W)]$. Pratt (1964) generalizes this result to the case of normally distributed wealth and an arbitrary (continuous and continuously differentiable) risk utility function.

2.3 Willingness to Pay for Certainty, Risk Aversion, and Prudence

▶ **Conclusion 2.4** The risk premium as a measure of the maximum willingness to pay for certainty is given by the product of subjective risk valuation (reflected by the coefficient of absolute risk aversion) and the objective variance of wealth.

The coefficient of *absolute risk aversion* R_A is a useful measure of risk aversion to the extent that it is invariant to linear transformations of the risk utility function. Such a transformation might affect the slope of the function. This problem is avoided because (2.18) contains a division by $v'[W_0]$, the slope. In view of the requirement that $v''(W) < 0$, there is a problem in case of increasing wealth since the risk utility function must continue to increase in W. This however implies that R_A must in its turn decrease with wealth beyond some point. As a consequence, very wealthy individuals are predicted to demand less insurance for their assets than less wealthy individuals.

This is an acceptable prediction as long as the variation of wealth remains constant and therefore loses importance relative to increasing wealth. After all, very wealthy individuals can self-insure to a greater extent than less wealthy ones. However, larger amounts of wealth usually are also subject to greater losses. In other words, absolute risk aversion is a plausible measure as long as the size of the risk does *not depend on wealth* (see Rothschild and Stiglitz (1970)).

Whenever wealth and risk vary in proportion, then the coefficient of *relative risk aversion* R_R is the more appropriate measure. To derive this measure, assume that the risky prospect is proportional to wealth such that $\tilde{Y} = W \cdot \tilde{X}$, with $\tilde{X}$ independent of wealth, as before. One then has $E\tilde{Y} = W \cdot E\tilde{X} = 0$. The indifference relation (in full analogy to (2.11)) reads

$$v[W_0 - W \cdot \rho^*] = EU[W_0 + \tilde{Y}], \quad \text{with } \rho = W \cdot \rho^*. \qquad (2.19)$$

Therefore $\rho^* = \rho/W$, with ρ^* denoting the proportional risk premium, i.e., the risk premium expressed as a share of wealth. Applying a Taylor approximation to the left-hand side of (2.19), one obtains

$$v[W_0 - W \cdot \rho^*] = v[W_0] - W \cdot \rho^* \cdot v'[W_0]$$
$$+ \text{ terms in } (W \cdot \rho^*)^2 \text{ and higher order.} \qquad (2.20)$$

For the right-hand side of (2.19), one has

$$EU[W_0 + \tilde{Y}] = v[W_0] + v'[W_0] \cdot E\tilde{Y} + \tfrac{1}{2}v''[W_0] \cdot E(\tilde{Y} - E\tilde{Y})^2$$
$$+ E(\text{remainder}), \qquad (2.21)$$

with the remainder containing terms in $\tilde{Y}^3$ and higher order. Equality of the two approximations and using $E\tilde{Y} = 0$ as well as $E(\tilde{Y} - E\tilde{Y})^2 := \sigma_{\tilde{Y}}^2$ yields

$$-W \cdot \rho^* \cdot v'[W_0] = \frac{1}{2} v''[W_0] \cdot \sigma_{\tilde{Y}}^2. \qquad (2.22)$$

Since $\tilde{Y} = W \cdot \tilde{X}$, its variance is $\sigma_Y^2 = W^2 \cdot \sigma_X^2$. Dividing (2.22) by $v'[W_0]$ as before, one obtains

$$\rho^* = -\frac{1}{2}\sigma_X^2 \cdot \frac{v''[W_0]}{v'[W_0]} \cdot W. \tag{2.23}$$

Evidently, a natural definition is

$$R_R := -\frac{v''[W_0]}{v'[W_0]} \cdot W = R_A \cdot W, \quad R_R : \text{coefficient of relative risk aversion}. \tag{2.24}$$

Using this definition in Eq. (2.23), one obtains

$$\rho^* = \frac{1}{2}\sigma_X^2 \cdot R_R, \tag{2.25}$$

in full analogy to (2.17).

The relative risk premium defined as a share in wealth again expresses the maximum willingness to pay for certainty. It is given by the product of

- (one half) the *variance of wealth* (objective component) and
- the coefficient of *relative risk aversion* (subjective component, which as before generally depends on initial wealth).

In accordance with Definition (2.24), the coefficient of relative risk aversion equals the coefficient of absolute risk aversion scaled up (multiplied) by the wealth of the individual.

An important advantage of the relative risk aversion measure is that its absolute value is equal to the "elasticity of the marginal value of wealth" (more precisely, of additional risky wealth),

$$e\{v'(W), W\} = -\frac{\partial v'(W)}{\partial W} \cdot \frac{W}{v'(W)} = -\frac{v''(W)}{v'(W)} \cdot W. \tag{2.26}$$

The measure of relative risk aversion therefore indicates by how many percent the slope of the risk utility function approximately decreases when risky wealth increases by one percent. It thus provides a measure of a function's curvature that is invariant to the choice of units when measuring wealth.

▶ **Conclusion 2.5** The relative risk premium ρ^* indicates the maximum willingness to pay for certainty expressed as a share of wealth.

For the sake of completeness, a third measure of risk aversion is presented, *partial risk aversion* (without proof). Partial risk aversion answers the question of how willingness to pay for certainty changes when only the size of the variation $\tilde{X}$ varies, while wealth remains constant. Note the difference from relative risk aversion, where

2.3 Willingness to Pay for Certainty, Risk Aversion, and Prudence

wealth is assumed to increase, causing the size of losses to increase as well. Let $\beta := \tilde{X}/W_0$ be a partial variation with wealth constant; therefore, $W_0(1+\beta)$ below is the same as $(W_0 + \tilde{X})$ in (2.11), motivating the definition,

$$R_P := -W \cdot \frac{v''[W_0(1+\beta)]}{v'[W_0(1+\beta)]}, \quad R_P : \textit{coefficient of partial risk aversion.} \quad (2.27)$$

Again without proof, the three measures of risk aversion can be shown to be related as follows:

$$R_P = R_R - \tilde{X} \cdot R_A. \quad (2.28)$$

Since the risk premium ρ as defined in (2.11) varies as a function of wealth and the risky prospect, one can prove the following statements (see Menezes and Hanson (1970), 485):

$$\frac{\partial \rho(W, \tilde{X})}{\partial W} \gtreqless 0, \quad \text{if} \quad \frac{\partial R_A}{\partial W} \gtreqless 0. \quad (2.29)$$

The absolute risk premium increases (decreases) with wealth when the coefficient of absolute risk aversion R_A increases (decreases) with wealth.

$$\frac{\partial \rho\left(\frac{\beta W, \beta \tilde{X}}{\beta}\right)}{\partial \beta} \gtreqless 0, \quad \text{if} \quad \frac{\partial R_R}{\partial W} \gtreqless 0. \quad (2.30)$$

The absolute risk premium increases (decreases) when both wealth and its variation increase (decrease) by the factor β if the coefficient of relative risk aversion R_R increases (decreases) with wealth.

$$\frac{\partial \rho\left(\frac{W, \beta \tilde{X}}{\beta}\right)}{\partial \beta} \gtreqless 0, \quad \text{if} \quad \frac{\partial R_P}{\partial W} \gtreqless 0. \quad (2.31)$$

The absolute risk premium increases (decreases) when the amount of variation but not wealth increases (decreases) by the factor β if the coefficient of *partial risk aversion* R_P increases (decreases) with wealth.

Arrow (1965) and Pratt (1964) hypothesize that *absolute* risk aversion *decreases* with wealth. An analogous statement is not available for *relative* risk aversion. The empirical evidence is discussed in Sect. 2.4.

2.3.3 Prudence and Higher Order Derivatives of the Risk Utility Function

Soon after the pathbreaking contributions by Arrow (1965) and Pratt (1964), researchers went beyond the concept of risk aversion. They considered insurance as one among several instruments of risk management such as stockpiling and the building of reserves (see Sect. 2.5). In the present context, the build-up of reserves is equivalent to accumulating more initial wealth through savings. However, there are many motives for savings. One is the smoothing of consumption over time, and another the desire to bequeath wealth to one's spouse or children. The motive most germane to risk management is precaution: prudent consumers accumulate wealth in order to be better prepared for a risky future. One may think of unexpected expenditure for treatment of an illness or planned outlays that turn out to be unexpectedly high (for instance, when moving house), or a shortfall of income due to unemployment or bad health. Therefore, precautionary saving is induced by the fact that future income and wealth is risky rather than predetermined. This type of risk is of concern for consumers who cannot diversify risks through the capital market; it therefore needs to be distinguished from the rate-of-return risk confronting someone who has to decide which assets to hold (see Sandmo (1970)).

Evidently, precautionary saving serves an insurance function, and demand for it should respond to an increase in future risk in a similar way. Typically, the analysis is couched in terms of a trade-off between current and future consumption. If total utility is additive in (discounted) per-period utilities, then Leland (1968) and Sandmo (1970) show that precautionary saving is a response to increased risk if the third derivative of the utility function is positive. Starting with Kimball (1990), this property is called *prudence*.

The importance of prudence can be seen when returning to the Taylor approximation (2.13) that permitted to derive the risk premium. There was no need to stop at the second-order term. Consider adding a third term to the Taylor series,

$$EU[W_0] = v[W_0] + \frac{1}{2!}\sigma_X^2 \cdot v''[W_0] + \frac{1}{3!}\sigma_X^3 \cdot v'''[W_0]$$
$$+ E(\text{remainder}), \text{ with} \tag{2.32}$$
$$\sigma_X^3 := E(\tilde{X} - E\tilde{X})^3. \tag{2.33}$$

Recall that willingness to pay for certainty is defined as the amount that can be deducted from initial wealth W_0 such that the individual is indifferent between the risky and the riskless alternative. However, this amount now also takes into account the possible skewness σ_X^3 of the variation in wealth rather than just its variance σ_X^2.

Using $v[W_0 - \rho^{**}] = EU[W_0 + \tilde{X}]$ as in (2.11), where ρ^{**} (again dropping its arguments for simplicity) denotes the risk premium including a prudence effect, one can solve in full analogy to Eq. (2.16) to obtain

$$\rho^{**} = -\frac{1}{2!}\sigma_X^2 \cdot \frac{v''[W_0]}{v'[W_0]} - \frac{1}{3!}\sigma_X^3 \cdot \frac{v'''[W_0]}{v'[W_0]}. \tag{2.34}$$

2.3 Willingness to Pay for Certainty, Risk Aversion, and Prudence

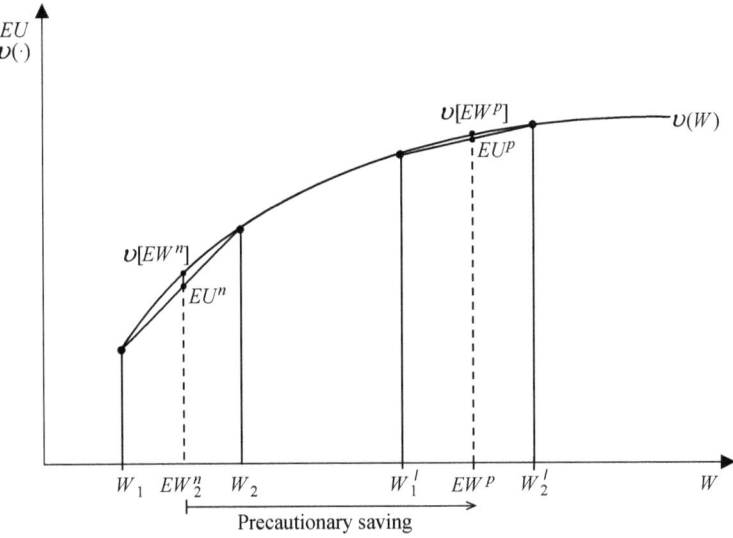

Fig. 2.10 Precautionary saving as a means to reduce the "Pain of Risk Bearing"

As before, the first term says that $v''[W_0] < 0$ is necessary for a positive risk premium, i.e., willingness to pay to avoid a wealth risk characterized by $\sigma_X^2 > 0$. The additional term first involves σ_X^3, i.e., the skewness of the distribution of wealth. It is natural to assume that there should be an increase in the risk premium if the consumer is able to avoid a *downside risk* characterized by $\sigma_X^3 < 0$, meaning that very low values of wealth can occur, albeit with low probability. Equivalently, the loss distribution is skewed in a way that extremely high losses can occur, but with low probability (see Fig. 2.1 again). Being symmetric, the variance σ_X^2 neglects this downside risk. Taking it into account should increase willingness to pay for certainty. However, in view of (2.34) and with $\sigma_X^3 < 0$, this requires that $v'''[W_0] > 0$. One therefore has

$$\rho^{**} > \rho \text{ if } \sigma_X^3 < 0 \text{ and } v'''[W_0] > 0. \tag{2.35}$$

A risk utility function with $v'''(W) > 0$ exhibits *prudence*.

If prudence characterizes human behavior, decision-makers are predicted to especially avoid downside risk. One way to mitigate downside risk is precautionary saving. The role of precautionary saving is illustrated in Fig. 2.10. The risk utility function depends on wealth only (strictly speaking, wealth in the subsequent period because savings take place during the current period; however, this detail is neglected for simplicity of notation). Without precautionary savings, let wealth take on the values W_2 or W_1 with probability 0.5 each. The difference between the certain utility $v[EW^n]$ and the expected utility EU^n reflects the disutility associated with risk bearing, or "pain" in the words of Eeckhoudt and Schlesinger (2006). Their basic insight (which will be crucial for deciding between the two state-dependent risk utility functions in Sect. 3.2.2) is that people seek ways to apportion risks in ways as to reduce this pain. One way to achieve this objective is to shift to a higher level of wealth in case the unfavorable state of the world materializes. Consider an amount of precautionary saving that moves the lower value of wealth from W_1 to W_1'. Since the risk to be faced remains the same because uncertainty of returns to savings is disregarded, W_2 shifts by the same amount, to W_2'. Now for this shift to "ease the pain" of risk bearing, the difference between $v[\overline{E W^p}]$ and $E U^p$ (with p denoting the effect of precautionary saving) must decrease. However, Fig. 2.10 shows this to be possible only if the curvature of the risk utility function diminishes with wealth. Since $v''[W_0] < 0$ indicates curvature, this implies $v'''[W_0] > 0$. Note that implicitly (actuarially fair) insurance was assumed to be available in both situations, permitting the consumer to reach riskless utilities $v[EW^n]$ and $v[EW^p]$, respectively.

▶ **Conclusion 2.6** In the presence of insurance, precautionary saving is predicted if the third derivative of the risk utility function is positive, i.e., if decision-makers exhibit prudence and downside risk aversion.

In the special case of constant relative risk aversion (CRRA), Kimball (1990) shows the coefficient of relative prudence (defined by $-v'''[W_0] \cdot W_0 / v''[W_0] > 0$; note the difference from (2.34)) is equal to the coefficient of relative risk aversion as defined in (2.24) plus 1. Therefore, prudence must result in a greater share of wealth being held in safe or insured forms than would be predicted from risk aversion alone.

Finally, note that the Taylor approximation in (2.32) could have been extended to include still more terms. For instance, the fourth-order term would read $-1/4! [\sigma_X^4 v^{IV}/v']$. If σ_X^4 is positive, there is excess probability mass in the tails (positive kurtosis), something people who are concerned about risk would want to eschew. This extra term augments willingness to pay for safety if $v^{IV} < 0$, i.e., if the fourth derivative of the risk utility function is negative. Eeckhoudt and Schlesinger (2006) call this property *temperance*, and they show that avoiding the pain of risk bearing implies that the signs of the derivatives alternate such that $\text{sign}[v^{(j)}] = (-1)^{j+1}$, where j denotes the order of the derivative.

2.4 Estimates of Risk Aversion

Since willingness to pay for certainty crucially depends on the strength of risk aversion (see Conclusions 2.4 and 2.5), demand for insurance is expected to also increase as a function of risk aversion. Indeed, this is an important prediction derived from the theory of insurance demand (see Sect. 3.3.2, notably Conclusion 3.5). For this reason, it is of considerable interest to know whether there are differences within a population, and how marked risk aversion is.

2.4.1 Microeconomic Evidence

Risk aversion is a characteristic that likely varies within a population. Three important influences that may lead to systematic differences are a person's wealth, age, and gender. They are considered in turn.

2.4.1.1 Risk Aversion and Initial Wealth

Final wealth is the result of stochastic shocks and risk management efforts (in particular, insurance coverage) and therefore endogenous. By way of contrast, initial wealth (denoted W_0 in Sect. 2.3) can be regarded as exogenous in many circumstances. The main hypotheses and findings follow.

- *Decreasing absolute risk aversion* (DARA, see end of Sect. 2.3.2): This is very intuitive and hardly disputed.
- *Increasing relative risk aversion* (IRRA): According to Conclusion 2.5, IRRA has the observable consequence that rich individuals devote a higher share of their wealth to keeping them safe than do poor ones. One can also say that protecting one's assets from losses is a luxury good, claiming an increased share of wealth among rich consumers. Conversely, the share of risky assets is predicted to fall with individuals' wealth. However, Friend and Blume (1975) frequently cannot reject the hypothesis of constant relative risk aversion (CRRA), depending on whether housing property is counted as a riskless asset or whether or not human capital is included.
- *Constant relative risk aversion* (CRRA): Across all wealth levels, this hypothesis could not be rejected by Siegel and Hoban (1982) as soon as they excluded housing property from total assets. However, this overall result conceals differences within subgroups. In the lowest wealth bracket, there was evidence of IRRA, whereas individuals in the highest bracket exhibited decreasing relative risk aversion (DRRA). Similar findings are presented by Morin and Suarez (1983).
- *Decreasing relative risk aversion* (DRRA): Evidence strongly supporting this hypothesis comes from Blake (1996), who investigated the demand of British households for risky assets. A risky asset is characterized by an expected rate of return that exceeds the risk-free rate of interest. The share of these higher yield assets then reflects relative risk aversion. It strongly increases with initial wealth, pointing to DRRA.

2.4.1.2 Risk Aversion and Individual Characteristics

Two characteristics that can be considered exogenous (unlike, e.g., education, which constitutes an investment that may be influenced by risk aversion) are a person's age and sex.

- *Risk aversion and age:* Riley and Chow (1992) were among the first to relate decisions regarding the allocation of assets to age. They concluded that risk aversion first decreases with age but increases after age 65. Other authors have used hypothetical gambles to measure risk preferences. Barsky et al. (1997) recorded responses by participants in the U.S. Health and Retirement Study. Relating them to one characteristic at a time, they find an inverted U-shaped relationship, with risk aversion attaining a maximum in the 55 to 70 age group. Confronting respondents with a hypothetical income gamble while controlling for other influences, Halek and Eisenhauer (2001) find a negative relationship between relative risk aversion and age, but with a marked increase after age 65, confirming the finding by Riley and Chow (1992). This is intuitive because people find it difficult to make up for a shortfall in income and wealth after retirement.

- *Risk aversion and gender:* One frequently encounters the stereotype that women are more risk-averse than men, especially when it comes to financial matters. Indeed, Riley and Chow (1992) found that females opted for less risky asset allocations than men. Similarly, Barsky et al. (1997) find men to be significantly less risk-averse than women. Moreover, highly risk-averse individuals tend to have more health insurance coverage; conversely, respondents without life insurance are substantially less risk-averse than those with it, confirming Conclusion 3.5 of Sect. 3.3.2. Controlling for several other influences, Halek and Eisenhauer (2001) also conclude that females exhibit stronger relative risk aversion than do males. However, findings of this type are subject to an important criticism voiced by Schubert et al. (1999). They argue that (as always in economics) observed phenomena are to be interpreted as optima that result from the matching of preferences and constraints limiting feasibility sets. Therefore, observed differences in risk aversion could be attributable to differences in *individual-specific feasibility sets*. In a controlled experiment, they can indeed relate differences between men and women in terms of risk aversion to differences in their financial situation. Conversely, when these differences are controlled, gender-specific differences in risk preference disappear.

2.4.2 Macroeconomic Evidence

One of the few studies using macroeconomic data is by Szpiro (1986a), seeking to test the constancy of *relative risk aversion* $[R_R(W) = W \cdot R_A(W)]$ with respect to wealth. Whereas the hypothesis of decreasing absolute risk aversion as a function of wealth $[R'_A(W) < 0]$ advanced by Arrow (1965) and Pratt (1964) is generally accepted, such a consensus is lacking in the case of relative risk aversion.

The point of departure is the formula

$$I = W - \lambda/R_A[W]. \qquad (2.36)$$

Here, I denotes claims payments by insurers, W is wealth, $\lambda > 0$ is the proportional loading for administrative expense and risk bearing (see Sect. 3.3), and $R_A[W]$, the coefficient of absolute risk aversion. Dividing this formula by W, one obtains

2.4 Estimates of Risk Aversion

$$\frac{I}{W} = 1 - \frac{\lambda}{W \cdot R_A[W]} = 1 - \frac{\lambda}{R_R[W]}. \tag{2.37}$$

Therefore, the *part of wealth covered by insurance* is

- = 1 at a maximum (if $\lambda = 0$, see Sect. 3.3.1);
- the more below 1, the higher the proportional loading λ; and
- approaching 1 with an increasing coefficient of relative risk aversion $R_R = W \cdot R_A$ (see Sect. 2.3.2).

If now R_R were independent of wealth, then insured wealth as part of total wealth would have to be independent of wealth as well, at least as long as the loading λ remains constant.

Szpiro uses aggregate data for I/W and λ in order to estimate a nonlinear regression linking R_R to W. In order to build this link in a simple way, he proposes the risk utility function,

$$v(W) = \frac{W^{1-\gamma}}{1-\gamma}, \quad \gamma \neq 1. \tag{2.38}$$

Indeed, (2.38) qualifies as a risk utility function because $v'(W) = W^{-\gamma} > 0$ and $v''(W) = -\gamma W^{-1-\gamma} < 0$, establishing concavity from below. Using these derivatives, one obtains

$$R_R := \{-v''(W)/v'(W)\} \cdot W = \gamma. \tag{2.39}$$

Therefore, $\gamma > 0$ equals the *coefficient of relative risk aversion*. Szpiro uses $R_R(W) = R_A(W) \cdot W$ again to solve for $R_A(W) = \gamma/W$. By introducing the parameter h, he is able to test for all the possibilities distinguished in Sect. 2.4.1.1,

$$R_A(W) = \frac{\gamma}{W^h}; \tag{2.40}$$

$$R_R(W) = \frac{\gamma \cdot W}{W^h} = \frac{\gamma}{W^{h-1}}. \tag{2.41}$$

The value of h thus determines the way absolute and relative risk aversion vary with wealth.

$h = 0$: Constant absolute risk aversion (CARA);
$h > 0$: Decreasing absolute risk aversion (DARA);
$h > 1$: Decreasing relative risk aversion (DRRA);
$h < 1$: Increasing relative risk aversion (IRRA);
$h = 1$: Constant relative risk aversion (CRRA).

Substituting (2.41) into (2.37), Szpiro (1986a) is able to relate observable I/W ratios to W and h in a nonlinear regression. For major insurance markets, he frequently found values around $h = 1$, confirming the hypothesis of *constant relative risk aversion*. For example, R_R varied between 1.2 and 1.8 for the United States. Using $R_R = 1.5$ as a representative value, the results can be interpreted in the light

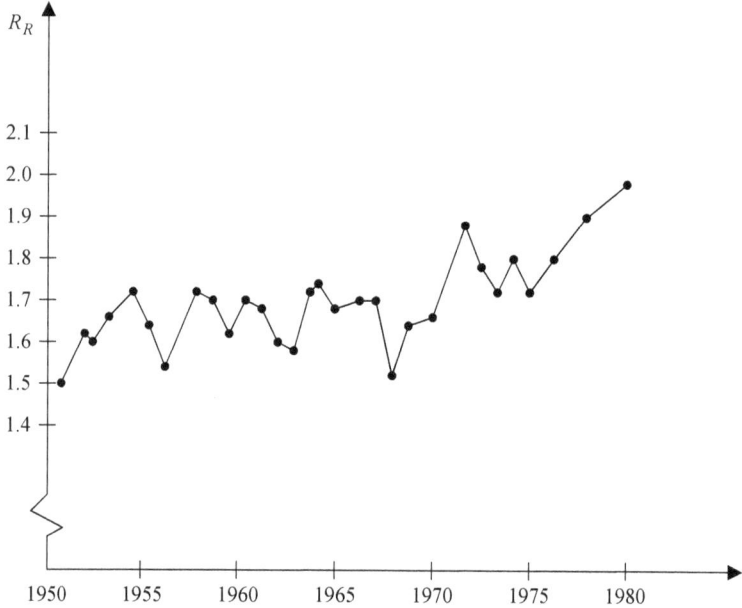

Fig. 2.11 Coefficient of relative risk aversion, Switzerland (1950–1980). *Source* Szpiro (1986b)

of (2.25) as follows. If the volatility of wealth σ_X^2 were to double (increase of 100%), relative willingness to pay for certainty ρ^* is predicted to increase by 75% ($= 1/2 \cdot 100 \cdot 1.5$), from 10% to 17.5% of wealth (say).

Note that estimated coefficients of relative risk aversion, while not systematically varying with wealth, still fluctuate over time. For instance, R_R varies between 1.5 and 2.1 during the period 1950–1972 in the case of Switzerland (see Fig. 2.11). More specifically, R_R is rather stable until 1972 but then jumps to the higher value in 1973 for the first time, a pattern observed in several countries. This jump may well be a statistical artifact, however. While assets lost value after the oil price shock of 1973, insurance payments I could not be adjusted quickly, being typically based on contracts struck in previous years. According to (2.37), R_R must increase in this case. The second increase of R_R in the years 1979 and 1980 can be explained in an analogous way as the consequence of the second oil price shock of 1979. In addition, however, risk aversion may have indeed increased because the two oil price shocks terminated a phase of continuous economic growth during the 1960s. With the transition to flexible exchange rates in 1973, economic agents might have become more aware of financial risks associated with international transactions (see Sect. 2.2.1). By 1979, capital markets had begun to develop at an unprecedented pace.

▶ **Conclusion 2.7** Risk aversion is not only a theoretical concept. It can be determined using insurance data and behavioral experiments. The evidence points to a constant coefficient of relative risk aversion with an overall value around 2.

2.5 Instruments of Risk Management

The different alternatives of action for dealing with risk can be viewed as instruments of risk management. In the literature, several measures are discussed that economic agents employ in an attempt to protect themselves against the unexpected occurrence of perils (for the risk management of insurance companies (IC), see Sects. 5.4–5.8).

If (new) information becomes available about such events (be they desired or undesired), decision-makers first of all *adjust their economic plans*—provided the expected benefit exceeds the cost of adjustment, of course. Thus, human planning usually is of the "flexible" or "rolling" rather than the "once-and-for-all" type. Indeed, optimization requires that the incoming information about uncertain future situations is taken into account when evaluating the consequences of a decision.

Besides this revision of plans, there are two types of risk management measures:

- Measures targeting the causes of the risk (etiological) and
- Measures targeting the amount of damage (palliative).

(1) *Etiological measures.* These amount to preventing risks and—in the extreme—avoiding them. *Risk avoidance* means renouncing to a risky activity, hence limiting one's choice to riskless activities. The use of sure and known technology rather than new, untested processes or products is instances in point. However, postponement of decisions until additional or more reliable information is available and incrementalism (making small changes only) also constitute risk avoidance. At the societal level, incrementalism corresponds to the "social piecemeal engineering" of Karl Popper (1945).
Risk prevention comprises measures designed to lower the probability of occurrence of a loss, such as the use of non-flammable materials in building and clothing, double walls of oil tanks, and measures against theft. If they constitute a behavioral adjustment rather than a material investment (such as driving at a reduced speed), they are called self-protection (see Sect. 3.5). Sometimes risk diversification is also counted as prevention; however, this leaves the probability of occurrence of an individual loss unaffected and is therefore considered a palliative measure.

(2) *Palliative measures.* These are designed to reduce the damage once a peril has materialized. Usually one distinguishes between risk taking, risk reduction, risk splitting, and risk transfer. *Risk taking*, i.e., the conscious assumption of a risk, is limited by the wealth of the individual, who moreover may stay away from this alternative because of risk aversion.
Risk reduction or self-insurance (see Sect. 3.5) contains all measures aiming to limit damages, e.g., sprinklers or alarm devices.
Risk splitting serves to relax the individual wealth constraint. Here, a further distinction has to be made between (a) individual and (b) collective measures. *Individual measures* comprise the holding of buffer stocks and reserves (e.g., stocking canned food to survive a failing harvest), assortment policy (e.g., holding more than one type of canned food to avoid malnutrition), diversification

of wealth (e.g., investing in several stocks and bonds rather than just one), and hedging (e.g., holding a foreign currency that tends to appreciate when the domestic currency loses value). *Collective measures* typically are undertaken by the government such as volume and price stabilization in agriculture and macroeconomic stabilization policy—and, of course, government-mandated social insurance.

The same distinction between *individual* and *collective measures* can be made with regard to *risk transfers*. At the individual level, contracts often are used to transfer risks from one party to another. For example, a seller may issue a warranty for the product, thus permitting the buyer to get rid of the risk of product failure. Or an employer may agree to pay wages during a layoff period, thus assuming the risk of low demand in a recession. However, such a clause in the employment contract could be the result of negotiations between unions and employers, making it a collective measure. Thus, the distinction between *individual* and *collective measures* is fuzzy here. This is due to the operation of markets that involve many people but are based on contracts struck between individuals. *Insurance markets*, of course, serve the purpose of shifting risks from consumers to insurers, who are able to neutralize them to a great extent by grouping together many clients (most of whom do not incur a loss during a given period). Other financial markets, ranging from currencies to securities, are discussed in Chap. 4, dealing with portfolio theory and the IC as a financial intermediary.

The difference between securities and insurance markets, in particular, can be characterized as follows: on *markets for securities*, shares of companies and hence risks associated with companies are traded. Owners of stock seek to transfer to other market participants those risks they do not want to bear themselves by structuring their portfolios accordingly. By way of contrast, non-diversifiable risks are generally traded on *insurance markets*. The ICs take these risks, divide them between themselves through co-insurance and reinsurance, and retain those they deem profitable in view of the losses they are likely to produce. To the extent these risks are uncorrelated or even negatively correlated, they can be aggregated to form an underwriting portfolio whose total loss dispersion is reduced. By specializing in these activities, ICs can thus diminish transaction costs and might even reap increasing returns to scale (see, in particular, Sect. 6.4).

Besides this, there always existed *"altruistic"*, voluntary risk transfers within the family, between friends, and through charitable institutions. They are induced and governed by social norms. Even more important is the *risk transfer by law*, notably through liability rules. For instance, consider a traffic accident. To the injured, this constitutes an externality caused by the driver responsible. However, liability makes the driver responsible for the damage. Thus, liability can be seen as an instrument for internalizing risky externalities. Yet, liability may in turn be limited. In particular, the owners of a limited liability company do not have to wager their fortunes in full should the company go bankrupt. Rather, they are allowed to shift part of this risk to third parties, e.g., creditors and other holders of claims against the company.

To prevent opportunistic behavior, legal supervision and creditor protection were introduced. Thus, the effects of liability rules vary depending on their precise form, which becomes especially evident in environmental and product liability.[15] Laws providing protection against layoffs, imposing minimum wages and gender or race quota also transfer risks; however, they are mainly motivated by social policy. This is especially true of social security and social insurance, which nowadays govern about one-quarter of the Gross National Product (GNP). Chapter 9 is entirely devoted to social insurance because of its importance.

2.6 Effectiveness of Risk Management and Risk Policy Measures

The use of instruments of risk management (at the individual or firm level) and risk policy (at the societal level) can be evaluated according to different criteria, both one by one and in combination.

Risk management instruments are employed by ICs. Therefore, they are referred to as *insurance technology* (which will be discussed in detail in Chap. 6). These instruments need to be evaluated with respect to (1) their goal conformity and (2) their cost–benefit ratio.

(1) *Goal conformity.* Instruments should be suitable to reach or foster a certain goal. Difficulties arise as soon as there are joint effects (effects of scope) and spillovers (externalities) from one measure to another.[16] A necessary condition for the solution of this problem is that the number of goals is not larger than the number of instruments. So it could be the case that—to take up the examples of Sect. 2.5— risk transfers motivated by social policy are not in conformity with certain goals. Indeed, they can lead to *adverse selection processes* (see Sect. 7.3) which hurt precisely the groups that should be favored (e.g., layoff protection for women, which may cause employers not to hire women in the first place). Sometimes, risks are simply shifted (e.g., from the labor market to society as a whole), resulting in diminished macroeconomic stability. Another instance is the possible side effect of minimum wage legislation, which transforms the risk of too low wages into one of unemployment.

---

[15]In Europe, e.g., (bi)annual exhaust controls are common. Car owners have to bear their cost as well as the risk of being sanctioned for non-compliance. In the United States, these controls are randomly performed on the road; moreover, car makers are obliged to recall and repair at their expense vehicles that fail to meet anti-pollution standards.

[16]In the theory of economic policy, this is discussed as the "assignment problem". The goals z_i can be seen as functions of the instruments t_j. Then for $\{z_1 = f_1(t_1, \ldots, t_n), \ldots, z_n = f_n(t_1, \ldots, t_n)\}$ the diagonal elements $\partial f_i/\partial t_i$ must be large while the off-diagonal elements $\partial f_i/\partial t_j$ for $i \neq j$ must be small. Furthermore, $\partial f_i/\partial t_i > \sum |\partial f_i/\partial t_j|$, i.e., the side effects of an instrument on the other goals must be smaller than the direct effects to ensure goal conformity.

(2) *Favorable cost–benefit ratio*. The criterion of goal conformity is not sufficient because it allows only statements about the suitability of an instrument in principle. A more refined evaluation of measures has to take into account their effectiveness. This calls for a *cost–utility* or *cost–benefit analysis* (CBA). The benefits (utility) of a measure are pitted against its cost. In the most simple case, benefits and costs can be measured in money, permitting the net gain to be calculated as the difference between benefits and costs. In CBA, not only the main effects but also the side effects (direct and indirect benefits, direct and indirect costs) need to be quantified.

However, benefits and costs frequently are not readily measurable in monetary terms. The leading case is the saving (or more precisely, prolongation) of human lives, especially in the areas of health, transportation, and environmental policy. At the level of an enterprise (specifically, an insurance company), the measurement problem is less relevant as long as management pursues the objective of expected profit (see Sect. 5.2.1 on the objectives of an IC). Therefore, the following examples relate to policies at the societal level.

In the *healthcare* sector, the development of new pharmaceuticals, the introduction of new surgical procedures, and the creation of medical facilities lead to enormous financial expenditures falling on health insurers or the government but ultimately on the insured or taxpayer. They have to be balanced against therapeutic advantages in the guise of a reduced risk of premature death, but increasingly also of an improved quality of life.

In *transportation policy*, the issues are the building of safer roads and freeways, the introduction of speed bumps in residential areas, or the mandate to wear safety belts for drivers and helmets for motorcyclists and bicyclists. These measures have costs in terms of money, but also loss of utility. Their benefit is the lowered risk of traffic accidents causing deaths and injuries.

Similar examples can also be cited for *environmental policy*. Installing filters in coal-fired power plants serves to reduce sulfur dioxide and other noxious emissions, to improve air quality and hence to lower the risk of respiratory disease. The removal of asbestos from schools and other public buildings helps to prevent asbestosis and lung cancer. However, depending on the safety standards applied, this can be a very costly way to save lives. If the norms of the U.S. Occupational Safety and Health Agency (OSHA) have to be satisfied, the estimate is U.S.$ 89.3 million for a statistical life saved. This even rises to U.S.$ 104.2 millions per life saved under the norms of the U.S. Environmental Protection Agency (EPA) (see Morall (1986)).

Estimates of this type raise a basic issue. Is it permissible to *value a human life* in monetary terms? And provided the answer to this question should not be "No" upfront, can the amount be anything else but infinite? This debate will not be taken up here (see, in particular, Zweifel et al. (2009, Sect. 2.3)); may it suffice to note that many private and public decisions simply do require a trade-off between the saving or the prolongation of statistical (rather than identified) human lives and other things of value. However, the examples cited have made clear this much: Given the necessity of such trade-offs, they should occur in a consistent manner, reflecting the preferences of the population (and not only of policy-makers).

To perform such an evaluation, there are three methods: (a) cost-effectiveness analysis, (b) cost–utility analysis, and (c) cost–benefit analysis. They can be ordered into the following hierarchy with respect to their contribution to decision-making.

(a) *Cost-Effectiveness Analysis* (CEA) does not require benefits to be measured in money. It thus allows to circumvent the difficulties connected with the valuation of human life. For instance, let the predetermined objective be to reduce the number of deaths from traffic accidents by x, and let there be different measures to attain this goal. Then, CEA advises the decision-maker to select the measure with minimum cost. Note, however, that minimizing $cost/x$ (average cost) need not coincide with minimizing $dcost/dx$ (marginal cost). Frequently, a higher degree of goal attainment comes at an increasing marginal cost. The problems with CEA are twofold. First, it cannot be applied to interventions that have effects beyond health. For instance, building a better road may not only reduce the number of traffic accidents but contribute to the economic development of a region. However, this additional benefit cannot be accounted for in a systematic way. Second, CEA does not tell the decision-maker where to stop. Maybe it would be worthwhile to reduce the number of deaths by more than x although the marginal cost $dcost/dx$ increases.

(b) *Cost–Utility Analysis* (CUA) goes one step further by distinguishing several objectives that may be advanced (or detracted from) by a measure. These effects are then weighted by marginal utilities and summed to obtain a utility value. The CUA criterion suggests that the decision-maker choose the alternative with the highest utility–cost ratio. The advantage of CUA is that its several steps (choice of objectives, effectiveness of measures considered, and utility weights applied) are made transparent. However, it still has two important weaknesses. First, utility is a subjective concept. The use of expert opinion does not solve the problem of lacking objectiveness because the preferences of an expert do not count more than those of any citizen (at least in a democratic society). Second, once more the CUA criterion fails to indicate where to stop. A well-known example is the rule adopted by the National Health Service of the United Kingdom stipulating that interventions costing more than £30,000 per Quality-Adjusted Life Year (QALY) are not to be paid for by the Service.[17] But possibly UK citizens would be willing to spend more (or less) money per QALY.

(c) *Cost–Benefit Analysis* (CBA) makes the biggest contribution to decision-making. Only CBA can give an answer to the question of whether a certain public project is worth the sacrifice of other goods and services. However, this great advantage can

---

[17] QALYs are calculated by assigning a weight of 1 to a year spent in perfect health and of less than 1 in less than perfect health. The weights are typically derived from expert opinion (from physicians, nurses, etc.).

only be attained if both costs and benefits are measured in money, the yardstick of those other goods and services. In many cases, this requires putting a monetary value on a (statistical) human life. The *human capital approach* uses foregone labor income as a measure of value. This implies that the rest of society (through the labor market) puts a value on a good (a person's life), whereas in economic theory, this is the individual. Apart from ethical concerns, this inconsistency makes the human capital approach unsuitable (see, e.g., Zweifel et al. (2009, Sect. 2.3)). *The willingness-to-pay (WTP) approach* avoids this difficulty. Note that it does not relate to the question "live or die", but rather to small changes in the probability of death during a certain time interval. The marginal willingness to pay can be defined as the marginal rate of substitution between a small change in wealth (or income) and a small reduction in the risk of dying.

For WTP measurement, economists have been trying to find evidence on actual choices. Table 2.4 contains a few examples. They relate to compliance with safety belt laws, purchase of smoke detectors, purchase of homes in areas with air pollution, cigarette smoking, purchase of new (safer) cars, equipping cars with a baby seat, living away from a waste disposal site with a health risk, and wearing a bicycle helmet. Each of these decisions concerns an activity that changes the probability of death by a fraction of a percentage point. For example, consider the study by Blomquist (1979) estimating the value of a statistical life from drivers' decision to wear the safety belt. Prior to the legal mandate, 17.2% of them were regular users, 9.7% frequent users, and 26% rare users. A full 46.6% never wore the belt. To obtain a cost estimate, Blomquist (1979) puts the time used for buckling up at 8 seconds. Valued at the average wage rate of drivers, the disutility cost of wearing the safety belt amounts to U.S.$ 45 annually. On the benefit side, buckling up is known to lower the annual probability of death in road traffic by 0.00375 percentage points (or $3.75 \cdot 10^{-5}$, respectively). Extrapolated to 100 percentage points for one statistical life, one obtains a value of U.S.$ 1.2 mn. Thus, one can say that at the time, U.S. drivers behaved in a way as though a human life was worth U.S.$ 1.2 mn.

The estimates of Table 2.4 show that the more recent studies yield higher WTP values than the earlier ones (all values in 1990 US$). This impression is confirmed by Table 2.5, which contains a survey of fatal occupational injuries in the United States covering years up to 2002. Whereas Table 2.4 exhibits values of US$ 1 mn. or even lower (at 1990 prices), the value of a statistical live can be said to be at least US$ 6 mn. (at 2012 prices; US$ 6.84 mn. at 2020 prices using the U.S. Consumer Price Index). The likely reason is income growth; when people get richer, they put a higher value on their chance of survival provided the marginal utility of wealth decreases with income (see, e.g., Zweifel et al. (2009), Sect. 2.4.3). This explanation is confirmed by the first entry of Table 2.5, where the estimate by Viscusi (2003) is as high as US$ 21.5 mn. for Whites but only US$ 10.3 (both at 2012 prices again) for Blacks, who historically have lower incomes. Similarly, the study by Hersch and Viscusi (2010) arrives at US$ 11.0 mn. for U.S.-born individuals but only US$ 6.6 mn. for (usually poorer) immigrants. However, at least among blue-collar workers, females are found to put a higher value on their lives than males in spite of their

2.6 Effectiveness of Risk Management and Risk Policy Measures

Table 2.4 Estimates of the value of a statistical life I

(Year)	Type of risk (year)	Financial trade-off	Average sample income	Implicit value of life (U.S.$ mn.)
Blomquist (1979)	Death due to car accident (1972)	Money value of disutility of wearing safety belt	U.S.$ 29,840	1.2
Dardis (1980)	Death due to fire without smoke detector (1974–1979)	Purchase price of smoke detector	n.a.	0.6
Portney (1981)	Death due to air pollution (1978)	Property values in Allegheny county, PA	n.a. (Value of life of a 42-year-old male)	0.8
Ippolito and Ippolito (1984)	Death due to cigarette smoking (1980)	Money equivalent of the effect of information	n.a.	0.7
Garbacz (1989)	Death due to fire without smoke detector (1968–1985)	Purchase price of smoke detector	n.a.	2.0
Atkinson and Halvorsen (1990)	Car accident (1986)	Purchase price of new car	n.a.	4.0
Carlin and Sandy (1991)	Baby seat in car (1985)	Purchase price of seat	U.S.$ 24,737	0.84
Dreyfuss and Viscusi (1995)	Car accident (1988)	Purchase price of new care	n.a.	3.8–5.4
Gayor et al. (2000)	Cancer risk near waste disposal site (1988–1993)	Property values in Grand Rapids, MI	n.a.	3.2–3.7
Jenkins et al. (2001)	Lethal injuries on bicycle (1997)	Purchase price of helmet	n.a.	1.4–2.9[a] 1.2–2.8[b] 2.1–4.3[c]

[a] Ages 5–9
[b] Ages 10–14
[c] Ages 20–59
Note All values in U.S.$ of 1990; n.a. = not available
Source Viscusi (1993, 1936); Viscusi and Aldy (2003, 25)

lower income; the figures reported by Viscusi (2004) are US$ 12.2 mn. and US$ 10.0 mn., respectively.

Implicitly, public authorities also value lives by spending on prevention or imposing costly life-saving regulation. For example, the resting requirements imposed on flight crews increase passenger safety—at the price of an increased cost of air travel. Viscusi (2014) calculates the value of a statistical life at US$ 9.1 mn. (at 2012 prices), which is toward the high end. The minimum estimated value is US$ 3.2 mn., implicit

Table 2.5 Estimates of the value of a statistical life II

Study	CFOI Measure	Worker Sample	Representative VSL Estimates ($millions)
Viscusi (2003)	Industry-race, 6-year average	CPS (1997) 20 equations	21.5 white (full sample) 10.3 blacks (full sample)
Leeth and Ruser (2003)	Occupation-gender-race, 3-year average	CPS (1996–1998) 28 equations	6.0 (risks by occupation, men)
Viscusi (2004)	Industry, Occupation, Industry-Occupation annual and 6-year averages	CPS (1997) 80 equations	6.7 (full sample) 10.0 (blue-collar males) 12.2 (blue-collar females)
Kniesner and Viscusi (2005)	Industry-occupation, 6-year average	CPS (1997) 6 equations	6.7 (full sample) 6.9 (male sample)
Kniesner et al. (2006)	Industry-occupation	PSID (1997) 10 equations	12.8 (base case with industry controls)
Viscusi and Aldy (2007)	Industry-age, 6-year average	CPS (1996) 4 equations	9.8 (nonsmokers) 9.7 (smokers)
Aldy and Viscusi (2008)	Industry-age	CPS (1993–2000) 8 equations	6.4 (full sample) 5.0 (age 18-24) 12.8 (age 35-44) 4.6 (age 55-62)
Kniesner et al. (2010)	Industry-occupation	PSID (1993-2001) 5 quantile equations	9.8 (median)
Evans and Schaur (2010)	Industry-age	HRS (1994–1998) 5 quantile equations, 1 OLS equation	20.7 (mean for 50 year olds)
Hersch and Viscusi (2010)	Industry-occupation-age-immigrant status, 3-year average	New Immigrant Survey (2003); CPS (2003) 22 equations	11.0 (native U.S.) 6.6 (immigrants)
Kochi and Taylor (2011)	Accident or homicide by MSA for drivers	CPS (1996–2002) 13 equations	6.1–8.4 range
Schoder and Zweifel (2011)	Industry-occupation, 6-year average	CPS (1996–1998) 9 equations	12.3 (undifferentiated deaths)
Kniesner et al. (2012)	Industry-occupation, annual and 3-year averages	PSID (1993-2001) 59 equations	11.4 (static first differences)
Kniesner et al. (2014)	Industry-occupation, annual and 3-year averages	PSID (1993–2001) 40 equations	13.0 (first difference for job changers, 3-year average risk)

Note CPS: Current Population Survey; PSID: Panel Study of Income Dynamics; estimates in 2012 U.S. dollars
Source Viscusi (2014, p. 412)

2.6 Effectiveness of Risk Management and Risk Policy Measures

in the cost of collecting advance information on private aircraft arriving in the United States incurred by U.S. Customs and Border Protection.

An alternative to WTP measurement derived from actual choices by individuals and public authorities is experimental evidence from stated choice. Here participants in an experiment are confronted with a series of choices, often of the yes/no type between an alternative and a status quo. For more details, see, e.g., Zweifel and Telser (2009).

Finally, note that the total value of a project usually is calculated as the *sum of individual WTP values*. This can be justified by the argument that one is dealing with public goods, whose use by one individual does not preclude the use by others.

▶ **Conclusion 2.8** Risk-averse individuals use a variety of risk management instruments to influence both the probability of occurrence and the severity of consequences. For the evaluation of an instrument at the societal level, cost–benefit analysis constitutes the gold standard. The often necessary valuation of a statistical life can be obtained by measuring marginal willingness to pay. Modern economics thus offers the tools to pursuit consistent decision-making under risk. Additional research is necessary, however, to determine the optimal mix of all available instruments.

2.A Appendix: Stochastic Dominance

2.A.1 First-Degree Stochastic Dominance

In the body of Chap. 2, only binary prospects are considered. In this appendix, arbitrary distributions, i.e., probability (density) functions are admitted.

Rothschild and Stiglitz (1970) proved the *equivalence* of the following three statements, which revolve about comparing two random variables X and Y with equal expected value (for simplicity, the tilde is dropped here). Higher values of X and Y are preferred, indicating that X and Y are goods or levels of wealth.

(a) *Y is the same as X plus random error.* Formally, let there be a random variable $Y = X + \epsilon$ such that Y has the same expected value as $X + \varepsilon$, with $E(\varepsilon \mid X) = 0$ for all X (the transformation from X to Y is often called "mean preserving spread"). Panel A of Fig. 2.A.1 illustrates. The probability density function $g(Y) := dG(Y)$ is derived from the function $f(X) := dF(X)$ by moving probability mass to the tails while keeping the expected value the same.

(b) *Y has more weight in its tails than X.* Formally, let $F(X)$ be the cumulative distribution of X and $G(Y)$ the cumulative distribution of Y. Let the domain of these functions be the closed interval $[a, b]$. Also, define T as the difference between realized values y and x. Therefore,

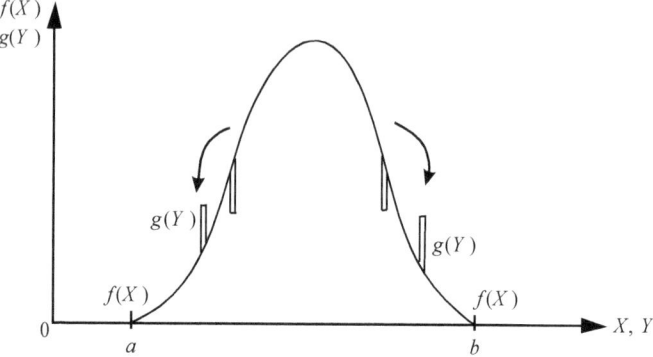

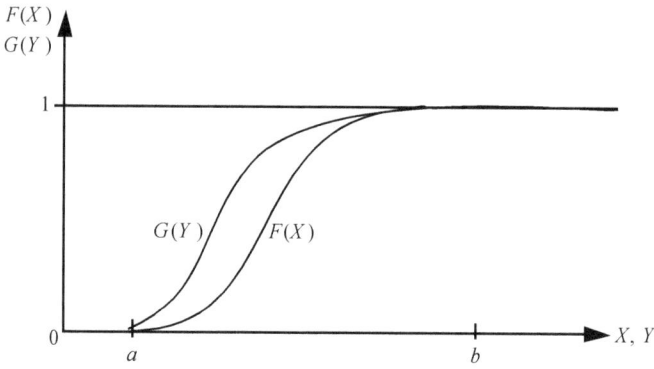

Fig. 2.A.1 Mean preserving spread resulting in first-degree stochastic dominance (FDSD)

$$T(X) := \int_a^b \{G(y) - F(x)\}\, dx, \quad \text{for } \{x, y\} \in [a, b], \text{ such that}$$
$$T[a] = 0, \quad T(x) \geq 0, \text{ and } T[b] = 0. \tag{2.A.1}$$

Panel B of Fig. 2.A.1 is derived from panel A by integrating the probability density functions $f(X)$ and $g(Y)$ to obtain the cumulative functions $F(X)$ and $G(Y)$, respectively. It shows that Y has a higher frequency of low values than X; on the other hand, its cumulative probability distribution $G(Y)$ approaches the upper limit of 1 faster than $F(X)$ when the argument moves from the lower bound a to the upper bound b of its domain. The crucial property, however, is that $F(X)$ has less probability mass in the domain of low values than does $G(Y)$.

(c) *All risk-averse decision-makers (i.e., with concave risk utility functions) weakly prefer X over the random variable Y.* Formally, $v(X) \succeq v(Y)$ for all concave $v(\cdot)$.

This third statement also is known as *first-degree stochastic dominance (FDSD)*. If a cumulative distribution function $F(\cdot)$ is weakly preferred compared to another cumulative density function $G(\cdot)$, then this constitutes stochastic dominance, a concept introduced by Hadar and Russell (1969).

Definition 2.A.1 The cumulative distribution function $F(X)$ exhibits first-degree stochastic dominance (FDSD) with regard to $G(Y)$ if $G(y) - F(x) \geq 0$ for all $x, y \in [a, b]$.

Note that the requirement of Definition 2.A.1 stating that $G(y)$ must be greater than or equal to $F(x)$ to be dominated is more stringent than Statement (2), which refers to the integral (basically, the sum) of the differences rather than each individual difference.

The connection between FDSD and the expected utility criterion is provided by a definition and a theorem.

Definition 2.A.2 The cumulative distribution function $F(X)$ is weakly preferred to the function $G(Y)$ if for an agent with risk utility function $v(x)$ it is true that $\int_a^b v(x)\, dF(x) \geq \int_a^b v(y)\, dG(y)$.

Clearly, the two integrals define expected values by summing the values of $v(\cdot)$ weighted by their probabilities. Now Hadar and Russell (1969) proved the following.

Theorem 2.A.1 *The cumulative distribution function $F(X)$ dominates the function $G(Y)$ to the first order if and only if $F(X)$ is weakly preferred to $G(Y)$ under the expected utility criterion.*

Therefore, FDSD and the expected utility criterion are equivalent.

2.A.2 Second-Degree Stochastic Dominance

As illustrated by Panel B of Fig. 2.11, the FDSD test is applicable only if the cumulative distribution functions do not intersect. A natural extension of the partial ordering by FDSD is provided by the following.

Definition 2.A.3 The cumulative distribution function $F(X)$ dominates the function $G(Y)$, with $Y = X + \varepsilon$ (second-degree stochastic dominance, SDSD), if $T(x) = \int_a^b \{G(y) - F(x)\}\, dx \geq 0$, $\forall \{x, y\} \in [a, b]$.

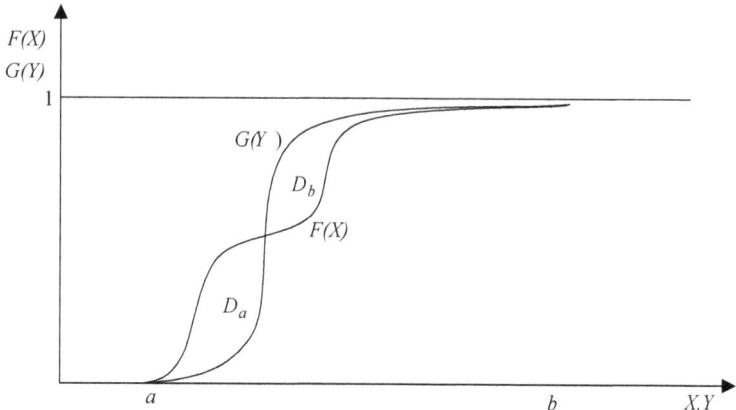

Fig. 2.A.2 Second-degree Stochastic Dominance (SDSD)

This definition states that contrary to panel B of Fig. 2.A.1, the two cumulative distribution functions may now intersect. For $F(X)$ to dominate $G(Y)$, it is sufficient that the sum of the areas between $G(Y)$ and $F(X)$ be positive when the summation is performed from left to right in the interval $[a, b]$ (see also Laffont (1990), Chap. 2.2). Graphically, one has the situation depicted in Fig. 2.A.2 ($D_a > D_b$).

Asymmetric (skewed) probability density functions can lead to intersecting cumulative distribution functions. Since FDSD does not permit such intersections, it establishes only a partial preference ordering of probability distributions. However, the SDSD criterion, while more general, still falls short of providing a complete preference ordering of arbitrary distributions. In other words, there are cumulative distribution functions that cannot be ordered using the SDSD criterion. This case is depicted in Fig. 2.A.3, where $(D_a + D_c) = (D_b + D_d)$, i.e., the summation of the areas between the limits a and b results in the same value.

Again, it is possible to link SDSD to risk aversion in the risk utility function. In the case of risk aversion ($v' > 0$; $v'' < 0$), one has

Theorem 2.A.2 *The cumulative distribution function $F(X)$ of second order dominates the cumulative distribution function $G(Y)$ if and only if $F(X)$ is weakly preferred to $G(Y)$ by an individual exhibiting risk aversion.*

The theory of stochastic dominance constitutes an interesting extension of the theory building on risky prospects of the binary type which is popular for analyzing decision-making under risk and the demand for insurance. By purchasing insurance, individuals are, in fact, capable of transforming an arbitrary distribution function of their wealth into another one. They buy insurance if and only if they prefer this new distribution function. However, the fact that these distribution functions cannot be completely ordered implies that one *cannot always predict* whether an individual will

2.A Appendix: Stochastic Dominance

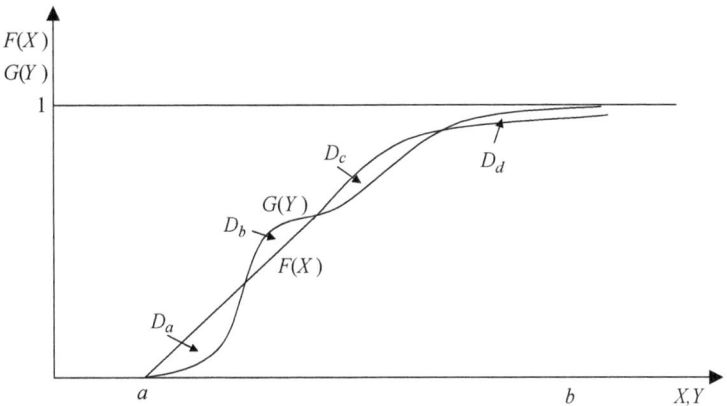

Fig. 2.A.3 The SDSD criterion fails to rank $F(X)$ and $G(Y)$

opt for insurance or not. For this reason, risky prospects of the binary type continue to be very common in theory of insurance demand (see Sect. 3.2).

There is, however, an interesting application of the SDSD concept. At least in Western societies, the ideal health profile is to live in perfect health (index set at $H = 100$ in Fig. 2.A.4) and to drop dead ($H = 0$) when the appropriate time has come. In the medical literature, the pertinent health profile BCD is said to exhibit "perfect rectangularization". When turned 90°, it becomes a somewhat peculiar cumulative distribution function (cdf). In fact, it shows that the entire probability mass defined over health states is concentrated at $H = 100$, a value that remains unchanged up to the age at death. However, Fig. 2.A.4 displays another profile BXE indicating that health status deteriorates slowly up to a rather high age, after which it drops quickly. This, therefore, reflects imperfect rectangularization; in return, it allows for a longer life. Once again, profile BXE is nothing but a cdf turned around as it shows the number of years that are spent enjoying health status H' (as an example) or better.

The question now becomes "Which profile is preferred by a risk-averse individual?" According to the SDSD criterion, profile BCD is preferred since the area BCX (corresponding to the probability mass of "more favorable health status") exceeds the corresponding area of BXE (XDE).

If the SDS criterion applies, there are two predictions:

1. Much on an individual's healthcare expenditure (HCE) occurs shortly before death. Therefore, the gap between the actual and the desired rectangular profile is greatest (see Fig. 2.A.4 in the neighborhood of point (C). However, at an advanced age, one's own efforts at improving health have low effectiveness compared to medical care. Attempts to move back toward the ideal profile BCD therefore call for the use of HCE.
Indeed, based on panel data of a Swiss health insurer, Zweifel et al. (1999) find that proximity to death is associated with a steep rise in HCE regardless of age

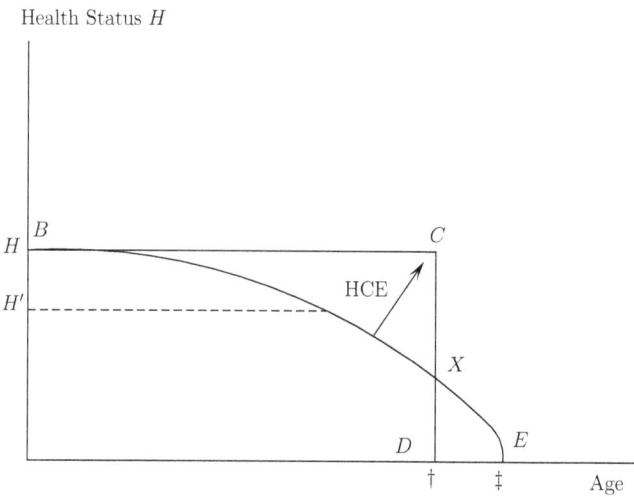

Fig. 2.A.4 Choice of health profile and its implications

(see Zweifel et al. (2009), Chap. 14.4.2 for an extended discussion of the so-called "red herring" hypothesis).

2. When living conditions improve, individuals attain an increasing degree of rectangularization of their health profile. In their younger years, efforts directed at maintaining and improving one's health have considerable effectiveness especially when undertaken in good health (see Zweifel et al. (2009), Chap. 3.4 on the state-dependent health production function). However, both this effort and HCE are costly in terms of (healthy) time and money, which are more abundantly available when living conditions improve.

   Evidence supporting this prediction is indirect: a necessary (but not sufficient) condition for attaining a more rectangular health profile is survival. Noting that perfect rectangularization of a population's survival curve implies that age at death is the same for everyone (i.e., a standard deviation of zero), Schoder and Zweifel (2011) test whether the standard deviation of age at death has decreased in OECD countries since the early 1960s. By 2005, it had fallen to 17.5 years (from 19) in the United States and even to 15 years (from 26) in Portugal, suggesting an improved control of health status. In addition, both per capita GDP and HCE are estimated to contribute in comparable measure to this decrease. Given a coefficient of risk aversion derived from U.S. choices of health insurance, further spending on health care could be justified even in that country—not counting the concomitant increase in longevity.

There have been tests of SDSD in other areas, in particular, portfolio choice where Bruner (2007) presents experimental evidence suggesting that strongly risk-averse individuals, in particular, act in accordance with SDSD.

2.A Appendix: Stochastic Dominance

Exercises

2.1

(a) A popular variant of the risk utility function is $v(W) = ln(W)$. Does it solve the St. Petersburg paradox? What are the properties of the coefficients of absolute and relative risk aversion?
(b) Another variant is $v(W) = aW - bW^2$, $a, b > 0$. Please answer the same questions.
(c) Still another variant is $v(W) = \frac{W^{1-\gamma}}{1-\gamma}$, $\gamma \neq 1$. Please answer the same questions.

2.2

(a) Measures of environmental and transportation policy often imply a valuation of human life. Is this ethically justifiable? And if so, should the value not be infinite?
(b) Using the purchase of a bicycle helmet as an example, show how the value of human life can be estimated.

2.3 Please answer the following questions related to the risk utility function:

(a) What data do you need in order to determine a certainty equivalent using the risk utility function $v(W) = ln(W)$?
(b) Please construct your own example, calculating not only the certainty equivalent but also the risk premium according to Arrow (1965) and Pratt (1964).
(c) Repeat these steps using the risk utility function $v(W) = 1 - e^{-W/\gamma}$.

References

ADAC (2009) (German Automobile Club, 2009). Versicherungsunfälle in Deutschland (http://www.adac.de/verkerhr/statistiken/unfalldaten, visited 5 Jan. 2009).
Aldy, L.E., & Viscusi, W.K. (2008). Adjusting the value of a statistical life for age and cohort effects. *Review of Economics and Statistics, 90*, 573-581.
Arrow, K.J. (1951). Alternative approaches to the theory of choice in risk-taking situations. *Econometrica, 19*, 404–437; In K.J. Arrow (Ed.), *Essays in the Theory of Risk Bearing* (pp. 1–43). Amsterdam: North-Holland.
Arrow, K.J. (1965). *Aspects of the Theory of Risk-Bearing*, Yrjö Johnsson Lectures, Säätiö, Helsinki Y.J.
Atkinson, S.E., & Halvorsen, R. (1990). The valuation of risks to life: evidence from the market for automobiles. *The Review of Economics and Statistics, 72*(1), 133–136.
Barsky, R.B., Juster, T.F., Kimball, M.S., & Shapiro, M.D. (1997). Preference parameters and behavioral heterogeneity: an experimental approach in the health and retirement study. *Quarterly Journal of Economics, 112*(2), 537–579.

Bernoulli, D. (1730/31). Specimen theoriae novae de mensura sortis. *Comentarii Academiae Petropolis, Petersburg, 5*, 175–192; English translation (Exposition of a new theory on the measurement of risk), *Econometrica, 22*(1), 23–46.

Blake, D. (1996). Efficiency, risk aversion and portfolio insurance: an analysis of financial asset portfolios held by investors in the United Kingdom. *Economic Journal, 106*(438), 1175–1192.

Blomquist, G. (1979). Value of life saving: implications of consumption activity. *Journal of Political Economy, 87*, 540–558.

Bruner, D.M. (2007). Risk Aversion and stochastic dominance. *Working Paper*, Calgary: University of Calgary.

Carlin, P.S., & Sandy, R. (1991). Estimating the implicit value of a young child's life. *Southern Economic Journal, 58*(1), 79–105.

Christiansen-Szalanski, J. & Bushyhead, J.B. (1981). Physicians' use of probabilistic information in a real clinical setting. *Journal of Experimental Psychology: Human Perception and Performance* (7), 928-935.

Dardis, R. (1980). Economic analysis of current issues in consumer product safety: fabric flammability. *Journal of Consumer Affairs, 14*(1), 109–123.

Dreyfuss, M.K., & Viscusi, W.K. (1995). Rates for time preference and consumer valuations of automobile safety and fuel efficiency. *Journal of Law & Economics, 38*(1), 79–105.

Eeckhoudt, L., & Schlesinger, H. (2006). Putting risk in its proper place. *American Economic Review, 96*(1), 280–289.

Evans, D. (2012). Risk intelligence. In S. Roeser et al. (eds.), *Handbook of Risk Theory*, Vol. 2, 604-620.

Evans, M.F., & Schaur, G. (2010). A quantile estimation approach to identify income and age variation in the value of statistical life. *Journal of Environmental Economics and Management, 59*, 260–270.

Friend, I., & Blume, M.E. (1975). The demand for risky assets. *American Economic Review, 65*(5), 900–922.

Garbacz, C. (1989). Smoke detector effectiveness and the value of saving a life. *Economic Letters, 31*(3), 281–286.

Gayor, T., Hamilton, J.T., & Viscusi, W.K. (2000). Private values of risk tradeoffs at superfund sites: housing market evidence on learning about risk. *Review of Economics and Statistics, 82*(3), 439–451.

Hadar, J., & Russell, W. (1969). Rules for ordering uncertain prospects. *American Economic Review, 59*, 25–34.

Halek, M., & Eisenhauer, G.M. (2001). Demography of risk aversion. *Journal of Risk and Insurance, 68*(1), 1–24.

Hersch, J., & Viscusi, W.K. (2010). Immigrant status and the value of statistical life. *Journal of Human Resources, 45*, 749-771.

Human Mortality Database (2010). http://www.mortality.org/cgi-bin/hmd/country.php?cntr=USA&level=1.

Ippolito, P.M., & Ippolito, R.A. (1984). Measuring the value of life saving from consumer reactions to new information. *Journal of Public Economics, 25*(1–2), 53–81.

Jackwerth, J.C. (2000), Recovering risk aversion from option prices and realized returns. *Review of Financial Studies* 13(2), 433-451.

Jenkins, R.R., Owens, N., & Wiggins, L.B. (2001). Valuing reduced risks to children: the case of bycicle safety helmets. *Contemporary Economic Policy, 19*(4), 408.

Kimball, M.S. (1990). Precautionary saving in the small and in the large. *Econometrica, 58*(1), 53–73.

Kniesner, T.J., & Viscusi, W.K. (2005). Value of a statistical life: relative position vs. relative age. *American Economic Review, 95*, 142-146.

Kniesner, T.J., Viscusi, W.K., & Ziliak, J.P. (2006). Life-cycle consumption and the age-adjusted value of life. *Contributions to Economic Analysis and Policy, 5*, 1-34.

References

Kniesner, T.J., Viscusi, W.K., & Ziliak, J.P. (2010). Policy relevant heterogeneity in the value of statistical life: new evidence from panel data quantile regressions. *Journal of Risk and Uncertainty, 40*, 15-31.

Kniesner, T.J., Viscusi, W.K., & Ziliak, J.P. (2014). Willingness to accept equals willingness to pay for labor market estimates of statistical life. *Journal of Risk and Uncertainty, 48*, 187-205.

Kniesner, T.J., Viscusi, W.K., Woock, C., & Ziliak, J.P. (2012). The value of statistical life: evidence from panel data. *Review of Economics and Statistics, 94*, 19-44.

Knight, F.H. (1921). *Risk, Uncertainty and Profit*. Chicago: Chicago University Press.

Kochi, I., & Taylor, L.O. (2011). Risk heterogeneity and the value of reducing fatal risks: further market-based evidence. *Journal of Benefit-Cost Analysis, 2*, 1-26.

Laffont, J.J. (1990). *The Economics of Uncertainty and Information*. Cambridge MA: Cambridge University Press.

Leeth, J.D., & Ruser, J. (2003). Compensating wage differentials for fatal and nonfatal injury risk by gender and race. *Journal of Risk and Uncertainty, 27*, 257–277.

Leland, H.E. (1968). Saving and uncertainty: the precautionary demand for saving. *Quarterly Journal of Economics, 82*(3), 465–473.

Lichtenstein, S., et al. (1978). Judged frequency of lethal events. *Journal of Experimental Psychology, 4*, 551–578.

Menezes, C.F., & Hanson, D.L. (1970). On the theory of risk aversion. *International Economic Review, 11*, 481–487.

Morall, J.F. (1986). A review of the record. *Regulation, 10*, 25–34.

Morin, R.A., & Suarez, A. (1983). Risk aversion revisited. *Journal of Finance, 38*(4), 1201–1216.

Murphy, A.H. & Winkler, R.I. (1977). Reliability of subjective probability forecasts of precipitation and temperature. *Journal of the Royal Statistical Society C, Applied Statistics* 26 (1), 41-47.

Popper, K. (1945). *The Open Society and Its Enemies*. London: Oxford University Press.

Portney, P.R. (1981). Housing prices, health effects and valuing reductions in risk of death. *Journal of Environmental Economics and Management, 8*(1), 72–78.

Pratt, J. (1964). Risk aversion in the small and in the large. *Econometrica, 32*, 122–136.

Riley, W.B., & Chow, K.V. (1992). Asset allocation and individual risk aversion. *Financial Analysis Journal* Nov./Dec.: 32–37.

Rothschild, M., & Stiglitz, J.E. (1970). Increasing risk: I. A definition. *Journal of Economic Theory, 2*, 225–243.

Sandmo, A. (1970). The effect of uncertainty on saving decisions. *Review of Economic Studies, 37*(3), 353–360.

Schoder, J., & Zweifel, P. (2011). Flat-of-the-curve medicine: a new perspective on the production of health. *Health Economics Review*1(2), https://doi.org/10.1186/2191-1-2.

Schubert, R., Brown, M., Gysler, M., & Brachinger, H.W. (1999). Financial decisionmaking: are women really more risk-averse? *American Economic Review, 89*, 381–385.

Setbon, M., Raude, J., Fischler, C., & Flauhault, A. (2005). Risk perception of the "Mad Cow Disease" in France: determinants and consequences. *Risk Analysis, 25*(2), 813–826.

Siegel, F.W., & Hoban, J.P. (1982). Relative risk aversion revisited. *Review of Economics and Statistics, 64*(3), 481–487.

Sinn, H.W., & Weichenrieder, A.J. (1993). The biological selection of risk preferences. In Bayerische Rückversicherung (Ed.), *Risk as a Construct*. Munich: Knesebeck, 67-73.

Slovic, P., Fischhoff, B., & Lichtenstein, S. (1980). Facts and fears – understanding risks. In Schwing, R.C., & Albers, W.A. (Eds.), *Societal Risk Assessment*. New York: Plenum Press.

Statistisches Bundesamt (2018). *Statistisches Jahrbuch für das Ausland (Statistical Yearbook for Foreign Countries)* Wiesbaden.

Szpiro, G.G. (1986a). Measuring risk aversion: an alternative approach. *Review of Economics and Statistics, 68*, 156–159.

Szpiro, G.G. (1986b). Über das Risikoverhalten in der Schweiz (On risk behavior in Switzerland). *Swiss Journal of Economics and Statistics, 122*(III), 463–470.

Szpiro, G.G. (1997). The emergence of risk aversion. *Complexity, 2*(4), 31–39.

Urquhart, J., & Heilmann, K. (1983). *Riskwatch: The Odds of Life*. New York: Facts on File.

US Department of Transportation, Bureau of Transportation Statistics (2011). http://www.bts.gov/data_and_statistics.

US Department of Transportation, Bureau of Transportation Statistics (2016). Traffic Safety Facts: Research Note. March.

US Department of Transportation, Bureau of Transportation Statistics (2020). Traffic Safety Facts: Research Note. October.

Viscusi, W.K. (1993). The value of risks to life and health. *Journal of Economic Literature, 31*, 1912–1946.

Viscusi, W.K. (2003). Racial differences in labor market values of a statistical life. *Journal of Risk and Uncertainty*, 27, 239-256.

Viscusi, W.K. (2004). The value of life: estimates with risks by occupation and industry. *Economic Inquiry*, 42, 29-48.

Viscusi, W.K. (2014). The value of individual and societal risks to life and health. In Mark J. Machina & W. Kip Viscusi (Eds.), *Handbook of the Economics of Risk and Uncertainty*, Chap. 7. Amsterdam: North-Holland.

Viscusi, W.K., & Aldy, L.E. (2003). The value of a statistical life: a critical review of market estimates around the world. *Journal of Risk and Uncertainty, 27*(1), 5–76.

Viscusi, W.K., & Aldy, L.E. (2007). Labor market estimates of the senior discount for the value of statistical life. *Journal of Environmental Economics and Management, 53*, 377–392.

Zweifel, P. (2001). Improved risk information, the demand for cigarettes, and anti-tobacco policy. *Journal of Risk and Unvertainty, 23*(3), 299–303.

Zweifel, P., & Telser, H. (2009). Cost-benefit analysis for health. In R.J. Brent (Ed.), *Handbook of Research on Cost-Benefit Analysis* (pp. 31–54). Northhampton MA: Edward Elgar.

Zweifel, P., Breyer, F., & Kifmann, M. (2009). *Health Economics*, 2nd ed. New York: Springer.

Zweifel, P., Felder, S., & Meier, P. (1999). Ageing of population and healthcare expenditure: A red herring? *Health Economics, 8*(6), 485-496.

Insurance Demand I: Decisions Under Risk Without Diversification Possibilities

Throughout this chapter, economic agents are assumed to have at their disposal two instruments of risk management only, viz., purchasing insurance coverage or exerting preventive effort. The possibility of coping with uncertainty through diversification of assets is, therefore, neglected. This alternative is available to enterprises and their owners and will be treated in Chap. 5. Section 3.1 refers to the risk-utility function derived in Sect. 2.2.2 and presents the expected utility maximization hypothesis which is used to resolve the decision-making problem under uncertainty. With this, the groundwork is laid for developing the basic model of insurance demand in Sect. 3.2. It predicts the choice of full coverage if the potential purchaser of insurance is charged the so-called fair premium, i.e., a premium that just covers the expected value of the loss insured. Accordingly, Sect. 3.3 addresses the issue of optimal insurance coverage in the more realistic case where the premium exceeds the fair value. Demand for insurance is also related to the premium rate and the value of the asset to be covered. Bringing the basic model closer to reality also motivates Sect. 3.4, which is devoted to the demand for insurance in the face of several risks. Sect. 3.5 deals with the relation between insurance and preventive effort on the part of the insured, an issue of great relevance for the insurance business. To be clear, expected utility theory is not the only decision-making rule. To that end, Chap. 4 contains a critical discussion of expected utility theory; it also introduces the reader to a few alternatives. However, since the predictions of these alternative decision-making rules do not essentially differ from those derived from expected utility theory, this theory will be retained for the remainder of this book.

3.1 The Expected Utility Maximization Hypothesis

Demand for insurance originates with households as well as enterprises. The two differ considerably in their demand behavior, however. Differences are caused by differing urgency of need, risk calculation capabilities, degrees of rationality, availability of information, risk management, and the structure of the insurance market they face. The most crucial difference is the following: large enterprises (more precisely, their owners) have diversification possibilities that go beyond those of a typical household. Shareholders usually own stock in many companies, which serves to protect them from the ups and downs in the value of a single asset. For this reason, demand for insurance by enterprises is relegated to Chap. 5.

Returning to the demand for insurance on the part of households, it will be shown to be the result of two interacting determinants. The *objective* component is the fact that an asset is exposed to risk; the *subjective* component is reflected by risk aversion (see Sect. 2.2.2 and specifically Sect. 2.3.2). This means that household demand for insurance changes over time. In particular, the growth of assets (e.g., due to an inheritance, purchase of a car or a home) can trigger additional demand for coverage. On the other hand, risk aversion may also change because certain events make people more aware of risks. In addition, there is always a replacement demand as existing contracts expire.

However, household demand for insurance also depends on the natural and social *environment*. First, someone living on the coast sea faces different perils from someone living in the mountains. Second, insurance policies cannot be written at will but must conform to general insurance regulation. Moral and ethical rules of conduct play an important role too (e.g., is it advisable to offer insurance for traffic tickets?).

All of these determinants are subsumed in the so-called *risk- utility function*. The risk-utility function, thus comprises the subjective and objective aspects of the decision-making situation. Individuals are assumed to have good knowledge of the probabilities of occurrence and the financial consequences of associated events. They, therefore, must be able to envisage a possible future situation in terms of their assets (i.e., health, wealth, wisdom; see Sect. 1.6) expressed in monetary terms. Probabilities of occurrence and financial consequences together describe the risk situation or the so-called risky prospect of an individual.

Example 3.1

Risky prospects
Assume for simplicity that events immediately lead to differences in wealth. The normal state (state 2) is characterized by final wealth W_2 of 100 TMU (TMU = thousand money units). If there is a loss of event (state 1), wealth drops to $W_1 = 80$ TMU. Let the individual have a (rather precise) notion of the probability (i.e., the relative frequency) of the occurrence of this event, e.g., 1:10 or 10%. The expected wealth (EW) of the individual is then given by

$$EW = 0.1 \cdot 80 + 0.9 \cdot 100 = 98 \text{ TMU}. \tag{3.1}$$

More generally

$$EW = \pi W_1 + (1-\pi)W_2, \text{ with } \pi : \text{ probability of loss.} \qquad (3.2)$$

Note that the theory is couched in terms of final wealth rather than changes in wealth. One can summarize the probabilities in a probability density function with

$$f(W) = \{0.9, 0.1\}, \text{ or more generally } f(W) = \{\pi, (1-\pi)\}. \qquad (3.3)$$

Then, a *risky prospect* is defined as a random variable $\widetilde{W}$ with

$$\{EW; f(W)\} \text{ or } \{W_1, W_2; \pi, (1-\pi)\}. \qquad (3.4)$$

∎

Frequently, a risky prospect can be modified by actions taken by the potential insurance buyer (IB) such as avoiding risks, reducing their probability of occurrence or severity of consequence by preventive effort, and by buying insurance coverage.[1] Since these actions (and non-actions) give rise to a set of prospects, the problem then becomes to find the one that is deemed preferable. In order to make a decision, the agent, therefore, must have a *preference relation* or a *decision rule* defined over risky prospects. Economic theory abounds with decision rules under uncertainty or risk. To name a few, there are the Minimax rule by Wald (1945), the Minimax-regret rule by Niehans (1948), the Focus-profit-and-loss rule by Shackle (1949), the Modus rank criterion by Lange (1943), the Criterion of equivalent gains and losses by Krelle (1959), the well-known (μ, σ)-criterion, the Criterion of minimum ruin probability, the Non-expected utility criterion by Machina (1995), and most prominent of all, the Expected Utility principle by Bernoulli (1738) [see equation (3.6) below].

However, one feature that is common to all these rules is that they take into account *risk aversion* (see Sects. 2.2.2 and 2.3.2): The individual is supposed to prefer a certain (i.e., safe) average of a distribution (often called a lottery) to the distribution (lottery) itself. In Example 3.1 above, this means that even if insurance were to cost 2 TMU or more, one is willing to buy it in order to have *certain*, rather than volatile, wealth. Evidently, one does not simply maximize the expected value of wealth but takes into account the amount of deviation from the expected value.

Example 3.1 (continued)

Assume now that the loss considered could amount to 66.666 TMU, causing wealth to drop to 33.334 TMU (rather than 80 TMU), with a probability of occurrence of 0.03 only (rather than 0.1). The expected value of wealth of this risky

---

[1] We use IB throughout the text to represent the consumer of insurance. The consumer is also often referred to as the policyholder.

prospect is given by

$$EW = 0.03 \cdot 33.334 + 0.97 \cdot 100 = 98.00002 \text{ TMU}. \tag{3.5}$$

Clearly, the two risky prospects (3.1) and (3.5) have the same (or almost the same) expected value. Nevertheless, almost everyone would prefer the first. The reason for this is risk aversion. ∎

Among all the decision rules cited above, the expected utility maximization principle by *Bernoulli* [see in particular Von Neumann and Morgenstern (1944)] has axiomatic properties that make many consider it as the rational one. The *Expected Utility* criterion (or *Bernoulli principle*) is an operator that permits the reduction of a decision-making situation under risk to the maximization of an unambiguously defined objective.

This can be shown rigorously with the help of Table 3.1 (copied from Sect. 2.2.2.2 for convenience). To construct it, one needs first to determine the risky prospects, viz., the states space (i.e., the possible states of nature that are mutually exclusive, S) and the consequences space C (the body of Table 3.1) which relates the action space A to consequences prior to valuation. For instance, action a_1 could be "accept the first risky prospect", with $EW = 98$, while a_2 could be "accept the second prospect", with $EW = 98.00002$. The second step is to value these consequences using a risk-utility function to arrive at a decision, e.g., "opt for the first risky prospect".

The relationship between the three components of a decision problem (actions, states, and consequences) can be formulated in the two methods that follow.

Formulation 1: An action a_i is a function which assigns to each state s_j a consequence c_{ij} for all $a_i \in A$ and $s_j \in S$. Note that "consequence" stands for all outcomes of relevance to the decision-maker; restricting them to assets, wealth, consumption, or any bundle of goods is not necessary. In the following, they are referred to as *contingent claims*.

Formulation 2: The space of actions A symbolizes the set of probability densities (objective or subjective) that are defined over the consequence space C. By choosing an action a_i, the agent chooses a *probability density function* defined over consequences, as in the case of the two risky prospects above.[2] Even if probabilities do not change, actions still modify consequences, shifting the distribution horizontally (see Fig. 2.1 of Sect. 2.1.1). Therefore, the choice of action does amount to the choice of a probability density.

Now, in order to make a decision, the individual must have a preference relation that imposes at least a weak ordering on outcomes (and hence actions). Under a set of plausible axioms, it is rational to *maximize expected utility* (see Sect. 2.2.2.2). This assumes that individuals act as though they

(1) can assign utility estimates to each of the actions in their possibility sets; and,
(2) select the action which maximizes their expected utility.

---

[2] In fact, preventive effort constitutes an action influencing probabilities of occurrence; see Sect. 3.5.

3.1 The Expected Utility Maximization Hypothesis

Table 3.1 Decision matrix

Alternative actions $a_i \in A$ ($i = 1, \ldots, n$)	States of nature $s_j \in S$ ($j = 1, \ldots, m$)				Valuation of consequences
A	s_1	s_2	$\ldots$	s_m	C
a_1	c_{11}	c_{12}	$\ldots$	c_{1m}	$v[c_{1j}]$
a_2	c_{21}	c_{22}	$\ldots$	c_{2m}	$v[c_{2j}]$
a_3	c_{31}	c_{32}	$\ldots$	c_{3m}	$v[c_{3j}]$
$\ldots$	$\ldots$	$\ldots$	$\ldots$	$\ldots$	
a_n	c_{n1}	$\ldots$	$\ldots$	c_{nm}	$v[c_{nj}]$
Probabilities of occurrence	π_1	π_2	$\ldots$	π_m	

An action then amounts to choosing a row in Table 3.1. If one takes into account that probabilities pertain to states, then the choice of an action is equivalent to the choice of a risky prospect (or again, a probability density).

The expected utility criterion connects the ordering over actions in terms of utility with the preference ordering over consequences. The *Expected Utility Theorem* (Bernoulli principle) states that consequences are ordered according to a cardinal[3] (order-preserving) utility function, the *risk-utility function*. The probability measure entering the calculation of expected utility can be objective or subjective. In the case of a finite number of consequences, the theorem says[4]

$$\text{Given } EU[c_{ij}] := \sum_j \pi_j v[c_{ij}; a_j] \equiv EU[a_i] \equiv \sum_j \pi_j v[a_j, s_j]$$

$$\text{with } \pi_j \geq 0 \text{ and } \sum_j \pi_j = 1,$$

$$c_1 := c[a_1, s_j] \text{ and}$$

$$c_2 := c[a_2, s_j] \text{ for all } s_j, \text{ then}$$

$$EU[c_1] \geq EU[c_2], \text{ if and only if } a_1 \succeq a_2, \tag{3.6}$$

where $\succeq$ means "is at least as preferred as".

That is, the expected utility of action a_i is the weighted average of the utilities of the pertinent consequences, with the probabilities serving as weights. The expected utility $EU[a_1]$ is greater or equal to the expected utility $EU[a_2]$, if and only if action a_1 is preferred to action a_2 (indifference included). Therefore, the individual considered chooses the action that has maximum expected utility.

After the choice of a specific action a_i, the utility values associated with the relevant states $\{v(c_{i,j})\}$ are fixed. Now the probability density function defined over the states of nature amounts to a probability density function defined over utilities,

---

[3] See method No. 1 for constructing the risk-utility function in Sect. 2.2.2.3.
[4] The notation with brackets means that a function is evaluated at a particular value of the argument, e.g., the risk-utility function $v(c)$ at a particular value c_{ij}.

hence $EU[a_1] > EU[a_2]$. The decision problem, therefore, can also be seen as a choice between different probability density functions, as in Formulation 2 above.

▶ **Conclusion 3.1** In order to resolve the decision-making problem under uncertainty, the individual needs to have a precise notion of (1) the alternative actions, (2) the probabilities with which possible states occur, (3) a function mapping the actions into consequences, and (4) a preference ordering or risk-utility function defined over consequences.

Using these elements, individuals are capable of valuing the alternative actions using the expected utility rule by opting for the one associated with the highest expected utility.

3.2 Theory of Insurance Demand

3.2.1 The Basic Model

In this section, the supply of insurance is considered as predetermined in order to focus on the demand side. A survey of contract types will follow in Sect. 3.3.4. To study the demand for insurance coverage (insurance demand for short), a representative individual facing a risky prospect of the binary type is considered. Specifically, with probability π, a loss of amount L will occur, and with the counter-probability $(1 - \pi)$, this will not happen. For example, one's home may be destroyed at least in part by fire with a certain known probability π. However, agents do not know whether or not they will belong to the ones suffering a loss.

More formally, let the risky prospect ($\tilde{W}$) associated with the action "Do not buy insurance" be sufficiently represented by two levels of wealth and their probabilities of occurrence. In the loss state (state No. 1 henceforth), final wealth is given by $W_1 = W_0 - L$, with W_0 denoting predetermined initial wealth. In the no-loss state (state No. 2 henceforth), wealth is unchanged such that $W_2 = W_0$. The probabilities are π and $(1 - \pi)$, respectively. Therefore, the initial prospect $\tilde{W}$ is given by

$$\tilde{W} = \{W_1, W_2; \pi, 1 - \pi\}. \tag{3.7}$$

This is very similar to the two-goods model of standard microeconomics; the only difference lies in the goods (W_1, W_2) being conditioned by probabilities $\{\pi, (1 - \pi)\}$. Accordingly, Fig. 3.1 exhibits a (W_1, W_2)-space in which points represent risky prospects. An indifference curve passes through each such point. In contradistinction to standard microeconomics, however, utility cannot be held constant along an indifference curve because probabilities are involved. Rather, it is only *expected utility* that is held constant, at $\overline{EU}$.

3.2 Theory of Insurance Demand

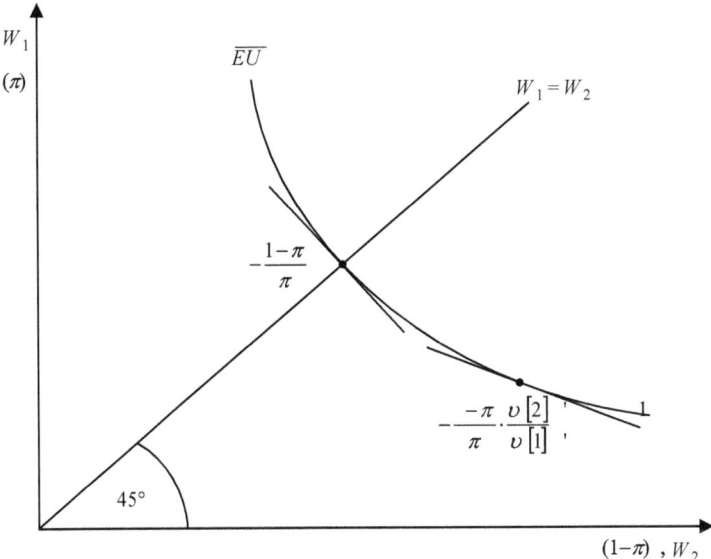

Fig. 3.1 Indifference curve for a risky prospect of the binary type

Note that the representation in terms of indifference curves does not differ from the representation in terms of the risk-utility function $v(W)$. In Sect. 2.2.2.3, $v(W)$ is constructed by using indifference between a risky prospect and a certain alternative (see Fig. 2.7). By varying the end points of $v(W)$ while leaving probabilities constant, risky prospects with the same expected utility can be constructed. These equivalent risky prospects form the indifference curve of Fig. 3.1.

The starting point for deriving the indifference curve in (W_1, W_2)-space is expected utility,

$$EU(\tilde{W}) = \pi v[W_1] + (1 - \pi)v[W_2]. \tag{3.8}$$

Now consider dEU, noting that $dEU = 0$ holds expected utility constant, thus defines an indifference curve. Given fixed probabilities, such a change can only come through changes in wealth levels. Therefore, the equation of an indifference curve reads

$$dEU(\tilde{W}) = \pi v'[W_1]dW_1 + (1 - \pi)v'[W_2]dW_2 = 0. \tag{3.9}$$

Frequently, the shorthand notation $v'[1] := v'[W_1]$ for the marginal utility of (risky) wealth in the loss state No. 1, and $v'[2] := v'[W_2]$ for the marginal utility in the no-loss state No. 2 will be used. Solving equation (3.9) for the slope of the indifference curve, one, therefore, obtains[5]

$$\left.\frac{dW_1}{dW_2}\right|_{EU} = -\frac{1-\pi}{\pi} \cdot \frac{v'[2]}{v'[1]}, \qquad (3.10)$$

with $v'[2] := v'[W_2]$ shorthand for the marginal utility of (risky) wealth in the no-loss state No. 2 and $v'[1] := v'[W_1]$ the marginal utility in the loss state No. 1.

Another important locus in Fig. 3.1 is the 45° line. Since it is defined by $W_1 = W_2$, points on it indicate equality of wealth across the two states, i.e., a situation of certainty. Following Hirshleifer (1965/66), the 45° line is, therefore, called the *certainty line*. However, given that the two wealth levels are equal, the marginal utility of a possible wealth increment must be the same as well, i.e., $v'[2] = v'[1]$ along the certainty line. Using this in equation (3.10), one obtains a specific value for the slope of the indifference curve where it intersects the certainty line

$$\left.\frac{dW_1}{dW_2}\right|_{W_1=W_2} = -\frac{1-\pi}{\pi}. \qquad (3.11)$$

Therefore, on the certainty line, the slope of the indifference curve boils down to the ratio of the two probabilities. It is steep when the probability of loss π is small, indicating that an individual would have to be compensated by a lot of wealth in the loss state for giving up wealth in the no-loss state.[6]

The next step consists of complementing the model with a restriction, permitting to identify an optimum. Turning to Fig. 3.2, let the initial situation be represented by point A (also called the endowment point). Point A lies below the certainty line because wealth W_1 in the loss state necessarily is below wealth $W_2 = W_0$ in the no-loss state. Now let there be an insurance company (IC) who offers an indemnity payment I in the case of a loss L for a premium $P(I)$ that of course increases with I. If the consumer decides to purchase this contract, wealth levels change to become

$$W_1 = W_0 - L + I - P(I) \text{ and } W_2 = W_0 - P(I), \text{ respectively.} \qquad (3.12)$$

At this point, a behavioral assumption concerning the IC is needed. Possible objectives of IC are discussed in Sect. 6.2; for the time being, assume that the insurer seeks to maximize expected profit $E\Pi$. This implies that the IC is risk neutral, in contradistinction to the potential insurance buyer (IB). With probability π, the IC has to come up with the promised payment I while receiving the premium $P(I)$.

---

[5] As to the curvature, the indifference curve is convex from the origin. It can be shown that (strict) convexity follows from (strict) concavity of the risk utility function $v(W)$. Therefore, the convexity of the indifference curve reflects risk aversion (specifically, R_A). For a proof, see, e.g., Eisen (1979b, 44) or Zweifel et al. (2009, Chap. 5).

[6] In conventional microeconomics, the precise slope of the indifference curve is not known. The additional information available here is due to the fact that the EU-function (3.8) is additive while conventional utility functions $u(W_1, W_2)$ can be of any form.

3.2 Theory of Insurance Demand

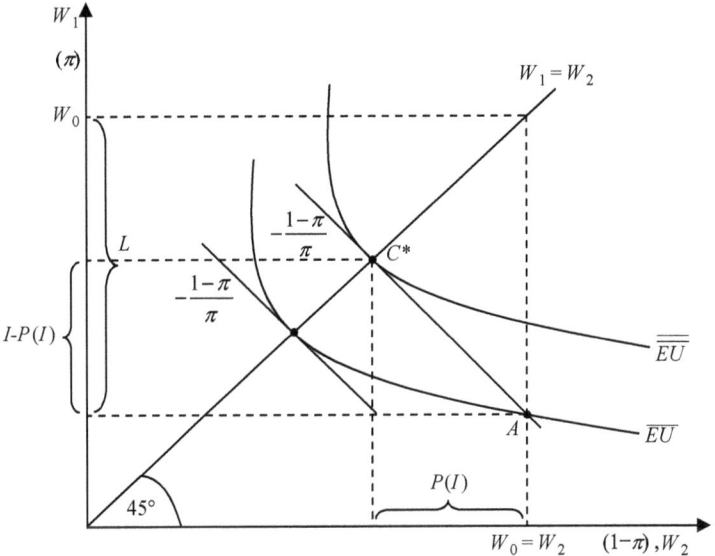

Fig. 3.2 Optimal insurance coverage given marginally fair premium

With probability $(1 - \pi)$, however, the IC cashes in the premium without having to pay the indemnity. In both states, there is administrative expense C that is assumed independent of I for simplicity. Therefore, one has

$$E\Pi = \pi\{P(I) - I\} + (1 - \pi)P(I) - C. \tag{3.13}$$

Now consider marginal changes of premium dP and of coverage dI (the fact that the two are functionally related will be taken into account below). With $dC = 0$, the extra indemnity is not loaded with administrative expense; the IC only takes into account that it has to pay the indemnity with probability π. Therefore, the premium can be said to be *actuarially fair at the margin*. In this case, the so-called iso-expected-profit or insurance line is given by

$$dE\Pi = \pi(dP - dI) + (1 - \pi)dP = 0. \tag{3.14}$$

This condition can be projected into the $[W_1, W_2]$-space of Fig. 3.2 by noting that $dP = -dW_2$ and $(dP - dI) = -dW_1$. Substitution in (3.14) yields

$$\frac{dW_1}{dW_2} = -\frac{1-\pi}{\pi}. \tag{3.15}$$

This is the slope of the so-called *insurance line*. It states that the insurer is able to transfer wealth from the no-loss state W_2 to the loss state W_1 while keeping its expected profit constant at a rate that reflects the ratio of the two probabilities. If the probability of loss π is small, it can transform a sacrifice of W_2 (i.e. the additional premium) into a great deal of W_1 (i.e., the additional indemnity minus the additional premium). The line through point A of Fig. 3.2 runs steep in this case. Conversely, it runs flat if π is high.

Note that equation (3.15) holds true at an arbitrary level of expected profit $E\Pi$. However, if insurance markets are competitive (i.e., not protected by barriers to entry), expected profits will be driven to zero by new competitors. In this case, the premium must be equal to the (expected) marginal cost of enrolling an additional IB, amounting to the so-called *actuarially fair premium*. Accordingly, the line through A is often referred to as the *fair odds line*.

The ingredients for determining the consumer's optimum are now ready. On the one hand, it is known that the indifference curves all have slope $-(1-\pi)/\pi$ when they intersect the certainty line. On the other hand, the insurance line which guarantees the IC a certain expected profit (of zero at the limit) has slope $-(1-\pi)/\pi$ as well. Since the optimum always is given by the tangency between the indifference curve and the constraint (the insurance line in the present context), it must lie on the certainty line (see point C^* of Fig. 3.2). Therefore, given a marginally fair premium, the optimal amount of coverage (provided the consumer buys coverage; see Sect. 3.3) is given by the equality of wealth levels between the two states, which means *full coverage* of the loss.

For this coverage, the consumer pays a premium $P(I)$, which can be read off directly from the W_2-axis of Fig. 3.2, obtaining in return coverage I while continuing to pay the premium in the loss state. Therefore, on the W_1-axis, only the net amount $I - P(I)$ can be read off, offsetting loss L. Of course, L is equal to the difference between the initial wealth levels.

An interesting result of some generality is that in the optimum, the marginal utilities of wealth are equal ($v'[1] = v'[2]$) since C^* lies on the certainty line [see the text leading up to equation (3.11) again]. However, as shown below, this obtains only if the premium is fair at the margin, permitting wealth to be transferred between the two states without a loading. This means no proportional loading but does not preclude a fixed loading $c > 0$. In the latter case, the marginal premium is still given by $P'(I) := \partial P/\partial I = \pi$. This can be used in the IB's optimization problem

$$\max_I EU = \pi \cdot v[W_0 - L + I - P(I)] + (1-\pi)v[W_0 - P(I)]. \quad (3.16)$$

The IB thus chooses the amount of insurance coverage that maximizes expected utility. The first-order condition reads

$$\begin{aligned}\frac{dEU}{dI} &= \pi \cdot v'[1]\{1 - P'(I)\} + (1-\pi)v'[2]\{-P'(I)\} \\ &= \pi v'[1](1-\pi) + (1-\pi)v'[2](-\pi) = 0.\end{aligned} \quad (3.17)$$

Division by $\pi(1-\pi)$ results in the condition

$$v'[1] = v'[2], \quad (3.18)$$

3.2 Theory of Insurance Demand

i.e., equality of the two marginal utilities of risky wealth. And with a strictly concave risk-utility function having wealth as its only argument, equality of marginal utilities can only be attained when the two wealth levels are equal.

Finally, a welfare statement can be gleaned from Fig. 3.2. By purchasing insurance, consumers reach a higher level of expected utility, for the indifference curve $\overline{\overline{EU}}$ through point C^* is higher-valued than the curve $\overline{EU}$ through point A. Note, however, that consumers may nevertheless regret having bought insurance if after paying a premium for many years, they never experience a loss triggering the payment of an indemnity. This illustrates the important difference between utility and expected utility. Under risk, the best one can hope to attain is an optimum in terms of expected utility (which can differ considerably from optimum utility under certainty).

▶ **Conclusion 3.2** Given a marginally fair premium, risk-averse individuals are predicted to opt for comprehensive insurance coverage, meaning that the indemnity equals the loss. On expectation, they are better off than without insurance coverage.

3.2.2 Insurance Demand in the Presence of Irreplaceable Assets

The basic model of insurance demand presented in the preceding section assumes that the assets to be insured can be replaced and have a market value. However, this is not always true. There are assets that exist only in certain states of nature (thus are totally irreplaceable) or whose subjective value depends on the state of nature. Finally, there are assets that can be partially but not fully recovered.

The value of such assets (life, health, and collector items) depends on personal preferences and is not easily expressed in monetary terms. Yet, their impact on the risk-utility function and hence the demand for insurance is amenable to economic analysis. For concreteness, let the irreplaceable asset be health. There are two health states, healthy (h) and permanently sick (s), e.g., due to a loss of limb. While the risk-utility function continues to be defined in terms of wealth, it typically becomes *state-dependent*. A natural assumption is

$$v_h[W] > v_s[W], \qquad (3.19)$$

stating that for a given level of wealth (and hence consumption goods), the risk-utility function has a higher value if the individual is healthy than when he or she is sick or disabled (see Fig. 3.3). However, an additional issue is the marginal utility of risky wealth. Here, two cases are possible. In panel A of Fig. 3.3, the marginal utility of (risky) wealth at a given wealth level is higher in the sick state (as compared to the healthy state), causing the utility difference between the two states to decrease with wealth. The two state-dependent risk-utility functions converge. A possible increase in wealth when sick may indeed be associated with a high marginal utility because it allows the patient to, for example, pay for healthy (and more expensive) food or benefit from a stay at a spa.

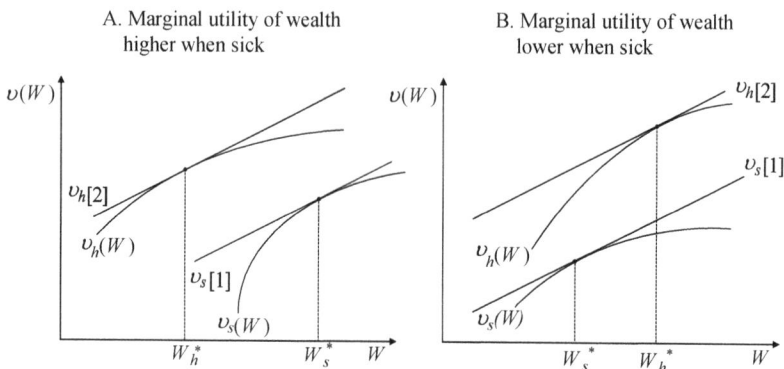

Fig. 3.3 Optimal wealth levels depending on the state of health

By way of contrast, panel B of Fig. 3.3 depicts the case where the marginal utility of (risky) wealth is higher in the healthy than the sick state, causing good health to become more important with a possible increase in wealth. The two state-dependent risk-utility functions diverge. The justification could be that a possible gain (e.g., a vacation trip) confers more utility if one is healthy than when one is sick.

The two cases generate precisely opposite predictions concerning demand for insurance coverage of the "normal" replaceable asset. This can be seen from considering again the optimization problem of an IB

$$\max_I EU = \pi \cdot v_s[W_0 - L + I - P(I)] + (1 - \pi) \cdot v_h[W_0 - P(I)]. \quad (3.20)$$

The analogy with (3.16) is evident. However, the functions $v_s(\cdot)$ and $v_h(\cdot)$ are state-dependent here while there is only one function $v(\cdot)$ in equation (3.16). Accordingly, the first-order condition for an optimum looks slightly different in the present case

$$v'_s[1] = v'_h[2]. \quad (3.21)$$

Thus, given a marginally fair premium, marginal utilities of risky wealth still should be equal for low and high wealth. However, contrary to condition (3.18), this does not imply any more that the two optimal wealth levels are the same.

Figure 3.3 illustrates. The optimality condition (3.21) is shown in both panels by equality of the slopes of the two state-dependent risk-utility functions. In panel A, the optimal wealth level W_s^* in the sick state exceeds the one in the healthy state. This means that insurance coverage should optimally exceed the loss in order to make up for pain and suffering (however; see the discussion of moral hazard effects in Sect. 8.2). In panel B, by way of contrast, wealth in the sick state should optimally fall short of that in the healthy state ($W_s^* < W_h^*$). Thus, insurance coverage should be less than complete, calling for some copayment in health insurance (say).

This analysis can be pushed a bit farther using the pioneering model by Cook and Graham (1977). Its crucial concept is the so-called ransom, i.e., the maximum amount of money one would be willing to pay in order to get the irreplaceable asset

3.2 Theory of Insurance Demand

back. In the present context, this is the amount that makes the healthy and the sick state equivalent in terms of utility. In Fig. 3.3, this amounts to the horizontal distance between the two state-dependent risk-utility functions (not shown). However, this distance decreases with wealth in panel A (due to convergence) but increases in panel B (due to divergence). Therefore, the ransom can be expected to depend on wealth

$$v_s[W_s] = v_h[W_h - r(W)], \quad r(W) : \text{ransom}. \tag{3.22}$$

Differentiating both sides of this equation w.r.t. W yields

$$v'_s \cdot \frac{\partial W_s}{\partial W} = v'_h \cdot \left(\frac{\partial W_h}{\partial W} - \frac{\partial r}{\partial W} \right). \tag{3.23}$$

This equation can be simplified by realizing that a change of wealth affects wealth both in the sick and healthy state one by one, implying $\partial W_s/\partial W = \partial W_h/\partial W = 1$. Therefore, (3.23) becomes

$$v'_s = v'_h - \frac{\partial r}{\partial W} v'_h. \tag{3.24}$$

This can be solved for $r'(W)$, writing pertinent wealth levels explicitly again as in (3.22)

$$r'(W) := \frac{\partial r}{\partial W} = 1 - \frac{v'_s[W_s]}{v'_h[W_h - r(W)]}. \tag{3.25}$$

This equation can be interpreted as follows, distinguishing three cases.

- The irreplaceable asset is a *normal* good. In this case, $r'(W) > 0$, which implies

$$v'_s[W_s] < v'_h[W_h - r(W)].$$

  The marginal utility of (risky) wealth is lower in the sick than in the healthy state. This calls for *partial coverage* of the insurable asset (see panel B of Fig. 3.3).
- The irreplaceable commodity is an *inferior good*. Then, $r'(W) < 0$, and one has from (3.25)

$$v'_s[W_s] > v'_h[W_h - r(W)].$$

  The marginal utility of (risky) wealth is higher in the sick than in the healthy state. This calls for *more than comprehensive coverage* of the insurable asset (see panel A of Fig. 3.3).
- The irreplaceable commodity *does not* have any wealth effect. Then $r'(W) = 0$, and

$$v'_s[W_s] = v'_h[W_h - r(W)].$$

The marginal utilities of (risky) wealth are the same across health states. This calls for *full coverage* of the insurable asset in the sense that the insurance benefit compensates the ransom as well.[7]

Clearly, these three cases can be related to Fig. 3.3. In panel A, the horizontal distance between the two risk-utility functions representing $r(W)$ decreases with W, implying $r'(W) < 0$. Therefore, the irreplaceable asset must be an inferior good. In panel B, the horizontal distance increases with W, pointing to the superiority of the irreplaceable asset. Finally, if the horizontal distance between the two risk-utility functions were to remain constant over the values of wealth considered, the irreplaceable asset would have no wealth effect (not shown in Fig. 3.3). In most cases, one would be led to conclude that the case depicted in panel B of Fig. 3.3 obtains because assets (irreplaceable or not) typically are normal goods.

However, note that the concepts of normal and inferior goods relate to a change in certain wealth. If the loss of an irreplaceable asset is interpreted as an event that is comparable to a change in risky wealth, it is not clear which of the two cases depicted in Fig. 3.3 actually obtains. A theoretical development by Eeckhoudt and Schlesinger (2006) resolves this ambiguity. The authors argue that risk aversion implies an aversion against the accumulation of losses. Recall that W is not certain wealth but risky wealth, motivating the concavity of the risk-utility function. The avoidance of cumulative risk can now be translated into a utility loss, i.e., the vertical distance between the two state-dependent risk-utility functions in Fig. 3.3. If wealth happens to be low, a simultaneous health loss should entail an especially high utility loss. Based on this argument, the relevant case is depicted in panel A of Fig. 3.3. Therefore, consumers are predicted to prefer more than comprehensive coverage for their health care expenditure if they are allowed to. They would like to fully insure what is insurable (their health care expenditure) and add a surcharge for what is non-insurable (loss of health, pain, and suffering).

▶ **Conclusion 3.3** Insurance demand for an insurable asset in the presence of a irreplaceable commodity importantly hinges on the impact of the irreplaceable commodity on the marginal utility of wealth (the replaceable asset). If the insurable loss is triggered by the loss of the irreplaceable commodity, the prediction is that consumers opt for more than comprehensive coverage.

The theoretical contribution by Eeckhoudt and Schlesinger (2006) can also be used to show that an increase in the probability of losing the irreplaceable asset induces an increase in the demand for insurance coverage. In panel A of Fig. 3.3, let the state of sickness occur with probability one-half (say) rather than probability one.

---

[7] $W_s^* = W_u^*$ in Fig. 3.3 requires that the ransom (i.e., the horizontal distance between the two risk-utility functions) be compensated.

3.2 Theory of Insurance Demand

In expected value terms, this moves the conditional utility function $v_s[W]$ up by half the vertical distance between it and $v_h[W]$. But this move toward the function $v_h[W]$ implies that the optimal wealth level in the sick state W_s^* approaches the lower value W_h^*, which is associated with a lower level of insurance coverage. Conversely, if one starts from a zero probability of sickness and hence the function $v_h[W]$, an increase in this probability creates a function $v_s[W]$ which entails a higher optimal wealth level W_s^* and hence a higher level of insurance coverage. Based on this argument, one can predict that the occurrence of a new health risk (such as the COVID-19 epidemic) causes the demand for insurance to increase.

3.3 Demand for Insurance Without Fair Premiums

3.3.1 Optimal Degree of Coverage Without Fair Premiums

There are many reasons for insurance premiums not be actuarially fair. For one, the insurance company (IC) must not only be able to cover the expected value of the claims to be paid, but also its administrative expenses. It could still end up in insolvency if it failed to charge a safety loading as well (see Sect. 7.1). Finally, premiums may have to be actuarially unfair in anticipation of moral hazard effects (see Sect. 8.2.2.2). In general, the loading comes in two variants.

(1) The IC may charge a *proportional loading* $\lambda > 0$ in excess of the fair premium,[8] resulting in a surcharge per monetary unit of coverage. Or put another way, the IB obtains *less coverage* in the loss state (state No. 1) for each monetary unit of premium paid in the no-loss state (state No. 2). In Fig. 3.4, the absolute value of the slope of the transformation line (insurance line) decreases. As a consequence, it is not optimal anymore to buy full coverage, with the only exception of an indifference curve that is not strictly convex, reflecting a local absence of risk aversion. Point C^{**} of Fig. 3.4 symbolizes the new optimum; it lies away from the certainty line, implying $I^{**} < L$.
(2) The IC could also charge a *fixed loading* $c > 0$ in excess of the fair premium, such that $P = \pi L + c$. Since this loading has to be paid in both states, it causes the endowment point A to shift to the left and down by the amount of c, to point A'. However, the premium is marginally fair because $P'(L) = \pi$. With an unchanged slope of the insurance line, the consumer is predicted to continue to buy full coverage (see point C^{***}) if he or she buys insurance at all.

However, if the surcharge moves the insurance line so far back to the origin as to cause the indifference curve through C^{***} to run below A, then no coverage will be purchased at all. This case is illustrated in Fig. 3.5. The consumer now compares

---

[8] In the jargon of insurance, the fair premium is often called "pure premium" or "risk premium", in contradistinction to the definition in Sect. 2.3.2.

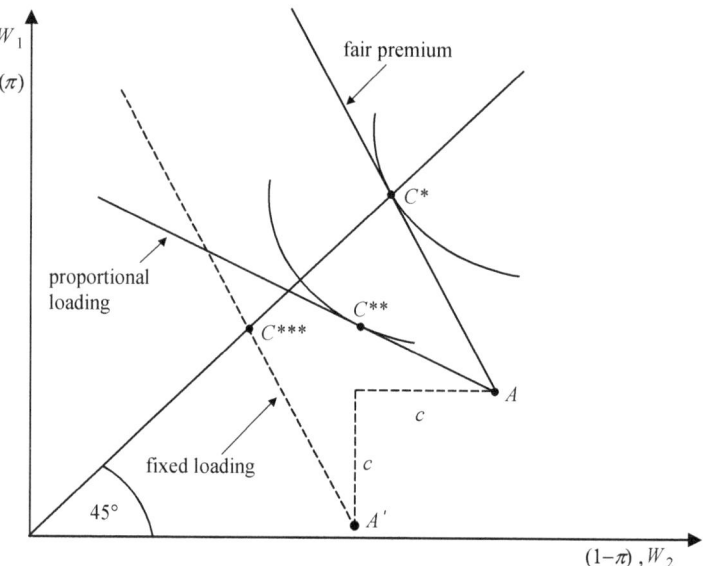

Fig. 3.4 Optimal insurance coverage without fair premium

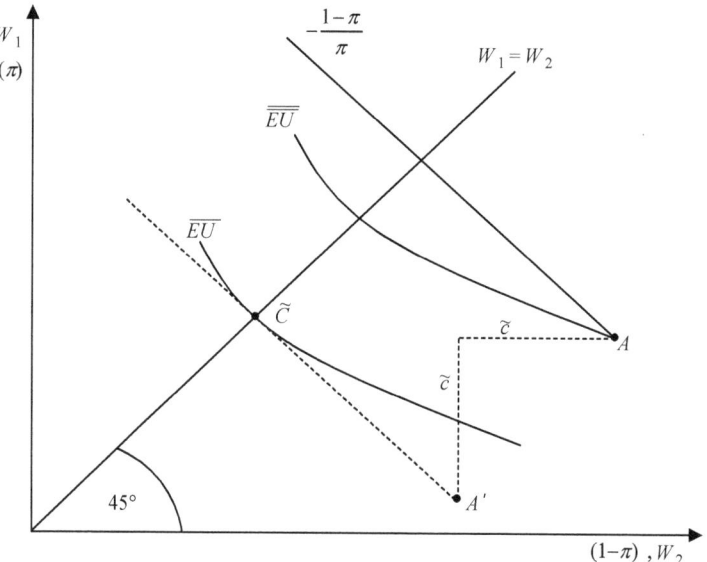

Fig. 3.5 No insurance coverage caused by an excessive fixed loading

point A (no insurance) with point $\tilde{C}$ (full coverage at a premium containing a very high fixed surcharge $\tilde{c}$). Here, A indeed lies on a higher-valued indifference curve $(\overline{\overline{EU}})$ than $\overline{EU}$ through $\tilde{C}$, indicating that *doing without insurance* is optimal.

3.3 Demand for Insurance Without Fair Premiums

Table 3.2 Premium functions and demand for coverage at the individual level

Type	Premium function[a]	Predicted insurance demand
Pure risk premium (fair premium)	$P = EL = \pi \cdot L$	Full coverage
Premium with proportional loading	$P = (1 + \lambda)\pi \cdot L$	Partial coverage; deductible and/or coinsurance
Premium with fixed loading	$P = \pi \cdot L + c$	Full coverage or no coverage
Premium with combined loading	$P = (1 + \lambda')\pi \cdot L + c'$	Partial or no coverage; deductible and/or coinsurance

[a]L: loss
$\lambda' < \lambda, c' < c$

For a combination of the two types of loading, the two results can be stated as follows. If the IB continues to buy coverage at all, it is optimal to purchase less than full coverage.

Table 3.2 provides an overview of commonly used premium functions and their impact on the demand for insurance. Where the prediction is less than full coverage, there are two variants. In the case of a *deductible*, an amount D is defined below which the IC does not pay at all. Above D, it pays a benefit net of D.[9] Alternatively, a *rate of coinsurance* $1 - \alpha$ defines a percentage rule, in that, the IC covers only αL, with $0 < \alpha \leq 1$ of a loss L.

Only the rate of coinsurance is considered here. After all, a deductible can be interpreted as a rate of coinsurance $1 - \alpha = 1$ below the threshold. Let the premium function and type of contract be predetermined, following work by Arrow (1963), Mossin (1968), and Smith (1968). Note that this precludes consumer choice concerning type of contract. Therefore, one is dealing with "second best" solutions at this point (however; see Sect. 3.3.4).

For simplicity, let the premium be without a fixed loading, causing it to vary in proportion with the degree of coverage α. Denoting P_0 as the fair premium for $\alpha = 1$, one, therefore, obtains for expected utility of the IB

$$EU(W) = \pi \cdot v[W_0 - L + \alpha L - \alpha P_0] + (1 - \pi) \cdot v[W_0 - \alpha P_0]. \quad (3.26)$$

In the case of a proportional loading λ, the premium becomes $\alpha P_0(1 + \lambda) = \alpha \pi L(1 + \lambda)$, and substituting in (3.26), one obtains for expected utility

$$EU(W) = \pi \cdot v[W_0 - L + \alpha L - \alpha \pi L(1 + \lambda)] + (1 - \pi) \cdot v[W_0 - \alpha \pi L(1 + \lambda)]. \quad (3.27)$$

The first-order condition for a maximum reads

$$\frac{dEU}{d\alpha} = \pi\{L - \pi L(1 + \lambda)\}v'[1] - (1 - \pi)(1 + \lambda)\pi L v'[2]$$
$$= \pi L\{1 - (1 + \lambda)\pi\}v'[1] - \pi L\{(1 - \pi)(1 + \lambda)\}v'[2] = 0, \quad (3.28)$$

---

[9]Doherty (1985, 451) in addition distinguishes the franchise, where the insurer pays the full indemnity without deduction if it exceeds the deductible.

with $v'[j]$ = marginal utility of risky wealth in state j ($j = 1, 2$) and $W_1 := W_0 - L + \alpha L - \alpha \pi L(1 + \lambda)$ and $W_2 := W_0 - \alpha \pi L(1 + \lambda)$. Now (3.28) can be divided by πL and multiplied by $(1 - \pi)/\pi$ to obtain the slopes of the indifference curve and the insurance line (see Fig. 3.4) at the optimum in absolute value

$$\frac{1-\pi}{\pi} \cdot \frac{v'[2]}{v'[1]} = \left[\frac{1-(1+\lambda)\pi}{(1-\pi)(1+\lambda)}\right] \cdot \frac{1-\pi}{\pi} = \left[\frac{1/(1+\lambda) - \pi}{1-\pi}\right] \cdot \frac{1-\pi}{\pi}, \quad (3.29)$$

after dividing numerator and denominator by $(1 + \lambda)$.

Condition (3.29) can be interpreted as follows:

- $\lambda = 0$ as the benchmark: This implies $v'[2]/v'[1] = 1$, i.e., equality of marginal utilities across the two states and hence *full coverage*. Given the fair premium, the first-best solution is attained (see C^* of Fig. 3.4).
- $\lambda > 0$: The proportional loading calls for $v'[2]/v'[1] < 1$ because the insurance line has a slope less than $(1 - \pi)/\pi$. Given a strictly concave risk-utility function, $v'[1] > v'[2]$ holds only if $W_1 < W_2$, which means *partial coverage*. In terms of Fig. 3.4, the right-hand side of (3.29) defines the slope of the insurance line, which is lower than $(1 - \pi)/\pi$ in absolute value, inducing C^{**} as the optimum in Fig. 3.4.
- $\lambda \gg 0$: Conceivably, an excessively high loading could turn the right-hand side of (3.29) negative. The critical value of λ is $\lambda^c = 1/\pi - 1$. The highest probability of loss is observed in health insurance, where up to 75% of the insured submit a claim for medical expenditure during a year. Setting $\pi = 0.75$, one has $\lambda^c = 0.33$, a value rarely reached in that line of insurance. However, in the extreme case of $\lambda > \lambda^c$, resulting in a negative slope of the insurance line, the optimum coincides with the endowment point A, i.e., *no coverage*.

If the loading is independent of the expected loss, the premium is given by $P = \alpha P_0 = \alpha \pi L + c$. Substituting this into (3.27), one obtains for an optimum

$$\frac{dEU}{d\alpha} = \pi(1 - \pi) \cdot L \cdot v'[1] - \pi(1 - \pi) \cdot L \cdot v'[2] = 0$$

$$\Rightarrow v'[1] = v'[2]. \quad (3.30)$$

This means *complete coverage*. However, as shown in Fig. 3.5, when c attains a high value such as $\tilde{c}$, an interior optimum may not exist. The fixed loading c evidently has some similarity to a lump-sum tax levied on the IB who, however, can avoid it by not buying insurance coverage at all.

▶ **Conclusion 3.4** For risk-averse consumers and in the presence of (necessarily) unfair premiums, partial coverage is optimal in the case of a proportional loading. In the case of a fixed loading, complete or zero coverage is optimal.

3.3 Demand for Insurance Without Fair Premiums

It is easy to see that although partial in the presence of a proportional loading, coverage does increase with the probability of loss. It suffices to recall that the left-hand side of Eq. (3.29) corresponds to the slope of the indifference curve in (W_1, W_2)-space and its right-hand side, the slope of the insurance line. Since $(1 - \pi)/\pi = (1/\pi) - 1$, a higher value of π is associated with both a flatter indifference curve and insurance line, which causes the optimum point C^{**} of Fig. 3.5 to move toward the certainty line, entailing a higher amount of coverage.

3.3.2 Risk Aversion as a Determinant of Insurance Demand

As shown in Eisen (1979b, 44), the curvature of the indifference curve mirrors the coefficient of absolute risk aversion. Equation (3.10) provides the intuition for this result. If the slope of the indifference curve is determined by the (probability-weighted) ratio of marginal utilities $(v'[2]/v'[1])$, then curvature as a change in the slope must contain at least one term $v''[1]$ or $v''[2]$. However, the second derivative of the risk-utility function appears in all measures of risk aversion (see Sect. 2.3.2).

The degree of risk aversion does not affect the optimum as long as the premium is marginally fair ($\lambda = 0$). In Fig. 3.6, the point C^* is determined by the equality of the respective *slopes* of the indifference curve and the fair odds line; the degree of curvature is irrelevant.

This changes as soon as the premium contains a proportional loading such that $\lambda > 0$. In Fig. 3.6, the indifference curve $\overline{EU_1}$ of slightly risk-averse individual No. 1 is associated with point C^{**}, entailing a marked reduction of coverage. By way

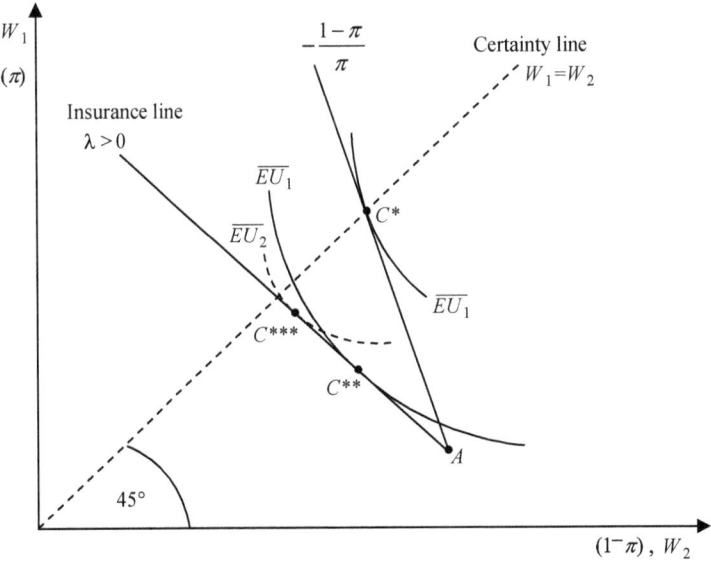

Fig. 3.6 Demand for insurance and risk aversion

of contrast, individual No. 2 is more risk-averse, with his or her indifference curve $\overline{EU_2}$ of equal slope as that of individual No. 1 on the certainty line but exhibiting more curvature. This means that $\overline{EU_2}$ becomes flattered quickly in the direction of point A, implying that it reaches the point of tangency with the insurance line quicker as well. Therefore, the optimum C^{***} of individual No. 2 lies closer to the certainty line than C^{**} of individual No. 1, implying more demand for coverage.

▶ **Conclusion 3.5** In the presence of a proportional loading, consumers with more marked risk aversion are predicted to opt for higher insurance coverage than those with weaker risk aversion, *ceteris paribus*.

3.3.3 Premium Rate and Wealth as Determinants of Insurance Demand

Two important questions of the theory of insurance demand revolve around the roles of the premium rate and wealth. The objective of this section is to derive demand functions with respect to these arguments. In the following, demand is represented by the rate of coverage α, wealth by W_A (denoting other assets than the insured one), and the premium rate p as introduced in Sect. 1.5.2. Note that p also reflects loadings λ and c.

The analysis is based on Mossin (1968). Let the IB have wealth $W = W_A + W_I$, with W_I denoting risky (and insurable) wealth, such as a home that may burn down. By way of contrast, W_A is neither subject to risk nor insurable. For instance, this could be bonds stored in the vaults of a bank which cannot burn and whose value cannot be insured. Let $L \leq W_I$ be the loss; with probability π, the house is destroyed, with probability $(1-\pi)$, it remains intact. Therefore, expected loss amounts to πW_I, which can be insured for premium $P_0 = pW_I$, with p symbolizing the premium rate. The IB has to decide the degree of coverage α. In case of $\alpha < 1$, the assumption is that the premium reduces to αP_0 while the indemnity reduces to αL in the loss state.

Final wealth of the IB amounts to $W_1 = W_A + W_I - \alpha pW_I - L + \alpha L$ in the loss state and $W_2 = W_A + W_I - \alpha pW_I$ in the no-loss state. The risk-utility function, therefore, reads

$$\begin{aligned} EU(W) &= \pi v[W_A + W_I - \alpha pW_I - L + \alpha L] + (1-\pi)v[W_A + W_I - \alpha pW_I] \\ &= \pi v[1] + (1-\pi)v[2], \text{ with} \\ v[1] &:= v[W_A + W_I - \alpha pW_I - L + \alpha L]; \\ v[2] &:= v[W_A + W_I - \alpha pW_I]. \end{aligned} \quad (3.31)$$

The first-order condition for the optimum value of α is given by

$$\frac{dEU}{d\alpha} = \pi v'[1](L - pW_I) + (1-\pi)v'[2](-pW_I) = 0, \quad (3.32)$$

3.3 Demand for Insurance Without Fair Premiums

with $v'[1]$ denoting the marginal utility of risky wealth in the loss state and $v'[2]$ marginal utility in the no-loss state, respectively. A solution $\alpha^* = \alpha^*(p)$ implicitly defines the demand for coverage as a function of price, i.e., the premium rate.

In order to obtain additional information about this function, so-called *comparative statics* analysis will be performed. It consists of subjecting the optimum condition (3.32) to an exogenous shock dp. This will entail an optimal adjustment $d\alpha^*$, resulting in the objective function $EU(\cdot)$ attaining its maximum somewhere else. However, the new maximum still must satisfy the condition, $dEU/d\alpha = 0$. Therefore, the equality to zero must hold before and after adjustment, resulting in the comparative-static equation

$$\frac{\partial^2 EU}{\partial \alpha^2} \cdot d\alpha + \frac{\partial^2 EU}{\partial \alpha \partial p} \cdot dp = 0. \tag{3.33}$$

The second term on the left-hand side shows the impact of the shock on expected utility, and the first term, the impact of the induced adjustment of α^*. Equation 3.33 can be solved to obtain

$$\frac{d\alpha}{dp} = -\frac{\partial^2 EU/\partial \alpha \partial p}{\underbrace{\partial^2 EU/\partial \alpha^2}_{(-)}} < 0, \quad \text{if} \quad \frac{\partial^2 EU}{\partial \alpha \partial p} < 0. \tag{3.34}$$

Since $v'' < 0$ is regularly assumed, the denominator is negative ($\partial^2 EU/\partial \alpha^2 < 0$, amounting to the sufficient condition for a maximum). Therefore, the sign of the numerator determines the sign of $d\alpha^*/dp$. The rate of coverage is predicted to decrease in response to a hike in the premium rate if the mixed second-order derivative is negative.

Differentiating (3.32) w.r.t. p and recalling that $W_1 = W_A + W_I - \alpha p W_I + \alpha L$ and $W_2 = W_A + W_I - \alpha p W_I + \alpha p W_I$, respectively, one obtains

$$\begin{aligned}\frac{\partial^2 EU}{\partial \alpha \partial p} &= \pi\{v''[1](L - pW_I)(-\alpha W_I) - v'[1]W_I\} \\ &\quad + (1-\pi)\{v''[2](-pW_I)(-\alpha W_I) - v'[2]W_I\} \\ &= -\pi v''[1](L - pW_I)(\alpha W_I) \\ &\quad + (1-\pi)v''[2](\alpha p W_I^2) - W_I Ev' \lessgtr 0,\end{aligned} \tag{3.35}$$

with $Ev' := \pi v'[1] + (1-\pi)v'[2]$ denoting expected marginal utility of risky wealth.

The definition of the coefficient of absolute risk aversion introduced in Sect. 2.3.2, $R_A[\cdot] := -v''[\cdot]/v'[\cdot] > 0$, can be used to obtain $-v''[1] = R_A[1] \cdot v'[1]$, and $v''[2] = -R_A[2] \cdot v'[2]$. Therefore, (3.35) can be rewritten

$$\frac{\partial^2 EU}{\partial \alpha \partial p} = \pi R_A[1] \cdot v'[1](L - pW_I)(\alpha W_I)$$
$$- (1-\pi) R_A[2] \cdot v'[2](\alpha p W_I^2) - W_I E v'[\cdot]$$
$$= \alpha W_I \cdot \{\underbrace{\pi R_A[1] \cdot v'[1](L - pW_I)}_{(+)}$$
$$\underbrace{- (1-\pi) R_A[2] \cdot v'[2] p W_I}_{(-)}\}$$
$$- W_I E v'. \tag{3.36}$$

The sum of the first two terms is the wealth effect of an increase in the premium rate; it can be positive or negative because risk aversion may vary with wealth. The last term is unambiguously negative, representing the substitution effect of the premium that has to be paid in both states. According to Sect. 2.4.1.1, three cases can be distinguished.

(1) *Constant absolute risk aversion (CARA)*. This means $R_A[1] = R_A[2] := R_A$. This allows R_A to be factored out, causing the bracket of the first term to become $\pi v'[1](L - pW_I) - (1-\pi)v'[2] p W_I$. According to the first-order condition (3.32), this expression is zero. Therefore, the sign of (3.36) is negative, implying $d\alpha^*/dp < 0$ in view of (3.34). The wealth effect drops out, leaving the substitution effect only. This constitutes the *normal response* since the degree of coverage is reduced when insurance becomes more expensive.

(2) *Decreasing absolute risk aversion (DARA)*. This implies $R_A[1] > R_A[2]$, since wealth in the no-loss state No. 2 is larger. The following four subcases can be distinguished.

- A *large loss L* compared to the initial premium pW_I causes the wealth effect to become positive and dominant such that $\partial^2 EU/\partial\alpha\partial p > 0$ and hence, $d\alpha^*/dp > 0$. This constitutes an *anomaly* since the degree of coverage α rises rather than falls when the premium rate increases. The reason is the wealth effect of a large loss that causes the marginal utility of wealth $v'[1]$ to be high in the loss state.
- *High insurable wealth* W_I causes the difference $(L - pW_I)$ in the first term of the wealth effect to go toward zero while boosting the negative terms, resulting in $\partial^2 EU/\partial\alpha\partial p < 0$ and hence $d\alpha^*/dp < 0$. This again constitutes a *normal response*.
- A *small probability of loss* π also causes the first component of the wealth effect to be dominated by the negative second component in (3.36), with $\partial^2 EU/\partial\alpha\partial p < 0$ and hence $d\alpha^*/dp < 0$. This again constitutes a *normal response* to an increase in the premium rate p.
- A *high initial premium rate p* serves to reduce $(L - pW_I)$ and to increase the second negative component of the wealth effect, resulting in $\partial^2 EU/\partial\alpha\partial p < 0$ and hence $d\alpha^*/dp < 0$. This again constitutes a *normal response*.

(3) *Increasing absolute risk aversion (IARA)*. In view of the discussion in Sects. 2.3.2 and 2.5, this case can be neglected.

3.3 Demand for Insurance Without Fair Premiums

▶ **Conclusion 3.6** The optimal degree of coverage and hence the demand for insurance decreases in response to an increased premium rate if the individual exhibits constant absolute risk aversion, has high insurable wealth, and is exposed to a low probability of loss or a high initial premium rate.

In order to obtain the relationship between the demand for insurance and wealth, the first-order condition (3.32) must be subjected to a change dW_A (note that insured wealth W_I remains constant). In full analogy to the comparative-static equation (3.33) and its solution, one obtains

$$\frac{d\alpha^*}{dW_A} = -\frac{\partial^2 EU/\partial\alpha\partial W_A}{\partial^2 EU/\partial\alpha^2} > 0, \quad \text{if} \quad \frac{\partial^2 EU}{\partial\alpha\partial W_A} > 0. \tag{3.37}$$

Taking the derivative of (3.32) w.r.t. W_A yields

$$\frac{\partial^2 EU}{\partial\alpha\partial W_A} = \pi v''[1](L - pW_I) + 0 - (1 - \pi)v''[2]pW_I + 0$$

$$= \underbrace{\pi v''[1](L - pW_I)}_{(-)} - \underbrace{(1 - \pi)v''[2]pW_I}_{(+)}. \tag{3.38}$$

Using $R_A[1] := -v''[1]/v[1]$ and $R_A[2] := -v''[2]/v[2]$ once more, one has

$$\frac{\partial^2 EU}{\partial\alpha\partial W_A} = \underbrace{\pi R_A[1] \cdot v'[1](L - pW_I)}_{(-)} + \underbrace{(1 - \pi)R_A[2] \cdot v'[2]pW_I}_{(+)} \gtrless 0. \tag{3.39}$$

Again, three cases can be distinguished.

(1) *Constant absolute risk aversion (CARA)*. This means $R_A[1] = R_A[2] := R_A$. Factoring this out results in the first-order condition (3.32) once more, causing (3.39) to become zero. Therefore, there is no wealth effect, and the prediction is *no response* in insurance coverage to a change in the amount of other, noninsurable wealth.
(2) *Decreasing absolute risk aversion (DARA)*. This means $R_A[1] > R_A[2]$. Four subcases need to be distinguished.

- A *large loss L* compared to the premium pW_I causes the first term to become dominant such that $\partial^2 EU/\partial\alpha\partial W_A < 0$ and hence $d\alpha^*/dW_A < 0$. This constitutes an *anomaly* since demand for coverage is predicted to decrease with increasing wealth, making insurance an inferior good.
- *High insurable wealth W_I* drives the negative first term of (3.39) toward zero (recall that loss L always exceeds premium pW_I, precluding a change of sign). At the same time, it boosts the second, positive term, tilting the expression toward $\partial^2 EU/\partial\alpha\partial W_A > 0$ and hence the *normal response* $d\alpha^*/dW_A > 0$.
- A *small probability of loss π* also makes the first negative term approach zero, tilting the expression toward $\partial^2 EU/\partial\alpha\partial W_A > 0$ and the *normal response* $d\alpha^*/dW_A > 0$.

- A *high premium rate* p also makes the first negative term small while increasing the positive second term. The predicted tendency is $\partial^2 EU/\partial\alpha\partial W_A > 0$ and hence the *normal response* $d\alpha^*/dW_A > 0$.

(3) *Increasing absolute risk aversion (IARA)*. In view of the discussion in Sect. 2.3.2, this case can again be neglected.

▶ **Conclusion 3.7** If the loss is large compared to the premium or insured wealth and absolute risk aversion is decreasing, then a reduction of demand for coverage in response to higher wealth cannot be excluded. In the remaining cases considered, demand for coverage is predicted to be unaffected or to increase with risky wealth.

Conclusions 3.6 and 3.7 are somewhat irritating because they point to the possibility of insurance being an *inferior* or even a *Giffen good*. However, one has to take into account that usually risk (here represented by the possible loss L) increases along with wealth. This consideration calls for making the coefficient of *relative* rather than absolute risk aversion the relevant subjective parameter. As shown in Sect. 2.3.2, constant risk aversion induces a constant share of insured assets in total wealth. Under plausible conditions, this constancy implies that the income elasticity of insurance demand is > 1, making insurance coverage even a *luxury good*.[10]

Ultimately, however, the purchase of insurance coverage revolves around the allocation of money to different forms of financial provision for an uncertain future, with savings constituting an alternative compared to insurance. Saving has the advantage of honoring less consumption today with more final wealth in the future in all states of nature. By way of contrast, the insurance premium paid guarantees a higher final wealth in the *loss state* only. Now it is quite possible that with increasing wealth, the opportunity cost of saving falls, causing insurance to be substituted by savings. Whether this in fact happens depends on how risk aversion changes compared to the opportunity cost of saving when wealth increases (see the role of prudence discussed in Sect. 2.3.3). In sum, financial provision for the future likely is a normal good. This, however, does not entirely preclude the component "insurance" from constituting an inferior good and possibly even a Giffen good [for more detail; see Eeckhoudt et al. (1997)].

---

[10] The relation between the elasticity of expenditure on insurance $P = pW_I$ w.r.t. wealth, denoted by $e(P, W)$, and the income elasticity $e(P, Y)$ can be established as follows. By expansion, one obtains $e(P, Y) = \frac{\partial P}{\partial Y} \cdot \frac{Y}{P} = \frac{\partial P}{\partial W} \cdot \frac{\partial W}{\partial Y} \cdot \frac{W}{P} \cdot \frac{Y}{W} = e(P, W) \cdot e(W, Y)$. If outlay on insurance as a share of wealth is to be constant, it must be true that $e(P, W) = 1$. On the other hand, the fact that the concentration of wealth exceeds that of income implies $e(W, Y) > 1$. In combination, these two statements result in $e(P, Y) > 1$.

3.3.4 Pareto-Optimal Insurance Contracts

The IB up to this point were seen as optimizing their degree of coverage (or conversely, of copayment) *given* the type of insurance contract. As pointed out already by Arrow (1963, Appendix) and Arrow (1974), this leaves open the question what the optimal type of contract would be if both parties were free to agree on it. This section, therefore, addresses the issue of Pareto-optimal insurance contracts.

Pareto-optimality calls for the modeling not only of the IB but of the insurance company (IC) as well. For simplicity, stochastic claims and nonstochastic administrative expense are amalgamated into a total cost function $C(I)$ which increases with claims covered $I \geq 0$, with $C'(I) > 0$. A Pareto-optimal insurance contract characterized by the amount of indemnity $I(L)$ and the premium $P(I)$ can then be derived by letting the IB optimize while the IC is guaranteed a certain level of utility,

$$\text{Max } EU(W) \text{ with } W = W_0 - P(I) - L + I(L),$$
$$\text{s.t.: (a) } EV(W_V^*) = V[W_V] \text{ and } W_V^* = W_V + P(I) - C(I(L));$$
$$\text{(b)} \quad 0 \leq I(L) \leq L. \tag{3.40}$$

As before, $EU(W)$ symbolizes expected utility on the part of the IB. In addition, $EV(\cdot)$ is the expected utility of the (potentially risk-averse) IC. It depends on its wealth W_V^* (i.e., assets) *after* concluding the optimal contract (i.e., after adding premium income and subtracting cost), whereas W_V stands for its assets without the contract. Accordingly, constraint (a) states that the IC must be at least as well off in expectation after the conclusion of the contract as before.[11] When cost is given by $C(I) = (1+\lambda)\pi I$, then constraint (a) requires that the premium be given by the standard formula $P = (1+\lambda)E[I(L)]$. Moreover, Arrow (1974) proves that the optimal indemnity function contains a deductible $D \in [0, L_{max}]$ such that

$$I^*(L) = \min\{L - D, 0\}. \tag{3.41}$$

Therefore, the optimal indemnity is zero as long as the loss falls short of the contractual deductible $(L - D < 0)$. Beyond that threshold $(L - D = 0, \text{ thus } L = D)$, the indemnity increases in step with the loss (after subtracting the deductible). This result was generalized by Raviv (1979) as follows.

(1) If the IC is *risk neutral* while its cost function is *linear* (with $\lambda > 0$), then the contract contains a positive deductible $D > 0$.
(2) If the IC is *strictly risk-averse* while its cost function is *linear* (with $\lambda > 0$), then the Pareto-optimal contract contains not only a non-negative deductible D, but also a copayment beyond D such that

$$I^*(L) = \begin{cases} 0 \text{ for } L \leq D, \ D \geq 0; \\ 0 < I^*(L) < L \text{ for } L > D, \text{ with } 0 < I'(L) < 1. \end{cases} \tag{3.42}$$

---

[11] This is the so-called zero-utility principle of premium calculation (see Sect. 7.1.3).

(3) If the IC is *risk neutral* but its cost function is convex from below, such that $C''(I) > 0$, then the Pareto-optimal contract again contains a strictly positive deductible D and a copayment beyond D.

▶ **Conclusion 3.8** The Pareto-optimal insurance contract has a deductible and no copayment beyond it in case of a linear total cost function and risk neutrality of the insurer. Risk aversion or a convex total cost function entail a deductible combined with copayment beyond it.

In fact, commonly observed insurance contracts do feature a copayment beyond a positive deductible. The probable reason is a convex cost function (case No. 3). The convexity is caused by moral hazard effects, which become increasingly important when the benefit I approaches the maximum loss L_{max} (see Sect. 8.2), because the probability of and/or the size of loss increases. Copayment thus serves to provide incentives to IB to limit their moral hazard.

3.4 Demand for Insurance with Multiple Risks

Economic agents often are exposed to several risks at the same time. A spell of illness may cause a loss to the asset "health", an accident gives rise to a liability claim affecting "wealth", and a new computer generation may devalue the asset "wisdom" (i.e., skills). In contradistinction to this fact, the models presented so far dealt with one risk only that could be mitigated or even neutralized through risk management, in particular insurance. However, an insurance contract protecting the IB from all possible perils (so-called all-risk policy) is not available. Usually, it takes different contracts to deal with different risks, e.g., covering the car, the home, healthcare expenditure, or the loss of life. Losses to other assets (notably "wisdom") are not insurable at all.

Doherty and Schlesinger (1983) model this incompleteness of insurance markets by positing a non-insurable *background risk* (N). There continues to be an insurable loss L. Both types of losses occur with a certain probability. Therefore, four states of nature can be distinguished, ranging from no loss at all (S_1) to the occurrence of both losses (S_4; see Table 3.3).

Let π_L and π_N be the probabilities of the occurrence of L and N, respectively. For deriving the probabilities of the four states of nature $\pi_1, \ldots, \pi_4$, they need to be expressed in terms of conditional probabilities. Specifically, $\pi_{N|L}$ denotes the probability of N occurring given that L happened. For instance, the probability of suffering both losses becomes $\pi_4 = \pi_L \cdot \pi_{N|L}$. Conditional probabilities are crucial because they permit expressing whether losses L and N are correlated or not. If, for

3.4 Demand for Insurance with Multiple Risks

Table 3.3 Probabilities of occurrence in the presence of an uninsurable loss

States of nature	Wealth (without insurance)	Probabilities
S_1 No loss occurs	W_0	$\pi_1 = 1 - \pi_N - \pi_L + \pi_L \pi_{N|L}$
S_2 Insurable loss occurs	$W_0 - L$	$\pi_2 = \pi_L - \pi_L \pi_{N|L}$
S_3 Non-insurable loss occurs	$W_0 - N$	$\pi_3 = \pi_N - \pi_L \pi_{N|L}$
S_4 Both losses occur	$W_0 - L - N$	$\pi_4 = \pi_L \pi_{N|L} = \pi_N \pi_{L|N}$

example, $\pi_{N|L} > \pi_N$, the two losses are positively correlated because the occurrence of L serves to increase the probability of N happening.[12]

The decision variable is the degree of coverage α as before, with αL the share of L falling on the IC and $(1-\alpha)L$, the share borne by the IB. With a proportional loading $\lambda > 0$, the premium accordingly amounts to $P = \alpha \pi_L L(1 + \lambda)$. As W_0 symbolizes initial wealth, expected utility can be written, using Table 3.3

$$\begin{aligned} EU = &\, \pi_1 v\{W_0 - \alpha \pi_L L(1+\lambda)\} \\ &+ \pi_2 v\{W_0 - \alpha \pi_L L(1+\lambda) - (1-\alpha)L\} \\ &+ \pi_3 v\{W_0 - \alpha \pi_L L(1+\lambda) - N\} \\ &+ \pi_4 v\{W_0 - \alpha \pi_L L(1+\lambda) - (1-\alpha)L\} - N\}. \end{aligned} \quad (3.43)$$

To find out whether full coverage (or even more than full coverage) is optimal in this situation, one needs to evaluate

$$\begin{aligned} \left.\frac{dEU}{d\alpha}\right|_{\alpha=1} = &\, \pi_1 v'[1]\{-\pi_L L(1+\lambda)\} + \pi_2 v'[2]\{-\pi_L L(1+\lambda) + L\} \\ &+ \pi_3 v'[3]\{-\pi_L L(1+\lambda)\} + \pi_4 v'[4]\{-\pi_L L(1+\lambda) + L\}. \end{aligned} \quad (3.44)$$

If the objective function has a positive or zero slope at the point of $\alpha = 1$, then full or "excessive" coverage is optimal; if it has negative slope at this point, some copayment is optimal. However, the case of full coverage permits two simplifications of (3.44). Since wealth is the same in states No. 1 and 2, it must be true that $v'[1] = v'[2]$, with $v'[\cdot]$ denoting marginal utility of risky wealth. Likewise, wealth in states No. 3 and 4 is equal, implying $v'[3] = v'[4]$. Therefore, using these two simplifications, one has

$$\begin{aligned} \left.\frac{dEU}{d\alpha}\right|_{\alpha=1} = &\, (\pi_1 + \pi_2) v'[1]\{-\pi_L L(1+\lambda)\} + \pi_2 L v'[1] \\ &+ (\pi_3 + \pi_4) v'[3]\{-\pi_L L(1+\lambda)\} + \pi_4 L v'[3]. \end{aligned} \quad (3.45)$$

---

[12]Conditional and nonconditional probabilities are related as follows. According to the Bayes theorem, the conditional probability is given by $\pi_{N|L} = \pi_{N,L}/\pi_L$. Solving for $\pi_{N,L}$, one obtains $\pi_{N,L} = \pi_{N/L} \cdot \pi_L$.

Moreover, from Table 3.3, it is evident that $(\pi_1 + \pi_2) = 1 - \pi_N$ and $\pi_3 + \pi_4 = \pi_N$. This results in

$$\left.\frac{dEU}{d\alpha}\right|_{\alpha=1} = (1 - \pi_N)v'[1]\{-\pi_L L(1+\lambda)\} + \pi_2 L v'[1]$$
$$+ \pi_N v'[3]\{-\pi_L L(1+\lambda)\} + \pi_4 L v'[3]. \quad (3.46)$$

Factoring out $\pi_L L$ and rearranging slightly, one obtains

$$\left.\frac{dEU}{d\alpha}\right|_{\alpha=1} = \pi_L L\{-(1 - \pi_N)v'[1](1+\lambda) + \pi_N v'[3](1+\lambda)\}$$
$$+ \pi_2 L v'[1] + \pi_4 L v'[3]. \quad (3.47)$$

Finally, π_2 and π_4 can be substituted as well by noting that according to Table 3.3, $\pi_2 = \pi_L - \pi_L \pi_{N|L}$ and $\pi_4 = \pi_L \pi_{N|L}$. Collecting terms involving π_N and $(1+\lambda)$ in the bracket results in

$$\left.\frac{dEU}{d\alpha}\right|_{\alpha=1} = \pi_L L\{\pi_N(1+\lambda)(v'[1] - v'[3]) + v'[1] - v'[1](1+\lambda)\}$$
$$- \pi_L \pi_{N|L} L v'[1] + \pi_L \pi_{N|L} L v'[3]. \quad (3.48)$$

After simplifying and factoring out $\pi_L L$ from the last two terms as well, this becomes

$$\left.\frac{dEU}{d\alpha}\right|_{\alpha=1} = \pi_L L\{\pi_N(1+\lambda)(v'[1] - v'[3]) - \lambda v'[1]\}$$
$$- \pi_L L\{\pi_{N|L}(v'[1] - v'[3])\}. \quad (3.49)$$

Finally, $\pi_L L$ can be factored out everywhere, resulting in

$$\left.\frac{dEU}{d\alpha}\right|_{\alpha=1} = \pi_L L\{[(1+\lambda)\pi_N - \pi_{N|L}](v'[1] - v'[3]) - \lambda v'[1]\}. \quad (3.50)$$
$$\phantom{\left.\frac{dEU}{d\alpha}\right|_{\alpha=1} = \pi_L L\{}(+/-)\phantom{\pi_N - \pi_{N|L}]}(-)(-)$$

Recall that full or "excessive" coverage is optimal if the value of (3.50) is nonnegative. Its value is positive if the objective function reaches its maximum at $\alpha > 1$ in principle, but the IC does not permit excess coverage to prevent moral hazard. Also, note that the effects to be discussed below are all leveraged by the expected value of the loss $\pi_L L$, which makes intuitive sense.

The following cases and subcases need to be distinguished.

(1) *Actuarially fair premium at the margin,* $\lambda = 0$. The multiplier of $(v'[1] - v'[3])$ boils down to $\pi_N - \pi_{N|L}$; moreover, the last negative term in the bracket vanishes.

- *Independence of the two losses.* This means $\pi_N = \pi_{N|L}$, causing the first term in the bracket to vanish. Therefore, (3.50) has the value of zero, and *full coverage* is optimal (see Table 3.4). This is an intuitive result. If the occurrence of the non-insurable loss N is independent of the insured loss L, then L can be dealt with in isolation. Therefore, the one-risk model applies, which predicts full coverage in the absence of a proportional loading (see Sect. 3.2.1).
- *Positive correlation of the two losses.* This means $\pi_N < \pi_{N|L}$ in equation (3.50). The occurrence of loss N makes that of loss L more likely.[13] Also, note that $v'[1] < v'[3] < 0$, since $v'' < 0$ and state No. 1 has highest wealth. Therefore, equation (3.50) takes on a positive value, indicating that *excessive coverage* ($\alpha > 1$) would be optimal in principle. Intuitively, consumers would like to over-insure loss L because they cannot obtain coverage for loss N. In this way, they could obtain at least partial coverage of loss N as well.
- *Negative correlation of the two losses.* Here, $\pi_N > \pi_{N|L}$, causing the first term to be positive and the value of (3.50), to be negative. Therefore, *partial coverage* is optimal. Indeed, the negative correlation between L and N serves as a "natural hedge" against the risk. Full coverage would reduce this negative correlation to zero, thus destroying this hedge.

(2) *Proportional loading, $\lambda > 0$.*

- *Independence of the two losses.* With $\pi_N = \pi_{N|L}$ as before, the first term in the bracket of (3.50) is now positive, multiplied by $(v'[1] - v'[3]) < 0$. Since the last term is negative, $dEU/d\alpha$ is negative when evaluated at $\alpha = 1$, indicating that *partial coverage* is optimal (see Table 3.4 again).
- *Positive correlation of the two losses.* This means $\pi_{N|L} > \pi_N$ or $\pi_N < \pi_{N|L}$. The first term thus has an ambiguous sign and is multiplied by $(v'[1] - v'[3]) < 0$. The last term is negative; therefore, the sign of $dEU/d\alpha$ is ambiguous. However, if the uninsurable loss N is small, $(v'[1] - v'[3])$ is small too, resulting in $dEU/d\alpha < 0$ at $\alpha = 1$, which makes *partial insurance* coverage optimal. Due to the loading, there is a tendency away from full coverage, which is counteracted by the desire to hedge against the loss N that is positively correlated with L. However, this counter-effect is unimportant when N is small enough.
- *Negative correlation of the two losses.* Here $\pi_{N|L} < \pi_N$ or $\pi_N > \pi_{N|L}$, causing the first term in the bracket of equation (3.50) to be positive. It is multiplied by $(v'[1] - v'[3] < 0)$, resulting in a negative quantity; the last term being negative, *partial coverage* is optimal again.

---

[13] The formula for a conditional probability reads, $\pi_{N|L} = \pi_{N,L}/\pi_L$ implying $\pi_{N,L} = \pi_{N|L}\pi_L = \pi_{L|N}\pi_N$. Substitution yields $\pi_{N|L} = (\pi_{L|N}/\pi_L) \cdot \pi_N$. Therefore, $\pi_{N|L} > \pi_N$ if $\pi_{L|N} > \pi_L$, i.e., L occurs with greater probability if N happens as well. However, this is the consequence of L and N being positively correlated.

Table 3.4 Insurance demand with two risks

Relationship between L and N^a	Loading	Optimal coverage
Independence	$\lambda = 0$	$\alpha^* = 1$
Positive correlation	$\lambda = 0$	$\alpha^* > 1$
Negative correlation	$\lambda = 0$	$\alpha^* < 1$
Independence	$\lambda > 0$	$\alpha^* < 1$
Positive correlation	$\lambda > 0$	$\alpha^* < 1$ if N small
Negative correlation	$\lambda > 0$	$\alpha^* < 1$

$^a L$: insurable loss; N: non-insurable loss

In sum, these results are in line with the prescriptions relating to optimal portfolio allocation. There, correlations play a crucial role as well, i.e., between stochastic rates of return on assets (see Sect. 5.1.3).

An interesting special case obtains if $\pi_1 = \pi_4 = 0$ and $L = N$. This defines perfect negative correlation between the two losses, since the states No. 1 (no loss) and No. 4 (both losses) cannot occur. The question to be answered is whether insurance coverage will be bought at all in this case. Therefore, the slope of the objective function is to be evaluated at $\alpha = 0$ this time. Noting that now $(\pi_2 + \pi_3) = 1$ and $\pi_2 = \pi_L$ and that equality of N and L with no insurance implies $v'[2] = v'[3]$, equation (3.44) becomes

$$\left.\frac{dEU}{d\alpha}\right|_{\alpha=0} = \pi_2 v'[2]\{-\pi_L L(1+\lambda) + L\} + \pi_3 v'[2]\{-\pi_L L(1+\lambda)\}$$
$$= \pi_L L \cdot v'[2]\{-(\pi_2 + \pi_3)(1+\lambda) + \pi_2/\pi_L\}$$
$$= \pi_L L \cdot v'[2]\{-(1+\lambda) + 1\} < 0. \quad (3.51)$$

Therefore, the sign of $dEU/d\alpha$ is negative at $\alpha = 0$, indicating that a negative value of α would be optimal in principle. The IB would want to turn into a supplier of insurance coverage. In the special case of $\lambda = 0$, Eq. (3.51) has the value of zero at $\alpha = 0$; *no coverage* is predicted.

Intuitively, a perfect negative correlation relieves the combined risky prospect of uncertainty. Purchasing insurance would introduce differences in wealth levels and hence risk while reducing the expected value of wealth or leaving it constant at best (fair premium). Therefore, no coverage is the best attainable solution.

▶ **Conclusion 3.9** In the presence of a non-insurable risk, demand for coverage of the insurable asset depends not only on the loading as the price of insurance, but on the correlation between the two risks.

Generalizing these results for the case of several insurable risks $(L_1, \ldots, L_n)$ turns out to be difficult because it is not realistic or reasonable to consider one of them at a time while neglecting the others. The exceptions are risks that are independent, causing demand for protection against a given risk L_i to depend only on the properties of L_i. In fact, the risk utility function must be quadratic or exponential to ensure separability [for more details; see e.g. Eisen (1979b, 106 f.)]. For instance, without specific assumptions concerning risk aversion, two independent risks may be undesirable when viewed in isolation but desirable when viewed in combination.

3.5 Relation Between Insurance Demand and Prevention

Up to this point, the probability of occurrence and the amount of loss were assumed to be predetermined for simplicity. This assumption is often inappropriate because the insured can exert preventive effort to lower both dimensions of risk. However, their incentive to undertake such an effort might be undermined by insurance coverage protecting them from the consequences of an incident. More generally, there is an extensive literature revolving around the relation between the demand for insurance and other instruments of risk management (among them preventive effort). After all, risk avoidance and risk reduction constitute *alternatives to insurance*.

Resources can be employed to reduce the probability of occurrence or the severity of consequences. Instances in point are sprinklers, fire walls, burglary alarms, and safe production processes. As always, the optimal intensity of such measures calls for equating their (certain) cost and their (uncertain) benefits at the margin, in short "marginal cost=E(marginal return)". However, expected marginal returns on the benefit side importantly depend on the structuring of insurance policies such as *ex-post* premium differentiation in the sense of experience rating, copayment, and deductibles. In addition, the price of insurance may reflect the amount of preventive effort, provided it is observable.

These considerations show that demand for insurance and prevention are related in an ambiguous way. If the price for insurance coverage increases, the amount of coverage falls, inducing more preventive effort which has become cheaper in a relative sense. This makes insurance and preventive effort *substitutes*. On the other hand, insurance coverage and preventive effort can be *complements*. If the insurance buyer (IB) increases preventive effort, the IC accounts for this by reducing the price for insurance, inducing a greater amount of coverage.

In the following, only the case of loss reduction (often called self-insurance) is considered [see Ehrlich and Becker (1972)]. The case of lowering the probability of loss occurrence through prevention will be dealt with in Sect. 8.2.2.1. As the IB can influence the amount of loss through preventive effort V only, expected utility is given by

$$EU(W, V) = \pi v[W_0 - C(V) - L(V)] + (1 - \pi)v[W_0 - C(V)], \quad (3.52)$$

with $L'(V) < 0$. The cost of this effort, $C(V)$, is assumed to be monotonically and progressively increasing in V, i.e., $C'(V) > 0$, $C''(V) > 0$. Accordingly, the first-order condition for a maximum w.r.t. V is[14]

$$\frac{dEU}{dV} = \pi\{-C'(V) - L'(V)\} \cdot v'[1] - (1-\pi)\underset{(-)}{C'(V)} \cdot v'[2] = 0, \quad (3.53)$$

with $v[1] := v[W_0 - C(V) - L(V)]$ and $v[2] := v[W_0 - C(V)]$. The second-order optimum condition then follows from the assumption $v''[\cdot] < 0$.

For the optimal amount of effort to be positive ($V^* > 0$), one must have $-C'[V^*] - L'[V^*] > 0$, or $-L'[V^*] > C'[V^*]$. This means that the marginal return, $-L'[V^*] > 0$, is at least as high as the marginal cost of additional effort for loss reduction, $C'[V^*]$. This is plausible because no one would want to spend an additional monetary unit (MU) for loss reduction if the potential return in the guise of loss reduction would be less than one MU. Using this inequality and assuming that the relevant indifference and transformation curves do not have kinks (see Fig. 3.7), condition (3.53) yields the sufficient condition for *positive* effort spent on loss reduction

$$\frac{L'[0] + C'[0]}{C'[0]} > \frac{(1-\pi)v'[2]}{\pi v'[1]}, \text{ or } -\frac{L'[0] + C'[0]}{C'[0]} < -\frac{1-\pi}{\pi} \cdot \frac{v'[2]}{v'[1]}, \quad (3.54)$$

with $L'[0]$ and $C'[0]$ indicating that $L'(V)$ and $C'(V)$ are to be evaluated at $V = 0$. In comparison to Fig. 3.6 of Sect. 3.3.2, the left-hand side of (3.54) is a new "insurance line" which represents the effort on loss reduction in (W_1, W_2)-space. This new "insurance line" must run at least as steeply as the fair odds line offered by the IC for preventive effort to be competitive. Failing this condition, prevention would be comparatively unproductive. There could even be an incentive to increase the loss ($V^* < 0$). In sum, market insurance and self-insurance turn out to be substitutes. However, this finding needs to be checked in a second step by letting the IB simultaneously optimize preventive effort (V) and insurance coverage (α) through maximizing expected utility,

$$EU(W, V, \alpha) = \pi v[W_0 - L(V) - C(V) - \alpha P_0 + \alpha L] + (1-\pi)v\{W_0 - C(V) - \alpha P_0\}. \quad (3.55)$$

As before, P_0 denotes the premium for full coverage ($\alpha = 1$). The crucial assumption is that due to lack of observability, the IC *does not honor* loss reduction effort

---

[14]If one simplifies by setting average and marginal cost of prevention equal to 1 [such that $C'(V) = C(V) = 1$], then the optimum condition (3.53) becomes $(1-\pi)v'[2]/\pi v'[1] = L'[V^*] + 1$. This condition is equivalent to maximizing expected utility if both the marginal utility of wealth and the marginal productivity of measures designed to reduce loss are decreasing [see Ehrlich (1972, 634)]. In the present context and referring to Fig. 3.7, indifference curves must be convex from, and the transformation curve TN concave to, the origin.

3.5 Relation Between Insurance Demand and Prevention

by lowering its premium. In this case, the first-order conditions pertaining to (3.55) read,

$$\frac{\partial EU}{\partial \alpha} = \pi v'[1](L - P_0) - (1-\pi)v'[2](P_0) = 0; \qquad (3.56)$$

$$\frac{\partial EU}{\partial V} = -\pi v'[1]\{L'(V) + C'(V)\} - (1-\pi)v'[2] \cdot C'(V) = 0. \qquad (3.57)$$

From (3.56), one has

$$\frac{(1-\pi)v'[2]}{\pi v'[1]} = \frac{L - P_0}{P_0}. \qquad (3.58)$$

Condition (3.57) yields

$$\frac{(1-\pi)v'[2]}{\pi v'[1]} = -\frac{L'[V^*] + C'[V^*]}{C'[V^*]} = \frac{-L'[V^*] - C'[V^*]}{C'[V^*]}. \qquad (3.59)$$

Equating these two conditions and recalling that $L'(V) < 0$, one sees that in the optimum the relative net "shadow price" of preventive effort must be equal to the true price of insurance (excess of premium over loss) in relative terms

$$\frac{C'[V^*] + L'[V^*]}{C'[V^*]} = \frac{P_0 - L}{P_0}. \qquad (3.60)$$

Thus, market insurance and self-insurance are *substitutes* in the sense that an increase of the relative price of insurance coverage serves to increase the demand for self-insurance (while of course reducing the demand for market insurance).

From (3.59), one might infer that strongly risk-averse individuals exert more loss-reducing effort than others. Prior to the influence of insurance, marked risk aversion implies a marked curvature of the risk-utility function, calling for a high value of $v'[1]$ pertaining to the loss state compared to $v'[2]$ pertaining to the no-loss state. This by itself would cause the left-hand side of (3.59) to be small. Therefore, the excess of marginal return over marginal cost of prevention on the right-hand side would have to be small as well, indicating a great amount of effort. However, according to Sect. 3.3.2, risk aversion entails a tendency toward full coverage. This causes $v'[1]$ to decrease relative to $v'[2]$; the left-hand side of (3.59) increases. Its right-hand side increases as well, pointing to reduced prevention effort (a moral hazard effect; see Sect. 8.3). The net influence of risk aversion on prevention, therefore, remains ambiguous in general. However, risk aversion ceases to be relevant if insurance coverage is available at fair premium ($\lambda = 0$). Then (because of $P = \pi L$), equation (3.60) reduces to

$$\frac{C'[V^{**}] + L'[V^{**}]}{C'[V^{**}]} = -\frac{(1-\pi)}{\pi}. \qquad (3.61)$$

Clearly, this condition does not contain any subjective parameter anymore, notably risk aversion. Individuals would, therefore, behave in *risk-neutral* manner when it

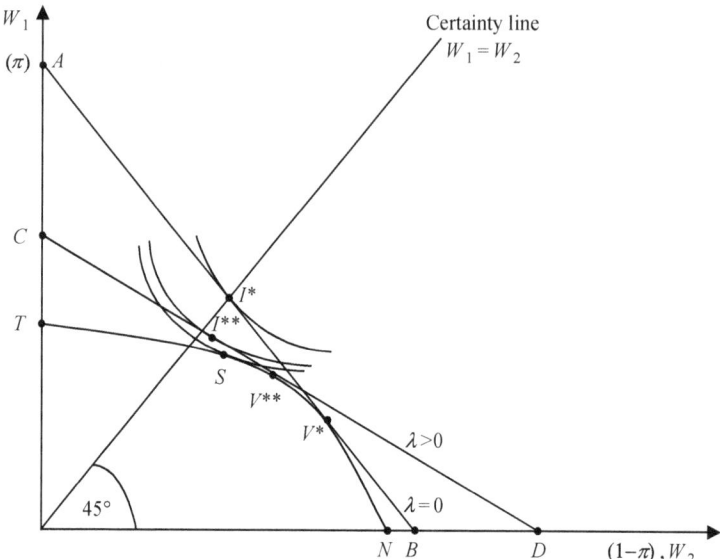

Fig. 3.7 Loss-reducing effort and market insurance

comes to reducing the amount of loss (since the marginally fair premium makes them buy full coverage). Even in the case of an unfair premium, it can be shown that the optimal amount of loss-reducing effort maximizes wealth, independently of the shape of indifference curves.

These findings are illustrated in Fig. 3.7. The transformation curve TN reflects the self-insurance possibilities of the individual in terms of transferring wealth from the no-loss state (W_2) to the loss state (W_1). In particular, the extreme point N corresponds to zero preventive effort, with the consequence that when the loss $L = W_2$ occurs, no wealth is available anymore such that $W_1 = 0$. The slopes of the insurance lines AB and CD reflect the conditions governing the transfer of wealth from the no-loss to the loss state through insurance. The insurance line AB runs relatively steep, reflecting a contract without proportional loading ($\lambda = 0$; see Sect. 3.3.1). This contract allows the individual to reach optimal point I^* which dominates in expectation the alternative S that can be attained "under autarky", i.e. without the availability of a market for insurance. Note that the curvature of the indifference curve (mirroring risk aversion) indeed does not matter, as indicated by equation (3.61). Moreover, I^* is attained by first moving from N to V^* on the production possibility frontier, entailing a reduction in no-loss wealth W_2 in favor of prevention. The distance between I^* and V^* is then covered by purchasing insurance coverage.

Now let the insurance line AB rotate to CD (reflecting a loading, $\lambda > 0$). This has three effects. First, the new optimum is given by I^{**}, indicating a lower expected utility. Second, the amount of preventive effort increases, from V^* to V^{**}. Third, the amount of insurance coverage decreases, from $(I^* - V^*)$ to $(I^{**} - V^{**})$. Therefore, the individual's own loss-reducing effort does decrease the demand for insurance coverage in this case (recall that the IC does not honor preventive effort by assumption).

3.5 Relation Between Insurance Demand and Prevention

▶ **Conclusion 3.10** An increase in the insurance premium induces an increased demand for self-insurance through loss-reducing effort and a decrease in market insurance, given that preventive effort is not honored by the insurer. Therefore, insurance and prevention are substitutes in this case.

It should be noted that when effort is directed at reducing the *probability of occurrence* $[\pi = \pi(V)$, with $\pi'(V) < 0]$, insurance and prevention need not be substitutes (see Sect. 8.2.2.1).

Exercises

3.1 Assume the following loss distribution $L = \{250; 500; 1,000\}$ MU with probabilities $\{0.4; 0.4; 0.2\}$.
Let there be insurance coverage available that (a) covers 50% of loss for a premium of 350 MU or (b) covers all losses but charges a premium of 550 MU. Your initial wealth is 2,000 MU and your risk- utility function is of the form $v(W) = ln(W)$, with W denoting final wealth. Which one of the policies (if at all) would you want to purchase?

3.2 In Sect. 3.5, the relation between prevention and insurance is characterized as ambiguous.

(a) Can you formulate this ambiguity in no more than four sentences?
(b) In one case, the initial change is an increase in insurance premium, in the other, an increase in preventive effort. Can there be an exogenous change in preventive effort?
(c) The analysis of Sect. 3.5 is limited to the case of nonobservable preventive effort. To complete the analysis, make P_0 depend on V in (3.54) and derive a new first-order condition w.r.t. V.
(d) Compare your new optimum condition with (3.56).

   (d1) Does the changed assumption of observability make a difference?
   (d1) In what sense could one say that insurance encourages prevention, establishing complementarity?

References

Arrow, K.J. (1963). Uncertainty and the welfare economics of medical care. *American Economic Review*, 53, 941–973.
Arrow, K.J. (1974). Optimal insurance and generalized deductibles. *Scandinavian Actuarial Journal*, 57, 1–42.

Bernoulli, D. (1730/31). Specimen theoriae novae de mensura sortis. *Comentarii Academiae Petropolis, Petersburg, 5*, 175–192; English translation (Exposition of a new theory on the measurement of risk), *Econometrica, 22*(1), 23–46.

Cook, P.J., & Graham, D.A. (1977). The demand for insurance and protection: the case of irreplaceable commodities. *Quarterly Journal of Economics, 91*(1), 143–156.

Doherty, N.A. (1985). *Corporate Risk Management, a Financial Exposition*. New York: McGraw-Hill.

Doherty, N.A., & Schlesinger, H. (1983). Optimal insurance in incomplete markets. *Journal of Political Economy, 91*, 1045–1054.

Eeckhoudt, L., & Schlesinger, H. (2006). Putting risk in its proper place. *American Economic Review, 96*(1), 280–289.

Eeckhoudt, L., Meyer, J., & Ormiston, M.B. (1997). The interaction between the demand for insurance and insurable assets. *Journal of Risk and Uncertainty, 14*, 25–39.

Ehrlich, J., & Becker, G.S. (1972). Market insurance, self-insurance and self-protection. *Journal of Political Economy, 80*, 623–648.

Eisen, R. (1979b). *Theorie des Versicherungsgleichgewichts (Theory of Equilibrium on Insurance Markets)*. Berlin: Duncker & Humblot.

Hirshleifer, J. (1965/6). Investment decisions under uncertainty. *Quarterly Journal of Economics, 79*, 509–536 and *Quarterly Journal of Economics, 80*, 252–277.

Krelle, W. (1959). A theory of rational behavior under uncertainty. *Metroeconomica, 11*(1–2), 51–63.

Lange, O. (1943). A note on innovations. *Review of Economics and Statistics, 25*, 19–25.

Machina, M.J. (1995). Non-expected utility and the robustness of the classical insurance paradigm. *Geneva Papers on Risk and Insurance Theory, 20*, 9–50.

Mossin, J. (1968). Aspects of rational insurance purchasing. *Journal of Political Economy*, 553–568.

Niehans, J. (1948). Zur Preisbildung bei ungewissen Erwartungen (On the formation of prices under uncertain expectations). *Schweizerische Zeitschrift für Volkswirtschaft und Statistik (Swiss Journal of Economics and Statistics), 84*, 433–456.

Raviv, A. (1979). The design of an optimal insurance policy. *American Economic Review, 69*, 84–96.

Shackle, G.L. (1949). *Expectations in Economics*. Cambridge MA: Cambridge University Press.

Smith, V. (1968). Optimal insurance coverage. *Journal of Political Economy, 76*, 68–77.

Von Neumann, J., & Morgenstern, O. (1944). *Theory of Games and Economic Behavior*. Princeton: Princeton University Press.

Wald, A. (1945). Statistical decision functions which maximize the maximum risk. *Annuals of Mathematics, 46*, 265–280.

Zweifel, P., Breyer, F., & Kifmann, M. (2009). *Health Economics*, 2nd ed. New York: Springer.

Insurance Demand II: Nontraditional Approaches to Decisions Under Risk

4

Up to this point, extensive use has been made of the expected utility theorem (i.e., the Bernoulli principle) for developing the theory of behavior under risk and of insurance demand in particular, which constitutes the traditional approach. However, mainly in response to the observed behavior that is found at odds with expected utility (EU), a number of nontraditional approaches have been developed.

This chapter starts by examining the axiomatic basis of EU theory in Sect. 4.1.1. It will become evident that while some of its axioms are widely accepted, others have been exposed to (sometimes severe) criticism. This criticism derives from anomalies in behavior (in the light of EU theory); partly it is more general, to be reviewed in Sect. 4.1.2.

Section 4.2 deals with a popular alternative to the EU hypothesis, so-called (*cumulative*) *prospect theory*, developed by Kahneman and Tversky (1979), two psychologists (in 2002, Kahneman received the Nobel Prize in economics for this work). Its crucial novelty is a kink in the risk-utility function around the status quo point: gains relative to the status quo are subject to risk-averse behavior (indicated by concavity from the origin), whereas losses are subject to risk-loving behavior (indicated by convexity from the origin). However, prospect theory has been shown to run into theoretical and empirical difficulties as well.

These difficulties have motivated additional nontraditional approaches. First, Quiggin (1982) proposed the concept of *rank-dependent utility*, to be presented in Sect. 4.3.1. This alternative retains the attractive features of EU theory; in particular, it avoids the kink in the risk-utility function as in prospect theory. Its innovation is that probabilities are weighted by the decision-maker, with the weights reflecting his or her optimism (or pessimism). For instance, an optimist would put a relatively high weight on the probabilities of favorable outcomes, assigning them a high "rank" as it were.

Another nontraditional approach is inspired by observed behavior in experiments, where participants exhibit the so-called *ambiguity aversion*. While in EU theory probabilities are assumed to be known, more often than not this does not hold true. In Sect. 4.3.2, the contribution by Gilboa and Schmeidler (1989) is sketched. This approach amounts to defining another probability distribution reflecting the degree of ambiguity concerning the original distribution of probabilities. The decision criterion turns out to be of the maxmin type, reminiscent of game theory.

A point of continued debate has been whether the probabilities entering a decision are subjective or objective. One way to bridge this gap is to use any available information about probabilities, regardless of whether it is objective or subjective, resulting in what Kofler and Menges (1976) call *linear partial information*. The price to be paid for exploiting even very weak information is the assumption that the decision-maker is engaged in a game against Nature (see Sect. 4.3.3). The decision criterion becomes maxEmin because Nature supposedly presents man with the urn of probabilities resulting in the worst *expected* outcome (rather than the worst outcome in every event, as a true adversary would).

Finally, the alternative that is closest to what has become to be known as Behavioral Economics is *regret theory*, reflecting the human tendency to regret bad decisions (and to rejoice in those that turn out to be good ones). Originally proposed by Niehans (1948, in German), it was further developed by Savage (1951) and popularized by Bell (1982) and independently by Loomes and Sugden (1982, 1983) for financial decisions. As argued in Sect. 4.3.4, a risk-averse decision-maker would seek to minimize regret.

The final Sect. 4.4 of this chapter presents some empirical evidence concerning the relative merits of the nontraditional approaches presented here. In its conclusion, it follows Hey, who started experimentally testing alternatives to EU theory in the early 1990s (Hey and Orme 1994). According to him and his co-authors, it appears that the choice boils down to the one between EU theory (with a great deal of error in decision-making) and rank-dependent utility theory (Conte and Hey 2013). Since the traditional alternative is far simpler, it will be retained in the remainder of this book.

4.1 Expected Utility Theory Revisited

4.1.1 The Axiomatic Basis of Expected Utility Theory

For a better understanding of both the strengths and weaknesses of expected utility (EU) theory, its essential axioms (which are also called rationality axioms) that form its basis are presented here. Note that the first two axioms are identical to those of conventional microeconomic theory, while the remaining three are specific to EU theory since they introduce probabilities and states of Nature.

1. *Complete ordering*. This means that all alternatives of action are compared to each other. Formally, a preference relation $\succeq$ is defined over the set of conse-

4.1 Expected Utility Theory Revisited

quences (c) and actions (a) that is complete, i.e., $c_1 \succeq c_2$ ($a_1 \succeq a_2$, respectively), or $c_2 \succeq c_1$ ($a_2 \succeq a_1$), or $c_1 \sim c_2$ ($a_1 \sim a_2$) for the case of two alternatives and two consequences, with "$\sim$" denoting indifference. Since this applies to any two alternatives within a given set, the ordering is indeed complete.

2. *Transitivity.* If $c_1 \succeq c_2$ ($a_1 \succeq a_2$, respectively) and $c_2 \succeq c_3$ ($a_2 \succeq a_3$), then it is true that $c_1 \succeq c_3$ ($a_1 \succeq a_3$).
3. *Continuity.* Given the ordering $a_1 \succeq a_2 \succeq a_3$, than there is a probability ρ (more generally, a probability distribution) with $0 < \rho < 1$ such that $a_2 \sim (a_1, a_3; \rho, 1 - \rho)$.
4. *Dominance.* If action a_1 results in a weakly better consequence [$c_1 = c(a_1, s)$] than the alternative a_2 [$c_2 = c(a_2, s)$] for all states $s \in S$, then it is true that $a_1 \succeq a_2$.
5. *Independence ("sure-thing" principle).* This axiom requires that the preference ordering of two alternatives must not depend on states of nature the occurrence of which results in identical consequences of the two alternatives. Formally, let w_1, w_2, and w_3 be probability distributions such that $w_1 \succeq w_2$. Then it must be true that $(w_1, w_3; \rho, 1 - \rho) \succeq (w_2, w_3; \rho, 1 - \rho)$, $0 < \rho < 1$, for any linear combination with $(\rho, 1 - \rho)$ defined over w_1 and w_3. In short, the ranking of R^* over R is unaffected by "mixing" a bit of R^{**} into both R^* and R, respectively.

The independence axiom can be formulated also as follows, "If the decision-maker prefers the prospect R^* over the prospect R or is indifferent between the two, then he or she must also weakly prefer the linear combination $\rho R^* + (1 - \rho) R^{**}$ to the linear combination $\rho R + (1 - \rho) R^{**}$ for all $\rho > 0$ and R^{**}".

An example of this is tossing a coin with $\rho = 1/2$. There are the prospects $\{R, R^*, R^{**}\}$, and the agent must choose between $\rho R^* + (1 - \rho) R^{**}$ (i.e., R^* with 50% of R^{**} mixed in) and $\rho R + (1 - \rho) R^{**}$ (i.e., R with 50% of R^{**} mixed in). If his or her preferences are such that, e.g., $R^{**} \succeq R^* \succeq R$ and the toss of a coin that is head up with probability $1 - \rho$, giving the right to take part in the lottery R^{**}, then the preference in favor of R^* over R should hold independently of the toss of the coin. If the coin is head up and the agent receives R^{**}, in which case the choices R^* and R are irrelevant. If the coin is tail up, the agent is thrown back to the choice between R^* and R. Rationality then requires to formulate one's preference for R^* over R from the beginning, implying that it holds regardless of whether the toss returns a head or a tail, in keeping with the independence axiom.

In empirical research [see especially MacCrimmon (1968)], most participants in experiments such as managers deem the axioms of completeness, transitivity, and dominance acceptable. However, this does not hold to the same extent for the independence or "sure-thing" axiom.

The importance of this axiom can be illustrated by two well-known counterexamples, the Ellsberg and the Allais paradoxon.

Example 4.1

The Ellsberg Paradox

In the decision situation developed by Ellsberg (1961), there is an urn containing a total of 90 balls. Of these balls, 30 are known to be red, while 60 are black or yellow, with shares unknown. One ball is randomly drawn from the urn. Depending on the color of that ball, you win 100 MU (or nothing otherwise; see exhibit below). You are asked to state your preference with regard to the respective lotteries.

Choice set I	Available choices	Your choice
a_1: Betting 100 MU on red	$a_1 \succeq a_2$	☐
a_2: Betting 100 MU on black	$a_1 \preceq a_2$	☐
Choice set II		
a_3: Betting 100 MU on red or yellow	$a_3 \succeq a_4$	☐
a_4: Betting 100 MU on black or yellow	$a_3 \preceq a_4$	☐

Now one often observes that respondents prefer a_1 over a_2 and at the same time, a_4 over a_3. However, this is inconsistent because a_3 and a_4 are the same as a_1 and a_2, respectively, save for the (irrelevant) addition of yellow balls. Clearly, observed behavior can contradict the independence axiom. ■

Example 4.2

The Allais Paradox

Another anomaly was discovered by Allais (1953), who developed the following decision situation. You are asked to choose twice between two risky situations.

Situation A

Prospect R_1 results in	Prospect R_2 results in
final wealth of 1 mn. MU with certainty	• final wealth of 5 mn. MU with 10% probability
	• final wealth of 1 mn. MU with 89% probability
	• final wealth to 0 MU with 1% probability.

Situation B

Prospect R_3 results in	Prospect R_4 results in
• final wealth of 1 mn. MU with 11% probability	• final wealth of 5 mn. MU with 10% probability
• final wealth of 0 with 89% probability	• final wealth of 0 with 90% probability.

Please mark your choices below (only one box at a time)

Situation A	Your choice	Situation B	Your choice
$R_1 \succ R_2$	☐	$R_3 \succ R_4$	☐
$R_1 \approx R_2$	☐	$R_3 \approx R_4$	☐
$R_1 \prec R_2$	☐	$R_3 \prec R_4$	☐

Frequently observed choice pairs are $R_1 \succ R_2$ and $R_4 \succ R_3$. However, one can show that there is *no traditional risk-utility function* that can represent these observed choices.

Specifically, in the *decision problem A*, the choice $R_1 \succ R_2$ can be evaluated as follows, with, e.g., $v[1]$ denoting utility associated with 1 mn. MU

$$R_1 \succ R_2: \quad \begin{array}{l} v[1] > 0.1 \cdot v[5] + 0.89 \cdot v[1] + 0.01 \cdot v[0], \text{ implying} \\ 0.11 \cdot v[1] > 0.1 \cdot v[5], \text{ since } v[0] \text{ can be} \\ \text{normalized to equal zero.} \end{array} \quad (4.1)$$

In the *decision problem B*, the choice $R_4 \succ R_3$ can be evaluated as follows:

$$R_4 \succ R_3: \quad \begin{array}{l} 0.1 \cdot v[5] + 0.9 \cdot v[0] > 0.11 \cdot v[1] + 0.89 \cdot v[0], \text{ implying} \\ 0.1 \cdot v[5] > 0.11 \cdot v[1] \implies 0.11 \cdot v[1] < 0.1 \cdot v[5]. \end{array} \quad (4.2)$$

The reversal of the inequality indicates a *contradiction*.

More generally, with probabilities ρ_1, ρ_2, and ρ_3, the preference of R_1 over R_2 implies for the risk-utility function $v(\cdot)$.

$$R_1 \succ R_2: \quad \begin{array}{l} v[1] > \rho_2 v[5] + \rho_3 v[1], \text{ therefore} \\ (1 - \rho_3) v[1] > \rho_2 v[5]. \end{array} \quad (4.3)$$

However, the preference of R_4 over R_3 implies (noting that $\rho_1 + \rho_2 = 90\%$)

$$R_4 \succ R_3: \quad \rho_2 v[5] > (\rho_1 + \rho_2) v[1]. \quad (4.4)$$

This is a contradiction (since $\rho_1 + \rho_2 = 1 - \rho_3$). Therefore, observed behavior in fact constitutes an *anomaly* in the sense of the Bernoulli principle.

The Allais Paradox initially was rejected as an "isolated example" [see Savage (1954, 101–103)]. However, it constitutes a special case of a very general phenomenon known as "common consequence effect". It seems that individuals exhibit stronger risk aversion in the case of a possible loss and weaker risk aversion in the case of a possible gain. By way of contrast, EU theory is couched in terms of final wealth rather than changes in wealth. Therefore, risk aversion should be independent of changes in wealth. ∎

4.1.2 Other Weaknesses of Expected Utility Theory

A vocal critic of EU theory is Taylor (2012), who starts with the "unknown unknowns" cited by the then-U.S. Secretary of Defense Rumsfeld in 2002. The usual solution has been to create a catch-all probability for the set of unforeseen outcomes. However, these unknown probabilities and their associated consequences may well change over time, without these changes being noticed. In experiments, it may be difficult to have participants neglect these unknowns. For instance, they may deem "winning 1 mn. MU with certainty" credible, while winning 5 mn. MU simply to be "too good to be true" (situation A in Example 2), inducing $R_1 \succ R_2$. However, when that same 1 mn. has a probability of just 11%, the "unknown unknown" may be that the bet is not honored after all regardless of outcome, so one might as well go for "5 mn. MU with 10% probability" (situation B), inducing $R_4 \succ R_3$ and with it, the Allais paradox.

Catch-all probabilities have been popular in financial models in an attempt to reduce their complexity. However, the contents of this category often change, perhaps most likely due to the emergence of new unforeseen events. The financial crisis of 2007/8 has been attributed to banks' and insurers' failure to recognize changes of this type (Brigo et al. 2010).

The first generalization of EU theory is to assign a measure of credibility to probabilities (and also consequences, although uncertainty regarding consequences can be dealt with by introducing additional probabilities defined over them). This amounts to defining a probability distribution over each of the original probabilities. But then, who is to know that this secondary probability distribution is the true one? Evidently, it is easy to argue that this generalization leads to an infinite regress. Moreover, once one leaves the realm of binary outcomes, a distribution law typically is assumed; otherwise, modeling a multitude of consequences becomes complicated. Yet the popular statistical distributions (the normal in particular) have too short a downside tail, thus failing to take very rare but extremely severe outcomes into account.

Another weakness of EU theory is that it is basically timeless. However, many decisions are (close to) irreversible, meaning that they can be corrected at a (very) high cost. One example is the purchase of a home, another, the choice of employment. Even reversing financial decisions until recently had a considerable cost in view of substantial fees charged by banks (financial services through the internet have changed this unless the provider is a fly-by-night). Conversely, when a course of action can be changed at a reasonable cost, decisions concerning the first period in a multiple-period context may be viewed as preliminary, to be adjusted in the light of experience.

In addition, EU theory neglects the fact that in many areas of public policy, the concept of acceptable risk plays an important role. Nuclear power provides a prominent example. Although most analysts put the risk of a maximum plausible accident (with thousands of deaths due to radiation) at 10^{-5} or less per year, the Fukushima (Japan) incident of 2011 caused the governments of Germany and Switzerland to stipulate an "exit from nuclear". For them, the risks associated with nuclear power became unacceptable, regardless of the fact that so far Fukushima has caused just one official death due to radiation. Likewise, a new pharmaceutical is allowed to be marketed by the U.S. Federal Drug Administration only if the risk (defined as the product of probability and severity of an adverse impact on health) does not exceed an acceptable threshold which, however, is not defined in the agency's guidance (FDA 2006, Fig. 4.1). By way of contrast, Solvency II regulation specifies that insurers must have sufficient capital to cover the expected value of possible losses (as measured by Value at Risk; see Sect. 6.9.1) with a probability of 99.5% during a given year.

It should be noted that the concept of acceptable risk often has the downside of being formulated regardless of any benefits. There is no doubt that nuclear power generates benefits, which are lost to the German and Swiss economies due to the government's exit decision. Patients suffering from a painful or lethal disease may be willing to try a new drug even if facing a considerable probability of being exposed to severe side effects. Finally, insurance managers are expected to act in the interest of shareholders, calling for underwriting risks that might result in a loss but promise a high return.

▶ **Conclusion 4.1** Expected Utility theory rests on five axioms, four of which are little disputed (Complete ordering and Transitivity correspond to standard Microeconomics; Continuity and Dominance are intuitive). However, Independence from irrelevant risks has been contradicted by experimental evidence. Additional weaknesses include the assumed completeness of the probability distribution (the "unknown unknowns"), an unspecified time horizon, and the neglect of acceptable risk, a concept of considerable importance in public policy.

4.2 Prospect Theory

Perhaps the most famous (and one of the earliest) explanations of individual behavior inconsistent with EU theory comes from Kahneman and Tversky. Though their early work was mostly of interest to psychologists, they eventually were published in leading economics journals. In their 1979 article in *Econometrica*, they slightly reframe the Allais paradox as choices in two distinct problems

		Problem 1: Choose between A & B			
A:	2,500 MU	with prob. 0.33	B:	2,400 MU	with certainty
	2,400 MU	with prob. 0.66			
	0 MU	with prob. 0.01			
		Problem 2: Choose between C & D			
C:	2,500 MU	with prob. 0.33	D:	2,400 MU	with prob. 0.34
	0 MU	with prob. 0.67		0 MU	with prob. 0.66

In Kahneman and Tversky (1979) 82% of survey respondents (out of 72, a significant proportion) are reported to prefer choice B to choice A. Assuming a standard concave utility function $u(\cdot)$ where $u[0] = 0$, this suggests that $u[2, 400] > 0.33 \cdot u[2, 500] + 0.66 \cdot u[2, 400]$. This preference ordering implies that $0.34 \cdot u[2, 400]$ is preferred to $0.33 \cdot u[2, 500]$, leading to the prediction that choice D in problem No. 2 will be preferred to choice C. However, the authors found the same survey respondents preferred C to D, in almost exactly the same proportion (83%).

They go on to point out a "reflection effect" whereby preferences are reversed when the outcomes are framed as losses (as opposed to gains as in the two problems outlined above). This reflection's effect and its implications essentially form the basis of what they label "prospect theory" [for a more complete discussion, including axioms, see Kahneman and Tversky (1979)]. A core element of prospect theory is the distinction between an editing phase and an evaluation phase in behavior under risk.

4.2.1 The Editing Phase

In the editing phase, decision-makers simplify risky options to make the decision-making process more straightforward using the following six steps:

(1) The *coding* operation introduces the notion of a "reference point"; outcomes are coded relative to a reference point (e.g., current wealth) as gains or losses. This reference point and the framing in terms of gains or losses form the foundation of prospect theory.
(2) The *combination* operation aggregates outcomes with the same probability into one outcome. For instance, consider two possible gains of 300 MU with probability 0.10 each. According to the combination operation, this becomes one possible gain of 300 MU with probability 0.20.
(3) In *segregation*, gains and losses that occur regardless of choice are netted out as a certain event. Consider two possible losses, one of -1,500 MU with a probability of 0.75 and one of -4,000 MU with a probability of 0.25. The segregation operation then results in a certain loss of -1,500 MU and a possible loss of -2,500 MU with probability 0.25.
(4) The *cancellation* operation reflects the observation that common outcomes seem to be ignored in practice. As an example, consider a choice between the following options

Choice set:

E		F	
5,000 MU	with prob. 0.40	5,000 MU	with prob. 0.40
2,000 MU	with prob. 0.15	3,000 MU	with prob. 0.15
-1,000 MU	with prob. 0.45	-2,000 MU	with prob. 0.45

4.2 Prospect Theory

Most decision-makers are found to cancel the component common to both choices (5,000 MU with probability 0.40), focusing on the other two possible outcomes.

(5) *Simplification* is the notion that decision-makers tend to "round" or "even out" outcomes. For instance, a loss of -2,001 MU with a probability of 0.095 is perceived as a loss of -2,000 MU with a probability of 0.10.

(6) *Dominance* means that dominated alternatives are discarded in contradiction with the completeness axiom of EU theory. Consider the following choice pairs,

Choice I clearly dominates choice H, which dominates choice G. Thus, choices H and G are removed from consideration, leaving choices I and J as the relevant ones.

Choice set:	
G	H
500 MU with prob. 0.40	500 MU with prob. 0.45
100 MU with prob. 0.60	100 MU with prob. 0.55

Choice set:	
I	J
1,000 MU with prob. 0.45	1,000 MU with prob. 0.50
100 MU with prob. 0.55	10 MU with prob. 0.50

4.2.2 The Evaluation Phase

Once decision-makers have edited the outcomes, they start the evaluation phase. In the evaluation phase, Kahneman and Tversky define the value of an edited outcome as being given by V which is a function of two parameters, σ and ν, where σ is a "decision weight" related to the probability (π) of the outcome. While the decision weights are a function of probability [i.e., $\sigma(\pi)$, with $\sigma'(\pi) > 0$], they do not necessarily sum to one. ν transforms the outcome x such that $\nu(x)$ represents the relative value of the outcome (again, usually relative to a reference point).

Assume two possible outcomes (both non-zero) x and y, occurring with probability p and q, respectively. Their evaluation (collectively called a "prospect") depends on their makeup. A prospect is "strictly positive" ("negative") if all individual outcomes $\{x, y\}$ are positive (negative) and the sum of p and q is one. A prospect is deemed "regular" if the prospect is neither strictly positive nor strictly negative.

A regular prospect is evaluated as follows:

$$V(x, p; y, q) = \sigma(p) \cdot \nu(x) + \sigma(q) \cdot \nu(y) \quad (4.5)$$

with $\nu[0] = 0$, $\sigma[0] = 0$, and $\sigma[1] = 1$.

Fig. 4.1 Possible value function, as shown in Kahneman and Tversky (1979, p. 279)

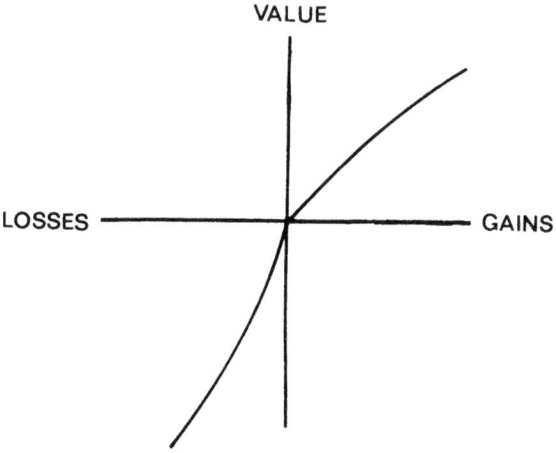

For strictly negative or strictly positive prospects, the evaluation is given as

$$V(x, p; y, q) = \nu(y) + \sigma(p) \cdot \{\nu(x) - \nu(y)\}. \tag{4.6}$$

In this formulation, the editing process (specifically, segregation) can be directly seen in the netting out of the risk-free outcome $\nu(y)$.

Two things should be noted here. First, though the notation and results may appear to be similar to expected utility theory, it is important to notice that under prospect theory, decisions are made (a) relative to a reference point and (b) using decision weights rather than probabilities. Second, while Kahneman and Tversky are generally credited with the codification of formal prospect theory, even they note that other economists (e.g., Markowitz 1952; Fellner 1961) introduced and discussed many of these concepts, including the core notions of evaluation relative to a reference point and the use of decision weights.

In representing the value function, ν, Kahneman and Tversky assume that the marginal value (utility) of wealth is decreasing in *both* the gain and loss dimension. Thus, prospect theory assumes that the value function is concave with respect to gains (as in EU theory) but is convex with respect to losses (contrary to EU theory). Further, the convexity in the loss domain is assumed to be greater than the concavity in the gain domain. This difference between valuing gains and losses is often referred to as *loss aversion*. The standard graphical representation of this relationship is shown in Fig. 4.1 below.

The value function of prospect theory can be written as

$$v(x) = \begin{cases} x^\alpha & \text{if } x \geq 0, \\ -\lambda(-x)^\beta & \text{if } x < 0, \end{cases} \tag{4.7}$$

with $\lambda > 0$ the loss aversion parameter. Further, since $\alpha > 0$ and $\beta > 0$ are not required to be the same, preferences are allowed to vary around the reference point (in this case, zero) in a way such that gains are treated differently from losses.

With regards to the decision weights, Kahneman and Tversky argue that decision-makers tend to overweight small probabilities such that $\sigma(p) > p$ for low values of p. Note that overweighting does not necessarily mean that the probability itself is overestimated. Rather, it simply means that the value associated with the event is given a higher weighting in the decision-making process (a subtle yet important distinction). Further, while prospect theory does not necessarily imply a similar underweighting of high probability events, Kahneman and Tversky do propose that the sum of the decision weights [e.g., $\sigma(p) + \sigma(1-p)$] is less than one, referred to as *subcertainty*.

In later works, the authors consider decision weights [based on Quiggin (1982)] as

$$\sigma^+(p) = \frac{p^\gamma}{(p^\gamma + (1-p)^\gamma)^{(1/\gamma)}} \qquad (4.8)$$

for the gain domain and

$$\sigma^-(p) = \frac{p^\delta}{(p^\delta + (1-p)^\delta)^{(1/\delta)}} \qquad (4.9)$$

in the loss domain. The combination of these updated decision weights, along with the value function given above, is referred to as *cumulative prospect theory*.

▶ **Conclusion 4.2** Prospect theory is highly descriptive of observed behavior in situations involving risk. Its primary downside is the kink in the risk-utility function at a reference point that need not be the status quo, a discontinuity that considerably complicates comparative-static analysis.

In Sect. 3.3.3, for instance, predictions concerning the impact of changes in wealth on the demand for insurance would depend on which side of the reference point the consumer is on.

4.3 Other Non-Traditional Approaches

4.3.1 Rank Dependent Expected Utility

In proposing *rank-dependent expected utility* (RDEU) (which he originally called *anticipated utility*), Quiggin (1982) offered an alternative to expected utility theory. RDEU was shown to explain some of the anomalies of EU theory. As stated in the preceding section, RDEU was ultimately incorporated into prospect theory, giving rise to cumulative prospect theory.

Quiggin's initial contribution was to apply decision weighting to the entire probability distribution. If individuals are known to over-weight extreme events (as discussed above), then they must also under-weight some "middle of the road" events so that the sum of the decision weights equals one, contrary to Kahneman and Tversky (1979). Thus, events with the same probability *do not* require the same decision weight.

Apart from the decision weighting, RDEU is quite similar to EU theory, positing the same concave risk-utility function without a kink at some reference point. Formally, all outcomes x_i occurring with probability p_i, associated with a probability vector $\mathbf{p} = (p_1, p_2, ...p_n)$, are ranked from the worst to the best (i.e., $x_1 < x_2 ... < x_n$). A decision-maker is then hypothesized to maximize the value function V which is given as

$$V = \sum_i h_i(\mathbf{p}) U(x_i). \tag{4.10}$$

The innovation lies in the weighting function, $h_i(\mathbf{p})$, which depends on the entire vector $\mathbf{p}$ and *not* the individual p_i. The probability weight assigned to an outcome x_i, therefore, is a function not only of the individual p_i but of all other probabilities, resulting in a ranking.

Specifically, Quiggin represents $h_i(\mathbf{p})$ as

$$h_i(\mathbf{p}) = f\left(\sum_{j=1}^{i} p_j\right) - f\left(\sum_{j=1}^{i-1} p_j\right). \tag{4.11}$$

Importantly, the function $f(\cdot)$ is the same for the p_j that make up the "gross rank" of p_i [in the first term of equation (4.11)] and p_j up to p_{i-1} which determines the "net rank" of p_i in the second term. This avoids a kink in the risk-utility function. The form of $f(\cdot)$ and hence $h_i(\mathbf{p})$ reflects the decision-maker's optimism (or pessimism, as appropriate). Concavity implies $h_i(\mathbf{p}) < p_i$ for probabilities associated with a high value of x_i and hence indicates pessimism; conversely, convexity implies $h_i(\mathbf{p}) > p_i$ for probabilities indicating optimism. Of course, it is possible that $h_i(\mathbf{p})$ has an inflection point (i.e., is "S"-shaped) or even multiple inflection points.

Ultimately, RDEU allows a decision-maker's choice to depend not only on the shape of the utility function (as in EU theory), but also on the shape of the decision weight function. This double dependence allows RDEU to more broadly explain observed individual choices than does EU theory. In particular, it retains the independence axiom only in a weakened form while having the same stochastic dominance properties that make EU theory attractive.

▶ **Conclusion 4.3** Rank-dependent expected utility accommodates features of prospect theory while retaining most of the axiomatic basis of EU theory. It does not posit a kink in the risk-utility function thus is amenable to comparative-static analysis.

4.3.2 Ambiguity Aversion

Ambiguity aversion (sometimes called *uncertainty aversion*) means that decision-makers prefer probabilities to be known rather than ambiguous. Reconsider the

4.3 Other Non-Traditional Approaches

Ellsberg paradox described as Example 4.1 in Sect. 4.1.1 (and reproduced below). There, the unknown quantity of yellow balls changed behavior, and in an irrational way according to EU theory.

Choice set I
a_1: Betting 100 MU on red (R)
a_2: Betting 100 MU on black (B)
Choice set II
a_3: Betting 100 MU on red or yellow (Y)
a_4: Betting 100 MU on black or yellow

According to Gilboa and Schmeidler (1989) a decision-maker maximizes the minimum expected outcome in a situation where ambiguity exists as a consequence of ambiguity aversion. Recalling that 30 balls are red and 60 are black or yellow, betting on red (a_1) has a known probability of success of 1/3, while betting on black (a_2) has a minimum probability of success of zero since the number of black balls (NB) might be zero.

Choice set I	Value of outcome
a_1: Bet 100 MU on R	$1/3 \times 100 = 33.33$
a_2: Bet 100 MU on B	$\min_{NB} NB/90 \times 100 = 0$
Choice set II	
a_3: Bet 100 MU on R or Y	$\min_{NY}(30 + NY)/90 \times 100 = 33.33$
a_4: Bet 100 MU on B or Y	$\min_{NY}(60 - NY)/90 \times 100 = 66.66$
	$= \min_{NB}(60 - NB)/90 \times 100 = 66.66$

Note: NB=number of black balls; NY=number of yellow balls

Using the maxmin criterion, the preferred outcome in Choice set I is a_1 over a_2. In Choice set II, opting for a_3 has a minimum probability of success of 1/3 since there are 30 red balls plus at least 0 yellow balls (NY–number of yellow balls). Yet a_4 has a minimum chance of success of 2/3 because regardless of whether the decision-maker focuses on black or yellow, there is a guaranteed probability of success of 2/3; the minimum number of the "other color" is $NY = NB = 0$. This makes a_4 the preferred action in Choice set II. As discussed in Sect. 4.1.1, this behavior is inconsistent with EU theory, but consistent with frequently observed choices. Ambiguity aversion helps to explain this behavior.

▶ **Conclusion 4.4** As with prospect theory and rank-dependent expected utility, ambiguity aversion retains the same basic properties as EU theory, but with the maxmin criterion that results in a weakening of the independence axiom and hence increased consistency with observed behavior.

4.3.3 Linear Partial Information

In traditional EU theory, the probabilities entering the risk-utility function are assumed to be known. In a gamble involving the toss of a dice, this assumption is credible. However, in most real-world situations, the mechanism determining probabilities is not known, allowing for estimates of objective probabilities at best. With these estimates, a subjective element enters decision-making, giving rise to the distinction between objective and subjective probabilities [Carnap (1950); the debate about their relative merits continues]. The best a decision-maker can do is use any information about probabilities, be they objective or ultimately subjective. This is the point of departure of Kofler and Menges (1976, in German), who note that the available information may well be partial, typically in the guise of linear restrictions.

These restrictions are called *linear partial information* (LPI). On the left-hand side of Figure 4.2 below, there are three possible outcomes with probabilities $\{\pi_a, \pi_b, \pi_c\}$. Therefore, each point in the cube delimited by the vertices $(1, 0, 0)$, $(0, 1, 0)$, and $(0, 0, 1)$ represents a probability distribution in principle. The summation restriction, $\pi_a + \pi_b + \pi_c = 1$ limits the set of admissible distributions to the points on the triangular hyperplane ABC. For instance, point D along AC reflects an admissible distribution because it is given by $(5/9) \cdot (1, 0, 0) + (4/9) \cdot (0, 0, 1)$ (see below). Now let the decision-maker know that outcome (a) is slightly more probable than outcome (c), resulting in the LPI: $4\pi_a \geq 5\pi_c \implies \pi_a \geq (5/4)\pi_c$, $\pi_c \geq 0$. The line BD reflects the equation, $\pi_1 = (5/4)\pi_2$ and thus limits the set of distributions satisfying this LPI to the shaded area ABD. Note that there may be no probability information regarding outcome (b), a lacking that precludes the application of EU theory from the outset.

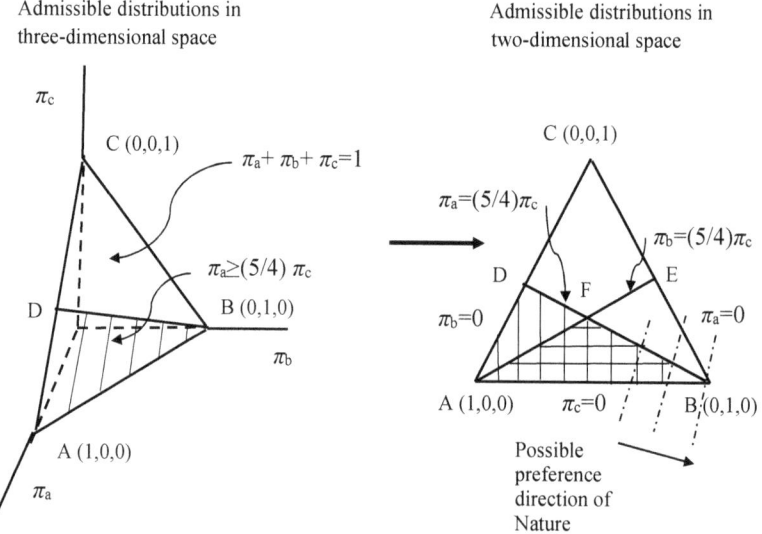

Fig. 4.2 The LPI: $\pi_a \geq (5/4)\pi_c$, the LPI: $\pi_b \geq (5/4)\pi_c$, and restriction, $\pi_a + \pi_b + \pi_c = 1$

4.3 Other Non-Traditional Approaches

On the right-hand side of Figure 4.2, the hyperplane ABC is shown as a triangle in two-dimensional space. As before, the original LPI: $4\pi_a \geq 5\pi_c$, $\pi_c \geq 0$ limits the set of admissible distributions to the shaded triangle ABD. In order to show how additional information serves to further reduce this set, the LPI: $4\pi_b \geq 5\pi_c$ is added creating the set ABE. Therefore, the two LPI combined give rise to the cross-hatched set ABF of admissible probability distributions. While it is smaller than the original unconstrained set ABC, it still contains an infinite number of points and hence probability distributions.

At this point, Kofler and Menges introduce their crucial assumption.

Assumption: The decision-maker is involved in a game against risk-neutral Nature which is zero-sum in expected value.

Under this assumption, Nature presents the decision-maker with an urn containing admissible distributions with a minimum expected value. Reflecting risk neutrality, her indifference curves are linear; on the right-hand side of Figure 4.2, let her preferred distribution be reflected by point B with $(0, 1, 0)$.

However, Nature's preferences could also call for A and C in principle, with C out of reach in view of the LPI. Thus, only the vertex points A, F, and B qualify as optima for Nature, who solves the dual of a linear program (Intriligator 1971). The decision-maker only needs to evaluate points A, F, and B; therefore, the infinity of admissible probability distributions boils down to a finite (usually small) number. These distributions are, $(1, 0, 0)$ for point A, $(5/14, 5/14, 4/14)$ for point F, and $(0, 1, 0)$ for point B [see Zimmermann et al. (1985) for more details].[1] There, the associated expected payoffs to the decision-maker are given by $(140.51, 140.51, 138.54)$ MU; since Nature is antagonistic by assumption, she opts for point B, as indicated in Figure 4.2. Knowing this, the decision-maker can now evaluate the two alternative actions a_1 and a_2. Since a_1 turns out to have a guaranteed expected payoff amounting to 138.54 MU, compared to 136.95 MU for a_2 [given the probability distribution $(0, 1, 0)$], the decision-maker is advised to opt for action a_1, applying the maxEmin criterion.[2]

It was noted in Sect. 4.1.2 that assigning probability distributions to the original probabilities might lead to an infinite regress in EU theory. This problem is at least mitigated herein that the credibility of original probability values can be modeled as another LPI. A rather practical example was investigated by Kofler and Zweifel (1988) in the context of economic policy in the face of the effect of the oil price hike on the U.S. economy in 1979/80. In 1979, President Carter had a choice between a restrictive and an expansionary policy. The Council of Economic Advisors (CEA) advocated a restrictive policy, designed to limit inflation, whereas Mork and Hall

---

[1] A vertex point is given by the condition that for an arbitrary number of probabilities, n, and $k \leq n$ restrictions (including the summation restriction), k are satisfied as equalities. Therefore, LPI theory is applicable quite generally.
[2] This criterion calls for maximizing over minimum expected utilities associated with the admissible probability distributions presented by Nature [again, see Zimmermann et al. (1985)].

(1979), based on a simulation model with an emphasis on the energy sector, called for monetary expansion. An LPI specifying a ranking of probabilities pertaining to central, favorable, and unfavorable outcomes could be derived from both the CEA report and Mork and Hall (1979). Outcomes were defined in terms of rate of inflation and rate of unemployment (which determine the chance of re-election according to a popularity function applicable to the United States).

Months before the 1980 election, President Carter had occasion to compare the actual figures for inflation and unemployment with the two competing forecasts. Since the CEA was farther off, it would have been appropriate to adjust their credibility compared to Mork and Hall (1979) (reflected by an *a priori* LPI) downward, to become an *a posteriori* LPI. Accordingly, the President's maxEmin-optimal policy in early 1980 would have been to shift from a restrictive policy (as advised by the CEA) to an expansionary one (as advised by Mork and Hall). Failure to do so cost him points in popularity which contributed to his defeat in the 1980 election [see Kofler and Zweifel (1988) for more details].

▶ **Conclusion 4.5** By accepting the assumption that the decision-maker is engaged in a game against risk-neutral Nature, any linear partial information (LPI) about probabilities can be exploited applying the maxEmin criterion. In addition, the credibility of information can also be modeled as an LPI, to be updated in view of experiences made.

It must be admitted, however, that the maxEmin criterion has never been tested for its empirical relevance. Although it does not posit that Nature acts like an actual adversary in a zero-sum game (who would seek to minimize the decision-maker's payoff in every instance), it is still somewhat pessimistic. Why should Nature present humans with the probability distribution resulting in their lowest possible expected utility but with the maximum expected utility to herself? In terms of the RDEU theory discussed in Sect. 4.3.1, this would mean that the weightings of probabilities are pessimistic without exception, a rather strong restriction.

4.3.4 Regret Theory

Regret theory was independently developed by Niehans (1948, in German) and Savage (1951), and popularized by Bell (1982) and Loomes and Sugden (1982, 1983) for financial decisions. According to the theory, decision-makers regret choices that turn out to be bad, while rejoice, when they turn out to be good. More specifically, regret theory assumes that individuals anticipate possible negative feelings if, once an event is realized, a differently chosen action would have resulted in a better outcome. These negative feelings are referred to as regret. Thus, a decision-maker's utility does not only depend upon the realized outcome but also on the outcomes that *would have been realized* had the choice been different.

As in Loomes and Sugden (1982, 1983), consider a decision-maker who is faced with choosing between two actions, a_1 and a_2. Action a_i results in outcomes

4.3 Other Non-Traditional Approaches

$x_{i,1}, x_{i,2}, \ldots, x_{i,n}$ that are dependent on states of the world, $s_1, s_2, \ldots s_n$ which occur with probabilities $\rho_1, \rho_2, \ldots \rho_n$. The choice of a_1 rather than a_2 with state s_j occurring returns outcome $x_{1,j}$ instead of $x_{2,j}$. Regret theory calls for the decision-maker's utility to be a function of both $x_{1,j}$ and $x_{2,j}$, represented as $M(x_{1,j}, x_{2,j})$.

The decision-maker is then assumed to maximize the expected value of utility and prefers a_1 to a_2 if and only if

$$\sum_{j=1}^{n} \rho_j M(x_{1,j}, x_{2,j}) = \sum_{j=1}^{n} \rho_j M(a_1) > \sum_{j=1}^{n} \rho_j M(x_{2,j}, x_{1,j}) = \sum_{j=1}^{n} \rho_j M(a_2). \tag{4.12}$$

If the converse is true, then a_2 is the preferred action to a_1 (it is also possible for the decision-maker to be indifferent).

Regret theory can be shown to explain the behavior observed in the Allais paradox. Recall Situation B from the Allais paradox presented as Example 4.2 in Sect. 4.1.1

Situation B	
Prospect R_3 results in	Prospect R_4 results in
• final wealth of 1 mn. MU with 11% probability	• final wealth of 5 mn. MU with 10% probability
• final wealth of 0 MU with 89% probability	• final wealth of 0 with 90% probability

Intuitively, it is easy to imagine that the anticipated "pang" of regret when R_3 returns 1 mn. MU while R_4 would have returned 5 mn. MU makes an individual prefer R_4 to R_3, even if the probabilities are not exactly the same.

Similar to other alternatives of EU theory, regret theory also satisfies the axioms of EU theory, although in a weakened form. Under certain assumptions, it also possesses first-order stochastic dominance (see Appendix 2.A.1).

4.3.5 Representative and Availability Heuristics

Representativeness or the *representative heuristic* refers to the behavioral response of individuals who tend to assume recent events (information) are "representative" of the true distribution of risk. That is, recent outcomes receive a disproportionate weight in the decision-making process. Tversky and Kahneman (1971, p. 105) first note that decision-makers may "view a sample randomly drawn from a population as highly representative." More recently, Volkman-Wise (2015) extends the representative heuristic for use in risk management, in particular, probability estimation and the demand for insurance. She uses Bayes' Rule to define the probabilities of losses occurring (or not) as follows:

$$ln\left(\frac{P(ND_{t+1}|D_t)}{P(D_{t+1}|D_t)}\right) = \delta_L ln\left(\frac{P(D_t|ND_{t+1})}{P(D_t|D_{t+1})}\right) + \delta_P ln\left(\frac{P(ND_{t+1})}{P(D_{t+1})}\right) \quad (4.13)$$

where $P(D_{t+1})$ $[P(ND_{t+1})]$ is the probability of a loss occurring (not occurring) in the next period and δ_L (δ_P) is the weight a decision-maker places on the posterior (prior) probability. A decision-maker without representative bias would place equal weights on the prior and posterior probabilities (i.e., $\delta_L = \delta_P$) while a decision-maker subject to representative bias will place a higher weight on recent outcomes (i.e., posterior probabilities), i.e. $\delta_L > \delta_P$.

For decision-makers operating under the representative heuristic, Volkman-Wise (2015) shows that the estimated probability of a future loss, given a recent loss, is higher than the true probability. She likewise shows the estimated probability of a future loss, in the absence of a recent loss, to be lower than the true probability.[3] These differences in probability estimates also affect the demand for insurance.

The *availability bias* or *availability heuristic* is similar to the representative heuristic in that a decision-maker estimates probabilities (frequencies) using information or examples that are readily "available" for analysis. Available information may come from events that are recent, numerous, or in any other way salient to the decision-maker. Tversky and Kahneman (1973) detected this bias in general decision-making while Kunreuther et al. (2013) note that the availability bias may affect the demand for flood insurance based on prior experiences.

The availability bias and representative heuristic are similar, but as discussed by Dumm et al. (2020), the availability bias is difficult to estimate empirically as one must know the full set of information available to an individual. Representativeness, however, only requires knowledge of recent events.

4.3.6 Other Biases

Meyer and Kunreuther (2017) note six additional behavioral biases that may affect a decision-maker's actions in the face of uncertainty. The *amnesia bias* is similar to that of the representative heuristic in that a decision-maker replaces painful memories (e.g., of large losses) quickly, especially when positive outcomes occur. As Meyer and Kunreuther (2017) note, this bias may lead to improper cancellation of risk management-related actions if losses are ignored or forgotten. *Herding* denotes decision-makers using the others' observable actions to make their own choices. With herding, a lack of preparedness may ensue if a risk is not broadly acknowledged and managed. *Inertia* (or *status quo bias*) simply suggests reluctancy to make changes in one's risk management decisions. Those who have obtained insurance, for example,

---

[3] The model of Volkman-Wise (2015) requires the probability of the true loss to be less than 50%. While this is a seemingly innocuous assumption, it does suggest that the difference in probability estimation may not apply to events with high likelihoods.

4.3 Other Non-Traditional Approaches

are likely to keep it, while those who have not, will not buy it. Decision-makers with *myopia* do not plan over long time horizons. Meyer and Kunreuther (2017) note that decision-makers subject to myopia may forgo investments in risk mitigation because they deem their present value too low to justify the initial outlay. *Optimism* refers to decision-makers with an irrational expectation that they are at a lower than average risk for an adverse outcome. Hence, they deploy insufficient effort to mitigate risk. Finally, *simplification* exists when decision-makers do not consider all causes that lead to a particular risk, resulting in an under-estimate. For instance, flooding may be the consequence of snow melting in the mountains combined with heavy rains in low-lying areas.

Meyer and Kunreuther (2017) suggest a "behavioral risk audit" for decision-makers and policymakers. By considering all behavioral biases (in particular, the six mentioned here), policymakers can design strategies to help mitigate decision-makers' tendencies to make improper investments in risk management.

4.4 Empirical Evidence

This section provides a non-exhaustive survey of the empirical evidence concerning the nontraditional theories presented above. While most of the empirical evidence is gleaned from experiments and quasi-experiments, some of it also comes from natural experiments in the guise of policy changes. Both sources have their deficiencies. Experiments allow researchers to focus on the phenomenon of interest; but participants often lack sufficient incentives to seriously consider their choices since they make too little difference in terms of a real payoff. Natural experiments, on the other hand, permit to examine the real-world decisions in a real-world environment, but are subject to confounding influences which cannot be adequately controlled. This results in favor of an agnostic view with regard to methodology.

4.4.1 Prospect Theory and Rank Dependent Expected Utility

Several studies have used experiments to estimate the parameters of the value function and probability weighting function of cumulative prospect theory. Tversky and Kahneman (1992) report the exponents of the value function [α and β in equation (4.7)] to be the same (median value of 0.88). Since $v'^{+} = \alpha x^{\alpha - 1}$ and $v''^{+} = \alpha(\alpha - 1)x^{\alpha - 2}$ on the "upside" while $v'^{-} = \lambda \beta x^{\beta - 1}$ and $v''^{-} = \lambda \beta(\beta - 1)x^{\beta - 2}$ on the "downside", the discontinuity in the marginal utility and in the curvature displayed in Fig. 4.1 rests on $\lambda \neq 1$ if $\alpha = \beta$, at least under the original version of prospect theory. Indeed, the authors further report a median loss aversion parameter λ of 2.25 which does suggest a kink in the risk-utility function. Still, the difference in its "upward" and "downward" slopes are estimated to be a constant, contradicting Fig. 4.1. This contradiction may be resolved by the weighting function in cumulative prospect theory.

The parameters in the weighting functions [γ and δ] were estimated as 0.61 and 0.69. Using these values and $p = 0.5$, one obtains $\sigma^{+}(p) = 0.42$ and $\sigma^{-}(p) = 0.45$;

for $p = 0.1$, $\sigma^+(p) = 0.19$ and $\sigma^-(p) = 0.18$. Allowing for their standard errors, these values are very likely to be equal, so the contradiction remains. Gonzalez and Wu (1999) examine the value function only in the gain domain finding lower values of α (median of 0.49). They also estimate a two-parameter weighting function

$$\pi^+(p) = \frac{\Delta p^\gamma}{(\Delta p^\gamma + (1-p)^\gamma)^{(1/\gamma)}} \quad (4.14)$$

designed to generalize Kahneman and Tversky's one-parameter function, with median values $\Delta = 0.77$ and $\gamma = 0.44$. They interpret γ as the "curvature" of the weighting function and Δ as the intercept, or "elevation". Note that this type of probability weighting also characterizes rank-dependent expected utility.

Abdellaoui (2000) provides a useful summary of parameterization strategies (and problems). He notes that despite also finding evidence consistent with probability weighting, "the data suggest a descriptive superiority of CPT [cumulative prospect theory] over RDEU [rank-dependent expected utility]" (p. 1498). His value function parameters ($\alpha = 0.89$ and $\beta = 0.92$), while consistent with Tversky and Kahneman (1992), still are in contradiction with Fig. 4.1. When using a one-parameter probability weighting function as in Kahneman and Tversky (1979), he estimates $\delta = 0.60$ and $\gamma = 0.70$, values which are quite close to those estimated by Tversky and Kahneman, though still potentially not different from each other. Estimates of the functional form as in Gonzalez and Wu (1999) result in values $\Delta = 0.65$ and $\gamma = 0.60$ in the gain dimension and $\Delta = 0.84$ and $\gamma = 0.65$ in the loss dimension. The first difference does suggest a kink in the risk-utility function (the second is again likely to be statistically insignificant). Nevertheless, Barberis (2012) in his survey claims that prospect theory "is still widely viewed as the best available description of how people evaluate risk in experimental settings" (p. 2).

▶ **Conclusion 4.6** Given that the kinks in the risk-utility function and the difference in its curvature between the two sides of the reference point constitute the features setting (cumulative) prospect theory apart from EU theory, the available experimental evidence is amazingly inconclusive.

4.4.2 Ambiguity Aversion and Regret

Several studies show the presence of ambiguity aversion. Camerer and Weber (1992) provide a useful summary of experiments documenting ambiguity aversion. Their summary table is replicated below in Table 4.1.

Camerer and Weber (1992) noted that several studies showed evidence of ambiguity aversion. However, the degree of ambiguity aversion varied. For example, Becker and Brownson (1964) found that approximately half of their experimental subjects were ambiguity averse. When the experiment involved business executives, MacCrimmon (1968) found less ambiguity aversion directly related to the

4.4 Empirical Evidence

Table 4.1 Summary evidence of ambiguity aversion (Camerer and Weber 1992, Table 3)

Stylized fact	Studies	Comments
Replication of Ellsberg	Becker and Brownson (1964, table 2)	Ambiguity = 70% of EV
	Slovic and Tversky (1974)	
	MacCrimmon and Larsson (1979)	= 20% of p
	Einhorn and Hogarth (1986, table 1)	
	Kahn and Sarin (1988)	
	Curley and Yates (1989)	= 5%–10% of EV
Strict aversion to ambiguity	Cohen et al. (1985)	Test by allowing indifference
	Curley et al. (1986, table 2)	
	Einhorn and Hogarth (1986, table 1)	
	Curley and Yates (1989)	
Aversion to partial ambiguity	Chipman (1960)	Subjects get samples from ambiguous urns
	Giglotti and Sopher (1990)	
Immunity to persuasion	MacCrimmon (1968)	Ambiguity aversion persists after exposure to written arguments
	Slovic and Tversky (1974)	
	Curley et al. (1986, table 4)	
Aversion to SOP	Yates and Zukowski (1976)	Ambiguity premium = 20% of EV
	Bernasconi and Loomes (1992)	20% of EV
Aversion to increasing range of probability	Becker and Brownson (1964)	
	Yates and Zukowski (1976)	
	Larson (1980)	
	Curley and Yates (1985)	
Ambiguity preference at low probabilities (gains) and high probabilities (losses)	Curley and Yates (1985)	$P_{low} = .4$
	Einhorn and Hogarth (1986)	$P_{low} = .001$
	Kahn and Sarin (1988)	$P_{low} = .1 - 3$, $P_{high} = .7 - .9$
	Curley and Yates (1989)	$P_{low} = .25$
	Hogarth and Einhorn (1990)	$P_{low} = .10$, $P_{high} = 90$
Extension to natural events	MacCrimmon (1968)	
	Goldsmith and Sahlin (1983)	
	Einhorn and Hogarth (1986, 1985)	
	Heath and Tversky (1991)	
	Keppe and Weber (1991)	
	Taylor (1991)	
Less ambiguity aversion for losses than for gains	Cohen et al. (1985)	
	Einhorn and Hogarth (1986) [table 1]	
	Kahn and Sarin (1988) (no difference)	
	Hogarth and Einhorn (1990, table 4)	
Independence of risk attitude and ambiguity attitude	Cohen et al. (1985)	Low correlations could be due to measurement error
	Curley et al. (1986, table 1)	
	Hogarth and Einhorn (1990, p. 797)	

Ellsberg paradox but more ambiguity aversion when they were confronted with a "real world" ambiguous situation [specifically, investments in foreign countries with known (risky) and unknown (uncertainty) frequencies of natural events].

Later, Camerer and Karjalainen (1994) "show a modest, but persistent, degree of ambiguity-aversion (p. 326); Eichberger et al. (2008) find their subjects to be ambiguity averse, as do Kelsey and le Roux (2015). Calford (2017) further shows that risk and ambiguity aversion are not separable. This is an important result since studies examining risk and ambiguity aversion separately (rather than jointly) likely have overestimated their effects.

4.4.3 Representative Heuristic

Dumm et al. (2017) and Dumm et al. (2020) identify a representative heuristic using data on the market for insurance for housing property in the state of Florida. The first study finds an uptick in aggregate insurance demand after hurricane losses. Consistent with the representative heuristic, aggregate demand increases immediately after hurricane losses and dissipates over time. Dumm et al. (2020) extend this study in several important ways. First, the authors show that this increase in insurance demand not only holds in the aggregate, but also at the individual level. Consumers tend to buy more insurance immediately after a loss, with the effect dissipating over time. This behavior is particularly marked for catastrophic losses as opposed to more common ones (e.g., fire). Finally, the authors provide an estimate of the degree to which individuals overestimate the probability of a subsequent catastrophic loss given a recent loss. Based on the model of Volkman-Wise (2015) they arrive at a value of almost 50%.

▶ **Conclusion 4.7** Allowing for variations in risk behavior adds nuance to individual decision-making. However, these modifications turn out to be rather complex and sometimes controversial. Since traditional expected utility theory is axiomatically well-founded and comparatively simple while providing important insights, it is retained in the remainder of the book.

Exercises

4.1

(a) Explain how the Allais and Ellsberg Paradoxes contradict Expected Utility Theory. Which axioms do they "violate"?
(b) In addition to the violation of the axioms in (a), discuss other apparent weaknesses of Expected Utility Theory.
(c) How does Prospect Theory "solve" the Allais Paradox?
(d) How does Ambiguity Aversion "solve" the Ellsberg paradox?

4.2

(a) Outline the basic tenets of Prospect Theory.
(b) Discuss the similarities and differences between Prospect Theory and Rank Dependent Expected Utility.

4.3 An insurance company considers launching a new product. If its reception by the market is favorable, the present value (PV) of predicted extra sales is 100 mn. MU, if it is somewhat favorable, 50 mn. MU, and if unfavorable, 10 mn. MU. Product development costs 20 mn. MU.

(a) Why is it important to express payoffs in terms of PV?
(b) With (π_a, π_b, π_c) denoting the probabilities of the favorable, somewhat favorable, and unfavorable reception, market research yields no more than the information, $\pi_a > \pi_c$. On the basis of this information, is it possible to determine whether it is worthwhile to launch the innovation using conventional criteria? Why (not)?
(c) Assume you are willing to use the maxEmin criterion of LPI analysis. Can you explain this criterion in no more than three sentences?
(d) What are the relevant probability distributions under the maxEmin criterion? What are the associated expected payoffs?
(e) What is the guaranteed minimum expected payoff? Is it sufficient to cover the cost of product development?
(f) Would you recommend to launch the innovation? Why (not)?

References

Abdellaoui, M. (2000). Parameter-free elicitation of utility and probability weighting functions. *Management Science, 46*(11), 1497-1512.

Allais, M. (1953). Le Comportement de l'homme rational devant le risque: Critique des postulats et axiomes de l'école américaine". *Econometrica, 21*, 503–546.

Barberis, N.C. (2012). Thirty years of prospect theory in economics: A review and assessment. *NBER Working Paper 18621*.

Becker, S.W., & Brownson, F.O. (1964). What price ambiguity? Or the role of ambiguity in decision-making. *Journal of Political Economy 72*, 62-73.

Bell, D.E. (1982). Regret in decision making under uncertainty. *Operations Research, 30*(5), 961-981.

Bernasconi, M., & Loomes, G. (1992). Failures of the reduction principle in an Ellsberg-type problem. *Theory and Decision 32*, 77-100.

Brigo, D., Pallavicini, A. & Torresetti, R. (2010). *Credit Models and the Crisis: A Journey into CDOs, Copulas, Correlations and Dynamic Models*. New York: Wiley.

Calford, E.M. (2017). Uncertainty aversion in game theory: Experimental evidence. *Purdue University Economics Department Working Paper No. 1291*.

Camerer, C., & Karjalainen, R. (1994). Ambiguity-aversion and non-additive beliefs in non-cooperative games: Experimental evidence. In B. Munier and M.J. Machina (eds.), *Models and Experiments in Risk and Rationality*, Springer Netherlands, 325-358.

Camerer, C., & Weber, M. (1992). Recent developments in modeling preferences: Uncertainty and ambiguity. *Journal of Risk and Uncertainty 5*, 325-370.

Carnap, R. (1950). *Logical Foundations of Probability*. Chicago: Chicago University Press.

Chipman, J.S. (1960). Stochastic choice and subjective probability." In D. Willner (ed.), *Decisions, Values and Groups*. Volume I. Oxford, England: Pergamon Press, 70-95.

Cohen, M., & Jaffray, J-Y., & Said, T. (1985). Individual behavior uncertainty: an experimental study. *Theory and Decision 18*, 203-228.

Conte, A., & Hey, J.D. (2013). Assessing multiple prior models of behavior under ambiguity. *Journal of Risk and Uncertainty, 46*(2), 113-137.

Curley, S.P., & Yates, J.F. (1985). The center and range of the probability affecting ambiguity preferences. *Organizational Behavior and Human Decision Processes 36*, 272-287.

Curley, S.P., & Yates, J.F. (1989). An empirical evaluation of descriptive models of ambiguity reactions in choice situations. *Journal of Mathematical Psychology 33*, 397-427.

Curley, S.P., & Yates, J.F., & Abrams, R.A. (1986). Psychological sources of ambiguity avoidance. *Organizational Behavior and Human Decision Processes 38*, 230-256.

Dumm, R.E., Eckles, D.L., Nyce, C., & Volkman-Wise, J. (2017). Demand for windstorm insurance coverage and the representative heuristic. *Geneva Risk and Insurance Review, 42*, 117-139.

Dumm, R.E., Eckles, D.L., Nyce, C., & Volkman-Wise, J. (2020). The representative heuristic and catastrophe-related risk behaviors. *Journal of Risk and Uncertainty, 60*, 2, 157-185.

Eichberger, J., Kelsey, D., & Schipper, B.C. (2008). Granny versus game theorist: Ambiguity in experimental games. *Theory and Decision, 64*, 333-362.

Einhorn, H.J., & Hogarth. R.M. (1985). Ambiguity and uncertainty in probabilistic inference. *Psychology Review 92*, 433-461.

Einhorn, H.J., & Hogarth, R.M. (1986). Decision making under ambiguity. *Journal of Business 59*, S225-S250.

Ellsberg, D. (1961). Risk, ambiguity, and the Savage axioms. *Quarterly Journal of Economics, 75*, 643–669.

Federal Drug Administration (2006). *Guidance for Industry. Q9 Quality Risk Management*. Rockville MD.

Fellner, W. (1961). Distortion of subjective probabilities as a reaction to uncertainty. *The Quarterly Journal of Economics, 75*(4), 670-689.

References

Gigliotti, G., & Sopher, B. (1990). The testing principle: a resolution of the Ellsberg paradox," working paper, Department of Economics, Rutgers University.

Gilboa, I., & Schmeidler, D. (1989). Maxmin expected utility with non-unique prior. *Journal of Mathematical Economics, 18*, 141-153.

Goldsmith, R.W., & Sahlin, N-E. (1983). The role of second-order probabilities in decision making. In. P. Humphreys, O. Svenson, and A. Vari (eds.), *Analysing and Aiding Decision Processes*. Amsterdam: North-Holland, 455-467.

Gonzalez, R., & Wu, G. (1999). On the shape of the probability weighting function. *Cognitive Psychology, 38*(1), 129-166.

Heath, C., & Tversky, A. (1991). Preference and belief: ambiguity and competence in choice under uncertainty. *Journal of Risk and Uncertainty 4*, 5-28.

Hey, J.D., & Orme, C. (1994). Investigating generalizations of expected utility theory using experimental data. *Econometrica, 62*(6), 1291-1326.

Hogarth, R.M., & Einhorn, H.J. (1990). Venture theory: a model of decision weights. *Management Science 36*, 780-803.

Intriligator, M.D. (1971). *Mathematical Optimization and Economic Theory*, Englewood Cliffs, New Jersey: Prentice-Hall.

Kahn, B.E., & Sarin, R.K. (1988). Modeling ambiguity in decisions under uncertainty. *Journal of Consumer Research 15*, 265-272.

Kahneman, D., & Tversky, A. (1979). Prospect theory: An analysis of decision under risk. *Econometrica, 47*(2), 263-292.

Kelsey, D. & le Roux, S. (2015). An experimental study on the effect of ambiguity in a coordination game. *Theory and Decision, 79*, 667-688.

Keppe, H.J., & Weber. M. (1991). Judged knowledge and ambiguity aversion, working paper no. 277, Christian-Albrechts-Universität, Kiel, Germany.

Kofler, E., & Menges, G. (1976). *Entscheidungen bei unvollständiger Information* (Decisions Based on Incomplete Information). Lecture Notes in Economics and Mathematical Systems 136. Berlin: Springer.

Kofler, E., & Zweifel, P. (1988). Exploiting linear partial information for optimal use of forecasts: with an application to U.S. economic policy. *International Journal of Forecasting, 4*(1), 15–32.

Kunreuther, H., & Pauly, M.V. & McMorrow, S. (2013). *Insurance and Behavioral Economics*. Cambridge: Cambridge University Press.

Larson Jr., J.R. (1980). Exploring the external validity of a subjectively weighted utility model of decision making. *Organizational Behavior and Human Performance 26*, 293-304.

Loomes, G., & Sugden, R. (1982). Regret theory: an alternative theory of rational choice under uncertainty. *The Economic Journal, 92*(368), 805-824.

Loomes, G., & Sugden, R. (1983). Regret theory and measurable utility. *Economics Letters, 12*(1), 19-21.

MacCrimmon, K.R. (1968). Descriptive and normative implications of the decision-theory postulates. In K. Borch, & J. Mossin (Eds.), *Risk and Uncertainty* (pp. 3–23). London: MacMillan.

MacCrimmon, K.R., & Larsson, S. (1979). Utility theory: axioms versus 'paradoxes.' In M. Allais and O. Hagen (eds.), *Expected Utility and the Allais Paradox*. Dordrecht, Holland: D. Reidel, pp. 333-409.

Markowitz, H.M. (1952). Portfolio selection. *Journal of Finance, 7*, 77–91.

Meyer, R. & Kunreuther, H.C. (2017). *The ostrich paradox: why we underprepare for disasters*. Wharton Digital Press.

Mork, K.A., & Hall, R.E. (1979). Energy prices, inflation, and recession, 1974-75. NBER Working Paper w0369. New York: National Bureau of Economic Research.

Niehans, J. (1948). Zur Preisbildung bei ungewissen Erwartungen (On the formation of prices under uncertain expectations). *Schweizerische Zeitschrift für Volkswirtschaft und Statistik (Swiss Journal of Economics and Statistics), 84*, 433–456.

Quiggin, J. (1982). A theory of anticipated utility. *Journal of Economic Behavior and Organization, 3*, 323–343.

Savage, L.J. (1954). *The Foundations of Statistics*. New York.

Savage, L.J. (1951). The theory of statistical decision. *Journal of the American Statistical Association*, 45, 57-67.

Slovic, P., & Tversky, A. (1974). Who accepts Savage's axiom? *Behavioral Science 19*, 368-373.

Taylor, K.A. (1991). Testing credit and blame attributions as explanations for choices under ambiguity, working paper, Department of Decision Sciences, University of Pennsylvania.

Taylor, P.R. (2012). The mismeasure of risk. In S. Boeser, R. Hillebrand, P. Sandin, M. Peterson (eds.), *Handbook of Risk Theory*. Vol. 1. New York: Springer, 441-475.

Tversky, A. & Kahneman, D. (1971). Belief in the law of small numbers. *Psychological Bulletin, 76*, 2, 105-110.

Tversky, A. & Kahneman, D. (1973). Availability: a heuristic for judging frequency and probability. *Cognitive Psychology, 5*, 207-232.

Tversky, A. & Kahneman, D. (1992). Advances in prospect theory: Cumulative representation of uncertainty. *Journal of Risk and Uncertainty, 5*, 297-323.

Volkman-Wise, J. (2015). Representativeness and managing catastrophe risk risk, *Journal of Risk and Uncertainty, 51*, 267-290.

Yates, J.F., & Zukowski, L.G. (1976). Characterization of ambiguity in decision making. *Behavioral Science 21*, 19-25.

Zimmermann, A., Zweifel, P., & Kofler, E. (1985). Application of the linear partial information model to forecasting the Swiss timber market. *Journal of Forecasting, 4*, 387-338.

Insurance Demand III: Decisions Under Risk with Diversification Possibilities

5

In Chap. 3, risk management was restricted to one of two alternatives: Either leave the asset in question without insurance protection or buy a certain amount of insurance coverage. This narrow view may be appropriate for the decision situation of a household who owns just one marketable asset (e.g., a house). Closer inspection shows that even in this case, two additional assets should be considered, namely health and human capital ("wisdom"). This gives rise to the question of whether the existence of these other assets might influence the decision to buy insurance coverage for the home. Consider a household whose human capital and hence labor income depend heavily on regional economic development. To a certain degree, it can diversify its assets by buying an apartment in a neighboring region that has different economic prospects. In this way, it can reasonably expect that its marketable asset does not lose value at the same moment when its wage income goes down. Obviously, risk can be reduced or mitigated through diversification, an additional means of risk management.

As shown in Sect. 5.1.1, diversification is already more useful for an enterprise that has different production units. It is even more beneficial to an investor who owns stock in enterprises. Therefore, in this chapter, the decision-making unit is no longer the household but the management of an enterprise or a firm who acts in the interest of investors. Section 5.1.2 takes up risk diversification through the capital market, with the capital asset pricing model (CAPM) providing the theoretical basis. In the course of the last twenty to thirty years, additional instruments for risk transfer have been developed that can also be used for risk management. Forward contracts, futures, and options in particular permit investors to transfer risks associated with marketable assets or liabilities to other participants of the capital market. They will be discussed in Sect. 5.2. Their transaction costs are compared to those of a risk transfer through insurance in Sect. 5.3. By examining the comparative advantages of insurance vis-à-vis the capital market, the demand for insurance by firms can be explained.

5.1 Risk Management and Diversification

Risk management comprises a set of instruments designed to measure and control the overall risk position of a firm with the aim to reduce risks (as part of the general goal of maximizing market value or expected profit). It, therefore, amounts to anticipating, assessing, and controlling the risks a firm is exposed to. Risk management employs all the measures cited in Sect. 2.3, including risk avoidance, prevention, reduction, splitting, allocation, and risk transfer. Short of perfect risk avoidance, *risk prevention* aims at reducing the probabilities of loss which, however, never reach zero. By way of contrast, *risk reduction* serves to limit the amount of damage or loss. *Mitigating measures* may be the building of buffer stocks, cash reserves, or goods. Mitigation may also involve a policy created in response to shifts in consumer demand as well as basic product (or service) diversification.

5.1.1 Risk Management and Internal Diversification

The essence of *diversification* is best understood by the proverb, "Never put all your eggs in one basket". Diversification helps in avoiding extreme results—or to make them less probable at least. This section is devoted to internal risk diversification. The possibilities of internal diversification derive from the fact that even a small firm can be perceived as a portfolio of different risk units. The identification of distinguishable risk units is mostly a question of judgment depending upon the degree of correlation between the risks [see Doherty (1985, 104)]. For instance, General Motors consists of plants and office buildings that may be considered as separate risk units in the case of fire. The probability of total damage or loss can always be reduced as long as the risks are not fully correlated between the units. The following example illustrates this assertion [see Doherty (1985, 128 f.)]. The possibilities of risk prevention are disregarded for simplicity, i.e., loss probabilities and loss amounts pertinent to the risk units are considered as given. These assumptions will be relaxed in Sects. 6.6 and 8.2, which deal with moral hazard.

Example 5.1

Internal diversification of fire risk
Firm A consists of 30 fast food restaurants. These restaurants are built the same way, are of similar value, and also similar with respect to the risk of fire. The risk

5.1 Risk Management and Diversification

units are locally separated, precluding any connection between a loss in restaurant i and restaurant j. The correlation coefficients, therefore, amount to $\rho_{ij} = 0$.[1]

Firm B produces synthetic articles. Its buildings are dispersed in a large area, leaving enough space between them to prevent conflagration. Although a large storm may damage more than one unit, the correlation coefficients between the risk units amount to a low $\rho_{ij} = 0.1$.

Firm C produces also synthetic articles, but production plants, warehousing, distribution, and offices are all situated in one building, which must, therefore, be considered as one risk unit when it comes to fire.

The values of the assets at risk, expected losses, and variations around expected losses (indicated by the standard deviation of the loss) are displayed below. Entries in bold are not assumed values but follow from calculations performed in the text. In terms of their value and their risk characteristics, these firms are comparable. All three have assets worth 9,000 TMU; the expected loss is 45 TMU; and the variability of the risk, measured, e.g., by the ratio of standard deviation to the expected value (mean), is equivalent at the level of the unit. It amounts to 2.5/1.5 $\approx$ 1.67 for one restaurant of firm A, 50/30 $\approx$ 1.67 for the plant of firm B, down to 3.333/2 $\approx$ 1.67 for the offices of B, as well as 75/45 $\approx$ 1.67 for firm C. However, the three firms differ with respect to their degree of diversification. It will be shown that this difference has considerable influence on the risk characteristic of the total loss and hence on the demand for insurance coverage to be expected. ∎

In general, the following relations hold. The expected value (mean) of a loss EL_i affecting unit i is given by

$$EL_i = \sum_k \pi_{ik} L_{ik}, \tag{5.1}$$

with π_{ik} denoting the probability of a loss in unit i amounting to L_{ik}. In the case of Firm B for instance $i = 1$ for the plant. The amount of loss L_{ik} can assume k values (depending, e.g., on whether expensive machinery is damaged or not) with probability π_{ik} such that the expected loss EL_i equals 3.0 TMU. In Table 5.1, only the result of this calculation is entered. Since damages to the plant may deviate from the expected value, they have a standard deviation σ_i. In general, the standard deviation of loss L_{ik} is given by

$$\sigma_i = \left[\sum_k \pi_{ik} (L_{ik} - EL_i)^2 \right]^{1/2}. \tag{5.2}$$

Again, only the result of the calculation is given in Table 5.1, with $\sigma = 50$ TMU in the case of Firm B and its plant ($i = 1$).

---

[1] Here, ρ is used to represent the correlation between the restaurants. In the previous chapter, ρ was used to represent compound lotteries.

Table 5.1 An example of internal risk diversification

Firm A: Risk unit	Value*	Expected loss*	Standard deviation*
Restaurant	300	1.5	2.5
Total (30 units)	9,000	45	**13.693** (see text)
Firm B: Risk unit	Value*	Expected loss*	Standard deviation*
Plant	5,000	30	50.000
Storehouse	2,000	10	16.667
Distribution	1,000	3	5.000
Offices	1,000	2	3.333
Total	9,000	45	**55.633** (see text)
Firm C: Risk unit	Value*	Expected loss*	Standard deviation*
Total (1 unit)	9,000	45	75

*in thousand monetary units (TMU)

Between the plant and the storehouse ($j = 2$), there is a distance, causing the coefficient of correlation between losses ρ_{12} to be small, with $\rho_{12} = 0.1$. Between two arbitrary losses, the coefficient of correlation $\rho_{ij} \in [-1, +1]$ is given by

$$\rho_{ij} = \frac{\sigma_{ij}}{\sigma_i \sigma_j}, \text{ with covariance given by } \sigma_{ij}$$

$$\sigma_{ji} = \sum_k \sum_l [\pi_{ik}(L_{ik} - EL_i)\pi_{jl}(L_{jl} - EL_j)]. \tag{5.3}$$

In Table 5.1, these coefficients of correlation are given, with $\rho_{12} = \rho_{21} = \rho_{23} = \cdots = \rho_{34} = 0.1$ in the case of Firm B.

Now, let L denote the sum of all losses L_1 (plant) to L_4 (offices). Its expected value is simply the sum of its components, i.e., $EL = 45$ TMU. To determine the standard deviation of this sum, i.e., of the portfolio of the risk units, recall first $Var(L_1 + L_2) = Var(L_1) + Var(L_2) + Cov(L_1, L_2) + Cov(L_2, L_1)$. More generally, and reverting to the notation used before, one obtains for the variance of a portfolio of random losses,

$$\sigma^2(L) = \sum_i \sigma_i^2 + \sum_i \sum_{i \neq j} \sigma_{ij} = \sum_i \sigma_i^2 + \sum_i \sum_{i \neq j} \rho_{ij} \cdot \sigma_i \sigma_j. \tag{5.4}$$

The standard deviation of the portfolio is, therefore, given by

$$\sigma_p(L) = \left[\sum_i \sigma_i^2 + \sum_i \sum_{i \neq j} \sigma_i \sigma_j \rho_{ij} \right]^{1/2}. \tag{5.5}$$

5.1 Risk Management and Diversification

Example 5.1 (continued)

Using Eq. (5.5), one can now calculate the standard deviation of the portfolio (i.e., the total loss) for each firm. The results are

For Firm A: $\sigma_p^A(X) = [30(2.5)^2]^{1/2} = 13.693$ TMU;
For Firm B: $\sigma_p^B(X) = [(50.000)^2 + (16.667)^2 + (5.000)^2 + (13.333)^2$
$+ 2 \cdot (0.1)(50)(16.667)$
$+ 2 \cdot (0.1)(50)(5)$
$+ 2 \cdot (0.1)(50)(3.333)$
$+ 2 \cdot (0.1)(16.667)(5)$
$+ 2 \cdot (0.1)(16.667)(3,333)$
$+ 2 \cdot (0.1)(5)(3.333)]^{1/2}$
$= 55.633$ TMU;
For Firm C: $\sigma_p^C(X) = 75$ TMU (by assumption).

Although the expected loss equals $EL = 45$ TMU for all three firms, the variability of total loss differs greatly between them. The most unfavorable risk situation is that of Firm C, which will be hit with some probability by a total loss of 120 TMU (45 + 75 TMU). At the other extreme, Firm A benefits from risk diversification to a high degree because of the local separation of its risk units. In most cases, its total loss will not exceed 59 TMU (45 + 13.693 TMU). Firm B is in between, with a likely total loss exposure of 101 TMU (45 + 55.633 TMU).
These differences will very likely influence the risk management strategy of the firm, and with it the decision to buy insurance coverage. This decision depends—according to the risk utility theory put forth in Sect. 2.3—in particular upon the variability of wealth, which in turn is determined by the amounts of damage and the probability with which they occur. ∎

Note that insurance does not relieve the firm from expected loss. Due to transaction costs and additional charges for risk bearing and administration, the insurance premium exceeds the expected loss. But insurance reduces, and in the extreme case of full coverage eliminates, the variability of wealth. It follows that Firm C benefits the most from insurance coverage, while Firm A benefits least. In general, one can expect enterprises with much dispersion of risk units to buy no insurance coverage and those with a concentration of assets in connected risk units to demand extensive insurance coverage. This expectation holds as long as any losses not covered fully fall on the owners of the enterprise; this condition will be qualified in the next section.

▶ **Conclusion 5.1** The demand for insurance coverage, in particular of enterprises which are managed by their owners, depends negatively on the possibilities of internal risk diversification. Demand is high if assets are concentrated in connected risk units.

This example can be applied more broadly (e.g., to burger chains, other multinational enterprises, or even local businesses). A particular application relates to mad-cow disease where suddenly consumers changed their demand away from beef toward chicken and pork. In order to cope with such an event, those businesses selling hamburgers are better positioned if they offer alternatives to beef such as chicken or pork products (e.g., chicken sandwiches, wings, etc.).

However, the analysis performed up to this point is only partial, for the objective of the firm is not risk minimization but maximization of its value. This is why all financial decisions have to be judged according to their contribution to the value of the firm. The interests of its owners are best served if the value of the capital invested (or the share price in the case of a stock company) is maximized. In fact, the maximization of equity value is equivalent to the maximization of the value of shares of the enterprise.

A more formalized notion of the benefit of diversification is given in the idea of *enterprise risk management (ERM)*. ERM is a holistic approach to risk management that suggests firms consider all risks together (instead of separately). When risks are not independent of one another, ERM will allow a firm to reduce its cost of hedging risk and, therefore, increase firm value.

Eckles et al. (2014) develop a simple model in which ERM is operationalized. Assume a single firm with two separate lines of business. Line B is arbitrarily set to have a higher expected loss and standard deviation of loss than line A. Let the loss distributions be given as follows:

	Line A		Line B
Loss	Probability	Loss	Probability
0	π_1	0	π_1
L_1	$1 - \pi_1$	L_2	π_2
		L_3	$(1 - \pi_1 - \pi_2)$

Each individual loss has the same expected value. Specifically, $L_1(1 - \pi_1) = L_2\pi_2 = L_3(1 - \pi_1 - \pi_2) = \mu_A$. The increase in their standard deviation follows from their increase in size (i.e., $L_3 > L_2 > L_1$). Also, let each line of business dispose of $L_1(1 - \pi_1)$ MUs for hedging risk [notice that the firm allocates $2L_1(1 - \pi_1)$ MUs in sum to hedge risk]. This allows for hedges of L_1 and L_3 in lines A and B, respectively. With hedging in place, the firm now faces the following loss distribution for each line:

Line A		Line B	
Loss	Probability	Loss	Probability
0	π_1	0	π_1
(hedged: $L_1 = 0$)	$1 - \pi_1$	L_2	π_2
		(hedged: $L_3 = 0$)	$1 - \pi_1 - \pi_2$

5.1 Risk Management and Diversification

The hedged means and variances (symbolized by subscript h) are then simply $\mu_{A,h} = 0 < \mu_{B,h} = L_2\pi_2$ and $\sigma^2_{A,h} = 0 < \sigma^2_{B,h} = (0 - \mu_{B,h})^2(\pi_1) + (L_2 - \mu_{B,h})^2(\pi_2) + (0 - \mu_{B,h})^2(1 - \pi_1 - \pi_2)$.

If the firm decides to manage the risks separately, (i.e., line by line, without ERM) it is exposed to a total loss with variance

$$\sigma^2_{firm,NoERM} = \sigma^2_{A,h,NoERM} + \sigma^2_{B,h,NoERM} + 2\rho\sigma_{A,h,NoERM}\sigma_{B,h,NoERM}. \quad (5.6)$$

Since $\sigma^2_{A,h} = 0$, which also reduces the covariances to zero, this variance reduces to

$$\begin{aligned}\sigma^2_{firm,NoERM} &= \sigma^2_{B,h,NoERM} \\ &= (0 - \mu_A)^2(\pi_1) + (L_2 - \mu_A)^2(\pi_2) + (0 - \mu_A)^2(1 - \pi_1 - \pi_2) \\ &= (0 - \mu_A)^2(1 - \pi_2) + (L_2 - \mu_A)^2(\pi_2). \quad (5.7)\end{aligned}$$

If the firm opts for the ERM alternative, it can give precedence to the highest overall risk. This calls for hedging of L_2 and L_3 confronting line B. With the same total allocation for hedging $[2L_2(1 - \pi_1)]$, the firm then faces the following loss distributions:

Line A		Line B	
Loss	Probability	Loss	Probability
0	π_1	0	π_1
L_1	$(1 - \pi_1)$	(hedged: $L_2 = 0$)	π_2
		(hedged: $L_3 = 0$)	$1 - \pi_1 - \pi_2$

The use of ERM, therefore, results in a mean and variance of zero for line B (i.e., $\mu_{B,h,ERM} = \sigma_{B,h,ERM} = 0$) and a mean and variance of line A the same as without hedging. With ERM, the variance of the losses confronting the firm, therefore, becomes

$$\begin{aligned}\sigma^2_{firm,ERM} &= \sigma^2_A + \sigma^2_{B,h,ERM} + 2\rho\sigma_A\sigma_{B,h,ERM} \\ &= \sigma^2_A \\ &= (0 - \mu_A)^2(\pi_1) + (L_1 - \mu_A)^2(1 - \pi_1). \quad (5.8)\end{aligned}$$

Recall the losses and probabilities were (arbitrarily) set such that line B had a higher standard deviation (variance) than line A. Thus, $\pi_1 < 1 - \pi_2$ and $(L_1 - \mu_A)^2(1 - \pi_1) < (L_2 - \mu_A)^2(\pi_2)$. It is easily seen now that $\sigma^2_{firm,ERM} < \sigma^2_{firm,NoERM}$. Thus, the firm utilizing ERM will have lower total risk exposure relative to the firm without ERM, *for the same total expenditure*.

Even if losses and probabilities are such that lines A and B contain risks of varying sizes, the firm can do no worse than achieving the same total risk exposure it would otherwise with a non-ERM focus. This is the benefit of considering the relative size and probability of risks *across the organization* rather than allowing the separate lines of business to adopt "tunnel vision" of the risk.

▶ **Conclusion 5.2** For firms with several lines of business which differ in terms of risk exposure, enterprise risk management calls for allocating funds available for hedging in a way to minimize overall risk exposure rather than allocating them line by line.

5.1.2 Risk Diversification Through the Capital Market

The value of the property right of an asset derives from the expected future returns of the asset. These returns are determined by operational decisions (e.g., which goods to produce and where to produce them) but also by financial risk management. Decisions that fail to take into account the relationship between risk management and the value of property rights to the enterprise, therefore, cannot be optimal. Likewise, decisions exclusively aiming at risk reduction and mitigation cannot be optimal either. Indeed, an inversion of this argument is possible: The quality of a firm's risk management is reflected in the share price. For instance, the expectation of an increased future probability of a damaging event has to show up in the quotation of shares today. Quite generally, share prices and their development reflect the changing expectations of market participants with respect to management policy. Hence, for the identification of an optimal risk management strategy, one has to know how share prices are determined, how the capital market functions, and what motivates investors to buy shares.

The starting point of the reasoning is the insight that the owners of a firm have to bear the residual risk after the cost-effective possibilities of internal risk diversification and enterprise risk management are exhausted. How does this residual risk impinging on the returns of the enterprise affect the quotation of its shares? One is tempted to argue that shareholders are risk-averse, hence, the risk will bring down share prices. This would imply that any risk management strategy (e.g., the purchase of insurance coverage) that serves to reduce the variability of the firm's value increases the value of its shares. However, such a conclusion would be premature, since shareholders themselves have diversification possibilities, which may substitute for risk management efforts by the firm. Specifically, they may be partial owners of several firms. By composing a portfolio of equities, they can reap an additional diversification effect. They may, therefore, be interested to a limited extent only in the risk management of a single enterprise. This argument is illustrated with an example.

Example 5.2

Diversification through the capital market
Let the shares of three firms exhibit the rates of return given in Table 5.2 over the last ten years.[2]
Evidently, with 8.1% p.a., Firm C has the highest average rate of return and, hence, the highest expected value of return. Firm B has the lowest expected rate of return, while Firm A lies in the middle. With respect to risk, A and C appear approximately equivalent; however, shares of Firm B again exhibit the worst performance. In Fig. 5.1, the three securities are depicted as points in a (μ, σ)-space, where $\mu_i := Er_i$ symbolizes the expected value and σ_i the standard deviation of the returns pertaining to shares of the i-th firm. An investor who can choose only one security would certainly avoid B. It depends on his or her risk attitude whether A or C is the optimal choice (the arrow pointing to the Northwest indicates the preference gradient in Fig. 5.1).
Now consider a fourth security D. With respect to expected return (7.7% p.a.) and risk (3.8% p.a.), it dominates A and B and is roughly comparable to C with respect to return, but better with respect to risk. However, this security D is nothing else but a simple portfolio composed of the securities B and C, with weights of 1/3 and 2/3, respectively (see Table 5.2). ∎

The original decision-making situation depicted in Fig. 5.1 (choice between A, B, and C) evidently ignores one important dimension: The risk of a portfolio depends not only on the individual standard deviations of its components but also on the covariances (correlations) between them. A negative covariance can reduce the risk of a portfolio considerably even though each component by itself may be characterized by a high risk.

A closer look at the values of Table 5.2 reveals that Firm B's stock tends to have a high rate of return when the values for A and C are low, while A and C vary in the same direction. Although A and C—considered in isolation—seem to be attractive, a portfolio consisting of the two would exhibit rather high variability (often called volatility). In contrast, the portfolio composed of shares B and C has a standard deviation which is much lower than that of B or of C but an expected return that lies between the values of B and C.

The calculation of the expected return and volatility of a portfolio of securities (symbolized by subscript p) is analogous to formulae (5.1) and (5.5), with rates of returns replacing monetary values. The only difference is that a portfolio involves a weighting of the different securities. In Example 5.2, the weights are 1/3 and 2/3,

---

[2]Let d_{t+1} denote the dividend and P_{t+1} the share price in period $t+1$. With P_t the share price in period t, the share's rate of return is given by $r_t = \frac{d_{t+1}+(P_{t+1}-P_t)}{P_t}$. The expected rate of return amounts to $Er_t = \sum \pi_i r_{it}$, with π_i symbolizing the probability of different values of r_{it} occurring. Its standard deviation is given by $\sigma(r_t) = [E(r_{it} - Er_t)^2]^{1/2}$. In Table 5.2, covariances and correlation coefficients are calculated using formulas (5.1) to (5.5), with $\pi_i = 1/10$ for simple averaging, using past observations to estimate (future) expected values.

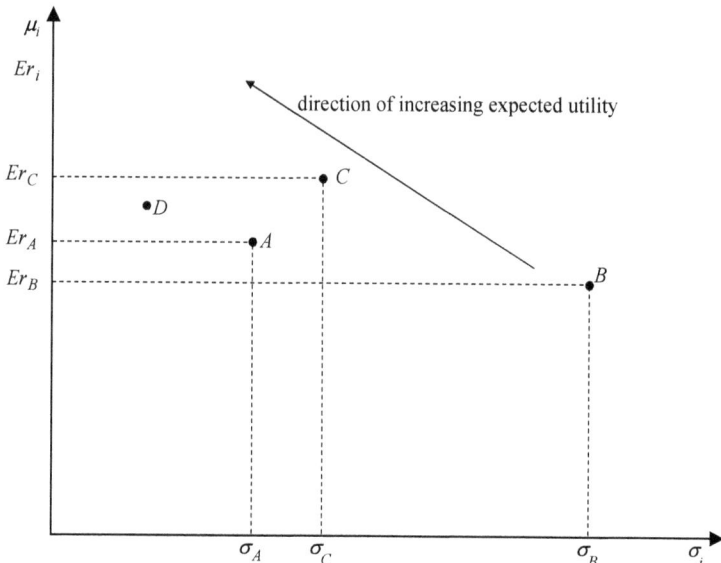

Fig. 5.1 Estimated expected returns (Er_i) and standard deviations (σ_i) ($i = A, B, C$)

Table 5.2 Rates of return of three firms and of a portfolio

Year	Shares of the firm			Portfolio*
	A	B	C	D
1	0.06	0.03	0.13	0.097
2	−0.03	0.33	−0.07	0.063
3	0.00	0.23	0.05	0.110
4	0.06	−0.12	0.13	0.047
5	0.20	−0.22	0.18	0.047
6	0.13	−0.15	0.13	0.037
7	0.10	0.19	0.08	0.116
8	−0.06	0.24	−0.10	0.013
9	0.05	0.16	0.10	0.120
10	0.23	0.00	0.18	0.120
Expected return	0.074	0.069	0.081	0.077
Standard deviation	0.089	0.179	0.091	0.038
Covariance	$\sigma_{A,B} = -0.0117$	$\sigma_{A,C} = 0.007$	$\sigma_{B,C} = -0.0131$	
Correlation coefficient	$\rho_{A,B} = -0.80$	$\rho_{A,C} = 0.86$	$\rho_{B,C} = -0.73$	

Note The portfolio is constructed according to the linear combination $D = 0.33B + 0.67C$

5.1 Risk Management and Diversification

respectively; more generally, they are denoted by w_i in the expressions below:

$$\mu := Er_p = \sum_i w_i Er_i, \tag{5.9}$$

$$\sigma(r_p) := [\sigma^2(r_p)]^{1/2}, \tag{5.10}$$

$$\rho_{i,j} := \sigma_{ij}/(\sigma_i \cdot \sigma_j), \text{ with} \tag{5.11}$$

$$\sigma_{ij} = \sum_i \sum_j w_i(r_i - Er_i)w_j(r_j - Er_j) \tag{5.12}$$

denoting the covariance between rates of return of securities making up the portfolio and

$$\sigma_p^2(r) = \sum x_i^2 \sigma_i^2 + \sum_i \sum_{\neq j} w_i w_j \sigma_{ij} \tag{5.13}$$

denoting the variance of the rates of return of the portfolio.

Throughout, the weights of the securities in the portfolio add up to $\sum_i w_i = 1$.

The two bottom rows of Table 5.2 contain the covariances and coefficients of correlation for each pair of the three shares, using Eqs. (5.9) to (5.13). In the case of the two securities B and C, the expected return of portfolio D is the weighted average of their returns ($0.077 = 1/3 \cdot 0.069 + 2/3 \cdot 0.081$), in keeping with Eq. (5.9). However, according to Eq. (5.13), the standard deviation of the portfolio is in general not equal to the weighted average of the standard deviations of the individual securities. Rather, there are mixed terms mirroring the covariances σ_{ij} (correlation coefficients ρ_{ij}). And with a correlation coefficient of -0.73, combining B and C serves to lower the variance of the portfolio considerably (see Fig. 5.1 again).

Three special cases can be analyzed easily and shown graphically in (μ, σ)-space. Here, σ stands for the standard deviation of the portfolio rate of return, shorthand for $\sigma(r_p)$ as defined in Eq. (5.9).

1. *Perfect positive correlation* ($\rho_{ij} = +1$). In Eq. (5.13), the terms σ_{ij} can be rewritten as $\sigma_{ij} = \rho_{ij} \cdot \sigma_i \cdot \sigma_j$ according to Eq. (5.11). With $\rho_{ij} = 1$, Eq. (5.13) becomes

$$\sigma^2(r_p) = \sum w_i^2 \sigma_i^2 + \sum_i \sum_{\neq j}(w_i \sigma_i)(w_j \sigma_j). \tag{5.14}$$

To illustrate, take the case of two securities A_1 and A_2. According to Eq. (5.9), the expected value of the portfolio's returns is the linear combination of the expected values of the individual securities with weights w_1 and $w_2 = 1 - w_1$. In addition, Eq. (5.14) can then be written

$$\sigma^2(r_p) = w_1^2 \sigma_1^2 + w_2^2 \sigma_2^2 + 2(w_1 \sigma_1)(w_2 \sigma_2) = [w_1 \sigma_1 + w_2 \sigma_2]^2. \tag{5.15}$$

Therefore, the standard deviation of the portfolio is also a linear combination of the standard deviations of the individual securities. In panel A of Fig. 5.2, the (μ, σ) values characterizing this portfolio must lie on the straight line connecting points A_1 and A_2. Combining the two securities does not serve to reduce overall risk beyond the lowest-risk component of the portfolio. There is no diversification effect.

2. *Perfect negative correlation* ($\rho_{ij} = -1$). In Eqs. (5.14) and (5.15), the plus sign must be replaced by a minus sign. In view of Eq. (5.14), there are combinations of w_i and σ_j that result in a value of zero. In the special case of two securities, the expression $[w_1\sigma_1 - w_2\sigma_2]^2$ in analogy to (5.15) reduces to zero if $w_1/w_2 = \sigma_2/\sigma_1$, i.e., if the weights are inversely proportional to the risks of the two securities. Such a zero-risk portfolio is represented by point B in panel B of Fig. 5.2. Now combining B with A_1 yields a portfolio with expected return $w_B Er_B + w'_1 Er_1$ and standard deviation $w_B 0 + w'_1 \sigma_1$. This is a linear combination represented by the straight line connecting B and A_1. Likewise, combinations involving B and A_2 give rise to the straight line connecting B and A_2. In sum, risk can be diversified away entirely by a judicious choice of weights.

3. *Perfect lack of correlation* ($\rho_{ij} = 0$). Here, the term involving the covariances disappears from Eq. (5.13). Because of the first term, however, $\sigma_p = 0$ cannot be reached. Still, there is a choice of weights w_i that minimizes σ_p. For the case of two securities, this combination can easily be deduced. Dropping the mixed term of Eq. (5.15), one obtains

$$\sigma^2(r_p) = w_1^2 \sigma_1^2 + (1 - w_1)^2 \sigma_2^2. \tag{5.16}$$

Differentiation with respect to w_1 yields the first-order condition,

$$\frac{\partial \sigma^2(r_p)}{\partial w_1} = 2w_1 \sigma_1^2 - 2\sigma_2^2 + 2w_1 \sigma_2^2 = 0.$$

Solving for w_1 results in

$$w_1^{min} = \frac{\sigma_2^2}{\sigma_1^2 + \sigma_2^2}.$$

Therefore, the weight of security No. 1 in a minimal variance portfolio is high if the other security is relatively volatile, contributing substantially to the total variance of the portfolio. This rule is similar to that found for the case of perfect negative correlation.

Panel C of Fig. 5.2 illustrates. Evidently, a zero-risk point such as B cannot be reached. The minimum variance portfolio is represented by point M. Combinations of securities A_1 and A_2 lie on the segment of an ellipse through M [suffice it to note that Eq. (5.16) contains the elements of an elliptic equation].

The general case of some degree of correlation between returns can be intuitively inferred from the special cases No. 1 to 3. For instance, high positive correlation must give rise to a portfolio combinations that lie between the two extremes of perfect

5.1 Risk Management and Diversification

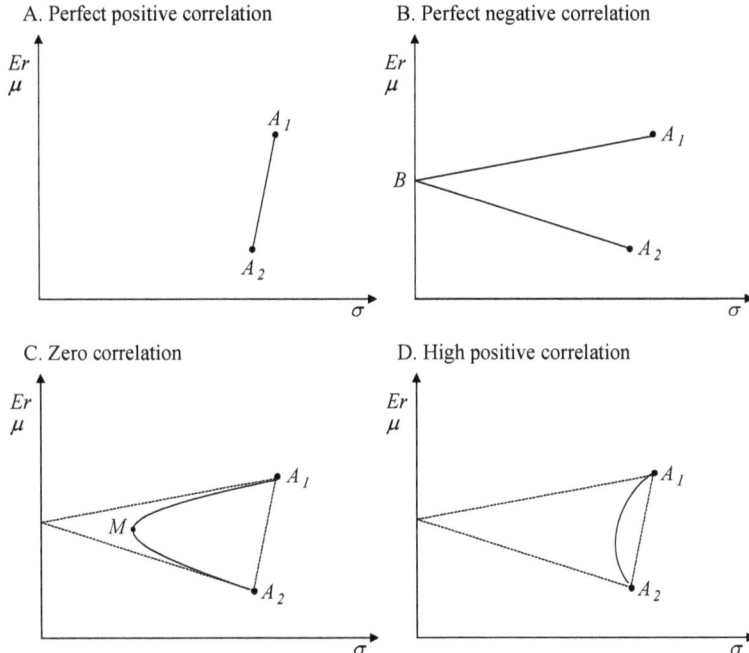

Fig. 5.2 Correlations between returns and portfolios in (μ, σ)-space

positive correlation as in panel A of Fig. 5.2 and lack of correlation in panel C with the preponderance of the first extreme. Panel D of Fig. 5.2 illustrates this case.

To sum up:

1. The lower the coefficient of correlation between two securities (the closer the coefficient is to 1.0), the more marked is the diversification effect;
2. Combinations of two securities have never more risk than in the case of perfect positive correlation. They are indicated by the straight line connecting the two positions in (μ, σ)-space; and
3. It is always possible to find a portfolio with minimum variance by the appropriate choice of portfolio weights.

With a little imagination, one can see that there exist additional possibilities of diversification when three or more securities are available for forming the portfolio. In principle, one could draw all possible combinations of risky securities in (μ, σ)-space. Doing this for a limited number of combinations, one arrives at Fig. 5.3.

Figure 5.3 displays four securities (A, B, C, D), each with its expected rate of return and standard deviation. For every pair of them, it is possible to construct portfolio curves by varying the ratio of the respective weights, with the curvature depending on the amount of correlation between the two components. A first possibility consists in selecting one security from each portfolio, e.g., A from AB and C from BC. This results in a set of new combinations symbolized by AC. Clearly, this new

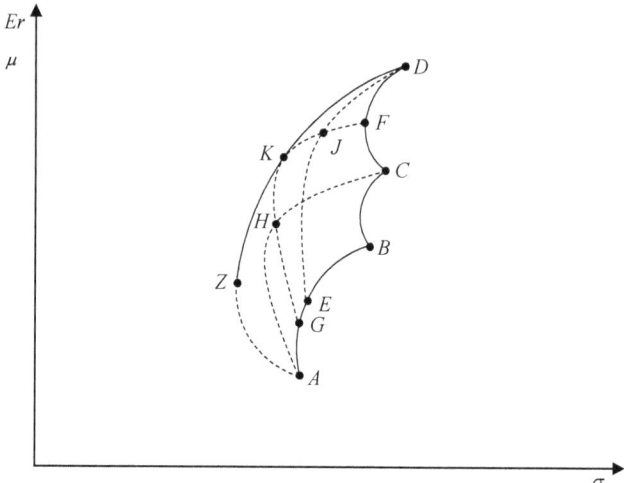

Fig. 5.3 Portfolios composed of portfolios and the efficiency frontier

frontier indicates a gain over AB due to an added diversification effect. The second step is to combine a portfolio with an individual security, such as E along AB with D to obtain ED. Finally, portfolios AB and CD can themselves be mixed, resulting in "portfolios of portfolios". The result is the boundary GF which contains a segment HJ dominating all possibilities considered up to this point (AB, CB, AC, ED).

Exhausting all possibilities gives rise to an area that is enclosed by $AZDCB$. Portfolios within this area are obviously dominated by those on the border between Z and D. For instance, point A can be excluded because there is a portfolio K that has less risk but offers a higher expected rate of return. In contrast, point D cannot be excluded because there is no portfolio with the same rate of return but less risk. Point D thus represents the portfolio (usually single security) which offers the highest expected rate of return. Conversely, point Z represents the minimum variance portfolio. Hence, all portfolios which are not dominated by any other portfolio lie on the envelope connecting Z and D of Fig. 5.3. This set of portfolios is, therefore, called the *efficient frontier*. Portfolios on it yield the maximum expected return for a given amount of risk. Conversely, they achieve minimum risk for a given expected rate of return. They are efficient with respect to mean and variance (or standard deviation).

These elements of portfolio theory go back to the pioneering work of Markowitz (1952). To determine the efficient frontier in (μ, σ)-space, one need not use the trial-and-error method sketched here. Rather, one can apply quadratic programming or the method developed by Merton (1972).

▶ **Conclusion 5.3** The efficient frontier in (μ, σ)-space is composed of individual securities or portfolios. It originates from the minimum variance portfolio and runs concave from below.

5.1 Risk Management and Diversification

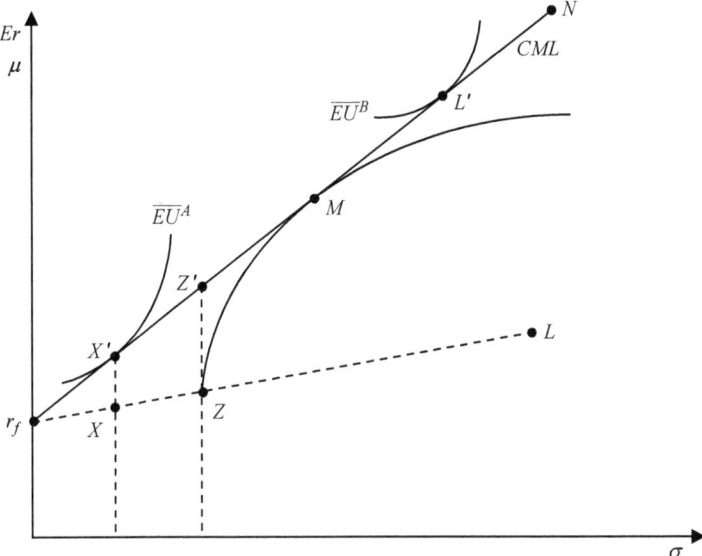

Fig. 5.4 Efficiency frontier and capital market line

If in addition to the risky securities there exists a *risk-free security* with rate of return r_f, then Fig. 5.3 needs to be complemented. Note that a constant such as r_f has no deviations from its expected value, causing covariance with any random variable to be zero [see the definition of covariance in (5.3)]. Therefore, special case No. 3 analyzed above (absence of correlation) applies. In Fig. 5.4, there accordingly is a straight line connecting the zero-risk point r_f, e.g., with point Z on the efficient frontier. Given this risk-free alternative, risk-averse investors will not want to invest all their capital in volatile securities. They could, e.g., put half of their funds in the minimum variance portfolio Z and the other half in the risk-free security (see point X of Fig. 5.4). Now assume that investors cannot only buy a security with an interest rate r_f but also sell such a security, which amounts to obtaining a loan. If credit is available at the same rate r_f, the dotted straight line through point X extends beyond point Z. A "leveraged" investor (with little risk aversion) might attain a loan at the rate r_f in order to scale up his or her portfolio indicated by L.

However, it is evident from Fig. 5.4 that none of the portfolios X, Z, L constitutes the best choice. The reason is that portfolios along the straight line $r_f MN$ can be reached as well. In particular, portfolios X', Z', L' offer a higher expected rate of return for the same amount of risk as X, Z, L. Therefore a strongly risk-averse investor (represented by the indifference curve $\overline{EU}^A$) opts for X' instead of X, a less risk-averse investor (represented by the indifference curve $\overline{EU}^B$), for L' instead of L. Accordingly, the line $r_f MN$ is called *capital market line*, with the segment $r_f M$ showing lending positions and the segment MN, borrowing positions or leveraged positions. It is also evident that the portfolio M plays a special role: It is optimal in the sense of contributing to a lending–borrowing position which dominates any other position which can be offered by any other risky portfolio on the efficient frontier.

5.1.3 The Capital Asset Pricing Model (CAPM)

Until now, optimizing behavior by a single investor was analyzed. The next step is to determine the equilibrium of the capital market from individual optimization. This can be achieved using the following argument. Provided a risk-free investment alternative exists, the *tangency portfolio M* of Fig. 5.4 is part of the optimum of all investors. Let a strongly risk-averse investor choose point X' on the capital market line (CML), a portfolio comprising a good deal of risk-free bonds. The risky component of that portfolio necessarily corresponds to point M on the efficient frontier. Let a less risk-averse investor borrow at the risk-free rate r_f to hold risky securities beyond his or her own wealth. Still, his or her optimum at point L' is nothing but a scaled-up version of the tangency portfolio M. This in turn implies that there is a single portfolio of securities that is *optimal for all investors*, independent of their risk aversion. This is portfolio M. All investors will, therefore, hold the same portfolio of risky securities; they differ only by their way of financing. The more risk-averse investors do not allocate their entire wealth to M, using the remainder to provide credit to others; the less risk-averse allocate more than their own wealth to M, borrowing the excess from others.

But if all investors want to hold the same portfolio of risky securities, then this portfolio can be nothing but the *market portfolio*, which contains all traded securities (i.e., securities that belong to the efficient frontier). Thus, every security is contained in the portfolio of each investor, with a weight equal to its market value relative to the market value of all securities.

The *equilibrium portfolio*, therefore, is the market portfolio. This fact suggests a reverse conclusion, from the risk-return characteristic of the market portfolio to the risk-return characteristics of the securities contained in it. Equations (5.9) to (5.13) of Sect. 5.1.2 established the relationship between a security and the portfolio. Now the problem is reversed: To find the relationship between the market portfolio and individual security. The question can be stated as follows: What combination of expected return and volatility ensures that security becomes part of the market portfolio?

Equation (5.9) shows that on the one hand the answer to this question will depend on the contribution of security i to the return of the market portfolio. On the other hand, according to Eq. (5.13), its contribution to the risk of the portfolio also plays a role. Besides its own variance, the covariances with the other securities j (i.e., with the market portfolio in the case of many securities) will be of importance. In this way, a security A can make it into the market portfolio because of two reasons. Due to a positive covariance with the return of the total market (r_M), it fails to contribute to risk diversification but stands out for its above-average expected return. Or it may have a below-average expected return, which, however, is compensated by a negative covariance with r_M bestowing it a strong diversification effect.

5.1 Risk Management and Diversification

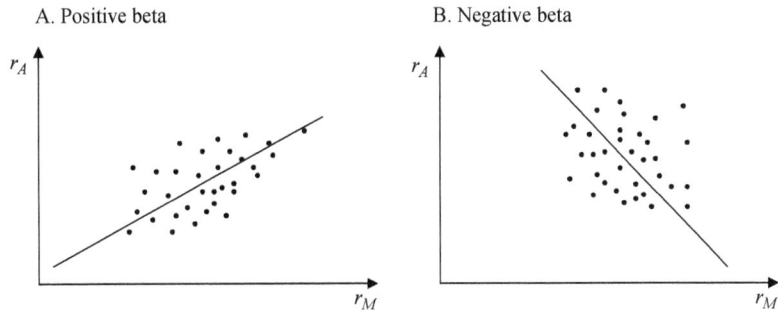

Fig. 5.5 Correlations between rates of return of two securities and the market portfolio

Now the relationship between the return of an individual security and the return of the market portfolio can be ascertained empirically. In Fig. 5.5, each point represents a pair of observed rates of return of a security and of the market portfolio M. In panel A, the returns of security A tend to be high when the market rates are high as well. For security B in panel B, the opposite is true, its rates of return being low when market rates are high.

While the relationship is not perfect, it is usually strong enough to permit estimation of a linear regression equation of the type

$$r_{i,t} = \alpha_i + \beta_i r_{M,t} + \varepsilon_{i,t} \quad i = A, B \text{ with} \tag{5.17}$$

α_i, β_i : regression coefficients to be estimated;
$\varepsilon_{i,t}$: residual error with $E\varepsilon_i = 0$ and $Var(\varepsilon_i) = $ const. for all values of $r_{M,t}$.

The parameter β_i (the *beta* of the security) shows the sign and the strength of the relationship between r_i and r_M. In panel A of Fig. 5.5, β_A has a positive value below 1, indicating a limited diversification effect. In panel B of Fig. 5.5, β_B is negative, pointing to a marked diversification effect. From regression analysis, it is known that β_i is nothing else but a scaled version of covariance:

$$\beta_i = \frac{Cov(r_i, r_M)}{\sigma^2(r_M)}. \tag{5.18}$$

Moreover, using $Cov(r_i, r_M) = \rho(r_i, r_M) \cdot \sigma(r_i) \cdot \sigma(r_M)$ in analogy to (Eq. 5.11), one can establish the relationship between β_i and the coefficient of correlation $\rho(r_i, r_M)$ between the individual and the market rate of return,

$$\beta_i = \rho(r_i, r_M) \frac{\sigma(r_i)}{\sigma(r_M)}. \tag{5.19}$$

Therefore, for a given ratio of individual to market volatility, a higher correlation coefficient goes along with a higher β_i.

Note that Eq. (5.17) decomposes the return of a security into a systematic and a non-systematic component (ε_{it}). This is also true of its variance. Again, the decomposition depends critically on the (absolute) value of β_i [see Doherty (1985, 150)]. According to Eq. (5.17), the expected value of security i's return is $Er_i = \alpha_i + \beta_i Er_M$, independent of time t. Therefore, one obtains for the variance of returns,

$$\begin{aligned}\sigma^2(r_i) &= E(r_i - Er_i)^2 = E[(\alpha_i + \beta_i r_M + \varepsilon_i) - (\alpha_i - \beta_i Er_M)]^2 \\ &= E[\beta_i(r_M - Er_M) + \varepsilon_i]^2 \\ &= E[\beta_i^2(r_M - Er_M)^2 + 2\varepsilon_i(r_M - Er_M \beta) + \varepsilon_i^2] \\ &= \beta_i^2 Var(r_M) + Var(\varepsilon_i) \quad \text{since } E\varepsilon_i = 0 \text{ for all values of } r_M. \end{aligned} \quad (5.20)$$

One part of the risk associated with security i emanates from the relationship between its return and the return of the market portfolio. This risk is reflected by the regression coefficient β_i (which in turn depends on the sign and amount of covariance or correlation). Therefore, $\beta_i^2 Var(r_M)$ is called *systematic risk* (*market risk*) because it relates systematically to the movements of the market portfolio. The higher β_i (the beta of a security), the more movements in the market rate of return translate into movements in the rate of return of security i.

The second component recalls the fact that the distribution of individual rates of return cannot be explained fully by the movements of the market portfolio. The points of Fig. 5.5 exhibit a stochastic variation around the regression line, pointing to unexplained variance in the error term $\varepsilon_{i,t}$. This unexplained component of the risk is called *non-systematic risk* (*idiosyncratic risk*).

▶ **Conclusion 5.4** The variance of the return of a security can be decomposed into a systematic component related to the capital market as a whole and a non-systematic component characterizing the security in question.

However, the relative importance of these two components changes systematically with an increasing degree of portfolio diversification. This can be easily shown for the special case of a portfolio consisting of N securities with equal weights $w_i = 1/N$ and equal non-systematic risk $Var(\varepsilon_i) = Var(\varepsilon)$. The non-systematic variance of the portfolio denoted by $Var(\varepsilon_p)$ is then given by

$$\begin{aligned} Var(\varepsilon_p) &= (\frac{1}{N^2})Var(\varepsilon_1) + (\frac{1}{N^2})Var(\varepsilon_2) + \cdots + (\frac{1}{N^2})Var(\varepsilon_N) \\ &= (\frac{1}{N^2}) \cdot N \cdot Var(\varepsilon) = (\frac{1}{N})Var(\varepsilon). \end{aligned} \quad (5.21)$$

For $N \to \infty$, $Var(\varepsilon_p) \to 0$, i.e., the non-systematic risk vanishes in a fully diversified portfolio. This result has three implications. First, because of (5.20), the non-systematic risk of any one security has no effect on the total risk of such a portfolio. This means that a diversified investor is indifferent to the non-systematic risk of an individual security. Hence, this component of risk cannot have an effect on the price of a security. Second, since the fully diversified portfolio coincides with the

5.1 Risk Management and Diversification

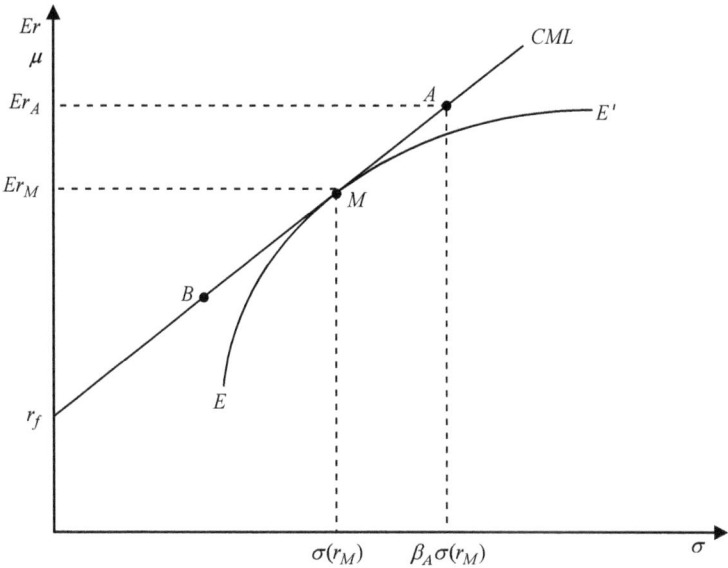

Fig. 5.6 Valuation of an individual security

market portfolio, investors attain a maximum degree of diversification by holding the market portfolio. Third, the price of a security does depend on its systematic risk. In Eq. (5.20), the first term does not vanish as $N \to \infty$. Each security contributes to the risk of the portfolio according to β_i, reflecting its covariance with the market portfolio. Hence, the value of β_i has a crucial effect on the price of the security.

Given these preliminaries, it is now possible to value an individual security. The idea is to examine the marginal variation of the market portfolio induced by a marginal variation of the weight of security i [see, e.g., Copeland et al. (2004, Sects. 7A–7C)]. Because of Eqs. (5.20) and (5.21), such a marginal variation of the market portfolio must satisfy the relationship

$$\sigma^2(r_i) = \beta_i^2 \sigma^2(r_M) \text{ and hence}$$
$$\sigma(r_i) = \beta_i \sigma(r_M). \tag{5.22}$$

The standard deviation of security i's return reduces to the systematic risk, i.e., the β_i-fold of the standard deviation of the market portfolio. The marginal variation of the market portfolio mentioned above is depicted in Fig. 5.6. In the neighborhood of point M, the efficient frontier EE' and the capital market line (CML) have the same slope. The movement caused by the variation of the weight of security A ($i = A$) in the portfolio is along EE'; however, because of the equality of slopes, it can also be shown along CML (in Fig. 5.6 only approximately so, since the changes are scaled up). The economic interpretation is clear: Whether the market portfolio is changed marginally by varying the weight of the risk-free asset or of a risky security, A does not make any difference to diversified investors.

Security A is, however, characterized by its expected return Er_A and by its standard deviation of returns given by Eq. (5.22). Since A lies on the CML, it follows from the theorem of intersecting lines

$$\frac{Er_A - r_f}{Er_M - r_f} = \frac{\beta_A \sigma(r_M)}{\sigma(r_M)} \quad \text{and hence}$$
$$(Er_A - r_f) = \beta_A (Er_M - r_f). \tag{5.23}$$

In general, for any security i contained in the market portfolio, this becomes

$$Er_i = r_f + \beta_i (Er_M - r_f). \tag{5.24}$$

This is the *core equation of the CAPM*. Since $\beta_i > 0$ and $Er_M > r_f$ as a rule, it states that on expectation, an individual security must exhibit a rate of return which contains a surcharge for its systematic risk in excess of the risk-free rate.

The relation (5.24) can also be expressed as a function of β_i. This is called the *security market line (SML)*. Because it is linear in β_i, one can construct the SML with the help of two values of β_i:

(1) $Er_i = r_f$ for $\beta_i = 0$. This is evident from Eq. (5.24).
(2) $Er_i = Er_M$ for $\beta_i = 1$. This also follows from Eq. (5.24). In addition, the beta of the market portfolio is one as well. Setting $i = M$ in (5.22) results in $\sigma(r_M) = \beta_M \sigma(r_M)$, implying $\beta_M = 1$.

In Fig. 5.7, the SML is drawn accordingly. It says that a security with a high (positive) β_i must achieve a high expected return. Furthermore, Fig. 5.7 shows two securities not lying on the SML. First, security X lies above the line, i.e., its expected return is high relative to its value of β. This means that X is under-valued. Investors will seek to buy this security, increasing its price and, hence, reducing future expected return until the equilibrium value is restored. By way of contrast, security Y lies below the SML. It is over-valued and will be sold, causing its price to decrease and its future expected return to increase. In this way, the price of an individual security is determined by its expected rate of return through the *SML*.

The Capital Market Line with point M as the equilibrium portfolio, Eq. (5.24), and the Security Market Line constitute the CAPM which was developed independently by Sharpe (1964), Lintner (1965), and Mossin (1966). It constitutes an extension of the portfolio theory of Markowitz (1952, 1959). While portfolio theory describes an individually optimal solution, the CAPM describes an equilibrium outcome on the market of securities (and hence derives from the demand for and supply of securities).[3]

---

[3] A critique of the CAPM will be provided in Sect. 7.2.2, where it is applied to insurance. It also has been subjected to empirical testing, with mixed results [see, e.g., Roll (1977)].

5.1 Risk Management and Diversification

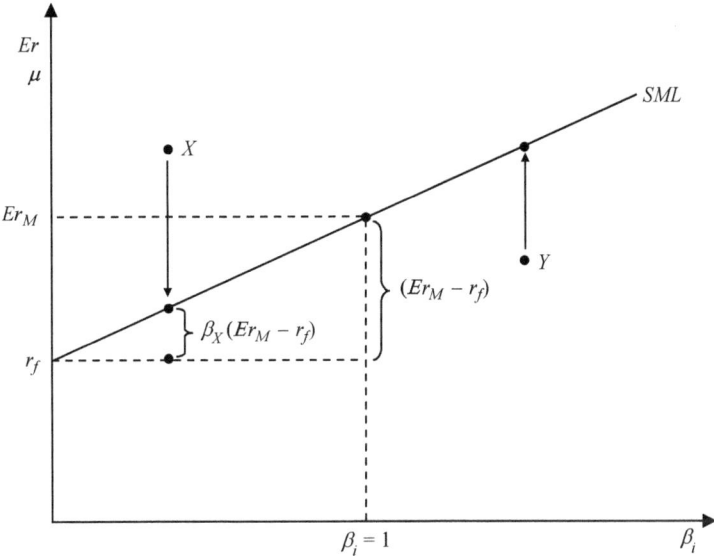

Fig. 5.7 The security market line (SML)

Equation (5.24) can be related back to the *risk management decision* of the firm. The equilibrium expected rate of return being Er_i, this is what investors ask at a given level of systematic risk. Since the enterprise must attain Er_i to attract investment capital, Er_i can be viewed as the cost of capital to this firm. Suboptimal decisions in risk management raise the risk borne by shareholders. A higher value $\sigma(r_i)$ is attributed to the securities of this firm, and according to Eq. (5.19), β_i increases for a given risk of the market portfolio $\sigma(r_M)$. Equation (5.24) states that this increase drives up the cost of capital. If on the other hand risk management achieves a reduction of β_i, it makes the shares of the firm attractive as a means of risk diversification, consequently lowering the expected rate of return and hence the cost of capital. In this way, risk management decisions are compatible with other financial decisions, contributing to the maximization of the value of the enterprise.

▶ **Conclusion 5.5** In order to be able to compete with the other securities of the market portfolio, an individual security must attain a certain rate of return on expectation. According to the CAPM, this benchmark importantly depends on the beta of this security, i.e., the slope coefficient of a regression of firm-specific returns on the returns of the market portfolio. Suboptimal decisions in risk management increase this beta and hence the cost of capital.

5.1.4 Empirical Asset Pricing Model (EAPM) and Arbitrage Pricing Theory (APT)

In Sect. 5.1.2, the possibilities of risk reduction through diversification were shown to depend mainly upon the degree of interdependence between the rates of return of the securities. In portfolio optimization, the degree of interdependence is measured by the covariance or the coefficient of correlation, respectively. In the CAPM of Sect. 5.1.3, it amounts to the beta of the security. The higher the (positive) beta, the more the returns of the security in question vary with that of the entire capital market and the larger is the risk surcharge by which the risk-free interest rate must be exceeded.

The linear regression $r_{i,t} = \alpha_i + \beta_i r_{M,t} + \epsilon_{i,t}$ forming the basis of the CAPM [see Eq. (5.17) again] generates two estimates of the constant α_i (the so-called alpha) and of the slope β_i (beta). The alpha indicates the extent to which the returns on stock i differ from the market return, r_M. A positive alpha means that stock i outperforms the market on average, while a negative value shows that this stock underperforms the market. A beta greater than one indicates that stock i has greater volatility than the market portfolio, while a beta less than one characterizes a stock that contributes to risk diversification through its relatively low volatility. The error term, $\epsilon_{i,t}$, captures unmeasured influences on the stock's rate of return, $r_{i,t}$, apart from the general development of the capital market.

These unmeasured influences motivated Fama and French (1992) to introduce two additional systematic risk factors, the size of the company issuing the stock (measured by the market value of its equity) and its value (measured by the ratio between its value according to its balance sheet and the market value of its equity, often referred to as the book-to-market ratio). Historically, small companies tend to outperform large ones (size effect), while high-value companies outperform low-value ones (value effect). This three-factor model constituted a significant improvement over the single-factor CAPM. The next step undertaken by Fama and French (2015) was to expand the CAPM even further, resulting in the empirical asset pricing model (EAMP). The authors included two additional risk factors, profitability (measured as profits to total assets) and investment (the change in total assets). In this way, they were able to explain an even greater share of the cross-sectional variance of companies' rates of return.

In the CAPM, the interdependence of returns is taken as given. By way of contrast, arbitrage pricing theory (APT) seeks to explain it. The APT attributes the rates of return of securities to a multitude of factors. As is usual in statistical factor analysis, these factors are assumed not to be directly observable (this will be qualified below), to be uncorrelated, and to have an expected value of zero. Some of them could be connected with the firm in question (e.g., profit expectations), some others with the branch of industry (e.g., cost-reducing technological change), and still others, with the capital market in general (e.g., interest rates). A positive correlation between rates of return results if the securities are influenced by one or several factors in the same way.

5.1 Risk Management and Diversification

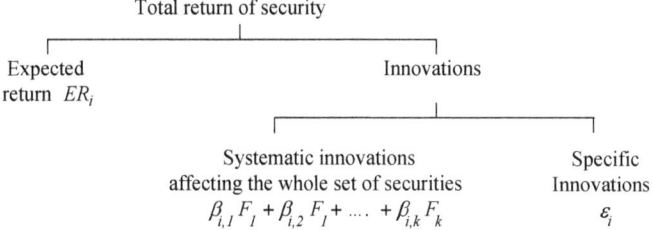

Fig. 5.8 Decomposition of the return of a security

The return of a security traded in the capital market is accordingly split into two components (see Fig. 5.8). The normal or expected return ER_i depends on all information the owners of this security have with respect to this security (or the above-mentioned factors, respectively). The second component is the uncertain or risky return of the security. It reflects additional, unexpected information, called *innovations*. Such innovations could be caused by reports on the most recent sales, inventions, or new products of the firm in question; new products of competitors; unforeseen changes in interest rates, inflation rates, and growth rates of the gross domestic product (GDP); or general political change in the country.

Innovations constitute the real risk associated with an investment decision. Conversely, as long as one always obtains the expected return, one bears no risk and no uncertainty.

The unexpected component can in turn be decomposed into systematic innovations impacting on a whole set of securities and a non-systematic or specific risk, which relates only to one particular security.

This description corresponds to the following factor model (see Fig. 5.8):

$$r_i = ER_i + \beta_{i,1}F_1 + \beta_{i,2}F_2 + \cdots + \beta_{i,k}F_k + \varepsilon_i. \tag{5.25}$$

Here, ER_i represents the expected return. The risky return is attributed to the k factors $\{F_1, F_2, \ldots, F_k\}$ and to the specific error term ε_i. The β coefficients $\{\beta_{i,1}, \beta_{i,2}, \ldots \beta_{i,k}\}$ are specific to the security or the firm i issuing it. This means that the firms are affected differently by the factors cited (interest rate, inflation rate, growth rate of GDP) or react to these influences with differing degrees of sensitivity. If a specific factor influences not only the return of the security i but also the return of all others in the same way, it causes positive covariance, resulting in a positive beta. Put another way, in the APT, the β_i of the CAPM is decomposed into several betas $\beta_{i,1}, \beta_{i,2}, \ldots \beta_{i,k}$ and the pertinent systematic innovations $F_1, F_2, \ldots, F_k$.

Since the relevant factors are often not known in detail, a *one-factor model* is popular in actual practice. Normally, the one factor used is an index of returns of securities or shares such as the Dow Jones, FTSE 100, Euro-Stoxx-50, or the S&P 500. However, these indices approximately mirror the return of the market portfolio r_M. Hence, the factor model of Eq. (5.25) simplifies to

$$r_i = ER_i + \beta_{i,1}F_1 + \varepsilon_i = Er_i + \beta_{i,1}r_M + \varepsilon_i. \tag{5.26}$$

The return r_i is related linearly to r_M, in a way similar to Eq. (5.17) which was used to derive the CAPM. Since the expected value of r_M is not zero, contrary to F_1, r_i has to be determined anew. To this end, the risk-free interest rate r_f is introduced. Eventually, the influence of a single security with its firm-specific risk (ε_i) disappears in a large, diversified portfolio owing to the law of large numbers. These considerations may be sufficient to see that under certain conditions the APT coincides with the CAPM [see, e.g., Copeland et al. (2004, Ch. 6L)]. In this event, $Er_i = r_f + \beta_{i,1}(Er_M - r_f)$, i.e., the expected rate of return of a security is equal to the risk-free interest rate plus a risk premium. The risk premium in turn is determined by the beta of the security and the excess of the expected return on the capital market over the risk-free rate.

▶ **Conclusion 5.6** The Arbitrage Pricing Theory extends the CAPM by attributing the beta of a security to factors leading to similar deviations from expected return both in this security and in the remainder of the market portfolio.

5.2 Risk Management, Forward Contracts, Futures, and Options

5.2.1 Hedging Through Forward Contracts and Options

While the focus in Sect. 5.1 was on risk management through diversification using ordinary securities like shares and bonds, it now shifts to so-called *derivatives* (forward contracts, futures, and options). Actors on the capital and foreign exchange markets may differ in their ability to bear the residual risk that is associated even with a diversified portfolio. To them, trading in derivatives provides an opportunity to transfer a risk to someone else. From the point of view of shareholders, buying a derivative amounts to an additional instrument of their own risk management, which may substitute for the risk management performed by the firm. Ultimately, trading in derivatives constitutes yet another alternative to the purchase of insurance coverage by the firm.

For several decades, the volume of trade in derivative instruments has been increasing fast. By now, trade in derivatives (interest futures, interest options, exchange futures, exchange options, share index futures, and share index options) is more important than trade in the so-called underlyings. In addition, there is a growing trade in over-the-counter (OTC) instruments.

Forward contracts and options are by no means new instruments. During the tulip crisis in the seventeenth century, there was considerable trade in options. However, modern business took off in the 1970s. During the 1950s and 1960s, the Bretton Woods exchange rate system with its fixed parities in combination with largely steady economic development had brought calm to the financial markets. This picture changed during the 1970s with the breakdown of the fixed exchange rate system of Bretton Woods. Since then, there has been almost free trade of the major currencies. The dramatic increase in oil prices in 1973/74 and the concomitant flows of capital

5.2 Risk Management, Forward Contracts, Futures, and Options

from the oil-producing countries as well as the economic crisis in industrialized countries paved the way to ever-increasing exchange rate fluctuations and hence the desire to hedge against exchange rate risk.

The motives for using derivatives are *arbitrage*, *hedging*, and *speculation*. In the following, the speculation motive will be disregarded. Arbitrage exploits differences in price between markets to make a profit. Prices of securities and exchange rates differ between stock exchanges worldwide. Arbitrageurs buy where the security is cheaper and sell where it is more expensive. Therefore, their activity is not related to risk transfer and risk management. This is not true for the hedging motive. Therefore, hedging is illustrated by an example.

Example 5.3

Hedging against exchange rate risk
Pursuant a delivery of goods, European firm F stands to receive payment in US$ in 90 days. However, the exchange rate in 90 days may deviate from the exchange rate of today. If the US$ rate (defined as the ratio of X Euro to 1 US$) increases from, e.g., 1.35 Euro to 1.40 Euro per US$, this results in a gain for F. If it decreases from 1.35 Euro to, e.g., 1.30 Euro, this is a loss for F [see panel A of Fig. 5.9]. To eliminate this risk of deviation from the current US$ rate, F can sell its dollars forward. Neglecting any transaction cost and assuming that the forward rate coincides with the current exchange rate, one obtains the payout profile of the forward contract as shown in panel B of Fig. 5.9. If the US$ spot rate of 90 days hence should be 1.40 Euro, A will have to give the dollars away too cheap, making a loss. With an exchange rate of 1.30 Euro/US$, F makes a profit. The combination of both transactions results in the perfect level profile of panel C of Fig. 5.9, indicating total independence from the spot rate prevailing in 90 days. Therefore, the combined position is risk-free. This "limiting" (in the extreme, elimination) of the risk is called "hedging". In this example, the exchange rate risk of an export business is hedged. To the extent that the firm could purchase export insurance, the capital market with its forward contracts again offers a substitute. ∎

Hedging, however, is not without cost. Hedging cost depends on the maturity of the contract and hence the difference between the domestic and foreign interest rate.

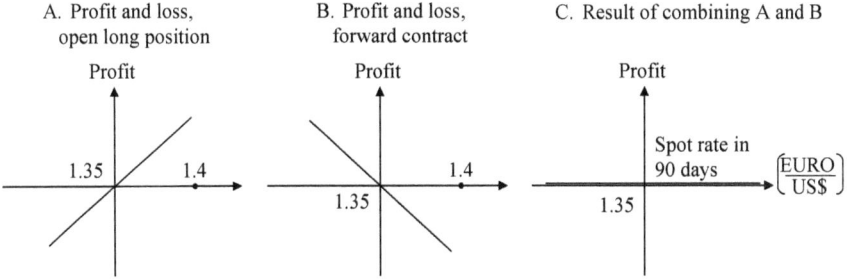

Fig. 5.9 Hedging through forward contract

Since the supplier ("writer") of the forward contract forgoes the opportunity of an investment in a US$-denominated security but obtains the opportunity of investing in a Euro-denominated one. In particular, the forward rate is lower than the spot rate (the current rate), if the rate of interest is lower than at home. In addition, hedging cost increases with the volatility of the exchange rate in question.

Note that forward contracts can be struck over other underlyings in addition to currencies. In 1848, American businessmen founded the Chicago Board of Trade (CBOT), where trade in standardized forward contracts (so-called *futures*) with agricultural produce and raw materials takes place. Oscillations of wheat and meat prices caused by unpredictable changes in weather make trade in futures profitable for both farmers and their customers. Because of the increasing dynamics of interest rates and exchange rates (and hence of security prices), participants in financial markets were faced with risks not known hitherto. Trade in forward contracts and futures involving financial capital, fixed-interest securities, shares, as well as stock and bond indices led to the creation of the International Money Market of the Chicago mercantile exchange (CME). Since then, forward stock exchanges were founded in financial centers around the world.

In addition to simple forward contracts, there exist also *swaps*. Here, the *quid pro quo* of the contract involves not only one but several future dates. Swaps can, therefore, be considered as a portfolio of forward contracts. From a risk management perspective, an interest rate swap is of particular interest. An interest rate swap amounts to an exchange of a loan with a variable interest ("floating leg") against one bill with a fixed interest ("fixed leg"). For example, enterprise B may have taken out a loan with a variable interest rate for six years. Hence, it must pay yearly interest of an unknown amount. If B anticipates a rising level of interest rates, it can transform this risky liability into a risk-free one using an interest rate swap. In effect, B exchanges its obligation to pay variable (and maybe increasing) interest for an obligation to pay a fixed interest, hedging its interest risk. Once again, B would rely on the capital market rather than insurance for its risk management.

Finally, an *option contract* is a special form of a forward contract. The buyer of an option has the right but not the obligation to ask from its seller (the "writer") a contractually defined service under conditions that are contractually fixed in advance as well. The object ("underlying") of options can be shares, bonds, loans, goods, currencies, stock indices, futures, etc. In Table 5.3, the rights and obligations are shown in connection with stock options (to be discussed in detail in Sect. 5.2.2 below).

The value of an option consists of an intrinsic value and a time value. Its *intrinsic value* reflects the advantage of exercising the option compared to performing the transaction on the spot market. For instance, if the agreed price of purchase ("strike price") of a call option on a stock is 50 monetary units (MU) while the stock has a spot price of 58, the intrinsic value is 8 MU per unit. If the spot price does not exceed the strike price, the intrinsic value of a call option is zero, implying that it will not be exercised. Note that sometimes the difference between the forward price and the strike price is also called intrinsic value.

5.2 Risk Management, Forward Contracts, Futures, and Options

Table 5.3 Call and put options

Type	Buyer		Seller	
	Right	Obligation	Right	Obligation
Call	Purchase of security at the strike price	Payment of option premium	Receipt of option premium	Sale of security at the strike price upon request by buyers
Put	Sale of security at the strike price	Payment of option premium	Receipt of option premium	Purchase of security at the strike price upon request by seller

The *time value* of an option is the difference between its price and its intrinsic value. It reflects the advantage of not having to perform the transaction right away but to be able to defer it. The time value of an option increases with the volatility of the underlying because the higher the volatility, the greater is the chance of the share price exceeding the strike price sometime before maturity, resulting in a gain in the case of a call option. The longer the time to maturity, the greater this chance, too. The possibility of a loss is not taken into account because in that event the option will not be exercised.

5.2.2 Hedging Through Stock Options

Among the alternatives to the purchase of insurance by firms provided by the capital market, stock options are of particular interest because they relate to the risk management of the insurance company (IC), a topic that will be taken up in Sect. 7.2. Indeed, shareholders of an IC will be shown to hold both a call and a put option on the value of the IC. This has implications for the financial risk management of an IC in the interest of its owners, motivating special emphasis on stock options.

In the case of stock options, the *underlying asset* ("underlying") is a share. A typical *call option* gives its holders the right to buy, e.g., 100 shares of the firm XY at (European option) or before (American option) the due date T (expiration date) at the strike price of 50 MU. Its value depends on the probability with which the price of the share at or before day T exceeds 50 MU. This probabilistic aspect will be addressed below. However, the hedging of options can be illustrated by simply examining conditional values at maturity.

Therefore, let S_T be the (unknown) market price of the underlying (the share) at date T, while the strike price is 50 MU. If the market price on this day is higher than the strike price, then the option is simply worth the difference $(S_T - 50MU > 0)$. Then the call is "in the money". If $S_T < 50$, which is possible, then the call is "out of the money". Its holder will not exercise the option in that event.

Figure 5.10 shows the payout profiles. As long as $S_T < 50$ MU, the call is out of the money and hence without value (panel A). If $S_T > 50$ MU, then the call is in the money, with its value increasing in step with the market price of the share. Note that

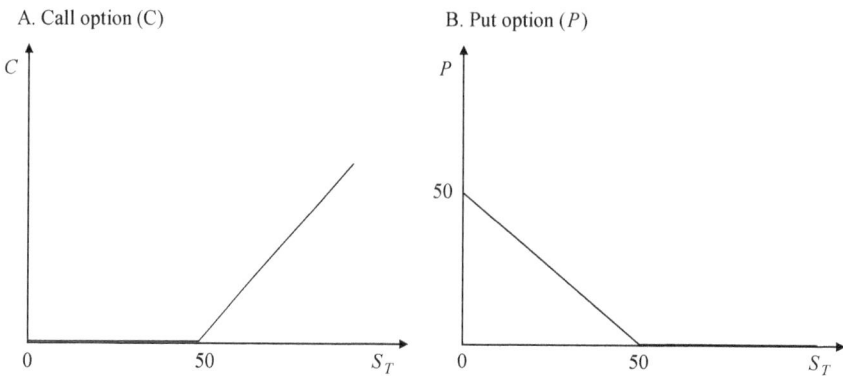

Fig. 5.10 Value of options at exercise time (strike price 50 MU)

at maturity, the option premium paid constitutes a fixed cost which is irrelevant to the value of the option. Therefore, at time T, a call option cannot have a negative value; it is an instrument with *limited liability*. For this reason, volatility of the underlying (measured by the variance or standard deviation of its market price) contributes to the value of the call option.

A put option constitutes the opposite of a call option (panel B). Whereas the call option gives its holder the right to buy the underlying at a predetermined price, the put option gives its holder the right to sell it at a previously fixed price. If the market price S_T of the underlying exceeds the strike price at time T, the put option has zero value hence will not be exercised. If, however, $S_T < 50$ MU, the put is in the money. It pays to buy the contracted number of shares at the price S_T and to sell them at the strike price. Again, the put option never has a negative value; therefore, its value also increases with increasing volatility of the underlying.

Put and call options can be combined to form complex option contracts. In particular, one option may offset the risk still inherent in the other, resulting in a risk-free outcome equivalent to insurance.

Example 5.4

Combination of a call and a put option
Let the strike price of a call option as well as a put option on the shares of firm XY be 55 MU. The price of the security today is 44 MU, at the expiration date (one month hence, so accrual of interest can be neglected), its price will be either 58 MU or 34 MU. Suppose the following strategy is chosen: Buy the security, buy the put, and sell the call. Payouts at the expiration date are shown in Table 5.4.
If the share price rises to 58, the call option is in the money. With regret, the investor has to deliver the shares at the price of 55 MU, realizing a loss of 3 MU. The put option has zero value. The total value of all transactions, thus, amounts to 55 MU. If the price of the security falls to 34 MU, then the put is in the money. Exercising it, the investor reaps a gain of 21 MU. The call option has zero value and is forfeited. The total value of all transactions amounts again to 55 ($= 34 + 21$)

5.2 Risk Management, Forward Contracts, Futures, and Options

Table 5.4 Payouts of a combination of options at expiration date

Original transaction	Price of the security increases to 50 MU	Price of the security falls to 34 MU
Buy a security	58 MU	34 MU
Buy a put	0 MU (forfeited)	21 MU (=55 − 34)
Sell a call	−3 MU (=55 − 58)	0 MU (forfeited)
Total value	55 MU	55 MU

Source Buckley et al. (1998, 405 f.)

MU. Thus, this strategy is risk-free. Also, note that the current share price of 44 MU does not enter calculations at all; what counts is but the price at time T. ∎

However, this risk transfer is not without cost. The purchase of an option calls for payment of the option premium (see Table 5.3 again). Example 5.4 makes clear that in general it is possible to combine the underlying asset (S_0), with the sale of a call option (C_0) and the purchase of a put option (P_0) to realize a risk-free investment. The subscripts indicate that the time these transactions are performed is one period before maturity T, calling for accrual of interest at the risk-free rate r_f. This rate is the appropriate one because there is a risk-free alternative, with a value equal to the strike price K. Since there must be equivalence between the two alternatives (otherwise there would be scope for arbitrage), one has

$$(S_0 - C_0 + P_0)(1 + r_f) = K. \tag{5.27}$$

From this, the difference between the two options can be derived,

$$C_0 - P_0 = \frac{S_0(1 + r_f) - K}{1 + r_f}. \tag{5.28}$$

This difference is called *put–call parity* [see Copeland et al. (2004, Sect. 8E)]. Since the variables on the right-hand side of Eq. (5.28) are known, it is easy to calculate the value of a call from the value of a put (and vice versa).

In the following, the discussion is limited to the so-called European option which permits holders to exercise only at a specified time but not before. It is analytically more tractable than the American option that can be exercised at any time up to expiration.

The value of a European call option depends on five fundamental determinants. They are listed in Table 5.5.

Table 5.5 Determinants of the value of a stock option

Determinant	Symbol	Call option	Put option
Value of the underlying (share price)	S_0	+	−
Strike price	K	−	+
Volatility	σ	+	+
Time to expiration	t	+	+
Interest rate	r_f	+	−

- Current share price or value of the underlying S_0 (because a higher current price increases the chance of the price exceeding the strike price at maturity);
- Strike price K (the higher, the smaller the value, for the same reason);
- Standard deviation or volatility of the share price σ (the higher, the more valuable is the call option; for an explanation, see below);
- Time to maturity t (more time increases the chance of the share price exceeding the strike price at expiration time); and
- The risk-free interest rate r_f (the higher, the more valuable is the call option, because payment for the security lies in the future, permitting investment in risk-free assets meanwhile; alternatively, the higher the interest rate, the lower the present value of the strike price).

The value of a put option depends on the same determinants, but not always in the same way.

- Current share price or value of the underlying S_0 (the higher, the lower the value because a higher current price decreases the chance of the price being below the strike price at maturity);
- Strike price K (the higher, the higher the value because this increases this chance);
- Volatility σ (the higher, the more valuable is the put option; for an explanation, see below);
- Time to expiration t (more time increases the chance of the share price being below the strike price at maturity); and
- Interest rate r_f (because the revenue from the sale of the shares lies in the future so needs to be discounted to present value; alternatively, the higher the interest rate, the lower the present value of the strike price).

At first sight, it may seem puzzling that according to Table 5.5, increased volatility in the value of the underlying causes the value of an option to increase. Figure 5.11 offers an explanation. In panel A, the payoff profile of a call option is complemented by two densities indicating the predicted distribution of share prices at the time of maturity. In the case of density $f(S_T)$, expected future share prices are tightly bunched around their expected value. Usually, this is a share whose price has not varied very much in the past. Note that there is limited probability mass to the right

5.2 Risk Management, Forward Contracts, Futures, and Options

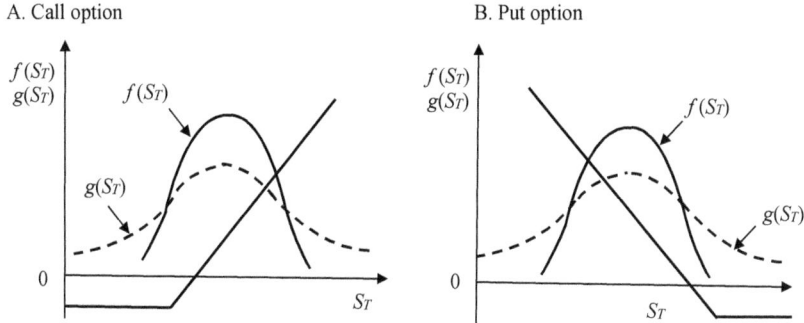

Fig. 5.11 The effect of volatility on the value of an option

where the call option is in the money, with payoff continuously increasing with S_T. Conversely, there is also limited probability mass to the left; however, it relates merely to the option premium to be paid, which is a constant rather than increasing.

In comparison, the density $g(S_T)$ displays higher volatility while having the same expected future value as $f(S_T)$. Of course, this means an increased downside probability mass; however, it again relates to the constant option premium. In contrast, the increased upside probability mass relates to the ever-increasing payoff of the call option once it is in the money. On balance, the value of the call option turns out to be higher for a share with substantial volatility than one with little volatility.

Panel B shows the case of a put option. Here, the investor has to pay the option premium if the share has a higher value at maturity than predicted, causing the option to be out of the money. Conversely, the payoff of a put option is higher, the lower the share price at maturity S_T. Given the "bunched" density $f(S_T)$, the risk of the option being out of the money is rather low; however, the investor only has to pay the constant option premium. Conversely, the chance of the put option being strongly in the money is rather low, too. Turning to the density $g(S_T)$ with its higher volatility, one sees that the higher downside probability mass (to the right this time) again relates to the constant option premium, while the higher upside probability mass relates to an ever-increasing payoff. Therefore, high volatility in the value of the underlying is associated with a high value of the put option as well.

Table 5.5 summarizes these relationships.

All these determinants are included in the famous *Black–Scholes formula* for the valuation of an option [see Hull (2018), Ch. 15]. In the example above, the share price at maturity took on only two possible values. It, therefore, followed a binomial distribution. This is a sufficient approximation for a short time to maturity. With a longer time, one must take into account that with many realizations the binomial distribution converges to the normal distribution. Therefore, the normal distribution plays a key role in the Black–Scholes formula,

$$C_0 = S_0 \cdot N[d_1] - K e^{-r_f t} N[d_2], \text{ where}$$

$$d_1 = \frac{\ln(S_0/K) + \left(r_f + \tfrac{1}{2}\sigma^2\right) t}{\sigma \sqrt{t}};$$

$$d_2 = \frac{\ln(S_0/K) + \left(r_f - \tfrac{1}{2}\sigma^2\right) t}{\sigma \sqrt{t}} = d_1 - \sqrt{\sigma^2 t}. \tag{5.29}$$

The variables are defined as follows: $N[d] :=$ probability that a standardized normally distributed random variable is smaller or equal to d; $S_0 :=$ current share price; $K :=$ strike price; $\sigma :=$ annualized standard deviation (volatility); $r_f :=$ risk-free interest rate p.a.; $t :=$ time (in years) to maturity.

Using the put–call parity [see (5.28)], it is possible to calculate the value of a (European) put option

$$P_0 = K e^{-r_f t} N[-d_2] - S_0 N[-d_1]. \tag{5.30}$$

The advantage of these fundamental equations is that four of the five parameters are observable: S_0, K, r_f, and t. Only one parameter, the variance of the returns σ^2, must be estimated. Note that d_1 and d_2 determine only a probability density; therefore, C_0 and P_0 must be interpreted as expected values of a random variable.

Equations (5.29) and (5.30) in fact postulate the equivalence of a random variable and a fixed quantity K without any adjustment for risk aversion (which would result in a certainty equivalent; see Sect. 2.3.1). This means that investors are assumed to be risk-neutral.

Quite generally, the assumptions underlying these formulae are very strong:

- There are no limits to short selling;
- Transaction costs and taxes are neglected;
- The option is of the European type;
- The security does not pay dividends;
- Development of the share price is continuous, and there are no jumps;
- Security prices are log-normally distributed;
- The market operates continuously; and
- The (risk-free) interest rate is known and constant.

For further discussion, see, e.g., Hull (2018).

▶ **Conclusion 5.7** According to the Black–Scholes formula of option price theory, the value of an option depends on five determinants, four of which (value of the underlying, strike price, time to expiration, and risk-free interest rate) are directly observable. Only the volatility of the underlying must be estimated; it contributes positively to the value of the option.

5.3 Corporate Demand for Insurance

5.3.1 Demand for Insurance in the Light of Capital Market Theory

When reviewing Sects. 5.1 and 5.2, readers will notice a stark contrast with the theory presented in Chap. 3. There, the difference between wealth levels (and hence, volatility of wealth) in combination with risk aversion was a crucial driver of the demand for insurance (see Figs. 3.2 and 3.6 again). Here, risk aversion does not play a role in the CAPM nor Option Pricing Theory; according to the latter, volatility even adds to the value of an option (see Table 5.5).

However, the contradictions are more apparent than real. As to volatility in wealth, it motivates both purchasers of insurance and of capital market instruments to consider paying a price (the insurance premium and the option premium, respectively) for transferring a risk. In both cases, the ideal outcome is full risk transfer, as stated in Conclusion 3.2 and illustrated in Example 5.4. As to risk aversion, it in fact ceases to be relevant if the insurance buyer buys full coverage in response to a marginally fair premium without an excessive fixed loading, as stated in Eq. (3.30). In the CAPM formula (5.23), investors' risk aversion does not figure explicitly; yet it is ultimately their risk aversion that causes them to demand a higher expected return on a share with a high beta, indicating high volatility of returns relative to that on the capital market in general. Finally, the Black–Scholes formula for option pricing is derived using an "objective" probability density function over future returns that does not reflect investors' risk aversion. However, investors act based on their subjective estimates of this density, which reflect their degree of risk aversion in the following way: the more marked risk aversion, the greater the probability mass in the tails of the estimated distribution. As shown, e.g., by Jackwerth (2000), the divergence between the theoretical, risk-free, and the observed densities of option quotes can be used to infer investors' degree of absolute risk aversion.

Still, it is important to distinguish between the enterprise and its shareholders as its owners. Although possible losses ranging from damage of property to liability claims may constitute important risks to the enterprise, they are of little or no concern to shareholders as the owners of the enterprise because they can be diversified away easily by their personal portfolio strategy. In Sect. 5.1.3, the variation in the firm-specific rate of return was divided into two components, the systematic and the idiosyncratic risk [see Eq. (5.20)]. The systematic risk (determined by β_i) of the firm cannot be removed by diversification; insofar as shareholders have an interest in the *reduction of the firm-specific beta*.

The purchase of insurance coverage by the firm serves to protect its assets against damages and claims that may arise notably from liability. It, therefore, can reduce the volatility of returns and decrease the specific β_i of the firm.[4] Shareholders again

---

[4]Since the variance of capital market returns cannot be influenced, β_i can only be lowered by lowering $Cov(r_i, r_M)$ according to Eq. (5.18). The covariance formula of Eq. (5.12) shows that this can be achieved by reducing the deviations of r_i from its expected value in a symmetric way (i.e., a reduction of its variance). However, volatility could also be diminished in an asymmetric way such

reap a gain from diversification because insurance distributes the risk over all policyholders of the insurance company (IC). Hence "diversification by oneself" (through one's individual portfolio of securities) and "diversification by insurance" (through its purchase by the enterprise) can lead to the same result and can be considered substitutes. For the investor, the purchase of insurance by the firm is beneficial if

- insurance *reduces* the amount of *systematic risk* of the firm's assets, i.e., its specific β_i;
- the *relative transaction costs* of insurance are *lower* than those of the individual portfolio strategy; or
- *other costs* are *reduced* by insurance, at least in expected value.

These possibilities are discussed in turn below.

5.3.1.1 Reduction of Systematic Risk Through Corporate Insurance

There is almost no direct evidence concerning the question of whether the purchase of insurance reduces the systematic risk of the firm. The reason is that enterprises do not single out insurance benefits received in their profit and loss statements. However, insured losses are reflected in the underwriting results of commercial business, which are published by the IC.[5] American data give rise to the conjecture that the β values of underwriting returns are slightly negative (see Sect. 5.3.3 below and Example 7.5 in Sect. 7.2.2). Denote this beta by $\beta_u = -0.1$. For the IC, the underwriting result has an almost perfect negative correlation with losses paid (which usually are not evidenced separately but only in combination with settlement expenses; see Sect. 6.1.2). Therefore, one can conclude that the beta pertaining to losses paid is slightly positive, e.g., $\beta_L = +0.1$ (to see this, replace r_B in panel B of Fig. 5.5 in Sect. 5.1.3 by $-r_B$). For the enterprise considering the purchase of insurance, this means that its (insured) losses have a tendency of being somewhat higher if returns on the capital market are high. Prior to insurance coverage, this positive correlation would be more pronounced because insured losses account only for part of effective damages. Hence, insurance has the potential of reducing the specific β_i of the firm, which is in the interest of its shareholders. In view of the CAPM formula of Eq. (5.24), one can say that the purchase of insurance coverage reduces the cost of capital to the corporation.

This argumentation is hardly applicable to firms in family ownership or, more generally, to firms that have only a few owners. Ownership of the few goes along with interest not only in value maximization but the pursuit of other aims as well such as maintenance of control over the enterprise. Furthermore, family members

---

that small values of r_i increasingly coincide with small values of r_M, causing covariance in fact to increase. Therefore, a reduction in the volatility of firm-specific returns can (but need not) result in a decrease of β_i.

[5]The distinction between risk underwriting and capital investment as the two core activities of an IC will be taken up in Sect. 7.2.1 below.

often work in the firm, causing their interest as owners to be aligned with those of an employee who values a safe job. Owners of this type are concerned about total risk, not only systematic risk. Therefore, the corporate demand for insurance can be modeled as that of (private) households in this case. Risk aversion of the owners can be considered as an important determinant of the corporate insurance demand of such a firm.

Even in the case of widely spread ownership (usually stock companies), a small holding of stock may constitute a substantial part of individual shareholders' wealth. The value of their private β can be reduced considerably by the purchase of corporate insurance.

5.3.1.2 Low Relative Transaction Cost of Insurance

The question concerning the transaction costs of insurance relative to those on the capital market can again be answered in general terms only. The transaction costs of insurance correspond to the loading contained in the premium, i.e., the surcharge over and above expected loss. This surcharge comprises administration and sales expense as well as a charge for risk bearing. The transaction costs of structuring a portfolio in the aim of diversification consist of the commissions for traders and banks, which depend in turn on the size of the market and the frequency of trading of the specific share or security in question. Hence, one can argue that diversification through insurance is relatively advantageous in terms of transaction cost if stock ownership is concentrated (resulting in a thin market) or if the shares are not or only rarely traded.

Furthermore, insurance coverage reduces the transaction cost associated with risk assessment, claims settlement, and presenting (insured) claims against third parties. Here, ICs have a comparative advantage. A good example is product liability cases where often unpleasant negotiations and settlement with claimants can be left to the IC.

Throughout, the critical relation remains the one between the relative effectiveness of insurance in reducing the risk of returns and its relative cost. Both effects operate through the cost of capital to the firm. But there are other costs that are of importance to the owners of the enterprise as well.

5.3.1.3 Other Advantages of Insurance

Up to this point, the discussion has been mainly in terms of the CAPM. However, the CAPM assumes a perfect capital market whereas capital market imperfections may lead to differences between the costs of raising capital internally and externally, to asymmetric information between contractual partners, and to bankruptcy costs. In this connection, Mayers and Smith (1988) discuss several reasons why firms may purchase insurance:

1. *Reduced probability of crowding out of owners.* The claims to the firm's assets held by its owners compete with those of other stakeholders such as employees, suppliers, customers, and injured parties in cases of third-party liability. These

claims can crowd out those of the owners. Corporate insurance coverage serves to lower the probability of crowding out. The larger the potential claims against the firm held by employees, customers, and suppliers, the higher will be pressure on management to buy insurance coverage in the interest of shareholders.

2. *Reduced probability of bankruptcy and associated cost.* Recall that ownership by the few usually goes along with a concern for total risk, which entails an interest in avoiding bankruptcy of that particular enterprise. Owners of this type face bankruptcy costs in the guise of a loss of reputation and control over a productive team (the cost to management will be discussed in Sect. 5.3.2.1 in the context of an imperfect agency). One way to lower the probability of bankruptcy is internal diversification, i.e., avoiding concentration of assets (as in Example 5.1 of Sect. 5.1.1, dealing with a fire risk). However, internal diversification may require physical separation of activities, which constitutes a very costly alternative in many cases. Therefore, the propensity to demand corporate insurance will be higher, the higher are the costs of possible bankruptcy, and the higher the geographic concentration of assets. Doherty (1985, 273) also cites possible disruption of business as a follow-up to a primary damage (e.g., fire). This constitutes an extra drain of funds making it difficult to finance remedial measures (e.g., rental of a building). In such a situation, insurance may actually avert bankruptcy.

3. *Protection of creditors against opportunistic behavior of owners.* As will be shown in Sect. 7.2.3, a share has the characteristics of a call option because its owners cannot lose more than its current value whereas they benefit without limit when the value of the firm increases. This creates a conflict of interest with the suppliers of credit (banks and bondholders in particular), who do not participate in increases of the firm's value but may lose their claim in part or entirely in the event of bankruptcy. Lucrative but risky projects are, therefore, in the interest of shareholders but not creditors. To protect creditors from this effect, loans often have stipulations obliging the firm to purchase different types of insurance.

4. *Reducing the tax burden of the firm.* If the marginal tax rate on internally generated funds exceeds that for external financing, there is an incentive in the case of a loss not to pursue self-financing (e.g., by liquidating reserves), but to revert to external financing by insurance. The purchase of insurance coverage is also advantageous when taxes are progressive in nature, income losses can be carried forward (or backward), or losses carried forward do not serve as a tax shield while loss reserves in the guise of insurance benefits are tax-privileged (indeed, the latter two categories have the effect of creating a more progressive tax structure). A simple analogy can be made here by considering the inverse of risk aversion. It was shown that insurance is advantageous for individuals *maximizing* a *concave* utility function. In the same way, insurance can be advantageous for individuals (or firms) *minimizing* a *convex* (e.g., progressive) tax burden.

5. *The existence of mandatory insurance.* The firm can potentially inflict damage on a third party; for internalizing these negative external effects, mandatory insurance is often required (e.g., by a regulator) because it also provides the injured party with compensation.

This multitude of arguments generally explains why firms with certain characteristics buy more insurance than others. But the widespread purchase of voluntary corporate insurance cannot be solely justified with them.

5.3.2 Reasons for Corporate Demand for Insurance Not Related to the Capital Market

5.3.2.1 Imperfect Agency

For many decisions, people use the advice of experts, and sometimes they delegate decision-making authority to them. In the same vein, owners of firms delegate competencies to management. Managers are hired to act on behalf of shareholders; however, their informational advantage provides them with leeway to pursue their own interests. Interests of management do differ from those of shareholders. When it comes to decisions involving risk, an important reason is that much of a manager's wealth consists of human capital that is tied to the firm they work for. This causes their portfolio to be much less diversified than that of most shareholders. This configuration has spawned principal–agent theory [see, e.g., Levinthal (1988)] which revolves around ways to pay management that induce an alignment of incentives with those of the principal (the shareholders in the present context; for an application to distribution systems in insurance, see Sect. 6.4.2). However, it may be less costly for the owners to mitigate the asymmetry of information by checking and exerting control. Yet exerting control comes at a cost, too, leaving room for imperfect agency. With regard to risk management, managers can benefit from this imperfection by purchasing corporate insurance to secure their own wealth at the expense of shareholders.

However, as argued by De Alessi (1987, 429), this explanation of corporate demand for insurance is unsatisfactory not only theoretically but also practically. In a competitive environment, possibilities of managers to shirk are limited by competition on output markets, for management positions, and control over the firm. Furthermore, choice of payment (e.g., profit sharing) can be used to rein in opportunistic behavior. Practical considerations also lead shareholders to let managers purchase an excessive amount of insurance coverage because this has a small impact on expected profits and the value of the firm compared to the cost of perfecting agency.

On the other hand, the existence of a binding profit constraint can result in excess corporate demand for insurance. Typically, public regulation imposes a constraint of this type, which reduces the benefits shareholders can reap from controlling management. Managers in turn have increased scope for opportunistic behavior. This argument predicts that regulated firms (e.g., public utilities) purchase more insurance than comparable non-regulated ones. But it can hardly explain the widespread demand for insurance also by firms who are not subjected to a binding profit constraint.

5.3.2.2 Sunk Costs

Enterprises normally represent a considerable amount of wealth in the form of assets which lead to sunk costs, such as specialized know-how and patent rights. These assets have in common that when sold they would fetch a price that is much lower than their value to the firm (i.e., their contribution to profit). For shareholders as the owners of these assets, it is impossible to distribute sunk costs over many firms in an attempt at risk diversification. Being subject to competing claims (e.g., from liability), shareholders also have an interest in purchasing insurance to protect their own claims. Note that this argument holds regardless of the degree of imperfection in the agency, i.e., the lack of control over managers.

Many types of assets lead to sunk costs. They not only comprise the teams at work but also, e.g., contracts that are signed to bind the team together and which govern its relations with third parties. These firm-specific values are called *quasi rents*, because the pertinent resources get paid more inside the firm than on the market. They differ from true rents, which are the consequence of barriers to competition [see De Alessi (1987)]. Events jeopardizing the existence of the firm mean the possible loss of all quasi rents, which were created by forming a successful team and its institutionalization. Bankruptcy is thus costly not so much because of the legal expenses associated with it but because it endangers the value of the team and of firm-specific assets in general.

On the other hand, sunk costs do not motivate corporate insurance demand for the interest of bondholders and creditors in addition to the considerations proffered in Sects. 5.3.1.3 and 5.3.3. Recall that the firm invests in assets with sunk costs since its owners expect to earn a much higher rate of return from their internal rather than external use. However, assets yielding higher returns in their present employment than in any other will not be liquidated by creditors in the event of bankruptcy; rather, they will be reshuffled with the aim of continuing business. Therefore, creditors do not stand to lose quasi rents in the event of bankruptcy. They do not have an interest in insurance coverage going beyond the one cited in Sect. 5.3.1.2 (preventing partial expropriation by shareholders).

5.3.2.3 Insurance-Specific Services

The last reason for corporate insurance demand revolves around services that ICs combine with their insurance product. Examples include services connected with loss prevention measures (such as risk and safety education) and with loss settlement (such as protection from and enforcement of liability claims, see Sect. 5.3.1.2). Obviously, these services could be purchased separately from insurance coverage. However, there are benefits of bundling (economies of scope), permitting these services to be offered at a lower cost when combined with insurance.

Prevention measures reduce the expected loss in the underwriting business of the IC. They usually derive from a risk assessment yielding information that can again be used for risk underwriting. In a similar fashion, the IC can use its own information gathering for an objective and efficient loss settlement process that has value for the insurance buyer.

▶ **Conclusion 5.8** Corporate demand for insurance can be explained using capital market theory insofar as it reduces the firm's beta through a positive correlation between paid losses and market returns, for which there is hardly empirical evidence, however. Insurance coverage can also reduce, through mitigating the risk of bankruptcy, costs that amount to lost quasi rents on firm-specific assets. Finally, insurance-specific services tied in with insurance coverage may be of value to customers.

5.3.3 Empirical Studies of Corporate Demand for Insurance

Despite the theoretical problems surrounding corporate demand for insurance and its importance in terms of premium volume, there have been few empirical investigations. The reason is that in corporate financial statements, insurance premiums paid are often summarized as "other expenses". Only one category of enterprise is legally obliged to detail its insurance coverage, namely the IC (and even then, insurers are only required to report their reinsurance usage). That is why Mayers and Smith (1988) analyze the demand for reinsurance by IC using the CAPM. They find clear statistical evidence to the effect that, e.g., an IC tightly held by few owners buys more reinsurance than others. Since several phenomena can be better explained by option pricing theory, this alternative will be emphasized in Sect. 6.7.3, dealing with the demand for reinsurance.

However, outside the insurance industry, the CAPM points to the following reason for the corporate demand for insurance. To the extent that losses impinging on enterprises tend to be high when capital market returns are low and low when capital market returns are high, they impart a positive covariance between firm-specific and capital market returns (a positive β_i). This constitutes a systematic risk that cannot be diversified away by investors but may be lowered by the purchase of corporate insurance. The crucial parameter, therefore, is the beta (in the following symbolized by β_L) resulting from the regression of the "return" (i.e., the relative change) of the firm's losses on the rate of return of the market portfolio. Even if these losses cannot be seen in the financial statements of the firms, they must appear—at least partially—in the books of the IC as paid claims.

This fact is exploited by Cummins and Harrington (1985). As the dependent variable, they use the rate of return from underwriting activity, relating it to the rate of return of the capital market (r_M). The corresponding beta is symbolized by β_u.[6] Based on a sample of 14 ICs and quarterly data, they found significant positive values for β_u. Since there is an almost perfect negative correlation between losses paid and the profitability of underwriting activity, this implies negative values for β_L. Thus, the percentage increase in losses tends to be high when capital market returns are low, thus accentuating the positive covariance between firm-specific rates of return and the capital market and motivating the purchase of insurance.

---

[6] This beta is the same as the β_u appearing in the insurance CAPM of Sect. 7.2.2 [see Eqs. (7.27) and (7.32)].

However, Davidson et al. (1992) argue that tests using the returns of underwriting activity may not be conclusive. Returns are calculated relative to premiums containing loadings of at least 20% for acquisition and administrative expense while profit imputation of these charges to the different lines of business is quite arbitrary.

The authors, therefore, favor two variables designed to reflect more accurately the (paid) losses of firms as purchasers of insurance:

1. *The quarterly change of net paid losses.* Here, a value of 0.05 means that the paid losses have increased by 5% compared to the previous quarter value. This can be interpreted from the viewpoint of the firm as a "rate of return on losses" amounting to 5%.
2. *The combined ratio.* The combined ratio is defined as the sum of paid losses including expenses for claims settlement and administration in relation to premiums written. A value of 1.05 is equivalent to a negative return on underwriting activity amounting to 5% for the IC whereas, for the insured firm, it constitutes a "return" of 5% net of premium paid.

In regression Eq. (5.31) below, these two quantities serve as indicators of the rate of return of losses $r_L(t)$:

$$r_L(t) = \alpha_K + \beta_{L,k} r_M(t-k) + \varepsilon(t), \quad k = 0, 1, \ldots, 4 \tag{5.31}$$

$r_L(t)$: return on losses in quarter t;
$r_M(t-k)$: rate of return of the market portfolio in the same quarter t ($k = 0$) or in one of the four previous quarters, respectively; and
$\varepsilon(t)$: normally distributed random variable with $E\varepsilon(t) = 0$ and $Var(\varepsilon(t)) = \sigma^2$ for all values of $r_M(t-k)$.

In this test of the relationship between r_L and r_M, lags of up to four quarters are introduced. In fact, portfolio diversification need not imply that an unexpected negative result of the firm considered should be balanced by a positive result of the market portfolio during the same quarter. Maybe investors are prepared to accept positive covariance during a short time period provided a negative correlation will result later. However, such reversals occur almost never beyond a lag of one quarter; therefore, estimates of $\beta_{L,k}$ are displayed for $k = 0, 1$ only in Table 5.6.

In Table 5.6, seven lines of commercial insurance are distinguished, ranging from fire to automobile. In column (1), estimated values of β_L are partly positive, partly negative but never significantly different from zero. Here, the relative change of (insured) losses of firms is related to the return of the market portfolio in the same quarter.

In column (2), the explanatory variable is the market return of the previous quarter. In two of the seven lines of business, the estimated value of β_L is significantly negative. However, regressions with time lags of two up to four quarters (not shown in Table 5.6) fail to confirm these two estimates. Since all the other coefficients

5.3 Corporate Demand for Insurance

Table 5.6 Estimated values of β_L, 1,800 American IC (1974–1986)

β_L estimated according to line of business (commercial)	Indicator of return on losses			
	Δ Losses paid		Combined ratio	
	$r_M(t)$	$r_M(t-1)$	$r_M(t)$	$r_M(t-1)$
	(1)	(2)	(3)	(4)
Fire	0.17	0.09	12.73	2.91
	(0.92)	(0.62)	(1.17)	(0.31)
Buildings (excluding fire)	−0.19	−2.01*	−5.83	12.24
	(−0.24)	(−2.85)	(−0.23)	(0.59)
Multiple peril	−0.08	−0.43	16.34	17.01
	(−0.41)	(−0.51)	(0.64)	(0.92)
Accident	−0.12	−0.25*	0.85	1.50
	(−1.23)	(1.99)	(0.08)	(0.14)
Liability (excluding auto)	0.04	0.31	37.33	48.32
	(0.19)	(0.94)	(1.23)	(1.75)
Auto liability	0.06	−0.05	26.82	24.30
	(0.35)	(−0.41)	(1.14)	(1.10)
Auto vehicle	−0.13	0.07	9.96	10.80
	(−1.28)	(0.31)	(0.71)	(0.98)

t-values in parentheses
*Statistical significance at the 5% level or better
Source Davidson et al. (1992)

continue to lack statistical significance, the two significant ones are best interpreted as a chance result.

In columns (3) and (4), regression Eq. (5.31) is estimated using the combined ratio as the second indicator of the rate of return associated with insured loss from the firm's point of view. Because of the different scaling of the dependent variable [with values around 1 rather than 0.05 to 0.1 in columns (1) and (2)], estimates of β_L are larger but again do not reach statistical significance in any one of the seven lines. The additional regressions with time lags up to four quarters (not shown) confirm this finding.

The study by Davidson et al. (1992) suggests that in the aggregate of American firms, returns of insured losses are uncorrelated with returns of the capital market. If, however, $\beta_{L,k} = 0$ is the best estimate, the CAPM cannot be used to explain corporate demand for insurance. It does not make sense for investors for a firm in their portfolio to insure possible damages that are not correlated with their remaining portfolio. The purchase of insurance coverage has no diversification effect in this situation, and it does not reduce the cost of capital for the enterprise.

▶ **Conclusion 5.9** Rates of return on insured losses appear to have a beta of zero for U.S. enterprises. Therefore, the CAPM cannot be used to explain the corporate demand for insurance, at least in the United States.

In view of their findings, the authors raise the question of whether there are other motives for corporate demand for insurance. They point to the conflict of interest between owners and creditors mentioned already in Sect. 5.3.1.3. By pursuing risky investment projects, owners can increase the value of their stock. Creditors do not share in the possible increase of the value of the firm while bearing an increased risk of losing at least part of their claims in the event of bankruptcy. One possibility to compensate them for this danger of quasi-expropriation is to offer a higher rate of interest on bonds; another, the purchase of insurance coverage. Creditors likely will prefer the insurance solution because the insurer also pays in case of bankruptcy of the firm, making funds available for covering their claims. By way of contrast, a bankrupt enterprise cannot fulfill the promise of paying higher interest.

These considerations point to the option pricing theory of Sect. 5.2.2. As in the case of an IC, the owners of a firm outside the insurance industry can raise the value of their call option. As explained in Sect. 7.2.3, creditors in fact supply a put option to the owner of the firm (who are relieved from any obligation to pay the net debt of their bankrupt firm). Any increase in the likelihood of bankruptcy serves to make this put option more valuable to the owners of the enterprise to the detriment of creditors, who can be said to be partially expropriated. Therefore, as soon as the risk of bankruptcy is considered, option pricing theory becomes the key building block for explaining the corporate demand for insurance in the light of capital market theory.[7]

Exercises

5.1

(a) Construct different portfolios from A, B, and C as given in Table 5.1. Display them in a (μ, σ)-diagram. Mix not only B and C, but also A and C.
(b) Calculate the portfolio with minimum variance.

5.2 Calculate the equilibrium returns of the following securities, assuming a risk-free rate of interest of 10% and an expected rate of return of the market portfolio of 0.18. What are the predicted adjustment processes if investors expect a rate of return of 0.17 for security C?

Security	Beta
A	1.5
B	1.0
C	0.5
D	0.0
E	−0.5

---

[7] The purchase of reinsurance coverage by the IC constitutes an application of option pricing theory to the insurance industry itself (see Sect. 6.7).

5.3 Demand for insurance comes from two main groups.

(a) Contrast the determinants of demand in the case of a typical household and a corporate enterprise.
(b) Consider the five enterprises listed in Exercise 4.2. For which one would you predict the greatest demand for insurance, *ceteris paribus*? Cite at least one *ceteris paribus* condition that might not hold true.
(c) "Buying insurance coverage is in the interest of shareholders". "Buying insurance coverage is in the interest of bondholders and creditors". Do these two statements contradict each other? Why (not)?
(d) A management consultant advises an IC to enhance demand for its products by offering "assistance in the event of loss". Is this a good idea?

References

Buckley, A., Ross, S.T., Westerfield, R., & Jaffe, J. (1998). *Corporate Finance Europe*. London: McGraw-Hill.
Copeland, T.E., Weston, J.F. & Shastri, K. (2004). *Financial Theory and Corporate Policy*, 4th ed. Reading MA: Pearson Addison Wesley.
Cummins, J.D., & Harrington, S.E. (1985). Property-liability insurance rate regulation: estimation of underwriting betas using quarterly profit data. *Journal of Risk and Insurance, 52*, 18–43.
Davidson, W.N.III, Cross, M.L., & Thornton, J.H. (1992). Corporate demand for insurance: some empirical and theoretical results. *Journal of Financial Services Research, 6*, 61–72.
De Alessi, L. (1987). Why corporations insure. *Economic Inquiry, XXV*, 429–438.
Doherty, N.A. (1985). *Corporate Risk Management, a Financial Exposition*. New York: McGraw-Hill.
Eckles, D.L., Hoyt, R.E., & Miller, S.M. (2014). The impact of enterprise risk management on the marginal cost of reducing risk: Evidence from the insurance industry. *Journal of Banking and Finance, 43*, 247-261.
Fama, E., & French, K.R. (1992). The cross-section of expected stock returns. *Journal of Finance, 47*(2), 427-465.
Fama, E., & French, K.R. (2015). A five-factor asset pricing model. *Journal of Financial Economics, 116*(1), 1-22.
Hull, J. (2018). *Options, Futures, and Other Derivative Securities*, 10th ed. New York NY: Pearson.
Jackwerth, J.C. (2000), Recovering risk aversion from option prices and realized returns. *Review of Financial Studies* 13(2), 433-451.
Levinthal, D. (1988). A survey of agency models of organizations. *Journal of Economic Behavior and Organization, 9*, 153–185.
Lintner, J. (1965). The valuation of risky assets and the selection of risky investments in stock portfolios and capital budgets. *Review of Economic Studies, 47*, 13–37.
Markowitz, H.M. (1952). Portfolio selection. *Journal of Finance, 7*, 77–91.
Markowitz, H.M. (1959). *Portfolio Selection - Efficient Diversification of Investments*. New Haven and London: Yale University Press.
Mayers, D., & Smith, C.W. (1988). Ownership structure across lines of property-liability insurance. *Journal of Law and Economics, 31*, 351–378.
Merton, R.C. (1972). An analytical derivation of the efficient portfolio frontier. *Journal of Financial and Quantitative Analysis, 7*(4), 1851–1872.
Mossin, J. (1966). Equilibrium in a capital asset market. *Econometrica, 34*, 768–783.

Roll, R. (1977). A critique of the asset pricing model theory's tests: part I. On past and potential testability of the theory. *Journal of Financial Economics, 4*, 129–176.

Sharpe, W.F. (1964). Capital asset prices: a theory of market equilibrium under risk. *Journal of Finance, 19*, 425–442.

The Insurance Company and Its Insurance Technology

6

Whereas Chaps. 3 and 5 revolve around demand for insurance, the focus of Chaps. 6 and 7 is on the insurance company (IC). Up to this point, the IC has been depicted as passive, its activity limited to charging a (fair) premium. However, an IC pursues objectives and has a host of instruments at its disposal for reaching them. The set of these instruments will be called insurance technology; it ranges from the design of products (for instance, exclusion of certain risks, "small print" in the contract) to providing services (advice regarding prevention, consumer accommodation, the settlement of claims) and on to the purchase of reinsurance and choice of strategy for capital investment.

Objectives and the means for attaining them are reflected in the financial statements published by the IC. Therefore, Sect. 6.1 contains a short introduction to the balance sheet and operational statement (also called income statement) of an IC. The concepts introduced there will be used repeatedly in the remainder of the chapter. In Sect. 6.2, the objectives of an IC are discussed. In particular, the question arises of whether the hypothesis of expected profit maximization is sufficient for describing the behavior of IC management or whether risk aversion must be accounted for. Although there is much to be said for taking risk aversion into account, the remainder of the chapter contains models that assume the management of an IC to be risk-neutral. A short survey of the instruments making up the insurance technology follows in Sect. 6.3, ordered according to the creation and termination of an insurance contract. The main stages are the acquisition through several channels of distribution, selection of risks offered for underwriting, control of moral hazard effects after the conclusion of the contract, purchase of reinsurance, and investment of surpluses and reserves on the capital market. The remainder of the chapter is devoted to these instruments. Distribution systems in insurance are compared in terms of their performance in Sect. 6.4; the optimal intensity of risk selection is studied in Sect. 6.5. In Sect. 6.6, the issue is the amount of control that should optimally be performed to limit the moral hazard exhibited by insurance buyers (IBs). This is followed by the primary insurer's demand for reinsurance (RI) in Sect. 6.7. Finally, Sect. 6.8 addresses issues revolving around the optimal investment policy of an IC, which could fill an entire textbook of its own. The one instrument that is not covered in this chapter is the pricing by the IC, a crucial aspect of the supply of insurance that deserves a separate discussion (Chap. 7).

6.1 Financial Statements of an Insurance Company

As with any business, a well-functioning insurer should maintain proper financial statements. Not only is this generally good practice, but financial statements are often required by various regulatory bodies. In the United States, for example, state-level regulators demand that most insurers file a comprehensive set of financial statements in a consistent way (usually on an annual basis). Further, any insurer that is publicly traded is required to provide financial statements to its various stakeholders (including owners). This section will explore some components of an insurer's financial statement and discuss at least one important nuance in reporting found in some countries (including the United States).

Before discussing details regarding the financial statements, one important observation regarding their use is in order. Financial statements submitted to regulators or to other stakeholders, by construction, represent a static picture. That is, while the insurer's business and financial position is changing in a dynamic way, the information gleaned from its financial statements constitutes a "snap shot" of its position at any one time. While this "snap shot" does provide information to the various stakeholders, it should be recognized that it is not a document that is updated in real time.

6.1.1 The Balance Sheet

The balance sheet is perhaps the most ubiquitous financial statement. Not only do businesses (including insurers) keep a balance sheet, indeed many individuals do as well. It simply summarizes the assets, liabilities, and equity value of a firm. For an insurer, the difference between assets and liabilities is often referred to as policyholder equity, capital, or surplus.

6.1 Financial Statements of an Insurance Company

		1 Assets	2 Nonadmitted Assets	3 Net Admitted Assets (Cols. 1 - 2)	4 Prior Year Net Admitted Assets
		Current Year	Current Year	Current Year	Prior Year
1.	Bonds (Schedule D)	49,150,589,953	0	49,150,589,953	57,792,865,347
2.	Stocks (Schedule D):				
	2.1 Preferred stocks	60,238,432	0	60,238,432	61,238,430
	2.2 Common stocks	1,646,846,098	33,027,310	1,613,818,788	1,252,354,303
3.	Mortgage loans on real estate (Schedule B):				
	3.1 First liens	5,715,358,788	0	5,715,358,788	4,999,790,414
	3.2 Other than first liens	9,885,150	0	9,885,150	12,010,987
4.	Real estate (Schedule A):				
	4.1 Properties occupied by the company (less $_____0 encumbrances)	0	0	0	0
	4.2 Properties held for the production of income (less $_____24,600,000 encumbrances)	11,158,154	0	11,158,154	9,658,609
	4.3 Properties held for sale (less $_____0 encumbrances)	0	0	0	0
5.	Cash ($_____(125,563,543), Schedule E - Part 1), cash equivalents ($_____249,777,427, Schedule E - Part 2) and short-term investments ($_____2,614,145, Schedule DA)	126,828,030	0	126,828,030	782,590,775
6.	Contract loans (including $_____0 premium notes)	0	0	0	0
7.	Derivatives (Schedule DB)	0	0	0	156,784,448
8.	Other invested assets (Schedule BA)	8,713,440,205	12,112,807	8,701,327,398	11,785,187,840
9.	Receivable for securities	211,541,395	0	211,541,395	165,708,953
10.	Securities lending reinvested collateral assets (Schedule DL)	0	0	0	0
11.	Aggregate write-ins for invested assets	0	0	0	0
12.	Subtotals, cash and invested assets (Lines 1 to 11)	65,645,886,206	45,140,117	65,600,746,089	77,018,190,106
13.	Title plants less $_____0 charged off (for Title insurers only)	0	0	0	0
14.	Investment income due and accrued	555,012,719	0	555,012,719	669,939,202
15.	Premiums and considerations:				
	15.1 Uncollected premiums and agents' balances in the course of collection	2,574,491,251	164,511,773	2,409,979,478	2,520,004,130
	15.2 Deferred premiums and agents' balances and installments booked but deferred and not yet due (including $_____0 earned but unbilled premiums)	1,053,558,394	2,766,852	1,050,791,542	1,193,372,324
	15.3 Accrued retrospective premiums ($_____1,621,212,479) and contracts subject to redetermination ($_____0)	1,657,141,776	35,929,297	1,621,212,479	1,701,197,232
16.	Reinsurance:				
	16.1 Amounts recoverable from reinsurers	775,475,116	0	775,475,116	794,371,417
	16.2 Funds held by or deposited with reinsured companies	620,248,042	0	620,248,042	491,632,615
	16.3 Other amounts receivable under reinsurance contracts	0	0	0	0
17.	Amounts receivable relating to uninsured plans	0	0	0	0
18.1	Current federal and foreign income tax recoverable and interest thereon	258,843,189	17,602,825	241,240,364	813,964,288
18.2	Net deferred tax asset	2,062,021,854	100,729,292	1,961,292,562	2,198,688,644
19.	Guaranty funds receivable or on deposit	24,654,617	0	24,654,617	23,068,726
20.	Electronic data processing equipment and software	103,814	341	103,473	276,889
21.	Furniture and equipment, including health care delivery assets ($_____0)	1,432,337	1,432,337	0	0
22.	Net adjustment in assets and liabilities due to foreign exchange rates	0	0	0	0
23.	Receivables from parent, subsidiaries and affiliates	53,342,057	13,321,403	40,020,654	436,985,876
24.	Health care ($_____0) and other amounts receivable	0	0	0	0
25.	Aggregate write-ins for other than invested assets	566,784,370	133,894,462	432,889,908	447,198,818
26.	Total assets excluding Separate Accounts, Segregated Accounts and Protected Cell Accounts (Lines 12 to 25)	75,848,995,742	515,328,699	75,333,667,043	88,308,910,267
27.	From Separate Accounts, Segregated Accounts and Protected Cell Accounts	0	0	0	0
28.	Total (Lines 26 and 27)	75,848,995,742	515,328,699	75,333,667,043	88,308,910,267

Fig. 6.1 Balance sheet (assets) of American International Group, 2017 (in US$, 000s omitted)

The asset side of a balance sheet shows the property owned by an insurer. Presumably, these assets are available for paying liabilities (e.g., claims). Of course, not all assets are liquid (meaning readily convertible to cash), causing the degree to which an asset can be used to pay claims to vary. The assets reported in 2017 by American International Group (AIG) to regulators are shown in Fig. 6.1. Notice that lines (1), (2), (3), and (5)–(7) of the statement of assets contain relatively liquid assets (mainly bonds, followed by loans, stocks, and some cash) amount to about US$56.4 billion (bn.), or 75% of total admitted assets. Also, note the US$75.3 bn. in total admitted assets (for a definition of admitted and nonadmitted assets, see below).

The liability side of the balance sheet represents the claims on the assets. The most important sources are policies that give rise to future claims ("Losses", line (1) of Fig. 6.2. Note that this suggests a precision of measurement that is far from realistic. Future claims can only be estimated on the basis of past experience; moreover, they need to be discounted to present value to make them comparable to other entries in the balance sheet [e.g., current surplus, line (37) of Fig. 6.2]. The choice of discount rate is a matter of continued debate; insurance practitioners typically prefer the internal rate of return (which makes the present value of future income and expenditure streams equal), whereas economists typically prefer the rate of return prevailing in the capital market adjusted for the IC's risk exposure [see the CAPM Eq. (5.24) again)]. In sum, the losses of US$35.5 bn. Shown on line (1) as well as the expected expenses US$4.2 bn. on line (3) are only rough estimates.

Further, AIG shows more than US$9.6 bn. shown in liability for premiums on policies written that have not been earned [line (9) of Fig. 6.2]. Under so-called accrual-based accounting, premiums on policies written are not "earned" if they can still be canceled on short notice. Therefore, part of the premiums written may have to be returned, constituting a liability. In sum, at least US$49.3 bn. of AIG's total liabilities of approximately US$54.2 bn. come directly from issuing insurance policies. This amounts to more than 90% of the total, exposing the IC to considerable risk. This can be seen by noting the change in line (1) from US$41.1 bn. to US$35.5 bn. within one year (imagine the reverse change unless matched by an increase in liquid assets).

The final primary data item found in the balance sheet is the difference between assets and liabilities. For insurers, this is referred to as capital, surplus, or even equity. This ambiguity of expression reflects another ongoing debate. While "surplus" suggests that policyholders can lay claim to these funds, "capital" and "equity" emphasize the rights of shareholders.[1] With admitted assets of US$75.3 bn. and liabilities of US$54.2 bn., AIG reports approximately US$21.1 bn., roughly 40% of its liabilities. This surplus is directly related to the financial health of the IC. The higher its surplus, the easier it will be for the IC to weather unexpected losses.

Though the example from AIG provides a specific insight into regulatory reporting in the United States, it is important to note that reporting standards are similar across the world. Common entries found on IC balance sheets are listed below.

1. *Debt securities.* Most of the investments contained in this category can be liquidated on short notice in order to finance loss payments. A significant share of this type of asset is typical of ICs outside the United States, whereas equities have more importance among U.S. insurers. The trade-off is between the high degree of liquidity and normally lower returns (see Sect. 6.8 for more details).

---

[1] Indeed, according to the option pricing theory discussed in Sect. 7.2.3, policyholders finance the put option held by the shareholders (who do not have to pay the debts of an IC; their shares cannot have a negative value). Arguably, this creates a claim against a joint part of the IC's accrued capital.

6.1 Financial Statements of an Insurance Company

		1 Current Year	2 Prior Year
1.	Losses (Part 2A, Line 35, Column 8)	35,477,244,285	41,117,561,678
2.	Reinsurance payable on paid losses and loss adjustment expenses (Schedule F, Part 1, Column 6)	659,751,325	717,510,696
3.	Loss adjustment expenses (Part 2A, Line 35, Column 9)	4,237,228,518	6,504,284,305
4.	Commissions payable, contingent commissions and other similar charges	(22,647,055)	(14,755,218)
5.	Other expenses (excluding taxes, licenses and fees)	55,806,642	64,934,125
6.	Taxes, licenses and fees (excluding federal and foreign income taxes)	439,053,573	603,673,032
7.1	Current federal and foreign income taxes (including $ _____530,242_____ on realized capital gains (losses))	4,951,942	7,045,385
7.2	Net deferred tax liability	(906,476)	(138,700)
8.	Borrowed money $ _____195,921,500_____ and interest thereon $ _____1,504,908_____	197,426,408	732,355,328
9.	Unearned premiums (Part 1A, Line 38, Column 5) (after deducting unearned premiums for ceded reinsurance of $ _____1,817,734,941_____ and including warranty reserves of $ _____142,742,245_____ and accrued accident and health experience rating refunds including $ _____0_____ for medical loss ratio rebate per the Public Health Service Act)	9,600,528,557	9,991,464,109
10.	Advance premium	0	0
11.	Dividends declared and unpaid:		
11.1	Stockholders	0	440,667
11.2	Policyholders	0	0
12.	Ceded reinsurance premiums payable (net of ceding commissions)	650,411,294	4,473,402,771
13.	Funds held by company under reinsurance treaties (Schedule F, Part 3, Column 19)	1,508,571,852	915,272,682
14.	Amounts withheld or retained by company for account of others	1,598,583	10,907,910
15.	Remittances and items not allocated	27,686,491	29,362,160
16.	Provision for reinsurance (including $ _____50,338_____ certified) (Schedule F, Part 8)	50,995,002	98,820,020
17.	Net adjustments in assets and liabilities due to foreign exchange rates	0	0
18.	Drafts outstanding	0	0
19.	Payable to parent, subsidiaries and affiliates	697,524,279	1,053,593,964
20.	Derivatives	39,114,701	2,174
21.	Payable for securities	73,587,001	38,432,306
22.	Payable for securities lending	0	0
23.	Liability for amounts held under uninsured plans	0	0
24.	Capital notes $ _____0_____ and interest thereon $ _____0_____	0	0
25.	Aggregate write-ins for liabilities	517,319,668	202,717,224
26.	Total liabilities excluding protected cell liabilities (Lines 1 through 25)	54,215,246,589	66,546,886,618
27.	Protected cell liabilities	0	0
28.	Total liabilities (Lines 26 and 27)	54,215,246,589	66,546,886,618
29.	Aggregate write-ins for special surplus funds	2,035,337,311	1,665,875,754
30.	Common capital stock	97,939,853	93,739,853
31.	Preferred capital stock	0	0
32.	Aggregate write-ins for other than special surplus funds	1,200,000	0
33.	Surplus notes	0	0
34.	Gross paid in and contributed surplus	17,803,913,191	17,672,632,745
35.	Unassigned funds (surplus)	1,180,030,099	2,329,775,297
36.	Less treasury stock, at cost:		
36.1	_____0_____ shares common (value included in Line 30 $ _____0_____)	0	0
36.2	_____0_____ shares preferred (value included in Line 31 $ _____0_____)	0	0
37.	Surplus as regards policyholders (Lines 29 to 35, less 36) (Page 4, Line 39)	21,118,420,454	21,762,023,649
38.	TOTALS (Page 2, Line 28, Col. 3)	75,333,667,043	88,308,910,267

Fig. 6.2 Balance sheet (liabilities) of American International Group, 2017 (in US$, 000s omitted)

2. *Equities.* Although the amounts are similar for life and non-life insurance, equities are more important in the life business relative to premium income. This reflects the fact that the loss event (survival to specified age or death) is comparatively easy to predict since life tables exhibit a great deal of stability over time. Moreover, the higher average return on equities compared to debt securities accumulates over a much longer time in life than in non-life insurance.
3. *Real estate.* Valuation of real estate depends crucially on future expectations. Changes due to revised valuations are often included in the income statement.
4. *Short-term investments.* In the main, these are government bonds.
5. *Other (Sundry) investments.* As an example, in life insurance, the IC can give a loan to a policyholder, accepting the policy as collateral.

6. *Total investments.* The stock of capital investments is markedly larger in life than in non-life business, although life premiums earned are lower. This is a consequence of the fact that most life insurance policies are of the so-called universal "whole life" type, i.e., they provide for a benefit both in the case of death (the event originally insured) and in the case of survival. These policies, therefore, have a savings component that is available for capital investment.
7. *Total assets.* Besides investments, this position also contains cash and cash equivalents to pay for losses, especially in non-life business. Substantial funds are required because non-life losses are deemed to be less predictable than in life business for two reasons. For one, the frequency of loss can vary suddenly. One may think of a hail storm damaging thousands of roofs and windows in a city. Although the IC may have guarded against insuring too many homeowners in the same area, losses can still accumulate unexpectedly in such a case. Second, the benefit is not determined as in a life insurance policy, where the amount to be paid is contractually fixed. Rather, benefits paid usually vary between a lower and upper limit, within a so-called layer.
8. *Total liabilities.* The liability side of the balance sheet shows the origin of the funds that are invested in the company's assets. This position contains the funds coming from outside the firm. The most important components are the insurance reserves. Liabilities also originate from an estimate of taxes not yet paid and from transactions with reinsurers. Often, the primary insurer wants to be able to draw on reinsurance benefits immediately after a loss event. The reinsurer (RI) makes a deposit which in turn amounts to a short-term liability on the part of the IC.
9. *Reserves for unearned premiums.* Especially in non-life business, there may be an unearned element of premium for two primary reasons. On the one hand, premiums pertaining to previous periods may be paid late; on the other hand, current contracts may be associated with losses that have not been reported yet to the IC (incurred but not reported, IBNR).
10. *Reserves for losses and expenses.* As soon as the IC is informed of a loss, it creates a reserve according to the estimated amount. The true benchmark is the present value of these claims, raising the issue of discounting to present value. Here, the Generally Accepted Accounting Principles (GAAP) of the U.S. differ from European Union (EU) norms. Under U.S. GAAP, discounting is disallowed (however, some U.S. states allow it for some lines of business). The EU requires claims settlement to have a maturity (mean lag on payment of premiums) of at least four years and stipulates a low rate of discount. It should be noted that these norms create an inconsistency because securities on the asset side are to be entered at fair value, which means that their valuation varies with market interest rates. By way of contrast, liabilities (mostly) cannot be adjusted when interest rates (and with them, discounting factors) vary.
11. *Future policy holders' benefits.* This entry is similar to the reserve for losses but is assigned to the life insurance business. The differentiation can be justified by noting that in life insurance, the lag between payment of premiums and benefits paid may amount to 20 or even 30 years. Since most contracts are of the universal type (also called "whole life"), containing a savings component, reserves must

also cover accrued interest. On the other hand, the corresponding liability should also be discounted to present value. Both in the United States and in European Union countries, the regulator prescribes a rate of interest (and hence discounting factor) to be used, along with standardized assumptions concerning mortality.

12. *Deposits and other outstanding claims of policyholders.* Especially in life insurance business, ICs let policyholders participate in future profits. Reserves need to be accumulated to be able to honor this commitment.
13. *Total of insurance reserves.* This is the sum of all reserves. It dominates the liability side of an IC that has a high degree of solvency. According to EU regulation, the IC could also enter separate reserves for catastrophic events; this is disallowed under U.S. GAAP.
14. *Minority shareholders.* In the course of past mergers and acquisitions, there may have been shareholders of the acquired company who preferred to retain their stock. Therefore, the IC considered here cannot be seen as the full owner of assets and liabilities, implying that its shareholders do not have a claim to the entire profit. Accordingly, current profits are in part credited to a special account.
15. *Total equity (policyholder surplus).* The difference between assets and liabilities.

▶ **Conclusion 6.1** The balance sheet of an insurance company is characterized on the asset side by a high share of capital investments, on the liability side by reserves for future claims held by buyers of insurance (reserves for future losses in the case of non-life business, reserves for future benefits in the case of life insurance).

6.1.2 The Income Statement

The income statement of the IC provides a summary of revenues and expenses. Revenue comes from two sources, premiums earned and investment income (dividends and capital gains). Insurers use the "float" provided by premium income to invest in assets (see Fig. 6.1 above) that earn income. These two sources can be seen in lines (1) and (11) of Fig. 6.3 below. AIG earned more than US$14.6 bn. in premiums and approximately US$2.9 bn. in investment income. The respective figures were US$17.2 bn. and US$4.2 bn. in the previous year, testifying once again to considerable volatility.

The income statement also shows the IC's expenses, mainly losses paid, loss-related expenses, and other underwriting expenses for policy acquisition. These expenses are seen in lines (2) through (5) of the income statement, totaling more than US$13.7 bn.

There are three commonly used indicators designed to characterize a year's insurance business: the loss ratio, the expense ratio, and the combined ratio. The *loss ratio* (LR) is given as:

$$Loss\ Ratio\ (LR) = \frac{Losses + Loss\ adjustment\ expenses}{Earned\ premiums}. \quad (6.1)$$

In 2017, AIG's LR was 0.94 (13.7/14.6). Similarly, the *expense ratio* (ER) is given as:

$$Expense\ Ratio\ (ER) = \frac{Underwriting\ expenses}{Earned\ premiums}. \qquad (6.2)$$

Since AIG spent more than US$4.7 bn. in underwriting-related expenses, its expense ratio was 0.32 (4.7/14.6). Together, the insurance component of the business netted AIG a loss of more than US$3.7 bn., often referred to as the *underwriting loss* (or gain, when appropriate). The loss ratio and the expense ratio together result in the *combined ratio* (CR):

$$\begin{aligned} Combined\ Ratio\ (CR) &= \frac{Losses + Loss\ adjustment\ expenses}{Earned\ premiums} \qquad (6.3) \\ &\quad + \frac{Underwriting\ expenses}{Earned\ premiums} \\ &= LR + ER. \qquad (6.4) \end{aligned}$$

A combined ratio greater (less) than one suggests that an IC earned a net underwriting loss (profit). It is not uncommon for insurers to report an underwriting loss, and AIG indeed shows a CR of 1.26. While an underwriting loss may seem unusual, ICs often make up for the loss with their investment income. In 2017, while AIG did earn approximately US$2.9 bn. in investment income, this was not quite enough to offset the underwriting loss of US$3.7 bn. In the preceding year, the IC's net loss even amounted to US$0.9 bn. $(= -5.1bn. + 4.2bn.)$; however, it is important to notice that items such as change in net unrealized capital gains [line (24) of Fig. 6.3] and change in provision for reinsurance [line (28)] are subject to substantial uncertainty and usually contain a great deal of hidden reserves.

In general, non-life business is relatively volatile, resulting in years with extremely high losses that are followed by others with favorable loss experience (see also Sect. 5.1). As stated above, a negative underwriting result need not indicate a threat to the solvency of the IC as long as it is balanced by returns from capital investment. Indeed, it is the overall return that determines the competitiveness of an IC when it comes to attracting funds in the capital market.

The remainder of the income statement shows how the operating loss (from underwriting and investment activities) creates a change in surplus. Other, less important items (e.g., dividends paid, deferred income tax) also affect the ultimate change in surplus [line (38)] causing it to differ somewhat from the net income shown [line (20)].

As before, the sample AIG statements are specific to the United States. Items universally found in income statements are briefly discussed below.

1. *Gross premiums written and policy fees.* These are the gross premiums generated by life, non-life, and reinsurance business before deduction of premiums paid to

6.1 Financial Statements of an Insurance Company

		1 Current Year	2 Prior Year
	UNDERWRITING INCOME		
1.	Premiums earned (Part 1, Line 35, Column 4)	14,641,925,157	17,193,317,434
	DEDUCTIONS:		
2.	Losses incurred (Part 2, Line 35, Column 7)	12,230,634,978	16,054,723,003
3.	Loss adjustment expenses incurred (Part 3, Line 25, Column 1)	1,511,509,737	1,111,157,419
4.	Other underwriting expenses incurred (Part 3, Line 25, Column 2)	4,661,583,883	5,115,715,635
5.	Aggregate write-ins for underwriting deductions	0	0
6.	Total underwriting deductions (Lines 2 through 5)	18,403,728,599	22,281,596,057
7.	Net income of protected cells	0	0
8.	Net underwriting gain or (loss) (Line 1 minus Line 6 plus Line 7)	(3,761,803,441)	(5,088,278,623)
	INVESTMENT INCOME		
9.	Net investment income earned (Exhibit of Net Investment Income, Line 17)	2,948,023,231	3,668,597,495
10.	Net realized capital gains or (losses) less capital gains tax of $ ____209,359,707 (Exhibit of Capital Gains (Losses))	(63,077,979)	543,725,067
11.	Net investment gain (loss) (Lines 9 + 10)	2,884,945,252	4,212,322,562
	OTHER INCOME		
12.	Net gain (loss) from agents' or premium balances charged off (amount recovered $ ____10,561,133 amount charged off $ ____63,670,379)	(53,109,246)	(63,624,061)
13.	Finance and service charges not included in premiums	0	883
14.	Aggregate write-ins for miscellaneous income	(296,590,941)	681,892,765
15.	Total other income (Lines 12 through 14)	(349,700,187)	618,269,587
16.	Net income before dividends to policyholders, after capital gains tax and before all other federal and foreign income taxes (Lines 8 + 11 + 15)	(1,226,558,377)	(257,686,474)
17.	Dividends to policyholders	0	0
18.	Net income, after dividends to policyholders, after capital gains tax and before all other federal and foreign income taxes (Line 16 minus Line 17)	(1,226,558,377)	(257,686,474)
19.	Federal and foreign income taxes incurred	(221,212,381)	(521,841,476)
20.	Net income (Line 18 minus Line 19)(to Line 22)	(1,005,345,996)	264,155,002
	CAPITAL AND SURPLUS ACCOUNT		
21.	Surplus as regards policyholders, December 31 prior year (Page 4, Line 39, Column 2)	21,812,221,709	23,596,010,067
22.	Net income (from Line 20)	(1,005,345,996)	264,155,002
23.	Net transfers (to) from Protected Cell accounts	0	0
24.	Change in net unrealized capital gains or (losses) less capital gains tax of $ ____(39,884,372)	382,990,757	(1,067,314,146)
25.	Change in net unrealized foreign exchange capital gain (loss)	182,140,385	(119,049,718)
26.	Change in net deferred income tax	(916,435,810)	786,948
27.	Change in nonadmitted assets (Exhibit of Nonadmitted Assets, Line 28, Col. 3)	899,386,818	(13,260,135)
28.	Change in provision for reinsurance (Page 3, Line 16, Column 2 minus Column 1)	47,825,018	10,242,707
29.	Change in surplus notes	0	0
30.	Surplus (contributed to) withdrawn from protected cells	0	0
31.	Cumulative effect of changes in accounting principles	0	0
32.	Capital changes:		
	32.1 Paid in	0	2,040,000
	32.2 Transferred from surplus (Stock Dividend)	0	0
	32.3 Transferred to surplus	0	0
33.	Surplus adjustments:		
	33.1 Paid in	10,802,598	363,061,720
	33.2 Transferred to capital (Stock Dividend)	0	0
	33.3 Transferred from capital	0	0
34.	Net remittances from or (to) Home Office	0	0
35.	Dividends to stockholders	(200,000,000)	(2,148,192,117)
36.	Change in treasury stock (Page 3, Lines 36.1 and 36.2, Column 2 minus Column 1)	0	0
37.	Aggregate write-ins for gains and losses in surplus	(95,165,025)	873,543,321
38.	Change in surplus as regards policyholders for the year (Lines 22 through 37)	(693,801,255)	(1,833,986,418)
39.	Surplus as regards policyholders, December 31 current year (Line 21 plus Line 38) (Page 3, Line 37)	21,118,420,454	21,762,023,649

Fig. 6.3 Income Statement of American International Group, 2017 (in US$ (000s))

reinsurers. U.S. GAAP emphasizes the distinction between contracts of short and long duration, which however broadly coincides with that between non-life and life business. Under U.S. GAAP, universal life premiums must not be fully entered as revenue because of their savings component, which belongs to the policyholder. However, expenses for acquisition and administration that would be charged to the policyholders who cancel ("buy back") the contract can be credited.

2. *Premiums ceded to reinsurers*. Reinsurance premiums are deducted to arrive at net premiums to the IC. They constitute an expense that reduces the profit of the IC. As soon as the regulatory authority (usually in the aim of consumer protection) limits IC profits, it, therefore, creates an incentive to cede premiums to the reinsurer (RI).

This incentive is especially marked if the primary insurer has a financial stake in the reinsurance company, permitting it to shift profits there. This is frequently the case in Germany, where shares ceded to RIs in excess of 20% of gross premiums are common (see Sect. 9.3.3). In comparison, the shares of 18% (=3.6/20.2) for non-life and 4% (=0.3/7.1) for life business as evidenced by this IC are not excessive. In particular, in view of the lower predictability of losses, it makes sense to cede a greater share of non-life business to RIs.

3. *Net premiums written and policy fees.* This net quantity is often used as an indicator of IC size.
4. *Net changes in reserves for unearned premiums.* Policies written in the past can result in losses during the current period. Reserves are accumulated for this on the balance sheet but revised estimates cause income in the current period to be adjusted.
5. *Premiums and policy fees earned.* This is the IC's premium income allocated to the period considered. It serves as the denominator for indicators such as the loss ratio or the combined ratio (see above).
6. *Net income from wealth management.* This position contains fees earned for wealth management services as well as income from selling securities and commissions earned for such sales. Note that the allocation of incomes and expenses may not always be easy because, e.g., an agent selling insurance may also provide wealth management services to clients.
7. *Net returns from capital investments.* For EU regulatory authorities, this position must currently be reported in greater detail, distinguishing between returns from investment (especially in affiliated companies), returns from the sales of securities, but also administrative expenses associated with these investments. However, future expectations again are important because securities may fall in value below their purchase price. Unless regarded as transitory, such a shortfall must be booked as a realized loss.
8. *Realized profits and losses from capital investments.* This entry contains realized values of securities and assets that were sold above purchase value or had to be sold below purchase value. Typically the amount in non-life is much higher than in the life business because short-term capital investments are more important there. Accordingly, deviations from purchase value are less likely to be considered transitory.
9. *Other income.* This contains ancillary income (e.g., revenue from services sold to employees).
10. *Total income.* This sum reflects the combined income of the IC.
11. *Losses paid, including expenses, non-life business.* This entry refers to losses paid by the IC itself, net of any RI contributions. The loss ratio (above) is a frequently used indicator of financial performance. From the point of view of policyholders as a group, a high loss ratio is beneficial because it shows that much of the premiums paid are returned to them in the guise of insurance benefits. From the point of view of the IC, a high loss ratio can be interpreted in two ways. On the one hand, it may reflect a generous consumer accommodation policy, which enhances the reputation of the IC as a reliable contractual partner. On the other hand, a high loss ratio can also be the result of careless past underwriting policy which

permitted to "earn" premiums in preceding years, with losses accumulating in the current period.

12. *Benefits paid and expenses, life business.* Again, this is net of RI contributions, which however are of less importance than in non-life insurance because less business is ceded to RI.

13. *Changes in technical reserves.* An increase in reserves is recognized as an expense for an IC. There are several reasons why such an increase may occur both in non-life and life business:

    - Increase in the number of policies written;
    - Inclusion of additional perils in existing policies;
    - Aggravated adverse selection effects (for instance due to a launch of new policies with higher benefits and higher premiums, see Sect. 8.3.1);
    - Aggravated moral hazard effects (for instance due to attempts by commercial IB to mitigate a liquidity crunch by presenting insurance claims);
    - Shortening of the time lag between loss occurrence and settlement of the claim; or,
    - Reduction of the discounting factor used for calculating the present value of future benefits (where discounting is allowed by the regulator).

    Evidently, this position provides IC with a great deal of leeway. For this reason, it has been used for profit smoothing which is of interest in view of progressive taxation [see Weiss (1995)].

14. *Participation in surplus and profit by policyholders.* From the point of view of the owners of the IC, commitments to let policyholders participate in the surplus (assets net of liabilities, i.e., reserves plus equity) and profit amounts to a liability. Note that profit-sharing provisions benefiting primary insurers are relatively even more important in reinsurance than in standard insurance contracts.

15. *Administrative expense.* Again, contributions from RI are excluded. The most important component is acquisition expense, be it for direct writers or independent agents and brokers. Much higher values are observed for younger ICs whose sales effort typically generates premiums only with a lag. Further, the expense is much lower in life insurance, reflecting its higher degree of product standardization.

16. *Other operational expense.* This entry reflects the need to run a back office. In non-life business, this comprises for instance specialized engineers who identify risks in buildings and production processes.

17. *Interest expense on debt.* Since returns are gross, they have to be corrected for interest paid on liabilities as evidenced in the balance sheet.

18. *Cost of restructuring, mergers, and acquisitions.* Merger and acquisition activity is not uncommon for ICs. The prices paid in these transactions must be considered.

19. *Amortization of goodwill.* In the course of earlier acquisitions and mergers, assets and liabilities were taken over and booked at purchase prices. However, there may have been an excess of assets over liabilities, resulting in a so-called goodwill.

The issue is whether this goodwill may be capitalized (i.e., be treated as an asset) or not. While both EU and U.S. GAAP regulations permit capitalization, the amortization of goodwill must occur within no more than ten years (U.S. GAAP) and five years (EU), respectively. This position reflects such an annual installment.
20. *Total losses, benefits, and expenses.* This sum represents the combined expenses of the IC.
21. *Net income before tax and minority shares.* Amounting to the net of revenues and expenses, this sum represents the income generated by the IC.

▶ **Conclusion 6.2** Ideally, the operational statement informs about the success in transactions associated with risk underwriting, reinsurance, and capital investment. However, published figures crucially depend on changes in loss reserves (non-life) and reserves for policyholder benefits (life).

6.2   Objectives of the IC

6.2.1   Theoretical Considerations

Financial accounting serves the purpose of informing owners and managers of an IC, but also the IB and regulatory authorities about the degree of goal attainment. However, what are the objectives pursued by the management of an IC?

In mission statements of companies, formulations such as "meeting the insurance needs of our customers" are common. However, such a need cannot be met if premium income falls short of the expected value of claims and expenses. Underwriting a risk of this type does not contribute to the economic survival of the IC. Therefore, the focus of economic analysis is on objectives that contribute to the economic survival of the firm. In the case of an IC, several objectives are postulated.

1. *Profit maximization.* Profit maximization may constitute an acceptable approximation for firms operating in a reasonably predictable environment. In a situation without regulation and a perfect capital market, it is indeed sufficient to maximize the present value of future profits (provided it is positive) to assure economic survival. A temporary loss could always be recovered under these conditions because in view of the positive net present value of the firm, there would be a lender of credit. However, in the case of an IC, one has to take into account that its core activity consists in the underwriting of risks. An IC specializes in bearing risks that other agents in the economy seek to transfer to it. With losses occurring with certain probabilities, the cost of insurance activity is not known *ex ante*, making a maximization of profit impossible.
2. *Maximization of expected profit.* This is what IC management at best can pursue in the absence of regulation and acting in the best interest of diversified shareholders. Strictly speaking, this objective requires that probabilities of loss are known,

which is not always the case (one may think, e.g., of environmental impairment liability insurance). However, the decisive benchmark is the capital market. Recall from Sect. 5.1.2 that if an individual firm's rate of return r_i is linked to the return on the capital market r_M through a regression equation $r_i = \alpha_i + \beta_i r_M + \varepsilon_i$, the variance of r_i can be decomposed to become $Var(r_i) = \beta_i^2 Var(r_M) + Var(\varepsilon_i)$. It was also shown that fully diversified investors can neglect $Var(\varepsilon_i)$, the volatility specific to the firm, focusing only on the first, systematic component, i.e., the β_i since $Var(r_M)$ is exogenous. If the shareholders of an IC are less than perfectly diversified, total volatility $Var(r_i)$ becomes relevant again. Still, being institutions specializing in the bearing and management of risk, the IC should excel in keeping the component $Var(\varepsilon_i)$ low through diversification and hedging. By way of contrast, consider a pharmaceutical company, where the attempt of creating a breakthrough new drug calls for an investment of hundreds of million dollars per year over several years [see DiMasi et al. (2003)]. Therefore, even an IC management acting on behalf of less-than-fully diversified shareholders may pursue the objective "maximization of expected profit".

3. *Maximization of expected utility.* There are two reasons why maximization of expected utility, where risk aversion becomes relevant, may be the appropriate objective for modeling IC behavior. First, a risk-averse IC management acts in the best interest of shareholders whose wealth is heavily concentrated in that single company, causing total variance $Var(r_i)$ to become relevant. Second, however, IC management could be pursuing its own objectives. This is possible when the owners are many (with only small shares in total stock outstanding), making it costly for them to organize in order to exert control. An important reason for IC managers to decide in a risk-averse manner is that they are far from perfectly diversified when it comes to their own assets. Usually, their most important component of total wealth is their human capital, which is know-how largely specific to the IC. In addition, they may be made to hold stock of the IC as a means to align their incentives with those of shareholders, resulting in a further lack of diversification. This leads to the expectation that IC managers often act under the influence of risk aversion (see also the analogous discussion in Sects. 5.3 and 5.3.2). This argument is strengthened by consideration of insolvency with its consequences for thousands of policyholders. A management responsible for such a large-scale failure would see its future prospects for employment and earnings damaged (Greenwald and Stiglitz, 1990).

4. *Growth.* In most industrial countries, public regulation of insurance entails a monitoring and often limiting of profits. This is against the interests of shareholders as the owners of the IC to begin with. Moreover, the performance of management cannot easily be judged in terms of profit if it cannot fully pursue that objective. Growth of premiums written or earned may serve as a substitute in this situation. For management, premium growth goes along with more power and prestige and often income as well. However, shareholders may benefit from premium growth, too. To see this, consider a rate-of-return regulation stating

$$\frac{\Pi}{K} \leq \bar{r}, \qquad (6.5)$$

with $\bar{r}:=$ maximum allowable rate of return on equity K, and $\Pi:=$ realized profit. Expanding by premium volume PV, this can be written as

$$\frac{\Pi}{PV} \cdot \frac{PV}{K} \leq \bar{r}, \text{ or } \Pi \cdot \frac{PV}{K} \leq \bar{r} \cdot PV. \tag{6.6}$$

By increasing PV, the constraint on profits Π can be relaxed (see also Sect. 9.3.3). This effect is particularly marked if premium growth goes along with an increase of PV/K, or so-called leverage. By focusing on premium growth, IC management thus indirectly acts in the interest of shareholders.

5. *Solvency.* The crucial service of an IC consists in paying benefits contracted under all circumstances specified in the policy. An IC failing to honor this commitment may suffer a loss of reputation that it possibly cannot recuperate. In addition, its insured population tends to increasingly consist of high risks, because consumers who suffered a loss do not migrate to a competitor for fear of losing their claim against the IC. Those without a loss leave it as long as they can. Since they are lower risk types on average, they initiate a process of adverse selection that may end up in insolvency of the IC. Once in insolvency, an IC has great difficulty becoming viable again, for the same reasons. The consequences of insolvency are deemed so serious that governments of countries with important insurance markets monitor IC solvency. Once a certain solvency (often also called solvability) level is enforced, solvency constitutes a constraint (which must be satisfied in all circumstances) rather than an objective (which may be violated to a degree). However, solvency ratios imposed by regulators are sometimes exceeded, suggesting that the degree of solvency constitutes an objective of the IC that can be traded off against other objectives. An exclusive emphasis on the ability to honor commitments under all circumstances, therefore, does not do justice to the decision-making situation of an IC.

6. *Generalized stakeholder approach.* The stakeholder approach maintains that not only the owners but several other groups hold stakes in the firm. These other groups comprise creditors, suppliers, employees, the government, and consumers of the goods and services produced by the firm. The objectives of these stakeholders differ, causing management to pursue them all to some degree.

The first thing to note is that the stakeholder approach makes it difficult to state how management should optimally respond to a change in the business environment. The analogy with the consumer optimization model makes this clear. This model has two components, a utility function reflected by indifference curves and a budget constraint. If one of the goods becomes more expensive, the standard prediction is that less of that good will optimally be purchased (the law of demand). However, this law may be violated even if there are only two goods (the so-called Giffen case). In the present context, the two goods can be interpreted as objectives (e.g., profits for shareholders, work satisfaction for employees). Indifference curves show the relative valuation of the two objectives by management. The budget constraint becomes a transformation curve stating that once management has reached it, pursuing one objective hurts the other. An increase in the "price" of an objective means that it has a higher opportunity cost in that its attainment entails more of a sacrifice of the other objective. The standard prediction states that the objective with the increased opportunity cost will be pursued to a lesser degree. But again, the analogy to the Giffen case cannot be excluded. Therefore, even with just two objectives rather than one, an outside observer cannot judge whether management optimally reacted to a change in the business environment. With three or more stakeholders, predicting behavior of IC management becomes virtually impossible. The stakeholder approach is, therefore, discarded here.

Conversely, one can argue that maximization of expected profit encompasses other goals. For example, if one defines the policyholders of an IC as one group of stakeholders, their preferences must be matched at least to the extent that they are willing to pay the premium; otherwise, the premium volume drops, and with its expected profit, *ceteris paribus*. Analogous arguments hold with regard to suppliers and employees because by honoring its contractual commitments, the firm can purchase services at favorable conditions, which contribute to keeping cost low and hence profit high. The argument may even be extended to public authorities as (implicit) contractual partners. A good standing with authorities can again help to keep the cost of business low. However, there is the crucial difference that authorities can impose legal norms without offering compensation (which is not possible for private contractual partners on equal terms). Therefore, there is no guarantee that good standing with public authorities indeed contributes to (expected) profit.

7. *Maximization of expected profit as a working hypothesis*. The preceding considerations can be summed up as follows. Simple profit maximization neglects risk as a crucial element of the decision-making situation of an IC. At the very least, it is the maximization of expected profit that should be postulated as the objective of an IC. In a situation with public rate-of-return regulation, the growth of premium volume is in the interest of shareholders. It may sometimes be appropriate to take risk aversion regarding expected profit into account, thus postulating a concave risk utility function for IC management. In this way, IC managers can be modeled as imperfect agents of shareholders as the owners of the IC. Insolvency and public regulation designed to avoid insolvency can also induce risk aversion. Finally, the stakeholder approach will not be pursued. The main reason is that every generalization of the objective function makes the derivation of empirically testable predictions difficult. For, any decision that does not seem compatible with maximization of expected profit (possibly, expected utility), could always be justified with reference to the interests of some other group of stakeholders besides shareholders.

▶ **Conclusion 6.3** Simple profit maximization is untenable as a hypothesized objective of an IC. Maximization of expected profit may serve in most cases; however, it neglects the role of risk aversion and growth orientation on the part of the management of an IC.

6.2.2 A Descriptive Study Concerning the Importance of IC Objectives

One way to find out about the importance of IC objectives is to simply ask IC management to rank them. However, the statement, e.g., "Ensuring solvency is most important" is not very informative because it fails to say how much management would be willing to sacrifice another objective for improving solvency. Ruminations of this type are reflected in the financial statements provided by the IC, at the very least with regard to (expected) profit and premium growth. Moreover, these performance measures can be related to efforts exerted to improve performance. If the management of an IC is committed to stated objectives, performance should exhibit a systematic relationship with its use of available instruments (the set of these instruments constitutes the so-called insurance technology, to be discussed in Sects. 6.3–6.8). Interestingly, this link does not seem to have been analyzed much. One exception is a study by the Association of German Insurers GDV (1996). The results of that investigation are reproduced in Table 6.1.

The entries of Table 6.1 are discussed assuming that the regression coefficients are all statistically significant. The original publication does not contain standard errors or t-ratios; moreover, it does not say whether there were additional explanatory

Table 6.1 Profitability and premium growth as objectives of German IC (1985–1994)

$PROF = -0.20CL - 0.82LR - 0.74EXP + 0.38EQ + 0.32NRC$
$GROW = 0.42DEG - 0.30EXP - 0.27AGE - 0.30PREM - 0.24DCG$

PROF:	Profitability (income before tax in percent of premiums earned, average value 1985–1994), in percentage points
GROW:	Premium growth (premiums earned, 1994 over 1985), in percentage points
AGE:	Age classification of the IC. $AGE = 1$ if the IC has been in existence for 15–20 years, $AGE = 2$ for 20–50 years, $AGE = 3$ for more than 50 years
CL:	Composition of lines, indicating homogeneity. Mean of the two first principal components, calculated from contributions to premium volume of 6 lines (accident, liability, auto, fire, contents, and residential buildings)
DEG:	$DEG = 1$ if the IC became active in former eastern Germany after reunification in 1989, $DEG = 0$ otherwise
DCG:	$DCG = 1$ if the IC targets its underwriting to particular customer groups (public employees, e.g.), $DCG = 0$ otherwise
EQ:	Equity in percent of premiums earned
EXP:	expense ratio, in percent of premiums earned
LR:	Loss ratio (including loss expenses and adjustments for final settlement), in percent of premiums earned
NRC:	Net return on capital investment, in percent
PREM:	Premium category, ranging from 1 (far below-average premium level) to 5 (far above average) calculated using the market shares of the three lines auto, private liability, and contents, as weights averaged over values of 1988 and 1993

N = 40, $R^2 = 0.76$
Source Association of German Insurers GDV (1996)

6.2 Objectives of the IC

variables in the regression equations. However, the source reports that the observation period was also split between 1985–1989 (prior to the reunification of Germany, with eastern Germany under the communist rule before 1989) and 1990–1994, without finding instabilities.

1. *Composition of lines (CL)*. The effect of this synthetic variable can be interpreted to show that concentration of underwriting of only a few lines of business is associated with a decrease in profitability while apparently not affecting premium growth (the variable is not included in the second equation). Therefore, there is no evidence of a trade-off between profitability and growth. During the period of observation, there was public rate-of-return regulation in Germany. Therefore, growth of premium volume may have indeed permitted higher profits Π, but not necessarily higher profitability Π/PV as defined in Table 6.1 [see Eq. (6.6) again].
2. *Activity in eastern Germany (DEG)*. The decision to enter the market of former communist eastern Germany ($DEG = 1$) apparently is not related to profitability. However, it goes along with additional growth of premiums amounting to 0.42 points. With the average in the sample attaining 2.12 (=212%) premium growth, this is an increase to 2.54 (254%).
3. *Loss ratio (LR)*. As was to be expected, a higher loss ratio goes along with lower profitability; the effect is estimated at 0.82 percentage points per additional percentage point in the loss ratio. Specifically, an IC with $LR = 0.96$ rather than the sample average of 0.95 (=95%) is estimated to have a profitability of 0.82 percentage points below the average value of 7.29% (i.e., some 6.47% relative to premiums earned; $0.0647 = 0.0729 - 0.0082$). According to the second equation, the loss ratio does not seem to affect premium growth. This is puzzling because a high loss ratio implies a low price of insurance to policyholders. Therefore, an IC with a high loss ratio should attract customers, resulting in premium growth. However, there are instruments of insurance technology not included in the two equations (e.g., control of moral hazard effects) that may allow the IC to keep its loss ratio low without jeopardizing customer satisfaction.
4. *Expense ratio (EXP)*. An increase of *EXP* by one percentage point is related to a decrease of 0.74 percentage points in profitability, e.g., as from the mean value of 7.29% to less than 6.6%. However, premium growth also suffers, with a reduction by 0.30 percentage points. At the sample mean, this amounts to a decrease from 2.12 (212%) to 1.82 (182%). The relationship with the insurance technology is evident. Through, e.g., the choice of the distribution system and risk-selection effort in underwriting policy, the expense ratio can be influenced.
5. *Equity relative to premiums (EQ)*. The variable *EQ* as defined here is an underestimate because it contains only the officially declared equity but no undisclosed reserves (which were admissible at the time). Additional equity reflects a stronger involvement of the owners of the IC in terms of risk bearing and may facilitate access to the capital market, with a beneficial effect on profitability. With 0.38, its estimated effect is on the high side, being almost one-half of that of *EXP* (which is also defined with reference to premium volume). Whether easier access to

the capital market conveys such a marked cost advantage is somewhat doubtful. Moreover, additional equity signals a lower risk of insolvency to potential IBs (Cummins and Sommer, 1996). It should, therefore, enhance premium growth *GROW*. However, such an effect is not evidenced in Table 6.1.

6. *Age of the company (AGE)*. The variable *AGE* does not belong to the insurance technology. Its coefficient indicates that an IC in the next higher age group attains a rate of premium growth that is 27 percentage points lower, amounting to (say) 185% rather than the mean value of 212%.

7. *Premium category (PREM)*. The fact that a higher premium level *PREM* is negatively related to the growth rate of the IC points to a negative price elasticity of demand for insurance. More cannot be said, for two reasons. First, *PREM* does not reflect premium differences in percentage terms, and second, the dependent variable is the growth of premiums rather than the premium volume itself. Therefore, it is not possible to compare its coefficient with the low estimated elasticity of demand with regard to premium rates presented in Sect. 1.5.2 for Germany.

8. *Net returns on capital investment (NRC)*. An IC who achieves a return on capital investment *NRC* that is one percentage point higher than average (7.9% rather than the mean of 6.9% nominal) can expect to have a profitability that is 0.32 percentage points higher than average. This underlines the importance of investment policy for the profitability of an IC, an element of insurance technology that will be discussed in Sect. 6.8 below. One could also expect an effect on premium growth because the IC can use returns from capital investment to offset administrative expense, permitting it to charge a low premium for a given value of expected loss (so-called cash-flow underwriting). Depending on the price elasticity of demand, this may induce more or less premium volume (and possibly premium growth). However, a reduction of the expense ratio (see item 4 above) likely is even more effective. This can be concluded from a comparison of elasticities. Evaluated at mean values (which happen to be 6.9% both for profitability *PROF* and net returns *NRC*), one obtains

$$e(PROF, EXP) = \frac{\partial PROF}{\partial EXP} \cdot \frac{EXP}{PROF} = -0.74 \cdot \frac{22.5}{6.9} = -2.41 \qquad (6.7)$$

$$e(PROF, NRC) = \frac{\partial PROF}{\partial NRC} \cdot \frac{NRC}{PROF} = +0.32 \cdot \frac{6.9}{6.9} = +0.32 \qquad (6.8)$$

$e(PROF, EXP)$: Elasticity of profitability w.r.t. the expense ratio;
$e(PROF, NRC)$: Elasticity of profitability w.r.t. the net returns on capital investment.

A reduction of the expense ratio by 10%, therefore, is estimated to increase profitability by about 24%,[2] whereas the same increase in the rate of return on capital investment serves to increase the profitability only by about 3%.
9. *Specialization in certain client groups (DCG)*. Some ICs concentrate their activity among particular segments of the population, such as public employees or independent workers ($DCG = 1$). Table 6.1 suggests that specializations of this type may substantially slow premium growth of the IC while leaving profitability unaffected (since DCG does not appear in the equation for $PROF$).

Comparing the two regression equations of Table 6.1, one sees that the two objectives are related to different elements of insurance technology. The one exception is the expense ratio EXP, which has a negative coefficient in both equations. Therefore, there is no trade-off for any element considered in the sense that profitability might be enhanced while premium growth might be slowed through its use. It seems that under the influence of rate-of-return regulation, the two objectives are not in conflict. However, note that the two equations of Table 6.1 are based on data also from inefficient ICs, who could attain both objectives to a higher degree by increasing efficiency.

▶ **Conclusion 6.4** Profitability (relative to premium volume) and premium growth are two objectives that are pursued by German ICs. This follows from statistically significant relationships with elements of the insurance technology, which are used to further these objectives.

However, there are objectives that cannot be measured easily, and some instruments of insurance technology such as "improved motivation of workers" can only be described using indicators. Therefore, the postulated relationships link latent quantities that cannot be observed directly. In this case, methods of so-called covariance analysis (in particular, LISREL) can be used, which also allow to verify whether the indicators reflect theoretical quantities well [for example; see Schradin (1994)].

6.3 Survey of Insurance Technology of an IC

In the previous subsection, elements of the insurance technology were directly related to the possible objectives of the IC. According to a more conventional definition, technology describes the set of processes available to generate outputs from inputs. In view of this definition, it makes sense to first clarify the concept of output in the context of insurance.

---

[2] In the case of the expense ratio, evaluation at the means of the sample can be problematic in case there are scale effects (see Sect. 7.4). In that event, profitability $PROF$ changes systematically with the size of the IC (indicated by premium volume PV). The same is true of the expense ratio EXP because it contains PV in the denominator as well. For this reason, estimated elasticities quite likely change with increasing PV.

6.3.1 What Is the Output of an IC?

For some time, there was a debate especially in countries with heavily regulated insurance markets (such as Germany) about how to define the output of an IC [see Müller (1981) for a survey]. The starting point was the view of a consumer who is likely to say that the IC does not provide any service as long as there is no loss event triggering payment of claims. To counter this view and to emphasize the continuity of output, the creation of an organization for pooling risks was defined as the output of an IC by some authors. Of course, this obviates measurement of output separately from input and of productivity and its development.

Another attempt was to define the provision of information as the output of an IC. The information in question is that the IC commits to cover damages to the extent stated in the contract. It serves the IB even though there may not be a payment of benefit, thus establishing continuity of service. However, information is also provided by economic agents that have nothing to do with insurance such as business consultants, lawyers, and journalists. Moreover, information has a public good property: Once created, it can be made available to third parties almost without cost. Excluding non-payers is, therefore, both difficult and often undesirable. In private insurance however, it is easy to exclude someone from coverage who does not pay the premium. Even in social insurance, extending coverage to non-payers is often deemed undesirable.

A definition that is in keeping with the theory of demand put forth in Chap. 3 is "commitment to pay contingent on occurrence of loss". This relates to the distinction of the two dimensions of risk, probability of occurrence and severity of consequences. The output of the IC then has two dimensions. First, the IC commits to pay regardless of whether the frequency of loss exceeds the estimated value. Rather, it calculates its premium on the best estimate of π. Second, the IC helps to reduce the severity of the consequences for the IB to the extent specified in the contract. To the IB, this certainly is information which is contained in the text of the contract and can be copied at a very low cost. In contradistinction to the case of a business consultant, a lawyer, or a journalist, however, this information is always combined with the commitment to pay a certain amount. The statement, "in case of the loss event of type X, the IC commits to pay an amount of Y" goes much further than, e.g., the information, "the probability of a loss event of type X is π". This commitment cannot be imitated at low cost but necessitates the holding of costly capital.[3]

This definition of output corresponds entirely with the theory of insurance demand of Sect. 3.2 based on contingent claims. There, it was shown that the purchase of insurance contributes to expected utility (and therefore, provides a service) although a loss payment may not occur. At the level of the insured population, the problem of discontinuity vanishes because the IC does pay for losses, and at guaranteed rates that are not adjusted to observed frequency of loss during the current period. Losses effectively paid then constitute an estimate of the expected value of commitments made, making them an appropriate measure of output [Doherty (1981)].

---

[3] Mutual insurers can rely on additional premiums or a loss reserve.

Finally, this definition of insurance output has the advantage of relating to the decision-making situation of the IB. To see this, one only needs to imagine that whoever drives a car would have to be liable with his or her entire wealth in the case of an accident because there is no auto liability insurance. Not very many people would dare to drive (assuming strict enforcement of the liability rule). Also, many product innovations could not be launched if their initiators were fully liable for any negative side effects, many of which can hardly be foreseen. Therefore, one indirect output of the IC is to permit the IB to pursue more risky, but also more lucrative alternatives of action [see Zweifel (2009)].

In this view, the IC has some similarities with other financial intermediaries such as banks. Also, moral hazard effects (to be analyzed in Sect. 8.2) turn out to be a necessary consequence of the intended extension of the IB's choice set made possible by the conclusion of the insurance contract.

Finally, the definition of an IC output is critical when studying economies of scope and scale as in Sects. 7.3 and 7.4. Early studies (e.g., Grace and Timme, 1992) used total premiums to measure output for the risk pooling function. Total premiums, however, represent the total revenue of the firm (e.g., price times quantity) and not only quantity. Later studies (e.g., Cummin and Zi, 1998) eschewed premiums in favor of losses paid as a proxy for output (i.e., quantity only). Of course, losses paid are not a perfect proxy as they are subject to variability, estimation, and manipulation. For the financial intermediation function, the use of invested assets or invested returns can be used.

▶ **Conclusion 6.5** The output of the IC can be defined as a commitment to pay contingent on the occurrence of loss in return for a premium that does not reflect current frequency of loss. This definition is in accordance with the theory of contingent claims shown in Sect. 3.2; it suggests losses paid as the observable output measure.

6.3.2 Instruments of Insurance Technology

In analogy to conventional production theory, one could relate an indicator of output (as defined in Sect. 6.3.1) to inputs such as labor of differing levels of qualification, capital goods, and materials. However, specifying a production function in this way turns out to be inappropriate because of reinsurance. On the one hand, one would have to define reinsurance coverage as an additional financial input. On the other hand, reinsurance coverage reduces the probability of insolvency, thus modifying the quality of output. This makes the analysis of the variation in output with quality held constant almost impossible.

For this reason, the alternative of formulating a production function reflecting the insurance technology is not pursued here. This has the disadvantage that one can say little about complementary and substitution relationships between the several instruments of insurance technology listed in Table 6.2 below. Also, statements about the optimal mix of these instruments cannot be derived. The analysis performed below

Table 6.2 Overview of insurance technology

– Distribution channels and remuneration systems
– Underwriting policy (amount of risk-selection effort)
– Composition of the underwriting portfolio
– Product design for controlling
– Risks inherent in the insured object
– Risks associated with IB behavior (moral hazard)
– Risks emanating from financial markets (loss of value in assets, increase of liabilities)
– Management of reserves and capital investment
– Purchase of reinsurance, coinsurance
– Statistical analysis of loss data; loss forecasting
– Pricing
– Experience rating
– Principles of calculating net premiums
– Premium adjustment clauses
– Premium reductions for deductibles and copayments
– Claims settlement (consumer accommodation policy)
– Contractual clauses governing the behavior of the IB in the case of loss
– Contract duration and conditions of termination of contract

focuses on one element at a time. It follows the life of an insurance contract, starting with acquisition effort and ending with a payment of benefits, usually years later. From Table 6.2, the following topics are selected.

- *Acquisition* (Choice of the distribution channels, Sect. 6.4);
- *Underwriting policy* (Risk selection, Sect. 6.5);
- *Settlement of claims* (Control of moral hazard, Sect. 6.6);
- *Purchase of reinsurance coverage* (Sect. 6.7);
- *Capital investment policy* (Sect. 6.8).

Because of its close connection to the supply of insurance, pricing is the topic of Sect. 7.1, where principles of premium calculation are discussed. Yet, Table 6.2 still omits several instruments, e.g., market research, methods of loss forecasting, details of experience rating, organizational structure of the IC, and possibilities of so-called alternative risk transfer (e.g., through securitization, but see Sect. 6.9.3).

6.4 Choice of Distribution Channel

6.4.1 Main Distribution Channels for Insurance Products

Insurance products are sold in different ways, each with its advantages and costs to the IC. Below, the five main variants are sketched.

1. *Direct writers*. The IC creates sales offices with employed sales personnel. This alternative calls for an important initial investment because the direct writers first have to introduce the IC to the local market. During this phase, they generate substantial cost but little premium volume. Incentives are usually biased toward premium growth rather than expected profit because the marginal cost of an additional contract is borne by the IC rather than the direct writer (see Sect. 6.4.4 below). On the other hand, the IC can monitor effort rather easily.
2. *Exclusive agents*. This type of sales agent is independent but is in exclusive dealing with the IC. This distribution form is more prevalent in the United States than in Europe. Since exclusive agents do business on their own accounts, they have to balance the additional revenue earned against the marginal cost. Still, the IC frequently shields them from investment outlay (especially for information technology), with the result that their incentive structure does not differ too much from variant (1). In return, monitoring effort is already more costly for the IC.
3. *Independent agents*. Since independent agents are already established, as a rule, they enable the IC to quickly attain a break-even premium volume in the local market. In return, monitoring of brokers meets with considerable difficulty because of their independence and their activity on behalf of several ICs. In particular, each IC fears to be disadvantaged in terms of risk selection, which may be performed by the broker in a way to subtly benefit some ICs to the detriment of others.
4. *Using the distribution network of another firm*. This variant is little known in the United States but somewhat popular in the United Kingdom for life insurance. By way of contrast, from the 1990s in France, more than one-half of life insurance contracts were sold through banks (*bancassurance*). Usually, an IC and a bank strike an exclusive dealing agreement with the aim of reducing the cost of distribution by avoiding brokers.
5. *Direct selling through the media or Internet*. The IC purchases advertisement time in the press, radio, and television, using the telephone, the Internet, or mail for contract applications. These efforts are rather costly; on the other hand, they allow for quick expansion of activity at the local and national level. Another advantage is that there is no need to monitor a sales agent. On the other hand, since there is no advice provided, this type of distribution is appropriate only for standardized products covering a single, well-defined risk. However, new electronic media increasingly provide low-cost alternatives that may tip the balance in favor of direct selling in the future.
6. *Brokers*. A broker is the agent of an IB. A broker will represent the IB in its negotiations with the IC with respect to premiums, coverage, and claims settlement. While brokers may be compensated by an IC via commission, their fiduciary responsibility is to the IB. Since brokers are agents of the IB, they can place insurance with any IC.

6.4.2 The Principal-Agent Relationship as the Underlying Problem

The distribution systems of the insurance industry provide an interesting case study of vertical integration [for a survey of the issues; see e.g., Carlton and Perloff (1999), Chap. 12]. At one end of the spectrum, direct writers represent full vertical integration, at the other, brokers no integration at all. A unifying theme for analyzing these differences in organizational structure is provided by the principal-agent relationship. A principal-agent relationship exists when one or more principals rely on the services of an agent [Jensen and Meckling (1976)]. Its crucial properties are the following [Grossman and Hart (1986); Levinthal (1988)]:

- The outcome of the agency relationship (income, profit) is a random variable that depends not only on the agent's effort but also on influences beyond the control of both principal and agent;
- Agent effort cannot be observed by the principal, creating the leeway for the agent to pursue his or her own objectives that differ from those of the principal (a moral hazard effect). Therefore, the contract between the two parties must be compatible with the agent's incentives to exert effort (incentive compatibility constraint);
- The agent cannot be forced to sign the contract, which means contract conditions must be sufficiently attractive to ensure voluntary participation (participation constraint); and,
- The principal can choose the type of contract. The objective for the principal is the maximization (on expectation) of the financial outcome net of payment for the agent.

Under certain conditions (Holmström, 1979), the optimal contract from the principal's point of view offers a bonus for an above-average outcome (which is observable) but a penalty for a below-average outcome. The bonus and penalty must be stronger the more the extra effort on the part of the agent contributes to the outcome. In addition, a fixed payment ensures the agent's participation. These general insights can be applied to the distribution channels for insurance as follows.

1. *Direct writers*. Through the employment contract, the IC acquires the right to monitor the activity of the agent, at least in principle. However, this right does not necessarily render the effort of the agent observable, apart from standardized processes (sales administration, claims settlement). Already for the managers of a writers' office, effort mainly consists in seeking out promising customers and assigning tasks to employees according to their capabilities—dimensions that are not easy to measure. Therefore, the IC may try to align the incentives of these managers with its own by paying bonuses for outstanding performance. The measure used may well be the growth of premium volume because this frequently corresponds to the objective of IC management, especially in a regulated environment (see Sect. 6.2.2).

2. *Exclusive agents.* In the absence of an employment contract, monitoring of effort is more difficult than in variant (1), suggesting an increased use of bonuses. Again, bonuses may be geared to premium growth.
3. *Independent agents.* For an individual IC dealing with this type of agent, observability of effort is further reduced because effort might also benefit a competitor. Therefore, it makes sense to especially honor premium growth as an indicator of extra effort exerted in favor of the specific IC. Indeed, contracts signed by a Swiss IC with both exclusive and independent agents show that independent agents received a special bonus for premium growth [Zweifel and Ghermi (1990); see also Sect. 6.4.4 below].
4. *Using the distribution network of another firm.* Here, the agency problem is twofold. First, the firm making its distribution network available to the IC must get its workers to serve as sales agents for additional products. Since this is not routine, observability of effort is not guaranteed in spite of an employment contract. The cooperating firm may, therefore, decide to also offer bonuses for the sale of insurance products. Second, its management in turn now acts as an agent of the IC. The solution may be to make the cooperating firm and its management participate in the equity and hence performance of the IC.
5. *Direct selling through the media or mail.* Here, the effort is measured rather easily, comparable to direct writers [alternative (1)].

▶ **Conclusion 6.6** The choice of distribution channel by the IC can be analyzed using principal-agent theory. In several cases, incentive compatibility calls for payment of a bonus for above-average performance.

The balance of advantages and downsides of distribution channels may also depend on the type of insurance regulation. In particular, Finsinger and Schmidt (1994) found marked differences in the market share of the so-called tied-advice channels [alternatives (1), (2), and (4)] among EU countries. Their market share in Germany was as high as 84%, compared to 35% in Belgium. The authors also find statistical evidence suggesting that countries with tightly regulated insurance markets (Germany and France in the 1980s) are dominated by tied forms of distribution (see Sect. 9.3.2). A possible explanation is that tied channels permit the IC to control the implementation of detailed regulations concerning prices and products at the point of sale.

6.4.3 A Comparison of the Cost of Distribution Channels Using U.S. Data

In an early investigation, Joskow (1973) found that ICs in the United States paid lower provisions and expense contributions to exclusive agents than to brokers. He concluded that brokers are a comparatively inefficient alternative for the distribution of insurance. However, this conclusion was criticized on three grounds: (1) The higher expenses of brokers could reflect better quality of advice; (2) a data set limited to just

Table 6.3 Expenses relative to premiums written by 46 ICs, United States (1978–1990)

Total expenses relative to	Independent agents	Mixed distribution	Exclusive agents	Direct selling	Average
– net premiums[a]	39.4%	37.5%	29.5%	26.3%	36.4%
– premiums earned	35.7%	34.8%	29.2%	25.6%	33.9%

[a]Premiums written by the IC net of provisions paid for distribution
Source Barrese and Nelson (1992)

1967 fails to reflect the fact that contrary to subsequent time periods, brokers served many small ICs at the time, precluding returns to scale; (3) the study measures only administrative expense but not total expense which also includes claims settlement.

A later contribution by Cummins and Van Derhei (1979) dealt with criticisms (2) and (3) by using time series data covering 1968–1976 and making total expense the dependent variable. Indeed, the cost disadvantage of brokers was reduced but there was no convergence to the cost level of the other distribution channels over time.

The investigation by Barrese and Nelson (1992) is discussed in greater detail below. It comprises the data of 46 ICs covering the years 1978–1990. It not only contains a re-estimation of Cummins and Van Derhei (1979) but also accounts for the fact that many ICs rely on several distribution channels rather than just one.

According to Table 6.3, direct selling (through the mail) appears to be the least costly alternative, with expenses amounting to 26% of net premiums. Again, independent agents and brokers perform worst, in part because their involvement in claims settlement (to the benefit of the IC) is neglected. However, there may be additional determinants of the cost that need to be controlled for. Panel A of Table 6.4 exhibits a re-estimation of the regression equation specified by Cummins and Van Derhei (1979), based on a longer period of observation. Total expenses of the IC for distribution, administration, and claim settlement are in logs [$Log(EXPENSE)$] and related to explanatory variables. Panels B and C introduce new explanatory variables (IA78, IA79; see below) reflecting the fact that an IC may employ a changing mix of distribution channels. In panel B, direct premiums written are the indicator of size, in panel C, losses paid. The estimation results can be interpreted as follows.

DPW: Direct premiums written constitute one of two alternative output indicators. Since both *DPW* and *EXPENSE* are in logs, the pertinent regression coefficients can be interpreted as elasticities. An increase of premium volume by 10%, therefore, is accompanied by an estimated increase of total expenses by 9.8%. This raises the question of whether a value of 1.00 (indicating absence of scale economies) is compatible with the estimated value of 0.98. Since the t-value of about 70 is equivalent to an estimated standard error of 0.014 (=0.98/70), the benchmark value of 1.00 still lies within the 95% confidence interval of 0.98 +/− 1.96 standard errors given by (0.953, 1.007). Expenses, therefore, increase in step with premiums written, with no evidence of either increasing or decreasing returns to scale.

LOSSES: Rather than premiums written, one can also use losses paid as an output indicator. This choice is in closer accordance with the theory of demand for insurance,

6.4 Choice of Distribution Channel

Table 6.4 Total expenses of 46 ICs using different distribution channels, United States (1977–1990)

	Re-estimation of Cummins and Van Derhei (1979)		Alternative estimations			
Dependent variables	Log(EXPENSE)		Log(EXPENSE/PI)			
	Panel A		Panel B		Panel C	
Explanatory variables	Coefficient	t-value	Coefficient	t-value	Coefficient	t-value
Constant	−1.0638	−11.11	−1.2409	−10.41	−9.3267	−2.30
$log(DPW)$	0.9803	70.23	0.9836	72.51	–	–
$log(LOSSES)$	–	–	–	–	0.9126	55.62
$1 - NPW/DPW$	−0.0068	−18.22	−0.0063	−15.91	−0.0027	−5.16
STK	−0.0038	−0.07	0.0487	0.99	0.1510	2.50
WC %	−0.0037	−2.94	−0.0009	−0.59	−0.0024	−1.22
AUTO %	–	–	0.0018	1.97	−0.0007	−0.66
MAIL	–	–	−0.2389	−2.87	−0.2044	−1.98
JCV78	0.2724	4.47	–	–	–	–
IA78	–	–	0.1980	3.07	0.187	2.28
JCV90	0.2680	4.39	–	–	–	–
IA90	–	–	0.2270	3.39	0.2561	3.01
HOME %	–	–	0.0067	2.71	0.0059	1.87
R^2		98%		96%		89%

Source Barrese and Nelson (1992)
EXPENSE: Cost incurred by the IC for acquisition, administration, and claims settlement; *PI*: Gross National Product deflator; *DPW*: Gross premiums written as a primary insurer, nominal values in panel A, adjusted for inflation using *PI* in panels B and C; *LOSSES*: Losses paid, adjusted for inflation; $1 - NPW/DPW = (DPW - NPW)/DPW$: Difference between gross and net premiums relative to gross premiums, i.e., share of premiums ceded to reinsurance; $STK := 1$ if stock company, $= 0$ otherwise; *WC%*, *AUTO%*, *HOME%*: Shares of premiums from workers' compensation, auto insurance and homeowners' insurance, respectively; $MAIL := 1$ if distribution by mail, $= 0$ otherwise. $CV78 := 1$ if the IC relied on independent agents in 1978, $= 0$ otherwise; *IA78*: Share of premiums acquired through independent agents in 1978; $JCV90 := 1$ if the IC relied on independent agents in 1990, $= 0$ otherwise; *IA90*: Share of premiums acquired through independent agents in 1990; $JCV79, \ldots JCV89$ and $IA79, \ldots, IA89$ not shown

as argued in Sect. 6.3.1 [see also Doherty (1981)]. An increase of loss payments by 10% goes along with an increase of expenses for acquisition, administration, and claims settlement of 9.1% only. With an estimated standard error of 0.016 (=0.91/56), the benchmark value of 1.00 is outside the 95% confidence interval pertaining to 0.91. Therefore, when loss payments are used as the output indicator, the estimation result points to weak scale economies.

$1 - NPW/DPW$: This can be rewritten as $(DPW - NPW)/DPW$. The difference between direct premiums written and net premiums written (*NPW*) is due to premiums ceded to reinsurers (RI). The higher, therefore, this relative difference, the more the IC purchases reinsurance. The significantly negative coefficient of this variable indicates that the use of reinsurance entails a relief from expense (frequently in the context of claims settlement).

STK: Stock companies ($STK = 1$) are expected to have lower expenses *ceteris paribus* than other ICs (in particular, mutuals). Especially in the United States, management of a stock company is under pressure to keep the cost low in the interest of profit. Failure to achieve this opens the door to a raider who promises better performance thanks to a new, more efficient management. However, the expected negative effect on expenses is not found, even to the contrary according to the estimates in panel C.

WC%, AUTO%, HOME% : Shares of premiums written in workers' compensation and auto insurance lines do not have a significant influence on expenses. However, the positive coefficients pertaining to homeowners' insurance (*HOME%*) suggest that this line of business may be more costly to operate than the others.

MAIL: Direct selling through mail ($MAIL = 1$) indeed goes along with lower expenses, making it the low-cost distribution channel.

JCV78, JCV90: These are two out of a set of 13 categorical variables that take on the value of 1 if the IC relied on brokers for distribution in the respective year, in analogy to Cummins and Van Derhei (1979). For both the beginning and the end of the observation period, brokers exhibit significantly higher expenses than the other channels (mainly direct writers). For the year 1990, the differential amounts to some 31% [$e^{0.268}/e^0 = 1.31$; for a more precise estimate, see Kennedy (1986)]. For the years 1979–1989 that are not evidenced in Table 6.4, a consistent cost disadvantage to the detriment of brokers amounting to 28% or more is found as well.

IA78, IA90: In panels B and C, the categorical variables *JCV78, ... JCV90* are replaced by the shares of premiums due to business acquired by brokers. The coefficients for the years 1978 and 1990 are significantly positive, pointing again to a cost disadvantage of independent agents. However, the estimates for the years in between (not evidenced in Table 6.4) fail to reach the usual threshold of statistical significance (their *t*-values are below 1.96).

▶ **Conclusion 6.7** Research from the United States suggest that distribution of insurance through independent agents and brokers is associated with higher total expense than tied-advice alternatives.

6.4.4 A Study Relating Performance to Incentives

During the 1980s, a Swiss IC distributed its products through dependent agents (exclusive agents and direct writers) as well as independent agents (brokers). Since the contracts were known to the authors in detail, Zweifel and Ghermi (1990) were able to model the incentive structure rather than relying on categorical variables or mixture indicators as in earlier studies. Income of dependent agents (D) comprised the following components.

D1. A so-called fixed remuneration, which however weakly depends on the growth of premiums written;
D2. A commission as a percent of the stock of premiums written; and,

6.4 Choice of Distribution Channel

D3. A bonus as a percent of premiums written that increases with the growth in premiums measured in percentage terms but decreases with the combined ratio of the agent.

Expenses for distribution and administration are not charged to dependent agents; however, higher expenses serve to increase their combined ratio, causing their bonus to be reduced. By way of contrast, the net income of brokers (B) has the following components:

B1. No fixed income;
B2. A commission as a percent of the stock of premiums written, lower than D2 of the dependent agent;
B3. A bonus as a percent of premiums written having the same structure as D3;
B4. A constant share in the absolute growth of premiums; and,
B5. Less: Cost, mainly wages for sales agents and administrative personnel employed by the broker.

Both remuneration systems contain elements designed to satisfy the participation constraint (D1 and D2; B2) and the incentive compatibility constraint, given that IC management aims at premium growth under the influence of regulation (D3; B3 and B4) as discussed in Sect. 6.4.2.

Both types of agents are assumed to maximize the present value of net income from their contract with the IC. Since premium growth enters the formula, the planning horizon must be extended to at least two periods. Dependent agents and brokers who achieve an increase in the stock of premiums written during the current year (see components D3, B3, and B4) are negatively affected in the following period since growth is measured from a higher baseline. This effect is part of the marginal cost of additional sales effort. The performance of the two distribution channels is evaluated with respect to premium growth (the apparent objective of the IC) and the expense ratio (for comparison with earlier studies).

- *Comparison of performance in terms of premium growth:* Component B4 of brokers happens to depend on premium growth in a similar way as the "fixed" component D1 of the dependent agent. The other components are quite similar as well, implying that the marginal returns to sales effort are about the same for dependent agents and brokers. Note that brokers bear a higher marginal cost because they have to pay for additional personnel (component B5) whereas dependent agents are protected by the IC. For this reason, brokers' premium growth is predicted to be below that of dependent agents. However, this expectation is not confirmed for the period 1981/82 to 1984/85 and a book of 50 contracts.
- *Comparison of performance in terms of the expense ratio:* Both dependent agents and brokers need to have administrative personnel, at the very least for claims settlement. The marginal return to personnel is more expeditious claims settlement and hence consumer satisfaction which contributes to the level and growth of premiums. This effect is similar for dependent agents and brokers. The difference

once more lies on the side of marginal cost, with brokers having to directly pay for the additional employment. Therefore, brokers are predicted to make do with less administrative personnel, resulting in a lower expense ratio. This expectation is confirmed in that their expense ratio *ceteris paribus* is some 11 percentage points lower than that of dependent agents, i.e., some 20% rather than 31% of premiums.

This finding is to be interpreted cautiously, however. It is quite possible that the two distribution channels differ systematically in the composition of their risk portfolio [as in Barrese and Nelson (1992), see Sect. 6.4.3]. However, the database lacks information about the break-down of premiums according to lines of underwriting.

▶ **Conclusion 6.8** In the case of a Swiss IC, contractual incentives lead to the expectation that (1) brokers contribute less importantly to premium growth and (2) exhibit a lower expense ratio than dependent agents. Prediction (1) is not statistically confirmed while (2) is confirmed. Contrary to evidence for the United States, distribution through brokers seems to be less costly than through dependent agents.

The contradiction between Conclusions 6.7 and 6.8 can possibly be resolved as follows. The U.S. data do not tell whether brokers are more costly because they are brokers or because incentives created by the IC cause them to run their business in a more costly way. The Swiss database contains information about these financial incentives, which however might be specific to the IC analyzed.

6.5 Underwriting Policy

6.5.1 Instruments of Underwriting Policy

Direct writers, exclusive agents, and brokers propose potential IBs to the IC for underwriting. The IC in turn seeks to avoid adverse selection effects (see Sect. 8.3). Adverse selection occurs if an insurance contract mainly attracts high, unfavorable risks. Underwriting policy is designed to prevent this from happening. Its main instruments are the following (see Table 6.2 again):

- *Structuring of contracts offered.* By creating a set of differentiated contracts, the IC may get the different risk types to select the contract designed for them. As shown in Sect. 8.3.1, high risks tend to select policies with a high degree of coverage but also a high premium. Low risks tend to prefer policies with substantial copayment in return for a low premium.
- *Collecting risk indicators.* In the case of commercial insurance, the type of the IB's activity already contains a good deal of information about risk. In the chemical industry, e.g., the use of toxic substances cannot be avoided in some production processes. This is an important consideration in liability insurance and workers' compensation. Detailed knowledge of the production process is also valuable. For

instance, insurance for fire risk in buildings and contents importantly depends on the materials used in construction.
- *Experience rating.* When signing an insurance contract, the IC often has but an imprecise estimate especially of the probability of loss. It, therefore, may retain the right to adjust the premium in the light of past loss experience. The most common variant of experience rating is the bonus-malus system. In actuarial science, devising a system with an optimal probability of misclassification of risks is the topic of so-called credibility theory [see Bühlmann and Gisler (2005)].
- *Selection of markets.* A multinational IC can decide not to do business in certain countries. A high expected value of losses per se is not a reason as long as it is matched by a high premium. However, premium regulation may make such a market unattractive. The IC thus performs *regulatory arbitrage*. This is an option even within a country when regional jurisdictions differ. For instance, ICs are known to avoid ("red line") U.S. states with courts that have a reputation of re-interpreting contracts in favor of IBs.

6.5.2 A Simple Model of Risk Selection

The measures listed in the previous section can be summarized as "risk-selection effort" (S). Such effort is costly. For example, the gathering of risk indicators in commercial insurance often requires on-site inspection. As to regulatory arbitrage, it cannot be performed without the help of lawyers who analyze the insurance jurisdiction of a country and its likely future development. For simplicity, the unit price of selection effort is set to 1, neglecting the fact that it entails activities that are expensive relative to those goods and services that make up losses paid. Also note that adverse selection (to be discussed in Sect. 8.3) is generally understood as involving effort on the part of the insured rather than the IC.

The issue at hand is to determine the optimal amount of risk-selection effort, S^*. The management of the IC is assumed to be risk-neutral in this section. This allows for the maximization of expected profit, $E\Pi$, as the objective. Let there be two types of risk, high (H) and low (L). Let $\mu(S) < 1$ be the share of low risks in the population insured; it increases with additional selection effort. The premium levels ($\bar{P}^L, \bar{P}^H$) are exogenous to the IC, reflecting regulation. In the case of low risk, the premium is higher than the expected value of loss, causing a positive contribution to the expected profit. In the case of high risk, this contribution is lower and possibly negative. Note that such a difference cannot exist in a competitive market with full information about risk types because the IC would want to achieve the same contribution to expected profit from all risk types. Moreover, contribution to expected profit would approach zero provided there is a free market entry. An alternative reason for differences in contribution to expected profit is imperfect information regarding risk types. In that situation, the IC may abstain from fully adjusting premium to risk even absent regulation because it ignores whether an IB who migrates to a competitor in response to a high premium can be replaced by another IB presenting a more favorable risk. Finally, the amount of loss L is an observable quantity; without loss of generality, it is assumed to be fixed and independent of risk type. The decision problem of the IC can then be written

$$\max_{S} E\Pi = \mu(S)\{\bar{P}^L - \pi^L \cdot L\} + \{1 - \mu(S)\}\{\bar{P}^H - \pi^H \cdot L\} - S, \quad \text{with} \quad (6.9)$$

$E\Pi$: Expected value of profit;
L: Loss payment including expenses, exogenous;
$\bar{P}^H, \bar{P}^L$: Premiums paid by the high- and low-risk types, exogenous;
S: Risk-selection effort (at the price of 1), $S \geq 0$;
$\mu(S)$: Share of low risks in the population insured, with $\partial\mu/\partial S > 0$ and $\mu(S) < 1$;
π^H, π^L: Probability of loss for high- and low-risk types.

For an interior optimum (associated with positive selection effort), one has the condition,

$$\frac{dE\Pi}{dS} = \frac{\partial\mu}{\partial S}\{\bar{P}^L - \pi^L \cdot L\} - \frac{\partial\mu}{\partial S}\{\bar{P}^H - \pi^H \cdot L\} - 1 = 0. \quad (6.10)$$

The first two terms correspond to the expected marginal return of additional selection effort due to the fact that the share of low risks increases while that of high risks decreases. These changes are to be weighted with the contribution to expected profit. The last term is the marginal cost of effort ($=1$). At the optimum, marginal returns and marginal cost must balance.

For ease of interpretation, it is worthwhile to perform a simple transformation, introducing $e(\mu, S) := (\partial\mu/\partial S) \cdot (S/\mu) > 0$. This defines the elasticity of the share of low risks with respect to selection effort. Note that $e(\mu, S)$ is assumed to be independent of S. Multiplying (6.10) by S^*/μ yields

$$e(\mu, S)\,[\{\underbrace{\bar{P}^L - \mu^L \cdot L}_{(+)}\} - \{\underbrace{\bar{P}^H - \pi^H \cdot L}_{(+/-)}\}] - \frac{S^*}{\mu} = 0. \quad (6.11)$$

$\phantom{e(\mu,S)[\{}\underbrace{\phantom{\bar{P}^L}}_{(+)}$

This can be interpreted as follows.

1. *Difference in contributions to expected profit.* This is the necessary condition for selection effort to pay off. If contributions to expected profit were equal, the term in square brackets of Eq. (6.11) would be zero, imparting a negative value to the whole expression. Additional selection effort would always reduce expected profit, calling for $S^* = 0$. Absent premium regulation, contributions would become equal at least in the long run due to experience rating, undermining the rationale for risk selection.

6.5 Underwriting Policy

2. *Positive contribution of high-risk types to expected profit.* Let the high-risk types contribute to expected profit. Intuition seems to suggest that the IC will refrain from risk selection in this situation. Yet the bracketed term of (6.11) may still be positive, indicating that optimal selection effort S^* is positive. It is the difference in expected profit margins that matters. As long as the margin of low-risk types exceeds that of high-risk types, the bracket in (6.11) has a positive value, implying $S^* > 0$.
3. *Negative contribution of high-risk types to expected profit.* In this case, the bracket in (6.11) necessarily has a positive value, inducing a positive value of S^*. The more strongly negative the contribution to expected profit from high-risk types, the more selection effort the IC is predicted to undertake.
4. *Influence of premium regulation.* As a rule, premium regulation seeks to reduce the premium paid by the high risks $\overline{P}^H$ in order to bring them closer to $\overline{P}^L$, the premium paid by the low risks. However, this induces a difference in profit margins that in turn triggers risk-selection effort [see item (1) again].
5. *Share of low-risk types in the population.* The higher the share of low risks μ, the higher must be S^* *ceteris paribus* to satisfy the condition (6.11). The intuition is that a substantial share of low-risk types means a small share of high-risk ones, permitting to target effort designed to avoid them. This serves to lower the marginal cost "per high risk avoided", justifying more risk-selection effort.
6. *Improved screening technology.* In the future, the IC might have access to genetic information of potential IBs (see Sect. 11.2.1). In the extreme, this would raise the probability that some IBs will suffer from a specific health condition to 100%, making them uninsurable. In all other cases, insurability is preserved but combined with increased selection effort. This follows from noting that genetic information would increase the effectiveness of the risk-selection effort. The elasticity $e(\mu, S)$ increases, implying that it takes a higher S^* to satisfy condition (6.11). Many observers indeed fear a boost of insurers' risk-selection effort in response to the availability of genetic information.

▶ **Conclusion 6.9** To the extent that management of an IC pursues expected profit maximization, an increase in risk-selection effort is predicted in particular when regulators reduce the premium of high-risk types or if access to genetic information is facilitated.

This model is simplistic because it neglects the fact that the IC could not only invest in risk-selection effort but also in product innovation. Product innovation also serves to attract favorable risks because the younger and better educated are likely to be those who first try a new (insurance) product. At the same time, it can reduce expected loss, e.g., by limiting moral hazard effects. In Zweifel (2007), innovation of this type is analyzed as an additional element of insurance technology. It turns out that regulation designed to limit risk-selection effort tends to make product innovation less attractive as well, in the main by lowering its contribution to expected profit [for a case study in health insurance, see Schoder et al. (2010)].

6.6 Controlling Moral Hazard Effects

Moral hazard is defined as the change in behavior of an IB induced by insurance coverage (for a precise definition; see Sect. 8.2). Moral hazard means that after signing the contract, the IC must consider an increased probability of loss as well as a higher amount of benefit to be paid once the loss has occurred. This causes not only expected loss but also the loading for administrative expense and risk-bearing to increase, undermining the IC's competitiveness. Specifically, unless its risk categorization is perfect, the IC charges customers who are characterized by little or no moral hazard with an excessive price causing them to migrate to a competitor.

These considerations establish an incentive for the IC to rein in moral hazard effects. In the literature, principal-agent theory has been applied to this issue [see e.g., Winter (1992)]. The objective is to determine the optimal payment function from the point of view of the principal (the IC) to ensure that the agent (the IB) acts in the interest of the principal to the greatest extent possible (as discussed in Sect. 6.4.2 in the context of distribution channels). Here, the payment function indicates how insurance benefits vary as a function of loss. Full coverage of the loss implies that preventive effort on the part of the IB would not be honored at all. Contrary to the optimal function in the absence of moral hazard (see Sect. 3.3.4), it is not optimal either to provide full marginal coverage beyond a deductible, except when there are other benefits to prevention (e.g., in terms of better health) while the cost of prevention to the IB is very low [see Winter (1992)].

While the standard assumption is that the IC cannot observe the preventive effort of the IB at all, it would clearly be going too far to claim that controlling moral hazard effects cannot be part of insurance technology. For instance, in commercial insurance, ICs do inspect fire prevention measures in some detail; in workers' compensation, they often retain the right to check the beneficiary's health status by a home visit. ICs can also verify that the IB has fulfilled duties of diligence as stipulated in the policy, with failure to perform permitting them to curtail or even cancel benefits. Evidently, these activities are rather costly. Checking for moral hazard in one hundred percent of cases is, therefore, out of the question. Rather, the issue is to determine an optimal frequency of checking.

Following the game-theoretic analysis by Borch (1990), there are two players, the IC and the IB. The decision variable on the part of the IC is the probability κ with which it exerts control. For instance, it can verify whether maintenance of fire extinguishers is performed as stipulated in the contract; in the case of health insurance, a physician commissioned by the IC may check whether the therapies applied were appropriate. Whenever the IC opts for "control" κ rather than "trust" $1 - \kappa$, it has to bear a cost b per contract monitored (see Table 6.5).

On the other hand, the IB also has a choice. With frequency ρ, he or she may decide to indulge in moral hazard, skimping on prevention. However, in case the IC finds out, it metes out a sanction in the guise of a reduction of benefit to the tune of Q.

The entries of Table 6.5 indicate the payoffs to the IB and the IC, conditional on the action of the other player. Both players are assumed to be risk-neutral with

6.6 Controlling Moral Hazard Effects

Table 6.5 Payoffs to the Insurance Buyer (IB) and the Insurer (IC)

		IC	
		Trust $(1-\kappa)$	Control (κ)
IB	Prevention $(1-\rho)$	(1) $EL - P - V$; $P - EL$	(2) $EL - P - V$; $P - EL - b$
	No prevention (ρ)	(3) $EL_0 - P$; $P - EL_0$	(4) $EL_0 - P - Q$, $P - E_0 + Q - b$

Note:
- EL: Expected loss ($=$ average benefit) given prevention
- EL_0: Expected loss ($=$ average benefit) in the absence of prevention
- P: Premium
- V: Preventive effort at the price of 1
- b: Cost of control per inspection
- Q: Reduction of benefit paid
- κ: Probability of checking
- ρ: Probability of performing prevention

regard to the implied variation of their wealth, which is disputable especially for the IB. However, risk neutrality permits the expression of payoffs in terms of money rather than utility. In cell (1) of Table 6.5, the IB invests in prevention while the IC does not check. On average, IBs obtain a benefit amounting to the expected loss EL (assuming complete insurance coverage), while having to pay premium P and bearing the cost of prevention V (effort valued at the price of 1). The second entry of cell (1) indicates the payoff to the IC. It consists of the premium P cashed in a net of EL, the expected value of the loss paid.

In cell (2), the IB again makes preventive efforts while the IC checks. The expected payoff to the IB remains unchanged, but the IC bears the additional cost b of the inspection. In cell (3), the IB skimps on prevention whereas the trusting IC refrains from checking. With no prevention, expected loss rises to $EL_0 > EL$, which the IB obtains as the benefit on average and which is paid by the IC. In cell (4), the IB is caught skimping on prevention. There is a sanction amounting to Q. In the one-period context of this model, this is a curtailment of benefits; in a multi-period context, it could also be a surcharge on the next period's premium. The IC receives Q but bears the cost of checking b.

In all, Table 6.5 contains the payoffs of a game in mixed strategies between the IC and the IB. Since the game can be interpreted as being of the repeated type, the probabilities ρ and κ become relative frequencies. The higher κ, the more frequently the IC performs its inspections and the greater are its effort and cost.

The expected value of the payoff for the IB is given by

$$\begin{aligned} EW^{IB} &= (1-\kappa)(1-\rho)(EL - P - V) + \kappa(1-\rho)(EL - P - V) + (1-\kappa) \\ &\quad \times \rho(EL_0 - P) + \kappa\rho(EL_0 - P - Q) \\ &= (1-\rho)(EL - P - V) + (\rho - \kappa\rho + \kappa\rho)(EL_0 - P) - \kappa\rho Q \\ &= EL - P - V + \rho(V + EL_0 - EL - \kappa Q). \end{aligned} \quad (6.12)$$

In a similar way, the expected payoff for the IC reads, after rewriting $(P - EL_0)$ as $(P - EL) - (EL_0 - EL)$,

$$\begin{aligned}
EW^{IC} &= (1-\kappa)(1-\rho)(P-EL) + \kappa(1-\rho)(P-EL-b) + (1-\kappa) \\
&\quad \times \rho(P-EL_0) + \kappa\rho(P-EL_0+Q-b) \\
&= (1-\kappa)(1-\rho)(P-EL) + \kappa(1-\rho)(P-EL) - \kappa(1-\rho)b \\
&\quad + (1-\kappa)\rho[(P-EL) - (EL_0 - EL)] + \kappa\rho[(P-EL) - (EL_0 - L)] \\
&\quad + \kappa\rho(Q-b) \\
&= P - EL - \rho(EL_0 - EL) - \kappa(b - \rho Q). \quad (6.13)
\end{aligned}$$

The game is assumed to be non-cooperative, meaning that the IB and the IC cannot exchange information in an attempt to attain a better outcome for both. This implies that the IC when choosing its frequency of checking κ must take optimization by the IB into account. Starting, therefore, with the IB, one has from (6.12)

$$\frac{\partial EW^{IB}}{\partial \rho} = V + EL_0 - EL - \kappa Q = 0. \quad (6.14)$$

In view of this condition, the IC sets

$$\kappa^* = \frac{V + EL_0 - EL}{Q}. \quad (6.15)$$

The IC is predicted to opt for a high frequency of checking (κ^* high) if

- V is large, i.e., if prevention is costly for the IB (implying that the incentive to indulge in moral hazard is marked);
- $(EL_0 - EL)$ is large, i.e., if moral hazard results in a marked increase in expected loss and hence benefits to be paid;
- Q is small, i.e., if the contract provides only for a small sanction in case moral hazard is detected.

In their turn, the IBs know that the frequency of checking by the IC must satisfy the condition,

$$\frac{\partial EW^{IC}}{\partial \kappa} = -b + \rho Q = 0. \quad (6.16)$$

This permits them to select their optimal propensity to skimp on prevention given by

$$\rho^* = \frac{b}{Q}. \quad (6.17)$$

6.6 Controlling Moral Hazard Effects

The IBs are, therefore, predicted to neglect prevention with high frequency (ρ^* high, see Table 6.5), if

- b is large, i.e., if the marginal cost of checking is high for the IC;
- Q is low, i.e., if the marginal return to checking in the guise of a reduced payment is low for the IC.

Note that the pair of values (ρ^*, κ^*) defines a Nash equilibrium because neither player has an incentive to deviate.[4]

▶ **Conclusion 6.10** IC effort at limiting moral hazard can be modeled as a game between the IB and the IC. The IC optimally checks on preventive effort employed by the IB with a frequency that depends positively on the cost of prevention, the effect of moral hazard on the expected loss, and negatively on the contractual sanction in case absence of prevention is detected.

Finally, Eq. (6.17) can be used to derive an optimal premium function. The reason is that the premium must not only cover the expected loss, but also the expected value of the additional payments and expenses that are induced by moral hazard. Since the IB will skimp on prevention with probability ρ^*, the premium is given by

$$P = EL + \rho^*(EL_0 - EL) = EL + \left(\frac{b}{Q}\right)(EL_0 - EL). \tag{6.18}$$

Clearly, the premium cannot be actuarially fair anymore in the presence of moral hazard but must contain a loading because $\rho^* > 0$ (a similar result is presented in Sect. 8.2.2.1).

6.7 Reinsurance

This section is devoted to reinsurance (RI) demanded by the primary insurer. The supply side of the RI market and equilibrium issues are only touched in passing (see also Sect. 10.4.2).

The question can be asked: Why should an IC buy reinsurance? As discussed at the beginning of Sect. 5.3, the CAPM predicts that the use of insurance is irrelevant

---

[4] At first sight, κ^* does not appear to be the best response to ρ^* because ρ^* does not figure in (6.15). However, by (6.17), ρ^* is governed by Q, causing κ^* to implicitly depend on ρ^* through Q in the denominator of (6.15). An analogous argument applies to ρ^* as the best response to κ^*.

in equilibrium. However, many arguments for the use of insurance are subsequently developed. These arguments are also relevant here with respect to the buying of reinsurance by an IC. What follows are related arguments for the reinsurance purchase decision.

6.7.1 Functions of Reinsurance

Generally, six functions of RI are distinguished [Outreville (1998), part V].

1. *Risk transfer*. Precisely as IBs decide to bear certain risks themselves or to transfer them to an insurer, the IC has the choice of retaining risk or transferring it at least in part to a reinsurer. In principle, the primary insurer could seek to also relieve itself from risks associated with capital investment; however, the focus of RI is on the underwriting business. Risks related to underwriting can be a "loss risk" (losses are on average higher than expected), a "probability risk" (the probability of loss is higher than calculated), and "distribution risk" (the cumulative density function used for calculating the premium was not the true one). RI relieves the primary insurer of all of these risks to an extent that depends on the type of contract (see Sect. 6.7.2).
2. *Increasing the underwriting capacity of the insurer*. The underwriting of risk could call for future payments that exceed the surplus (i.e., the sum of equity and reserves) of the primary insurer. If the IC seeks to keep the probability of insolvency at the level it (or the regulatory authority) deems optimal, there are three alternatives. First, the IC can renounce the business. Second, it can take only a share of the business, along with other ICs (this is often called coinsurance). The downside of this alternative is the costs of contract preparation, monitoring, and execution. The third alternative is to purchase RI coverage designed to cap loss payments at an amount that does not jeopardize the solvency of the IC.
3. *Substitute for equity*. The commitment to pay losses arguably is not reflected in the total of premiums but only in net premiums written after deductions made for premiums ceded to RI, showing the degree to which the IC is relieved of this commitment. Regulators, therefore, tend to relate equity (and reserves) to net premiums when judging the solvency of an IC. In this event, RI coverage has the same effect as equity capital. For instance, let the purchase of RI coverage reduce net premiums from 100 to 90 MU (monetary units). Given a required solvency margin of 10%, minimum equity falls from 10 to 9 MU. This permits the IC to underwrite more risk for a given amount of equity, i.e., to increase its leverage.
4. *Reserve smoothing*. The underwriting of losses that are known to vary little around expected value calls for a small increase in reserves, to be financed by the loading for risk-bearing included in the premium. However, a loss can be several standard deviations above the expected value (for instance, because of positive correlation due to a major storm). It would entail a safety loading so high that the IC might lose the business. Rather than running this risk by seeking to immediately adjust

its reserves, it can rely on RI for covering extremely high losses, permitting to avoid the hike in reserves and premiums.
5. *Services.* The reinsurer can support the primary insurer in its underwriting activities because it has a very comprehensive information base (especially concerning high and rare losses). Apart from the assessment of risks, services provided by a reinsurer comprise advice concerning loss prevention and pricing, help with claims settlement, and protection from unjustified claims.
6. *Financing.* The reinsurer may advance the provisions that must be paid to distributors when creating a new line of business. In return, the IC agrees to purchase RI coverage from its sponsor.

Traditionally, functions (1) and (4), risk transfer and reserve smoothing, are regarded as the core functions of RI. However, from the point of view of shareholders as the owners of an IC, this justification of demand for RI is not very convincing. They hold a claim against the net value of the firm, i.e., assets minus liabilities $(A - L)$ of the IC.[5] By buying RI coverage, the IC lowers both A and L. If the RI premium is fair, the expected value $E(A - L)$ does not change. There is a reduction in variance, $Var(A - L)$; however, this should be irrelevant to shareholders who are fully diversified themselves [see Doherty (1981)]; the argument is in full analogy with the one advanced in Sects. 5.1 and 5.2 for firms outside the insurance industry.

However, there are other groups of claimants in addition to shareholders. In the case of an IC, these are the policyholders (IBs). Most of them are not fully diversified through the capital market (otherwise they presumably would not have bought insurance). For them, the insolvency of the IC is of far greater importance than for a diversified investor, who is little affected by the loss of value of a stock that makes up a tiny share of his or her portfolio. The concern of IBs for a solvent insurer can be allayed by a risk transfer to a reinsurer. This, in turn, enhances the IBs' demand for coverage, and hence the premium volume of the IC. Through this demand effect, $E(A - L)$ does increase, causing the purchase of RI coverage to be in the interest of the owners of the IC. Indeed, Doherty and Garven (1986) found empirical evidence suggesting that premium rates of ICs in the United States are positively related to the degree of their RI coverage.

6.7.2 Types of Reinsurance

In a series of papers, Borch developed a "theory of reinsurance" [see, e.g., Borch (1961, 1962)]. He included a discussion on the optimal use of "stop loss" reinsurance (Borch, 1960a), "quota share" treaties (Borch, 1960b), and "reciprocal" treaties (Borch, 1960c). He argued a quota share treaty where the shares add to one "is superior to any other form of reinsurance because it gives ...the smallest possible variance" (Borch 1960c, p. 46). This suggests that ICs "should cede their entire portfolio to

---

[5]For simplicity, the same symbol L is used for payment of losses and liabilities more generally.

a pool, and then agree on some rule as to how payment of claims against the pool should be divided among the companies" (Borch 1960b, p. 170). However, as shown by Eisen (1979a), these treaties result in a situation which is not Pareto optimal.

Here, only the most important variants of reinsurance (RI) can be described. A comprehensive survey is provided by Strain (1989); for a concise introduction, see Outreville (2002). Formulas for loss distributions relevant to both primary insurers and reinsurers are available in Daykin et al. (1994, Sect. 3.4).

If there is a general agreement stipulating that the primary insurer automatically cedes risks underwritten in part to the reinsurer, this is called *mandatory RI* or *automatic RI*. As soon as the primary insurer can decide on a per-case basis whether or not it wants to buy RI coverage, this is called *facultative RI* (also known as treaty reinsurance). Facultative RI is associated with a problem of adverse selection because the primary insurer has an incentive to cede business with a high amount of loss risk, probability risk, or distribution risk.

Another important distinction is between proportional (pro-rata) and nonproportional (excess) RI.

1. *Proportional reinsurance.* In this case, a share α of a loss is paid by the reinsurer. It can be shown that the pro-rata RI is the alternative with the lowest premium provided the reinsurer charges a loading on the fair premium that increases with the variance of the loss to be covered [Von Eije (1989), p. 54, see also Beard et al. (1984), p. 174 f. and Sect. 7.1.3]. The RI premium P^{RI} is then given by

$$P^{RI} = EL^{RI} + f\{Var(L^{RI})\}, \text{ with } \partial f/\partial\{Var(L^{RI})\} > 0, \qquad (6.19)$$

P^{RI}: Reinsurance premium;
EL^{RI}: Expected value of loss covered by the reinsurer.

The loss to be borne by the reinsurer is given by the gross amount L minus the contribution of the primary insurer, I. One, therefore, has $L^{RI} = L - I$ and hence

$$Var(L^{RI}) = Var(L - I) = Var(L) + Var(I) - 2 \cdot Cov(L, I)$$
$$= Var(L) + Var(I) - 2 \cdot \rho_{LI} \cdot \sigma_L \sigma_I, \text{ with} \qquad (6.20)$$

$$\rho_{LI} := \frac{Cov(L, I)}{\sigma_L \cdot \sigma_I},$$

L^{RI}: Losses paid by the reinsurer;
I: Losses retained by the primary insurer; $I = L - L^{RI}$;
ρ_{LI}: Correlation coefficient between total losses and losses retained by the primary insurer.

For simplicity, assume that the variances $Var(L)$ and $Var(I)$ on the right-hand side of Eq. (6.20) are not affected by the type of RI contract. In that case, only the correlation between gross losses and retained losses ρ_{LI} is left as a choice variable. In accordance with Eq. (6.19), $Var(L^{RI})$ should be minimized for minimizing the RI premium. This implies that in Eq. (6.20) the correlation coefficient ρ_{LI} should be maximum, i.e., $\rho_{LI} = 1$. However, this is achieved if L and I move in a perfect linear relationship. Proportional RI with $I = (1 - \alpha)L$ is perfectly linear.

This result is intuitive from the point of view of the reinsurer. In the event of a very large loss, it has the guarantee that the primary insurer bears its share, permitting the reinsurer to offer coverage at a low premium. However, for the primary insurer, it may still be the case that RI affords an insufficient reduction of variance in liabilities. In view of this disadvantage, it would be inappropriate to discard other types of RI on the grounds that they are dominated by proportional RI.

2. *Aggregate-excess contract*. Here, the primary insurer pays the total loss of a year up to a limit, while the excess is covered by the reinsurer. This contract is of the so-called stop-loss type. Providing for full marginal coverage beyond a deductible, it is equivalent to the Pareto-optimal contract between a risk-averse IB and a risk-neutral IC in the presence of administrative expenses (see Sect. 3.3.4). Compared to proportional reinsurance, both components of the formula (6.19) change. First, the reinsurer is relieved of small losses, causing its expected value EL^{RI} to increase. A counteracting effect is that the reinsurer may also be involved in claims settlement [function (5) in Sect. 6.7.1]. In that event, an aggregate-excess contract prevents many small claims from being presented to the reinsurer that would be reported in a proportional contract. However, $Var(L^{RI})$ also changes. As long as claims remain below the primary insurer's retention limit, this term is equal to zero. Beyond the retention limit, it is the conditional variance $Var(L^{RI}|L > I)$ that affects the reinsurer. Note that a conditional variance is analogous to the explained variance in regression analysis, and therefore, cannot be greater than the unconditional variance. Therefore, for losses above the retention limit, the aggregate-excess contract would be less costly than proportional reinsurance. It is the net effect on EL^{RI} below the retention limit that is decisive.[6] The reason may be that loss risk and distribution risk strongly affect the reinsurer. For example, the rate of inflation may be higher than expected. This causes the gross amount of loss to exceed the retention limit with a higher frequency than anticipated, to the detriment of the reinsurer.

3. *Per-risk excess contract*. This contract is also of the stop-loss type, with the retention defined not in terms of an aggregate monetary value but in terms of an individual risk underwritten (i.e., an individual contract). This permits primary insurers to transfer part of their risk of cumulative loss to the reinsurer. For instance, let a storm damage several large buildings. In this case, the reinsurer

---

[6]This conclusion holds only as long as variance is used as a measure of risk. For all its advantages for the primary insurer, aggregate-excess contracts are not very common.

participates in each loss exceeding the retention limit. The fact that damages to buildings in other parts of the country possibly are below expected value (a diversification effect that would be to the benefit of the reinsurer in an aggregate-excess contract) is irrelevant here.

4. *Per-occurrence excess contract.* In this type of contract, the retention limit is defined in terms of a single event. In the example above, the risk would not be defined in terms of the single insured building but in terms of a storm or an earthquake. For this reason, per-occurrence excess contracts are often called catastrophic contracts. The definition of "occurrence of loss" can be very crucial. For example, do the two attacks on the World Trade Center of New York on 11 September 2001 constitute one or two events? If it is defined as one event, is the relevant retention limit simply the sum of the two limits applying to each of the buildings? RI payments depend strongly on the answers to questions of this type.

▶ **Conclusion 6.11** The primary function of reinsurance is risk transfer. It benefits even fully diversified owners of the primary insurer if a reduction of the insolvency risk results in a higher premium volume. Although proportional reinsurance has a minimum premium, it need not dominate other types of contract from the point of view of the primary insurer.

6.7.3 A Model of Demand for Reinsurance Based on Option Pricing Theory

This section is devoted to a study that models the functions No. (1) to (4) of reinsurance cited in Sect. 6.7.1 while also taking into account the effect of taxation [see Garven and Lamm-Tennant (2003)]. The basic assumption is that management of the primary insurer acts in the interest of the shareholders when deciding about the purchase of RI coverage.

Recall that the owners of an IC hold a call option on the value of the company. If the IC has a positive net value, they exercise their option by keeping their stock. In the case of a negative net value, the IC is insolvent. Rather than having to pay their share of the net debt, the owners of the IC have the right to simply not exercise their call option. As argued in Sect. 7.2.3, this right amounts to a put option that has positive value when the IC is insolvent. However, let the probability of insolvency be so low that this put option can be neglected. Taxation is introduced by noting that the government also has a call option. If the IC has a positive value at the end of the period, the government claims its tax share; conversely, if it has a negative value, the government does not share in the loss (by paying back taxes) but rather lets go of the option with a payoff of zero.

The call option of shareholders has a gross (before tax) value given by

$$C(A \cdot R_p; -U) = R_f^{-1} \int_{-\infty}^{\infty} \int_{-\infty}^{\infty} max[(A \cdot R_p + P_n - (1-\alpha)L), 0] \, \hat{f}(r_p, L) dr_p dL. \quad (6.21)$$

6.7 Reinsurance

Table 6.6 A model of reinsurance demand based on option theory

$$C(A \cdot R_p; -U) = R_f^{-1} \int_{-\infty}^{\infty}\int_{-\infty}^{\infty} max[(A \cdot R_p + P_n - (1-\alpha)L), 0]\,\hat{f}(r_p, L)dr_p dL \qquad (6.21)$$

$$C(A \cdot R_p; -U) = R_f^{-1} \int_{-P_n}^{\infty} (Y + P_n)\hat{f}(Y)dY \qquad (6.27)$$

$$\tau C(A \cdot \theta \cdot r_p; -U) = \tau R_f^{-1} \int_{-\infty}^{\infty}\int_{-\infty}^{\infty} max[(A \cdot \theta \cdot r_p + P_n - (1-\alpha)L), 0]\,\hat{f}(r_p, L)dr_p dL \qquad (6.28)$$

$$\tau C(A \cdot \theta \cdot r_p; -U) = \tau R_f^{-1}(Z + P_n)\hat{f}(Z)dZ \qquad (6.30)$$

$$V = C(A \cdot R_p; -U) - \tau C(A \cdot \theta \cdot r_p; -U) = R_f^{-1} \int_{-P_n}^{\infty}(Y + P_n)\hat{f}(Y)dY - \tau R_f^{-1} \qquad (6.31)$$

$(Z + P_n)\hat{f}(Z)dZ$

Note:

$C(\cdot)$:	Value of a call option written on the net value of the IC
α:	Share of losses ceded to RI
$P_n(\alpha)$:	Net premiums written by primary insurer, $P_n'(\alpha) < 0$
$A(\alpha)$:	Initial assets of primary insurer, $= S_0 + kP_n(\alpha)$ with $A'(\alpha) < 0$
k:	Fund-generating factor; share of premium revenue that is available for capital investment
θ:	Share of assets subject to taxation
τ:	Rate of taxation applied to profits of IC
$f(r_p, L)$:	Bivariate normal distribution defined over rates of return on capital investment and loss payments
$\hat{f}(r_p, L)$:	Bivariate normal distribution that makes IC management risk-neutral
r_f:	Risk-free rate of return
r_p:	Rate of return on capital investment by the IC
R_i:	$= 1 + r_i$, $i = f, p$ (f: risk-free), p: portfolio of the IC
S_o:	Initial surplus
U:	Underwriting result before tax, $= P_n - (1-\alpha)L$
V:	Net value of call option held by IC shareholders
Y:	Final surplus before net premium income, $= AR_p - (1-\alpha)L$

Source Garven and Lamm-Tennant (2003)

For simplicity, only proportional reinsurance is considered here; therefore, the decision variable is α, the share of aggregate loss ceded to RI (see Table 6.6). The gross value of the call option can vary in the following domain. On the upside, existing assets (A) increase by the interest factor $R_p = (1 + r_p)$ to become $A \cdot R_p$ by the end of the year, with r_p denoting the rate of return on the capital market. On the downside, $-U$ denotes the exercise price of the option, i.e., the value of the pre-tax underwriting result where the owners of the IC declare bankruptcy because their shares have zero value. This (negative) underwriting result is given by $-U = (1-\alpha)L - P_n$, i.e., the negative of the pre-tax underwriting result, with P_n denoting net premiums written and $(1-\alpha)L$, losses to be paid net of reinsurance by the IC. Net premiums are defined by

$$P_n(\alpha) = \bar{P} - \pi(\alpha) - \alpha P^R, \quad \text{with } \pi'(\alpha) < 0, \; P'_n(\alpha) < 0. \tag{6.22}$$

Here, $\bar{P}$ symbolizes the (exogenous) premium income that would be attainable if the IC had absolutely no risk of insolvency. This risk is denoted by $\pi(\alpha)$; it decreases with the share α of losses covered by RI. For simplicity, $\pi(\alpha)$ is expressed in monetary terms and includes costs related to insolvency (e.g., lawyers' fees and court expenses). Moreover, full reinsurance coverage would cost P^R; the reinsurance premium, therefore, amounts to αP^R. The overall effect of additional RI coverage on net premiums is assumed to be negative.

The right-hand side of Eq. (6.21) states that stockholders can choose (opt for the maximum) between the value of the IC (if positive) and zero. A positive value of the IC is given by the sum of the underwriting result $P_n - (1 - \alpha)L$ and final assets given by $A \cdot R_p = A(1 + r_p)$ due to capital investment at a rate of return r_p. This leaves initial assets A to be determined, which amount to

$$A(\alpha) = S_0 + k P_n(\alpha), \quad \text{with } A'(\alpha) < 0 \text{ because of (6.22)}. \tag{6.23}$$

Initial assets are, therefore, given by the surplus inherited from the previous period S_0 augmented by a multiple k of net premiums P_n that is not claimed as losses yet during the current period, and therefore, is available for investment. The constant k is called the funds generating factor; it reflects the time lag between premiums cashed and losses paid (if measured in years, $k = 1.5$ is a typical value in non-life).

Returning to Eq. (6.21), one sees that the pre-tax value of the option for shareholders depends on two stochastic variables, the rate of return in capital investment r_p and the amount of loss L. Let $f(\cdot)$ indicate the probability with which possible pairs of values $\{r_p, L\}$ occur. It is assumed to be bivariate normal although in particular losses in non-life business are characterized by rare outliers causing the density function to be skewed with regard to L. However, the normal distribution has the great advantage that linear transformations of its arguments are also normally distributed (see below). Finally, note that the true density function $f(\cdot)$ is replaced by the less disperse function $\hat{f}(\cdot)$. This serves to take risk aversion by IC management into account without having to introduce a risk utility function. Therefore, let IC management confront a density function $\hat{f}(\cdot)$ such that the *max* operator renders the certainty equivalent of the option value. The use of certainty equivalents can also be justified by noting that risk aversion does not show up in the pricing of securities in the capital market (see Sect. 5.1.3). For the same reason, the option is discounted to present value using the risk-free rate of return, with $R_f = (1 + r_f)$.

The integral over all possible value of the option must be formed to obtain an expected value. In the case of losses L, the range of values should in principle be $[0, \infty)$ rather than $(-\infty, \infty)$, but this difficulty can be solved through a transformation of variables. This transformation reads

$$Y = A \cdot R_p - (1 - \alpha)L, \tag{6.24}$$

6.7 Reinsurance

combining the two stochastic quantities into one. The variable Y is equal to the surplus (before net premium income) at the end of the period, with the liabilities of the IC reduced by RI coverage. For the hypotheses to be tested, the variance of this quantity is of interest,

$$\sigma_Y^2 = A^2 \sigma_p^2 + (1-\alpha)^2 \sigma_L^2 - 2A(1-\alpha)\sigma_{pL}, \qquad (6.25)$$

with σ_{pL}: covariance between r_p and L.

For future reference, the relation between the variance of Y and the amount of RI coverage α is of interest,

$$\frac{\partial \sigma_Y^2}{\partial \alpha} = 2A \cdot A'(\alpha)\sigma_p^2 - 2(1-\alpha)\sigma_L^2 - 2\sigma_{pL}[A'(\alpha)(1-\alpha) - A] < 0 \text{ if } \sigma_{pL} \leq 0] \qquad (6.26)$$

since $A'(\alpha) < 0$ (for the sign of σ_{pL}, see H2 below).

Using (6.24), one can simplify Eq. (6.21) to become

$$C(A \cdot R_p; -U) = R_f^{-1} \int_{-P_n}^{\infty} (Y + P_n) \hat{f}(Y) dY. \qquad (6.27)$$

Rather than two variables of integration, there is now only one since Y combines R_p and L. Equation (6.27) involves the expected value of the final surplus $(Y + P_n)$; note that net premiums P_n are considered to be nonstochastic. The final surplus is zero (implying a zero value of the option), if $Y = A \cdot R_p - (1-\alpha)L = -P_n$. In this case, the negative underwriting result $-U = -(1-\alpha)L + P_n$ is exactly balanced by the assets available after capital investment AR_p. Accordingly, the lower limit of integration changes from $-U$ to $-P_n$. The max operator can also be dropped because above $-P_n$, the value of the option is always positive.

Equation (6.27) determines the gross value of the call option held by the owners of the IC. From this, the value of the call option held by the government reflecting its tax claim needs to be deducted. It is given by

$$\tau C(A \cdot \theta \cdot r_p; -U) = \tau R_f^{-1} \int_{-\infty}^{\infty} \int_{-\infty}^{\infty} max[(A \cdot \theta \cdot r_p + P_n - (1-\alpha)L, 0] \cdot \hat{f}(r_p, L) dr_p \, dL. \qquad (6.28)$$

Compared to the expression (6.21) for the shareholders, there are three adjustments.

- The tax claim of the government does not concern the assets of the IC but its income. For this reason, the argument of the *max* operator contains investment income Ar_p rather than final assets AR_p;

- Tax usually is levied on investment income depending on its source. In particular, government bonds are often exempted from taxation. The share of assets whose returns are taxed is denoted by θ; and,
- The final value of the option for the government is higher, the higher the rate of taxation τ on the profit of the IC.

Again, the two stochastic variables r_p and L are combined into one normally distributed variable Z denoting the stochastic component of profit,

$$Z = A \cdot \theta \cdot r_p - (1-\alpha)L. \tag{6.29}$$

Substitution of (6.29) into Eq. (6.28) yields

$$\tau C(A \cdot \theta r_p; -U) = \tau R_f^{-1} \int_{-P_n}^{\infty} (Z + P_n) \hat{f}(Z) dZ. \tag{6.30}$$

Again, the limit of integration is modified, permitting the dropping of the *max* operator.

Finally, the net value V of the call option held by the owners of the IC is the difference between the expressions (6.27) and (6.30), resulting in

$$V = R_f^{-1} \int_{-P_n}^{\infty} (Y + P_n) \hat{f}(Y) dY - \tau R_f^{-1} \int_{-P_n}^{\infty} (Z + P_n) \hat{f}(Z) dZ. \tag{6.31}$$

For the derivation of the hypotheses H1 to H4 stated below, one would have to differentiate Eq. (6.31) w.r.t. the decision variable α to obtain the first-order condition for an optimum. This first-order condition is then subjected to a shock. Using comparative statics, one can qualitatively determine how the optimal rate of RI coverage has to adjust. Based on such an analysis (not detailed here), Garven and Lamm-Tennant (2003) state the following hypotheses.

H1: Other things equal, the demand for reinsurance will be greater the higher the leverage of the IC, i.e., the higher its net premiums compared to surplus.

The intuition is as follows. In terms of Eq. (6.23), high leverage means that for a given net premium P_n, there is a small amount of surplus S_0. This serves to reduce assets $A(\alpha)$ available for capital investment, with the consequence that the probability mass of Y in Eq. (6.27) shifts toward lower values, causing the value of the call option to decrease. This shift can be counteracted by buying additional RI coverage serving to reduce the variance of Y; this follows from Eq. (6.24). In addition, since the option held by the government depends much less on assets [$0 < \theta < 1$ in Eq. (6.28)] but rather on net premiums P_n, changes in its value have comparatively little influence on the net option value V.[7] On the whole, purchasing additional RI coverage serves to counteract the reduction in shareholders' option value due to increased leverage.

H2: Other things equal, the demand for reinsurance will be lower the more marked the (negative) correlation between the investment returns and claims costs of the IC.

The starting point is to recall from Sect. 5.2.2 that high volatility in the value of the underlying asset serves to increase the value of an option. Now Eq. (6.26) shows how the relationship between σ_Y^2 and α is affected by the change in σ_{pL}, the parameter in question. The authors argue that a positive covariance between rates of return on capital investment and losses ($\sigma_{pL} > 0$) would provide a natural hedge shareholders are *not* interested in. They, therefore, posit $\sigma_{pL} \leq 0$. Considering now a shift of σ_{pL}, the last term of (6.26) indicates that this serves to make the relation between σ_Y^2 and α more strongly negative. Purchasing more RI coverage would reduce the volatility of the surplus even more than otherwise. Therefore, a negative relationship between RI coverage and (negative) σ_{pL} is predicted.

H3: Other things equal, the demand for reinsurance will be greater for firms that write "longer-tail" lines of insurance.

In "longer-tail" lines claims come in with a considerable time lag on premiums. Therefore, a given surplus serves as equity for a greater amount of premiums written. The IB see this as an increased risk of insolvency, causing P_n in Eq. (6.27) to drop. As in H1, $Y + P_n$ shifts toward smaller values, diminishing the value of the option to shareholders. Increasing the share α ceded to RI serves to increase Y thus counterbalancing this effect.

H4: Other things equal, the demand for reinsurance will be greater for firms that concentrate their investments in tax-shielded assets.

The starting point here is that in equilibrium, after-tax returns must be the same across all investment alternatives. Adjusting returns for risk is not necessary because the density function $\hat{f}(\cdot)$ inducing risk neutrality is used. Fully taxed investments ($\theta \to 1$), therefore, must have higher rates of return r_p and hence $R_p = 1 + r_p$. An IC who reduces θ by investing in tax-shielded assets achieves a reduction of its tax burden but also a lower value of R_p. This causes Y in Eq. (6.31) to take on lower values, resulting in a reduced value of the option held by the shareholders of the

---

[7]Quite generally, the values of the options held by shareholders and the government move in parallel. However, the effect of the exogenous changes is smaller on the government's option, with the exception of a change in θ (see H4). Therefore, it suffices to examine the gross value of the option.

IC. As shown above, such a shift can be counterbalanced by purchasing more RI coverage.

6.7.4 Empirical Testing of the Model

Garven and Lamm-Tennant (2003) obtained data for 128 U.S. insurance companies covering the years 1980–1987 and resulting in a total of 1,350 observations. Some 60% come from mutuals, the remainder from stock companies. The evidence presented in Table 6.7 indicates that the ordinary least squares estimate is able to explain 32% of the variance in the dependent variable. However, the high t-ratios of *SIZE* and *LICENSE* (see below) show that much explanatory power is contributed by regressor variables that are not directly related to the theoretical quantities of the model. The results can be interpreted as follows.

REINS: Ratio between premiums paid for reinsurance coverage and gross premiums written. This is the dependent variable, reflecting the theoretical quantity α. However, the correspondence is imperfect because the data also contain other types of RI besides proportional reinsurance. Since some companies are net writers of reinsurance, *REINS* has a minimum value of -0.09. The mean value is 0.27, i.e., 27% of gross premiums are ceded to RI.

SIZE: Admitted assets of the IC, in log form. This is a first regressor that is not directly related to any of the theoretical variables. Nevertheless, there is the presumption that a large company has more possibilities of internal risk diversification (see Sect. 5.2), which may serve as a substitute for reinsurance. Therefore, *SIZE* is predicted to have a negative coefficient, which is strongly confirmed.

PSRATIO: Direct premiums written over surplus, with an average value 2.48. It indicates the degree of leverage. Since a high degree of leverage should increase the demand for RI (hypothesis H1), the predicted relationship is positive, which is clearly confirmed. Calculated at the means, an increase in this ratio by 10% (from 2.48 to 2.73, i.e., by 0.25 points) is associated with an increase of *REINS* by 0.0060 points ($= 0.0239 \cdot 0.25$). Compared to the mean value of 0.27, this amounts to an increase in *REINS* of some 2.2%. The estimated elasticity, therefore, is a non-negligible 0.22.

RHO: Coefficient of correlation between the rate of return on investment and loss payments, averaged over as many as 14 investment categories and 17 lines of underwriting. Each time, the correlation between the rate of return and the amount of loss is calculated, weighted by the respective shares, and summed up to obtain the average value.[8] This variable corresponds to σ_{pL}, and according to hypothesis H2, its partial relationship with *REINS* should be negative. The regression coefficient is indeed negative, supporting H2.

---

[8]The average value of the correlations amounts to 0.11, pointing to a preponderance of positive correlations. This contradicts the theoretical development that is based on $\sigma_{pL} \leq 0$ [see explanation of hypothesis H2 and Eq. (6.26)].

STDP: Standard deviation of returns on investment by the IC. In the model, this corresponds to σ_p^2; no hypothesis was derived for this variable. Yet according to Eq. (6.26), the variance of Y depends negatively on the share α devoted to RI (provided $\sigma_{pL} \leq 0$). This effect is stronger the higher σ_p^2, implying that there should be a negative relationship with RI coverage (in analogy with hypothesis H2). In fact, the regression coefficient is significantly positive, suggesting that IC management may buy RI coverage in its own (rather than shareholders') interests. This would constitute a sign of imperfect agency as discussed in Sect. 5.3.

STDL: Standard deviation of loss payments. The pertinent theoretical quantity is σ_L. Although again no hypothesis was derived for this variable, Eq. (6.26) shows that additional RI coverage serves to reduce the variance of Y when σ_L^2 is high. By purchasing reinsurance, IC management would especially hurt shareholders (who benefit from a high variance of Y). Therefore, the predicted sign of the coefficient is negative. In fact, it is positive but not significantly different from zero.

SCHEDP: Share of the premium volume that comes from schedule P (property/liability) lines. These lines are known for their long time lags between premiums cashed and losses paid; therefore, *SCHEDP* is an indicator of the theoretical variable k. According to hypothesis H3, it should be positively related with *REINS*. This prediction is clearly confirmed.

THETA: Share of returns from capital investment subject to tax, corresponding to the theoretical quantity θ. According to hypothesis H4, there should be a negative relationship with the demand for RI. The regression coefficient is negative but fails to attain statistical significance.

HERF: Herfindahl index of concentration applied to the underwriting activity of the IC. A high degree of concentration in underwriting means a lack of internal risk diversification and hence more volatility in the underwriting result Y. Purchasing reinsurance would reduce $Var(Y)$, hurting shareholders' interest. If an IC were to obtain premium income only from one line, its concentration index would be maximum with $HERF = 1$; if it obtained $1/17$ from each of the 17 lines of underwriting, $HERF = 1/17$.[9] Accordingly, *REINS* should depend negatively on *HERF*. The estimated coefficient is indeed negative but does not quite attain the 5% level of significance.

LICENSE: Number of states where the IC has a license to operate, multiplied by (-1). If an IC has a license in many states, its regional concentration of underwriting activity likely is low. The multiplication by (-1), therefore, makes *LICENSE* a direct indicator of concentration, analogous to *HERF*. More regional concentration likely causes $Var(Y)$ to increase. Shareholder interests would be hurt if this increase were counterbalanced by purchasing RI. Therefore, the partial relationship is predicted to be negative, which it definitely is.

---

[9] This index of concentration is given by $HERF = \sum m_i^2$, with $m_i :=$ share of the ith line of business in total premiums written. In the case of a uniform distribution over the n lines, one has $HERF = \sum(1/n^2) = m(1/n^2) = 1/n$.

Table 6.7 Demand for reinsurance by U.S. primary insurers[a] (1980–1987)

Regressor	Theoretical quality	Expected sign	Regression coefficient	Standard error	t-value
INTERCEPT			1.7587	0.1046	16.187
SIZE		−	−0.0777	0.0050	−15.417
PSRATIO	P_n/S_0	+	0.0239	0.0032	7.386
RHO	σ_{pL}	−	−0.1077	0.0461	−2.333
STDP	σ_p	−	0.8948	0.2333	3.835
STDL	σ_L	−	0.3177	0.3232	0.983
SCHEDP	k	+	0.0986	0.0285	3.458
THETA	θ	−	−0.0303	0.0261	−1.162
HERF[b]		(−)	−0.0460	0.0240	−1.916
LICENSE[b]		(−)	−0.0069	0.0004	−16.434
MUTUAL[b]		(+)	−0.0049	0.0112	−0.436
+ 7 dummy variables for years					

$R^2 = 0.32$, N= 1,350, OLS
Note [a]Dependent variable is *REINS*, the ration between RI premiums paid and gross premiums written. [b]Predicted signs not directly related to the model are in parentheses
Source Garven and Lamm-Tennant (2003)

MUTUAL: Categorical variable indicating whether an IC is organized as a mutual association (*MUTUAL* = 1) or stock company (*MUTUAL* = 0). While there is no direct relationship with the variables of the model, one can expect the management of a mutual to be less tied to the interests of owners, who cannot sell their shares to a challenger promising better performance. This fact speaks in favor of additional RI coverage since managers (whose assets are less diversified, see Sect. 5.3) have a personal interest in hedging against risk. However, this prediction is not borne out in Table 6.7.

T1 – T7: Seven dummy variables, not shown in Table 6.7. Each takes on the value of 1 in the respective year and of 0 otherwise (e.g., $T1 = 1$ for 1981, $= 0$ otherwise, $T2 = 1$ for 1982, $= 0$ otherwise, etc.), with 1980 constituting the benchmark year. They are all insignificant, with the exception of *T7* for 1987 whose coefficient is positive, likely reflecting the U.S. Tax Reform Act of 1986. This Act amounts to a hike in the tax rate τ serving to increase the value of the claim option held by the government. By increasing the purchase of RI beyond its original optimum, IC management can lower profit subject to taxation in the interest of shareholders. This calls for a positive partial relationship with *REINS* [see also Berger et al. (1992)].

▶ **Conclusion 6.12** Demand for reinsurance can be explained to a considerable degree with reference to the interests of the shareholders as the owners of the primary insurer who hold a call option on the net worth of the company, with deduction of the value of the call option held by the government in the guise of its tax claim.

6.8 Capital Investment Policy

The capital investment policy of an IC constitutes a crucial element of its insurance technology. As shown in Sect. 6.1, in the context of the financial statements of an IC, claims payment and expenses related to it can exceed the current premium income, making successful capital investment the condition determining whether an IC can meet its future liabilities.

The elements of portfolio theory examined in Sect. 5.1.2 can be used for the IC as an investor. In principle, however, the analogy is imperfect because in the main the IC does not invest its own funds but funds that result from reserves accumulated for meeting its liabilities. Therefore, an exclusive focus on assets invested is inappropriate; the development of liabilities matters too. The appropriate target quantity is the surplus i.e., the excess of assets A_t over liabilities L_t, discounted to present value because notably in life and pension insurance, liabilities have a maturity of 20 years or even more. Denoting by S_t the surplus amounting to equity and accumulated reserves of the IC, one has,[10]

$$S_t = A_t - L_t. \tag{6.32}$$

For simplicity, expenses related to the implementation of a capital investment policy are neglected. They typically constitute a fixed, nonstochastic deduction not directly relevant for optimization.

Change over time in surplus can be expressed as the change of assets minus the change of liabilities,

$$\begin{aligned} S_t - S_{t-1} &= (A_t - L_t) - (A_{t-1} - L_{t-1}) \\ &= (A_t - A_{t-1}) - (L_t - L_{t-1}). \end{aligned} \tag{6.33}$$

The next step is to express this change in percentage terms. One could use the surplus of the previous period S_{t-1} as the reference quantity; however, this would be problematic because S_{t-1} can assume both negative and positive values. It is more convenient to use assets of the previous period A_{t-1}. As to the second term of (6.33), it is multiplied by $1 = (L_{t-1}/L_{t-1})$ to become

$$\begin{aligned} \frac{S_t - S_{t-1}}{A_{t-1}} &= \frac{A_t - A_{t-1}}{A_{t-1}} - \frac{(L_t - L_{t-1})/L_{t-1}}{A_{t-1}/L_{t-1}} \\ &= \frac{A_t - A_{t-1}}{A_{t-1}} - \frac{L_t - L_{t-1}}{L_{t-1}} \frac{1}{A_{t-1}/L_{t-1}}. \end{aligned} \tag{6.34}$$

This equation states that the surplus rate of return achieved on the assets invested depends on three factors:

---

[10]The exposition follows Ezra (1991). Therefore, L_t does not represent current loss payments but total liabilities in the sense of present value of future payments.

1. The relative change in the value of assets. This change is nothing but the usual rate of return on a portfolio, a stochastic quantity;
2. The relative change in the value of liabilities. Note that this quantity also can be stochastic. This becomes especially clear in the context of a pension fund. For instance, changes in interest rates impact the present value of liabilities, and changes in the inflation rate have an influence as soon as the insured are guaranteed at least partial protection from inflation. Even in non-life insurance, inflation can unexpectedly shift the entire distribution of claims toward higher values; and,
3. The ratio between assets and liabilities in the previous period A_{t-1}/L_{t-1}, the so-called funding rate of the previous period. The higher this rate, the smaller is the impact of liabilities on net return.

Notation using the tilde serves to indicate that all quantities of Eq. (6.35) below are random variables, with the exception of the funding ratio $F:=A_{t-1}/L_{t-1}$, the value of which was determined in the previous period. Therefore, Eq. (6.34) can be rewritten as

$$\tilde{r}_S = \tilde{r}_A - \frac{1}{F} \cdot \tilde{r}_L, \quad \text{with } \tilde{r}_A := \frac{A_t - A_{t-1}}{A_{t-1}}, \quad \tilde{r}_L := \frac{L_t - L_{t-1}}{L_{t-1}}, \quad F := \frac{A_{t-1}}{L_{t-1}}. \tag{6.35}$$

The expected value of surplus return is given by

$$E\tilde{r}_S = E\tilde{r}_A - \frac{1}{F} E\tilde{r}_L. \tag{6.36}$$

As to the variance of $\tilde{r}_s$, note that $(1/F)$ enters as a scaling factor,

$$Var(\tilde{r}_S) = Var(\tilde{r}_A) + \frac{1}{F^2} Var(\tilde{r}_L) - \frac{2}{F} Cov(\tilde{r}_A, \tilde{r}_L). \tag{6.37}$$

The volatility of the surplus rate of return, therefore, coincides with the volatility of the rate of return on capital investment only if the funding rate is extremely high ($F \to \infty$). In all other cases, volatility in liabilities $[Var(\tilde{r}_L)]$ causes an increase in surplus volatility. This effect can be mitigated by choosing capital investments whose rate of return correlates positively with the change in liabilities $[Cov(\tilde{r}_A, \tilde{r}_L) > 0]$.

At this juncture, the formulation of the objective function deserves attention. According to the option pricing theory employed in the preceding section, diversified shareholders are interested in volatility in the value of the IC at least as long as it does not drive up the cost of capital provided by policyholders or bondholders. In the present context, however, the influence of public regulation and reputation are assumed to dominate, forcing IC management to act in a risk-averse manner. The even more extreme assumption is made that the objective function is also appropriate for capital investment on behalf of pension plan beneficiaries. Since both the expected return on the surplus $E\tilde{r}_S$ and its variance $Var(\tilde{r}_S)$ depend on the weights w_i of the different types of assets making up the investment portfolio, one can write the

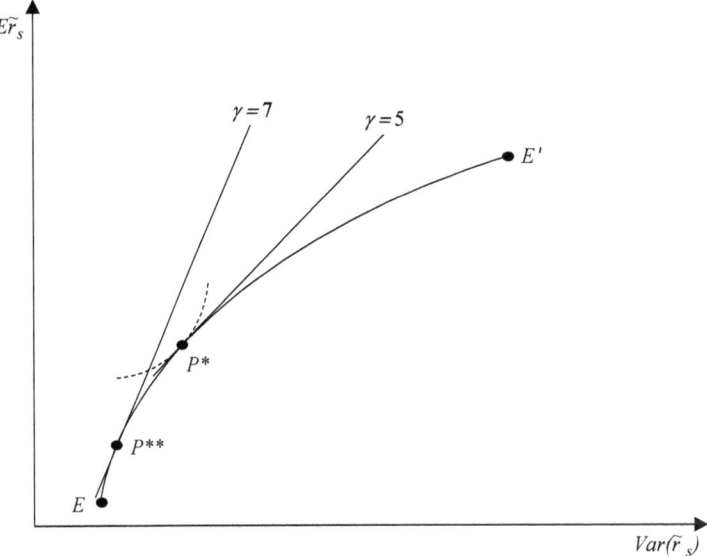

Fig. 6.4 Application of portfolio optimization to surplus S

optimization problem of the IC, with $\gamma > 1$ reflecting beneficiaries' subjective trade-off between expected returns and risk,

$$\max_{w_i} \left\{ E\tilde{r}_S + \frac{1}{2}(1-\gamma)Var(\tilde{r}_S) \right\}, \quad \text{s.t.} \quad \sum w_i = 1. \tag{6.38}$$

Figure 6.4 illustrates the efficient frontier EE' in the absence of a risk-free asset (which is realistic once the planning horizon is extended to several years). It has the same general shape as the efficient frontier discussed in Sect. 5.1.2. The line labeled $\gamma = 5$ indicates the slope of an indifference curve in the case where returns and (the absence of) volatility have equal subjective weight at the optimum point P^* (see the dotted indifference curve). For a more risk-averse clientele ($\gamma = 7$), point P^{**} constitutes the optimum. It contains higher shares w_i of low-risk, low-return assets.

The problem with this formulation is that it treats expected returns and their variances and covariances as time-independent. But with a time horizon of up to 25 years, the properties of the different assets and liabilities may change considerably, and with them the variances and covariances appearing in Eqs. (6.38) and (6.37). In addition, concentrating on conventional asset classes such as treasury bills, bonds, and stocks is insufficient to provide an effective inflation hedge over these long time periods. One contribution addressing this issue is by Hoevenaars et al. (2008). However, the authors note that with the presence of as many as seven asset classes, efficient portfolio weights become very sensitive to errors especially in the estimation of covariances, potentially triggering a repeated restructuring of the portfolio at a high cost. For this reason, they assume the IC's investment policy to remain unchanged for the planning period, which is $T = 1, 5, 10,$ and 25 years, respectively. Rather than using $\tilde{r}_S$ as defined in Eq. (6.35), they argue that the variable of interest for policyholders is the funding rate F_t because it shows whether benefits promised are sufficiently funded by assets. From the definition of (6.35), one has

$$lnF_t = lnA_t - lnL_t. \qquad (6.39)$$

The change in $\ln F_t$ over time $d\ln F_t/dt$ amounts to a percentage change that can be interpreted as a (stochastic) rate of return, given by

$$\tilde{r}_{F,t} = \tilde{r}_{A,t} - \tilde{r}_{L,t}, \qquad (6.40)$$

with $\tilde{r}_{A,t}$ and $\tilde{r}_{L,t}$ defined in Eq. (6.35). At decision time t for T periods ahead ($T = 1, 5, 10, 25$ years) and γ denoting risk aversion again, the optimization problem would become[11]

$$\max_{w_{iT}} \{E_t \tilde{r}_{F,t+T} + \frac{1}{2}(1-\gamma) Var_t(\tilde{r}_{F,t+T})\}. \qquad (6.41)$$

Note that both expected value and variance now are conditioned on the information available at time t. This information is assumed to be valid for the following T years. The authors first calculate a set of expected values based on quarterly U.S. data covering 1952: II to 2005: IV. Decision time, therefore, is 2006. Extrapolation into the future is performed by so-called vector autoregression; seven rates of return (six assets and one liability) are related to their own lagged values as well as five exogenous explanatory variables. For example, the dividend-price ratio of stocks can be considered an exogenous predictor of stock returns. Of course, the number of relationships must be cut down to make a vector autoregression system estimable. The authors test whether these exclusions are acceptable. For simplicity, changes in mortality, effects of aging, and other demographics are neglected. The maturity of liabilities is assumed to be 17 years; they are inflation-indexed. Beneficiaries, therefore, are guaranteed a rate of return on their contributions equal to that of an inflation-indexed bond.

Table 6.8 shows the optimal composition of portfolios $\{w_{iT}^*\}$ in a few cases. First, on the left-hand side, the solutions to the asset allocation problem are shown [$\tilde{r}_{L,t}$ is neglected in (6.40)]. The Global Minimum-Variance portfolio is selected as the optimum to reflect an investment policy on behalf of strongly risk-averse beneficiaries (see point E of Fig. 6.4). The right-hand side contains the solution to the full surplus optimization problem, similar to the one defined in (6.40) and (6.41), again assuming a very high degree of risk aversion. For each alternative, a comprehensive and a restrictive investment strategy are distinguished. In panel A, the set of investment categories comprise not only the three conventional ones (treasury bills, bonds, and stocks), but also four non-conventional ones (lending credit, investing in commodities, real estate, and hedge funds). In panel B, asset choice is restricted to the three conventional alternatives.

---

[11] Actually, the authors posit a risk utility function with constant relative risk aversion having slightly different properties than (6.41).

6.8 Capital Investment Policy

Table 6.8 Global minimum variance and liability hedge portfolios

Time horizon T (years)	Global minimum variance (Assets only)				Liability hedge (Assets and one liability)			
	1	5	10	25	1	5	10	25
A) Unrestricted portfolios								
Treasury bills (w_{tb})	1.06	0.99	0.92	0.83	0.62	0.52	0.41	0.29
Bonds (w_b)	−0.01	0.03	0.04	0.01	0.34	0.39	0.38	0.28
Stocks (w_s)	−0.03	−0.03	−0.01	0.12	−0.11	−0.11	−0.06	0.20
Credits (w_{cr})	−0.03	−0.02	0.01	0.01	0.12	0.16	0.21	0.16
Commodities (w_{cm})	0.02	0.02	0.02	0.04	0.03	0.04	0.04	0.06
Real estate (w_{re})	0.00	0.01	0.01	0.01	0.00	0.01	0.01	0.01
Hedge funds (w_h)	0.00	0.00	0.00	0.00	0.00	0.00	0.00	0.00
B) Restricted portfolios								
Treasury bills (w_{tb})	1.08	1.01	0.94	0.86	0.67	0.57	0.48	0.37
Bonds (w_b)	−0.05	0.02	0.06	0.01	0.44	0.53	0.56	0.39
Stocks (w_s)	−0.03	−0.03	−0.01	0.13	−0.11	−0.10	−0.04	0.23

Note The left-hand side shows the global minimum variance-portfolio for the asset-only. The right-hand side shows the liability hedge portfolio for the asset-liability problem. Weights may not add up to one due to rounding.
Source Hoevenaars et al. (2008)

- *Asset-only optimization:*

With a time horizon of just one year and admitting all seven asset classes (panel A), it would be optimal to hold 106% of the original portfolio in treasury bills. This is made possible by short-selling bonds, stocks, and credits. Among the non-conventional assets, only commodities enter the allocation, with a mere 2%.

When the investment horizon is extended to $T = 25$ years, treasury bills become slightly less preponderant. When extrapolating their estimates using vector autoregression, Hoevenaars et al. (2008) find that stocks exhibit less volatility while they increasingly provide a hedge against treasury bills. For this reason, their optimal share attains 12% at $T = 25$.

In panel B and on the left-hand side of Table 6.8, optimal allocation in the minimum-variance portfolio does not change much in spite of restricted asset choice. Again, there should be short-selling of stocks up to a planning horizon of $T = 10$ years which turns into a positive share of 13% at $T = 25$. However, these results are all partial, in that, they are asset-only, neglecting the need to hedge against future liabilities.

- *Optimal hedging against the liability:*

On the right-hand side of Table 6.8, pensions are assumed to provide protection against inflation while contributions earn accrued interest. Therefore, the liability amounts to an inflation-indexed bond. Treasury bills do not offer much of a hedge against inflation, causing their weight to drop to 62% right away ($T = 1$, panel A). Investing in bonds (but not stocks, surprisingly) and the provision of credit enter the optimal allocation with weights of 34% and 12%, respectively. The contrast with an asset-only approach is even more striking when the time horizon is extended to $T = 25$ years. Now treasury bills have a weight of just 29%, down from 83%. Conversely, investing in credit provision and commodities become important with 16 and 6%, respectively, because they are good inflation hedges.

Finally, portfolio choice is again restricted to the three conventional alternatives in panel B on the right-hand side of Table 6.8. Taking the pension liability into account again makes a difference regardless of planning horizon. For instance, for $T = 1$ bonds optimally should have a share of 44% rather than 34% in the portfolio; for $T = 25$, it is still 39% rather than 1%. When it comes to the impact of restricted asset choice, it remains optimal to short-sell stocks up to a time horizon of $T = 10$ years. In comparison with panel A, the suggestion is to invest more in bonds in order to make up for the non-conventional asset categories.

The authors also calculate the amount that would compensate for the loss of performance caused by limiting asset choice to the three conventional categories. It depends positively on risk aversion but may attain as much as 2 dollars per 100 dollars initially invested ($T = 1$) and still 1 dollar per 100 dollars invested with the long planning horizon ($T = 25$).

▶ **Conclusion 6.13** Capital investment by the IC needs to take into account that not only its assets but also its liabilities are subject to stochastic shocks impinging on interest and inflation rates. In the context of pensions, the appropriate objective is to reach the efficiency frontier in terms of expected returns and volatility of the funding rate.

Portfolio optimization based on the approach pioneered by Markowitz (1952) has not been without criticism, however. The most important are the following (Ramaswami, 1997):

- Variance and standard deviation are symmetric, reflecting positive and negative deviations from expected value in the same way. However, investors are almost exclusively concerned about negative deviations;
- Rates of return are asymmetrically distributed as soon as options are included in the set of assets. In particular by purchasing a put option on a stock, one exclusively retains the positive deviations from the strike price; and,
- The underwriting business also has on option characteristic. It in fact amounts to the writing of a put option since the IBs can buy their policy back at a predetermined price (which in the main makes up for the acquisition expense). This imparts an asymmetric downside risk to the premium income of the IC and ultimately, its assets.

Investment policy must also take into account that there are two groups of stakeholders. On the one hand, there are those who are receiving (or are about to receive) benefits such as retired employees in the case of a pension fund. On the other hand, there are the sponsors (the employers and current employees who contribute to the fund) whose interests differ. In fact, it may not be possible to find an investment policy that simultaneously satisfies the interests of both groups (Zweifel and Auckenthaler, 2008).

6.9 New Elements of Insurance Technology

6.9.1 The Value at Risk Concept

Catastrophic risk has the property of occurring extremely rarely but attaining a severity that could easily cause the insolvency of an IC. Figure 6.5 illustrates this property. It shows the surplus, i.e., the difference between assets (A_T) and liabilities (L_T) at the end of the planning period. The surplus amounts to the sum of equity and insurance reserves. There are two density functions [$f(A_T - L_T)$], one without reinsurance coverage [$f_0(\cdot)$], the other, with reinsurance coverage [$f_1(\cdot)$]. Both are extremely skewed to the right because of the skewness of the loss distribution (see Sect. 7.1.1.3). In the case of catastrophic risks, there is a very high probability of the IC having to make small or zero payments; on the other hand, the IC (at least without reinsurance coverage) has to come up with payments that can jeopardize its existence. The additional symbols appearing in Fig. 6.5 are defined as follows,

M_0: Most probable value of net assets ($= A_T - L_T$) without reinsurance;
M_1: Most probable value of net assets with reinsurance;
Z_0: Threshold value of net assets at time $T (= A_T - L_T)$ without reinsurance, to be detailed in the text;
Z_1: Threshold value of net assets at time T with reinsurance.

Therefore, reference values are not the expected values (not shown in Fig. 6.5) but the modal values $\{M_0, M_1\}$ reflecting the very likely normal situation without a loss.

An instrument increasingly used by insurers for the assessment of catastrophic risks is Value at Risk (VaR), a concept first developed for banks. VaR is the amount that can be lost during a certain period of time with a predetermined probability. In Fig. 6.5, let this probability be one percent; it is indicated by the shaded areas delimited by the threshold values $\{Z_0, Z_1\}$ which are set in a way that the probability with which net assets ($A_T - L_T$) fall below these values is precisely one percent.

In the case of the density function $f_0(\cdot)$, value at risk can be interpreted as the loss amounting to ($M_0 - Z_0$) that occurs with a certain small probability (one percent in Fig. 6.5). Such a loss could be caused by a catastrophic event and would lead to insolvency of the IC in the case depicted ($A_T - L_T < 0$).

By purchasing reinsurance, the IC can transfer large risks to the reinsurer. It moves to a new density function such as $f_1(A_T - L_T)$. This density function indicates that very high payments become sufficiently unlikely for the critical value Z_1 to shift

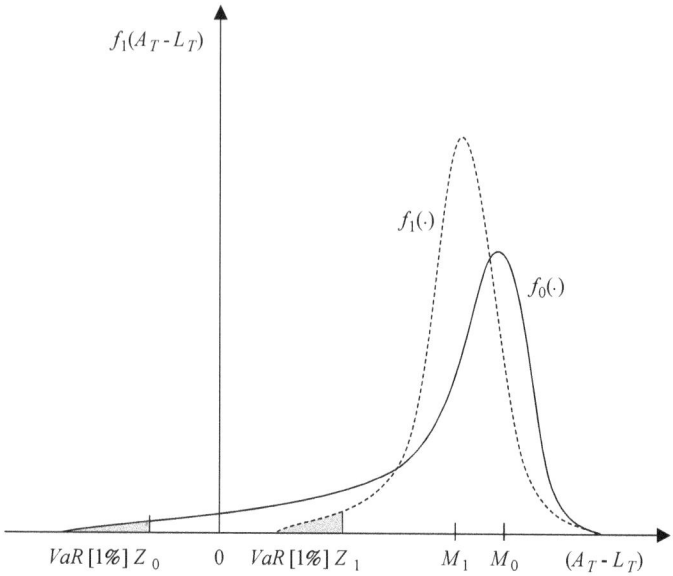

Fig. 6.5 Surplus density functions incorporating catastrophic risks and Value at Risk (VaR)

to the domain of positive surpluses ($A_T - L_T > 0$). This suggests that the density function $f_1(A_T - L_T)$ is to be preferred. However, its modal value M_1 is lower than M_0 (without reinsurance) since the IC must pay the reinsurance premium under all circumstances. It is, therefore, not evident that a risk-averse management, applying the criterion of second-order stochastic dominance (see Appendix 2.A.2 to Chap. 2) should change in favor of density $f_1(\cdot)$ by purchasing RI coverage.

In addition, Fig. 6.5 shows that VaR requires knowledge of the density functions $f_0(\cdot)$ and $f_1(\cdot)$. However, as argued in Sect. 7.1.1.2, there are hardly known distribution laws for densities that are typical in particular for property-liability insurance. The threshold values Z_0 and Z_1 can nevertheless be estimated by applying, e.g., the Normal Power Approximation that transforms the density functions $f_0(\cdot)$ and $f_1(\cdot)$ into standard normal distributions as an approximation (see Sect. 7.1.1.3). The threshold values Z_0 and Z_1 that delimit 1% (say) of the probability mass of a $N(0, 1)$ random variable can then be determined from Table 7.A.1 in the Appendix to Chap. 7.

Still, for applying, e.g., the Normal Power Approximation, the parameters $E(A_T - L_T)$, $Var(A_T - L_T)$ and $\sigma^3(A_T - L_T)$ must be estimated, i.e., the expected value, the variance, and the skewness of the surplus. Estimating these parameters meets with considerable difficulty because

- catastrophic risks are so rare that these three parameters can hardly be inferred with precision;
- there may be a positive correlation between risks induced by catastrophic events, causing tail dependence (see Sect. 6.9.2);

- the density of catastrophic risk may change over time since extreme losses seem to become more frequent (see Sect. 1.2); and,
- correlations between catastrophic risks and the components of the risk portfolio which are crucial for possible diversification effects may also change (see Sect. 5.2).

In sum, Value at Risk is a possibility of describing catastrophic risks in a simple way. This advantage must be weighed against the downside that the concept is not linked to stochastic dominance and hence the degree of risk aversion of the decision-maker. It is not clear that risk-averse decision-makers should choose the distribution with a smaller value at risk [for more details, see Dowd (1998)]. In spite of these weaknesses, insurance regulation both in the United States and the European Union have in fact adopted the VaR standard by prescribing a maximum probability of insolvency after taking reinsurance into account. In the case of the United States, this is the Risk-based Capital approach (see Sect. 9.3.2); in the case of the European Union, this is the Solvency II regulation (see Sect. 9.4.4).

6.9.2 Copulas for Dealing with Tail Dependence

Approximations such as Normal Powers cited in the preceding section serve well as long as only one loss distribution is analyzed at a time. But a multi-line IC underwrites several types of risks, and as is argued in Sect. 7.2, even a single-line insurer is exposed to two risks, namely the one emanating from its underwriting and the one from its capital investment activities. Of course, it could resort to convoluting the random variables, but this is practical (if at all) only if there is independence (see Sect. 7.1.1.2 again). The challenge to risk management becomes acute if there should be a positive correlation between highly skewed loss distributions of the type shown in Fig. 6.5. In that event, extremely high losses that individually are highly improbable could strike the IC simultaneously, reflecting so-called tail dependence. Reserves that are sufficient in a normal situation would then fail to prevent the insolvency of the IC.

Basically, one would want to estimate the joint density of distribution functions of two or more random variables in their tails. The traditional approach has been to calculate a correlation coefficient to check for dependence. However, correlation coefficients only indicate linear relationships that hold on average, over the whole sample space. One could try to limit their calculation to data in the tails, but these are usually too few to permit valid inference. Moreover, extrapolation from the overall sample to the tails can be misleading if the losses form a non-linear relationship that for some reason becomes more accentuated for high values. For example, consider two storm damage portfolios in two countries on the same continent that usually have different weather conditions. Therefore, the two risk portfolios can almost always be regarded as independent. However, let there be a storm so violent that it does affect the two countries at the same time, causing major damage in both. In that event, the two portfolios exhibit a positive correlation in the tails, which cannot be detected in the data generated by more regular losses.

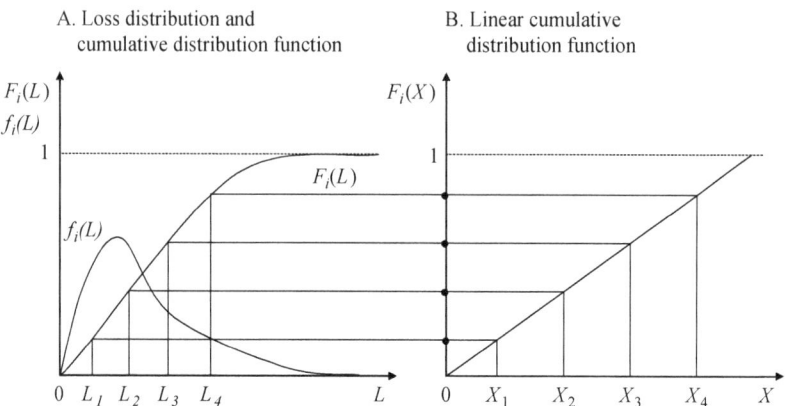

Fig. 6.6 Transformation of loss values to obtain a uniform distribution

The difficulty with combining arbitrary random variables is that each of them may have its own cumulative distribution (i.e., cumulated density) function. In panel A of Fig. 6.6, the skewed density $f_i(L)$ characterizing component i of an insurer's underwriting portfolio is shown. The corresponding cumulative distribution function amounts to the sigmoid function $F_i(L)$. However, by decreasing and increasing the values of the argument [the X values in panel B of Fig. 6.6], it is always possible to obtain a linearly increasing distribution function (which corresponds to a uniform density function). In the example, the probability $F_i(X)$ is increased by steps of 0.2. The corresponding values are $\{L_1, L_2, L_3, L_4\}$ for the original sigmoid function $F_i(L)$. They become $\{X_1, X_2, X_3, X_4\}$ for the linear distribution function $F_i(X)$, which must reach the maximum value of 1 at the same loss as $F_i(L)$. It should be much easier to combine these transformed, uniform distribution functions $F_i(L)$ to form an estimated joint distribution function $F(\cdot)$. In the following, let $X_1, , X_n$ denote these transformed random variables. The first thing to note is that if the value of the joint distribution (i.e., cumulated density) function $F[X_1, X_2, , X_n]$ of losses were known at some point $[x_1, x_2, , x_n]$, the values of the partial distribution functions $F_1[x_1], F_2[x_2], , F_n[X_n]$ could be determined. For instance, in the case of a bivariate distribution function $F(X_1, X_2)$, by setting $X_2 = 0$ (i.e., neglecting the second risk), the value $F_1[x_1]$ is known for $X_1 = x_1$. By choosing different values for X_1, the partial distribution function $F_1(X_1)$ can be determined.

This argument can be reversed. It should be possible to approximate the joint distribution function $F(X_1, \ldots, X_n)$ from a sequence of values $F_1[x_1], F_2[x_2], \ldots, F_n[x_n]$. Indeed, a theorem stated by Sklar (1973, originally 1955 in French) states that

$$F[x_1, \ldots, x_n] = C[U_1[x_1], \ldots, U_n[x_n]], \quad \text{with } U_i := F_i[X_i] \quad (6.42)$$

In words, a particular value of the joint distribution $F(\cdot)$ where the transformed random variables take on some values x_i can be determined through a combining function $C(\cdot)$, the so-called copula. In effect, the copula indicates the probability of these random variables being jointly below or equal a set of threshold values,

$$C[u_1, \ldots, u_n] = Pr[U_1 \leq u_1, \ldots, U_n \leq u_n]. \quad (6.43)$$

6.9 New Elements of Insurance Technology

Therefore, it is possible to recognize tail dependence from the properties of the copula function. In the bivariate case (with U and V replacing U_1 and U_2 and $u_1 = u_2 = u$ as a common threshold), one can define two types of tail dependence. If it is true that

$$\lim_{u \to 0} \frac{Pr[U \leq u, V \leq u]}{u} = \lim_{u \to 0} \frac{C(u, u)}{u} = \underline{b}, \quad 0 < \underline{b} \leq 1, \qquad (6.44)$$

then the two distributions exhibit so-called lower tail dependence. Even if one were to choose a value of u close to zero (hence low values of the two losses considered), the copula would still indicate some probability mass in the joint distribution function. Conversely, one has so-called upper tail dependence if

$$\lim_{u \to 1} \frac{Pr[U \leq u, V \leq u]}{1 - u} = \lim_{u \to 1} \frac{1 - 2u + C(u, u)}{1 - u} = \overline{b}, \quad 0 < \overline{b} \leq 1. \qquad (6.45)$$

Here, even if one lets u approach 1, there is again some probability mass left in the copula, pointing to the possibility of two very high losses occurring simultaneously in the case of insurance.

Given that one has discovered tail dependence, the question still remains how a copula can be chosen that best reflects it. The most general class is the so-called Archimedean copula, in which the u_i values are related back to the original values of losses by inverting the transformation function $\varphi(\cdot)$ such that

$$C[u_1, \ldots, u_n] = \varphi^{-1}[\varphi(u_1), + \ldots + \varphi^{-1}(u_n)]. \qquad (6.46)$$

The $\varphi(\cdot)$ function involves a parameter θ that indicates the degree of tail dependence. The choice of copula is still a matter of debate. At any rate, copulas contribute to a successful implementation of the Value at Risk approach described in Sect. 6.9.1 by a multi-line IC [for more detail; see, e.g., Embrechts et al. (2001) and Schölzel and Friederichs (2008)].

6.9.3 Alternative Risk Transfer (ART) Through Capital Markets

As discussed in Sect. 6.7, reinsurance (RI) is an important element of a primary insurer's technology. However, the underwriting capacity of a reinsurer sometimes proves insufficient in the face of losses amounting to US \$100 bn. and more. In addition, RI premiums can easily comprise loadings of 60% in excess of fair premiums (Doherty, 1997). Loadings of this size point to considerable transaction costs burdening the relationship between primary insurers and reinsurers. Indeed, there is scope for a great deal of moral hazard on both sides [for a survey of explanations, see Froot (1999)].

- *Ex-ante moral hazard on the part of the primary insurer*: In Sect. 8.2, *ex-ante* moral hazard is defined as a reduction of preventive effort on the part of the IB. Here, the IB is the primary insurer, whose preventive effort can be equated to a careful selection of risks underwritten. Reinsurance coverage of large risks undermines the incentives for this type of effort since in the event of a very high loss, it is almost certain that part of it will be paid by RI. As shown in Sect. 8.2.2.1, an IB with weak risk aversion (an IC with a diversified portfolio in the present context) is predicted to exhibit marked moral hazard. To neutralize this, the reinsurer counters by charging a high loading that usually increases progressively with the sum insured (see Sect. 8.2.2.1).

- *Ex-ante moral hazard on the part of the reinsurer*: The option pricing theory shown in Sect. 7.2.3 can also be applied to the shareholders of a RI company. This means that they are liable for losses only to the point where their shares have zero value; they do not have to foot a negative surplus. Therefore, the reinsurer is tempted to skimp on efforts to preserve its solvency. For instance, in the wake of Hurricane Andrew, several U.S. reinsurers became insolvent [see Swiss Re (1998)].

- *Ex-post moral hazard on the part of the primary insurer*: Ex-post moral hazard refers to the IB's influence on the amount of damage given a loss event occurred (analogous to the choice of medical treatment in the case of health insurance, discussed in Sect. 10.5.2). Here, it is the primary insurer who does not necessarily have an interest in limiting damage; generosity in consumer accommodation pays off in the media and through the grapevine. One would expect this effect to be especially marked if the RI contract is of the stop-loss type (see Sect. 6.7.2).

These considerations provide an explanation for the fact that primary insurers (and even RI companies themselves) have increasingly used capital markets to transfer risks. Known as Alternative Risk Transfer (ART), these instruments have become increasingly important since the 1990s when insurers were hit by several catastrophic losses (see Table 1.1). The Chicago Board of Trade was first to create a standardized contract that can be traded at low cost in the guise of PCS Cat Insurance Options. Moral hazard effects on the part of the primary insurer are absent because the amount of benefit to be paid in the case of loss does not depend on the information provided by the IB but on the value of an index that the two contractual parties have previously agreed upon. For example, the index may be based on the amount of precipitation during 24 hours in a defined area of the United States as a predictor of damage due to flooding. A value in excess of the predetermined threshold automatically triggers payment.[12]

Following the example of banks in the 1980s, ICs also seek to securitize their risk portfolios. One of the early variants (launched by AXA Winterthur in 1997) was the issue of a bond that paid interest inversely related to the amount of hail damage. In the event of losses due to hail exceeding a certain (high) limit, the IC is relieved from its obligation to pay interest; in addition, it cannot be made to buy back its bonds. While providing for a partial risk transfer, insurance-linked securities of this type

---

[12]This parametric trigger is but one option used in catastrophe bonds. Other triggers, such as an indemnity trigger, are not necessarily devoid of moral hazard effects.

fall short of a full hedge, which would also meet the need of the IC or RI company for new capital replenishing its reserves after an extreme loss. The solution to this problem could be the purchase of put options on the company's own shares, thus obtaining the right to sell them at a price that was set prior to the occurrence of the loss [for a survey of ART products currently in use; see, e.g., Ben Ammar et al. (2015)].

All of these variants of ART have in common that the estimated size of the loss and the amount of benefit to be paid cannot be influenced (i.e., manipulated) by either the primary insurer or the reinsurer. This serves to suppress moral hazard effects on both sides, thus lowering the cost of contract monitoring. Since the RI benefit is fixed, the primary IC has to bear any increase in the size of loss due to a careless underwriting policy. Conversely, a reinsurer who underwrites risks jeopardizing its solvency is sanctioned by the capital market because it has to pay a higher rate of interest on securities issued.

Finally, ART also benefits consumers, who are individuals and companies with limited diversification possibilities (see Chap. 3 again). In the absence of ART, they must take the risk of non-performance on the part of their IC into account: With a certain probability, the primary insurer cannot pay the promised benefit because the reinsurer ended up insolvent. To these IBs, the alternative to a risky initial situation, therefore, is not perfect security but another risky situation with a lowered but still positive probability of loss (rather than $\pi = 0$), combined with a lower level of wealth due to the premium paid. This means a reduced (maximum) willingness to pay for insurance (indicated by the risk premium, see Sect. 3.2.1) reflecting a reduced value of coverage. With ART, this risk of non-performance is practically zero, thus increasing consumer surplus since the actual premium almost never increases in step with consumers' maximum willingness to pay.

The advantages in terms of contract monitoring have to be weighed against the transaction costs associated with the writing of the contract. The securitization of a risk portfolio usually lacks the standardization characterizing an option (say), forcing potential buyers to undertake an in-depth analysis of the portfolio's properties. However, standardization has been occurring, e.g., in the guise of PCS Cat Insurance Options.

Another advantage of ART products is their excellent risk diversification properties [see Swiss Re (1996)]. They are illustrated by Fig. 6.7 for the observation period 1985-1995 with two efficient frontiers in (μ, σ)-space. For investments excluding PCS Cat Insurance Options, the minimum-variance endpoint of the frontier corresponded to U.S. government bonds. The maximum expected return endpoint was determined by French stocks (F), while Japanese stocks (J) constituted a high-volatility alternative that entered a globally efficient portfolio thanks to their diversification effect. When PCS Cat Insurance Options were included in the portfolio, the frontier moved upward, indicating a gain in efficiency, with the maximum expected return portfolio consisting of Cat options. The efficiency gain is due to the fact that the occurrence of natural catastrophes is hardly related to a fall in the rate of return of other securities, resulting in (almost) zero correlation and an excellent hedge.

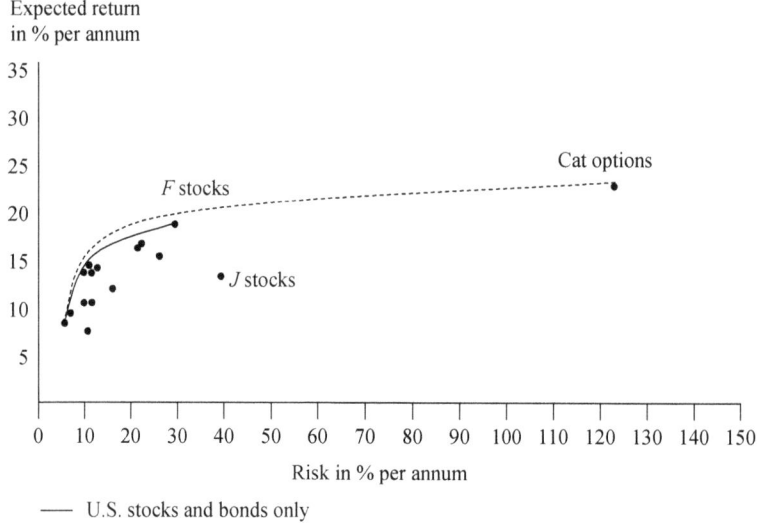

Fig. 6.7 Globally efficient portfolios excluding and including cat options, 1985–1995; *Source* Swiss Re (1996)

In order to reap this gain, however, investors had to accept a volatility that was almost six times higher than the next-best alternative, provided by French stocks (marked F stocks in Fig. 6.7). Evidently, they must be willing to go way beyond the customary risk-return combinations. In addition, the points shown in Fig. 6.7 represent estimates that are subject to considerable uncertainty. Since instability in estimated covariances (or correlations) is reflected in the beta of the CAPM (see Sect. 5.1.3), research has focused on the instability of beta values over time. For instance, Ebner and Neumann (2005) find considerable instability of betas even in the "mature" German stock market, prompting them to propose econometric methods more advanced than OLS for estimating them. In emerging markets such as Mexico, this instability is even more pronounced (Nieto et al., 2011).

Nevertheless, ART has become increasingly popular, including lines beyond non-life insurance. For instance, unexpected advances in medical technology boosting the cost of medical care constitute a challenge to health insurers. Since new medical technology spreads globally, this may create tail dependencies for ICs with activity in several countries, calling for the use of copulas (see Sect. 6.9.2). Depending on the result, the IC may decide to securitize its health insurance portfolio for placing it on the capital market. This is an option social health insurers do not have because their (usually small) capital investments are typically limited to bonds issued by domestic governments. Therefore, this addition to the insurance technology helps private ICs cope with challenges that would otherwise likely benefit social insurance (see Chap. 11).

▶ **Conclusion 6.14** In order to insure extreme risks in future, both primary insurers and reinsurers increasingly rely on Alternative Risk Transfer instruments provided by capital markets. Their main advantage is that they mitigate moral hazard effects burdening conventional reinsurance contracts.

Exercises

6.1

(a) Several entries in both the balance sheet and the income statement of an IC are affected by a change in interest rates. Which ones, and how?
(b) Consider *reserves for losses and expenses* as a balance sheet item in the case of life and non-life insurance. In which line of business do you expect interest sensitivity to be greater? Why?
(c) Usually, regulatory authorities either disallow discounting to present value or impose a fixed rate of discount. Does this modify your answer to (b)?
(d) As soon as non-insurance firms have possibilities of risk diversification in addition to purchasing insurance, interest rates may influence also their demand for insurance. Why? What happens to the demand for insurance in the case of decreasing interest rates? Does this modify your answer to (a)?
(e) Describe the difficulties for an IC to satisfy legal requirements in terms of reserves for future claims in a situation of falling interest rates while the rate set by regulatory authorities remains unchanged. Conversely, what happens when interest rates in the market increase? Do you see a possibility of explaining the existence of so-called insurance cycles?

6.2

(a) In the text, it is emphasized that instruments of insurance technology should in principle be viewed simultaneously. Why?
(b) As an example, consider paying provisions to direct writers not only according to the stock of premiums written but also to additional premium volume acquired. Under what type of objective function is it appropriate to create such an incentive?
(c) Describe the likely implications of such a payment rule

(c1) on the "quality" of additional risks acquired (in terms of their probability of loss);
(c2) on the homogeneity of total population at risk.

(d) What are the consequences for

  (d1) the expected value of profit of the IC?
  (d2) the variance of profit of the IC?
  (d3) the position of the IC on the capital market?
  (d4) the future cost of purchasing capital for this IC?

(e) Is paying provisions according to additional premiums in accordance with the interests of shareholders?

References

Association of German Insurers GDV (1996). *Wettbewerbsfaktoren der Kompositversicherungsunternehmen in Deutschland (Determinants of competitiveness of multiline insurers in Germany)*. Karlsruhe: Verlag für Versicherungswirtschaft.

Barrese, J., & Nelson, J.M. (1992). Independent and exclusive agency insurers: a reexamination of the cost differential. *Journal of Risk and Insurance, 59*(3), 375–397.

Beard, R.E., Pentikäinen, T., & Pesonen, E. (1984). *Risk Theory. The Stochastic Basis of Insurance*. London: Chapman and Hall.

Ben Ammar, S., Braun, A. & Eling, M. (2015). Alternative risk transfer and insurance-linked securities: Trends, challenges, and new market opportunities. University of St. Gallen: Institute of Insurance Economics.

Berger, L.A., Cummins, J.D., & Tennyson, S. (1992). Reinsurance and the liability insurance crisis. *Journal of Risk and Uncertainty, 5*, 253–272.

Borch, K.H. (1960a). An attempt to determine the optimum amount of stop loss reinsurance, in XVIth International Congress of Actuaries, Brussels 1960, Vol. 1, 597 - 610.

Borch, K.H. (1990). The price of moral hazard. *Scandinavian Actuarial Journal*, 173–176 (1980), reprinted. In K.H. Borch (Ed.), *Economics of Insurance* (pp. 346–362). Amsterdam: North-Holland.

Borch, K.H. (1960b). The safety loading of reinsurance premiums. *Scandinavian Actuarial Journal*, 3-4, 163-184.

Borch, K.H. (1960c). Reciprocal reinsurance treaties seen as a two-person cooperative game. *Scandinavian Actuarial Journal*, 1-2, 29-58.

Borch, K.H. (1961). Some elements of a theory of reinsurance. *Journal of Insurance, 28*(3), 35-43.

Borch, K.H. (1962). Equilibrium in a reinsurance market. *Econometrica, 30*, 424-444.

Bühlmann, H., & Gisler, A. (2005). *A Course in Credibility Theory and its Applications*. New York: Springer.

Carlton, D.W., & Perloff, J.M. (1999). *Modern Industrial Organization*, 3rd ed. Reading MA: Addison-Wesley.

Cummins, J.D., & Van Derhei, J. (1979). A note on the relative efficiency of property-liability distribution systems. *Bell Journal of Economics and Management Science, 10*, 709–719.

Cummins, J.D., & Sommer, D.W. (1996). Capital and risk in property-liability insurance markets. *Journal of Banking and Finance, 20*, 1069–1092.

Cummins, J.D., & Zi, H. (1998). Comparison of frontier efficiency methods: an application to the US life insurance industry. *Journal of Productivity Analysis, 10*(2), 131–152.

Daykin, C.D., Pentikäinen, T., & Pesonen, M. (1994). *Practical Risk Theory for Actuaries*. London: Chapman & Hall.

DiMasi, J.A., Hansen, R.W., & Grabowski, H.G. (2003). The price of innovation: new estimates of drug development costs. *Journal of Health Econonmics, 22*(2), 151–185.

Doherty, N.A. (1981). The measurement of output and economies of scale in property-liability insurance. *Journal of Risk and Insurance, 48*(3), 391–402.
Doherty, N.A. (1997). Innovations in managing catastrophic risks. *Journal of Risk and Insurance, 64*(4), 713–718.
Doherty, N.A., & Garven, J.R. (1986). Price regulation in property-liability insurance: a contingent claims approach. *Journal of Finance, 41*, 1031–1050.
Dowd, K. (1998). *Beyond Value at Risk. The New Science of Risk Management*. New York: Wiley.
Ebner, M. & Neumann, T. (2005). Time-varying betas of German stock market returns. *Financial Markets and Portfolio Management, 19* (1), 29-46.
Eisen, R. (1979a). Equilibrium in risk bearing: The principle of equivalence–different implications of alternative interpretations. *Geneva Papers*, No. 11 (January), 14–33.
Embrechts, P., Lindskog, F., & McNeil, A. (2001). Modelling dependence with copulas and applications to risk management. In S. Rachev (Ed.), *Handbook of Heavy Tailed Distributions in Finance* (pp. 329–384). Dordrecht: Elsevier.
Ezra, D. (1991). Asset allocation by surplus optimization. *Financial Analysts Journal* Jan./Feb.: 51–57.
Finsinger, J., & Schmidt, F.A. (1994). Prices, distribution channels, and regulatory intervention in European insurance markets. *Geneva Papers on Risk and Insurance Issues and Practice, 70*, 22–36.
Froot, K.A. (1999). Introduction. In K.A. Froot (Ed.), *The Financing of Catastrophe Risk* (pp. 1–22). Chicago: University of Chicago Press.
Garven, J.R., & Lamm-Tennant, J. (2003). The demand for reinsurance: theory and empirical tests. *Insurance and Risk Management, 7*(3), 217–237.
Grace, M.F. & Timme, S.G. (1992). An examination of cost economies in the United States life insurance industry. *Journal of Risk and Insurance, 59*, 72-103.
Greenwald, B.C., & Stiglitz, J.E. (1990). Asymmetric information and the new theory of the firm. *American Economic Review Papers and Proceedings, 80*(2), 160–165.
Grossman, S.J., & Hart, O.D. (1986). The costs and benefits of ownership: theory of vertical and lateral integration. *Journal of Political Economy, 94*, 691–719.
Hoevenaars, R.P., et al. (2008). Strategic asset allocation with liabilities: beyond stocks and bonds. *Journal of Economic Dynamics & Control, 32*, 2939–2970.
Holmström, B. (1979). Moral hazard and observability. *Bell Journal of Economics, 10*(1), 74–91.
Jensen, M.C., & Meckling, W.H. (1976). Theory of the firm: managerial behavior, agency costs, and ownership structure. *Journal of Financial Economics, 3*, 306–360.
Joskow, P.L. (1973). Cartels, competition and regulation in the property-liability insurance industry. *The Bell Journal of Economics and Management Science, 4*(2), 375–427.
Kennedy, P.E. (1986). Interpreting dummy variables. *Review of Economics and Statistics, LXVIII*, 174–175.
Levinthal, D. (1988). A survey of agency models of organizations. *Journal of Economic Behavior and Organization, 9*, 153–185.
Markowitz, H.M. (1952). Portfolio selection. *Journal of Finance, 7*, 77–91.
Müller, W. (1981). Theoretical concepts of insurance production. *Geneva Papers on Risk and Insurance, 6*(21), 63–83.
Nieto, B., Orbe, S. & Zaraya, A. (2011). Time-varying beta estimators in the Mexican emerging market. Working Paper, Bilbao: Universidad del País Vasco.
Outreville, J.F. (1998). *Theory and Practice of Insurance*. Dordrecht: Kluwer.
Outreville, J.F. (2002). Introduction to insurance and reinsurance coverage. In D.M. Dror, & A.S. Preker (Eds.), *Social Reinsurance. A new Approach to Sustainable Community Health Financing*, (pp. 59–74). Washington: The World Bank.
Ramaswami, M. (1997). Value at risk and asset-liability based asset allocation for a pension fund, a foundation, and an insurance company, *Working Paper*, Bankers Trust, New York.
Schoder, J., Sennhauser, M., & Zweifel, P. (2010). Fine-tuning of health insurance regulation – unhealthy consequences for an individual insurer. *International Journal of the Economics of Business, 17*(3), 313–327.

Schölzel, C., & Friederichs, P. (2008). Multivariate non-normally distributed random variables in climate research – introduction to the copula approach. *Nonlinear Processes in Geophysics, 15*, 761–772.

Schradin, H.R. (1994). Kritische Erfolgsfaktoren in der Versicherung. Untersuchungsansätze und Methodische Grundlagen für die Analyse organisatorischer Teileinheiten (Critical determinants of profitability in insurance: approaches and methods for the analysis of organizational units). *Zeitschrift für die gesamte Versicherungswissenschaft, 83*(4), 531–561.

Sklar, A. (1973). Random variables, joint distribution functions and copulas. *Kybernetika, 9*, 449–460.

Strain, R.W. (1989). *Reinsurance*. New York: The College of Insurance.

Swiss Re (1996). Risikotransfer über Finanzmärkte: Neue Perspektiven für die Absicherung von Katastrophenrisiken in den USA (Risk transfer through capital markets: New perspectives for hedging catastrophic risks in the USA). *sigma* 5/1996.

Swiss Re (1998). Der globale Rückversicherungsmarkt im Zeichen der Konsolidierung (The global reinsurance market in consolidation). *sigma* 9/1998.

Von Eije, J.H. (1989). *Reinsurance Management. A Financial Exposition*, Dissertation, Erasmus University (Foundation for Insurance Science), Rotterdam.

Weiss, M.A. (1995). A multivariate analysis of loss reserving estimates in property-liability insurers. *Journal of Risk and Insurance, 52*(2), 199–221.

Winter, R.A. (1992). Moral hazard and insurance contracts. In G. Dionne (Ed.) *Contributions to Insurance Economics* (pp. 61-96). Dordrecht: Kluwer.

Zweifel, P. (2007). The theory of social health insurance. *Foundations and Trends in Microeconomics, 3*(3), 183–273.

Zweifel, P. (2009). Technological change and health insurance. In J. Costa-Font, C. Courbage, & A. McGuire (Eds.), *The Economics of New Health Technologies. Incentives, Organization and Financing* (pp. 93–107). Oxford: Oxford University Press.

Zweifel, P., & Auckenthaler, C. (2008). On the feasibility of insurers' investment policies. *Journal of Risk and Insurance, 75*(1), 193–206.

Zweifel, P., & Ghermi, P. (1990). Exclusive vs. independent agencies: A comparison of performance. *The Geneva Papers on Risk and Insurance Theory, 15*(2), 171–192.

The Supply of Insurance 7

In this chapter, several dimensions of the supply of insurance coverage are examined. The first dimension is the pricing of insurance products. The objective is to calculate a minimum premium at which a single insurance product breaks evenly (provided the market accepts it). Section 7.1 introduces the reader to traditional premium calculation, where pricing depends upon the characteristics of the loss distribution and an exogenously given ruin probability (i.e., one minus the probability of solvency), applying elements of probability theory. On the other hand, for the determination of the market price of an insurance product, the alternatives which are available to investors and insurance buyers (IBs) in the capital market must be evaluated. Accordingly, in Sect. 7.2 elements of capital market theory are applied to derive the premium the insurance company (IC) must obtain to be sufficiently attractive to investors and can charge while still attracting IBs.

Besides pricing, determining the product spectrum is of great importance for supply, to be addressed in Sect. 7.3. IC management can decide to add a line of business to its existing array of products—or to drop an activity. This issue goes beyond the limits of insurance. Are economies of scope (often called synergies) sufficiently strong to call for combining insurance products with other financial services, in particular, in connection with *bancassurance*?

For a given product spectrum, the choice of scale in underwriting constitutes the third dimension of supply, taken up in Sect. 7.4. Economies of scale ultimately determine the size of the IC. Here, probability theory gives rise to the presumption that a higher volume of contracts goes along with a decreasing reserve requirement per contract, causing unit cost to fall. However, other types of costs may neutralize this effect. Empirical research is needed to come up with an answer with respect to economies of scale in insurance.

The results of Sects. 7.3 and 7.4 are of considerable importance for assessing the changing structure of insurance markets. If a tendency toward concentration reflects economies of scope or scale, it has to be judged differently from the case where size is a means to gain or maintain market power. However, maintenance of market power (and hence the ability to charge high premiums) requires barriers to entry that keep away potential competitors. Whether barriers to entry exist in the insurance industry has mainly to do with public regulation, which is discussed in detail in Chap. 9.

7.1 Traditional Premium Calculation

Premium calculation belongs to the most important activities in insurance. Up to the 1980s, price competition was strongly restricted by so-called material regulation of the insurance industry (see Sect. 9.1). National insurance associations typically worked with authorities to set a common level of premiums. To this end, several competing principles of premium calculation were developed, to be explained below.

7.1.1 Claims Process and Loss Distribution

The risk theory of insurance traditionally focuses on the underwriting activity of the IC, neglecting capital investment, the other primary activity of an IC. It revolves around the description and forecasting of the liabilities resulting from the underwriting of risks. Claims from the insurance portfolio arise irregularly and in different amounts. They form a stochastic (i.e., chance-determined) process in time, consisting of two components:

1. Uncertain number of claims and
2. Uncertain amount of a claim.

Parallel to this claims process, there is a premium income process which is considered as nonstochastic for simplicity.

Figure 7.1 illustrates at the start of the planning period, there is a surplus (equity capital plus insurance reserves) amounting to S_0. Until the first claim occurs, only the premium income process is active, causing the surplus to increase according to the premium income per time period given by $(1 + \lambda)\Pi$, where Π symbolizes the fair premium income.[1] The fair premium just covers the expected value of the claims (still to be determined; see below). The surcharge λ per monetary unit (MU) of premium is a safety loading designed to control the risk of insolvency (administrative expense is neglected at this point). After T periods and in the absence of any claim, the surplus would be $S[T] = S_0 + (1 + \lambda)\Pi \cdot T$. The fact that it can be invested in the capital market will be taken into account in Sect. 7.2.2 below. When the first claim occurs, surplus decreases by the amount of loss payment; from there, the surplus process continues with the same slope until another claim is presented. The vertical distance

---

[1] In this chapter, Π denotes the premium because the symbol P is used for the put option.

7.1 Traditional Premium Calculation

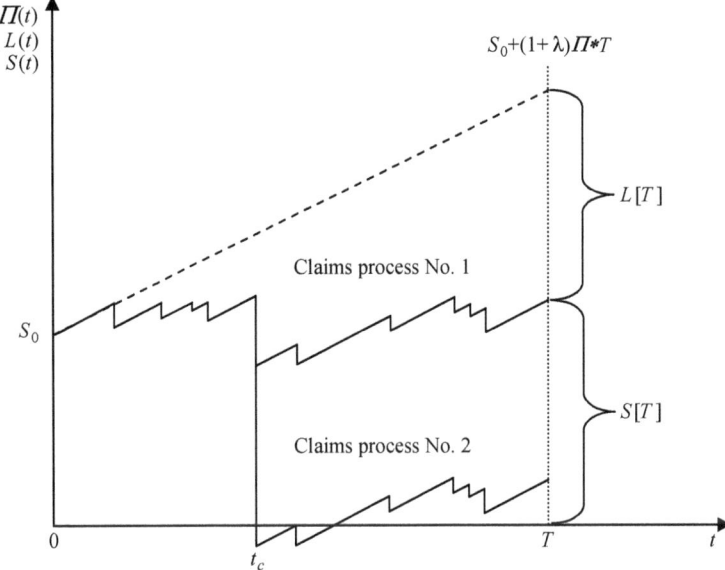

Fig. 7.1 Development of premium income, claims, and surplus over time

from the premium income process indicates the cumulative value of losses paid. At the end of the planning period, gross surplus $S_0 + (1 + \lambda)\Pi*T$, the total amount of losses paid $L[T]$ as well as resulting surplus $S[T]$ can be read off from the vertical line (shown for claims process No. 1 only).

However, the claims process could have been less favorable, with an extremely high claim hitting the IC at time t_c (claims process No. 2). This would cause surplus to be negative for a few periods, indicating insolvency or ruin of the IC. Evidently, the probability of ruin (or insolvency) can be reduced by

- increasing the surplus S_0 in the starting period through more equity capital (resulting in a parallel upward shift of the premium and surplus processes), or
- increasing the surcharge λ on the fair premium (resulting in a steeper slope of the premium and surplus processes).

However, success is *not guaranteed* for the second alternative especially in deregulated insurance markets because the increase in the price of insurance coverage will induce IBs to cancel their contract or at least to reduce the amount of coverage. The decisive parameter obviously is the price elasticity of insurance demand. If it is above one (in absolute value), premium income will not increase in response to a rise in λ and hence price, causing the premium process in Fig. 7.1 to run flatter rather than more steeply. Empirical research with respect to insurance demand points to a substantial price elasticity at least in less-regulated markets (see Sect. 1.5).

Note the importance of *timing of claims* for the risk of insolvency. For instance, let the extreme claim of process No. 2 occur around the middle of the planning period. This would be sufficient to avoid insolvency. Indeed, in Fig. 7.1 the duration of insolvency is so short that the IC should be able to obtain credit using its expected premium income as collateral. The traditional approach neglects this possibility, maintaining that the claims process No. 2 gives rise to a positive probability of ruin during the planning period, typically one year. In fact, it disregards the timing of claims as a first approximation, collapsing claims into a so-called loss distribution that holds for the planning period. Its typical shape is indicated by the solid density function of Fig. 7.2 in Sect. 7.1.1.3, reflecting the fact that most of the losses are small while a few are extremely high (as depicted in Fig. 7.1). It is intuitive that a known distribution law at best can provide an approximation to observed frequencies and sizes of losses (see Sects. 7.1.1.2 and 7.1.1.3 below). In some special cases, however, the amount of loss can be taken as fixed, leaving the number of claims as a stochastic quantity. One such special case is discussed in Sect. 7.1.1.1 below before turning to the problem of estimating a general loss distribution in Sects. 7.1.1.2 and 7.1.1.3.

7.1.1.1 Number of Claims Uncertain with Fixed Claim Severity

Let the claims (i.e., losses) be of (approximately) equal size. In whole life insurance, this assumption is acceptable for a cohort (i.e., individuals of the same age) with a fixed capital paid in the event of survival. Another example is a benefit fund which pays a fixed amount to cover the cost of funeral services. With the amount of loss fixed, the probability of ruin (i.e., insolvency) can be determined in a straightforward manner.

Denote by L_n total losses to be paid, which depend on the number of loss events n, the stochastic quantity. With the amount of loss per insured x fixed, one has

$$L_n = n \cdot x. \tag{7.1}$$

Let a regulatory authority (or the members of the fund if it is a mutual) require that the probability of ruin (i.e., the risk of insolvency) be no larger than ε during one year (say). Conversely, the probability of the initial surplus S_0 plus the earned premium income $\Pi_\lambda = (1+\lambda)\Pi$ exceeding total losses L_n must be $(1-\varepsilon)$ or higher. This can be written as

$$Pr\{S_0 + \Pi_\lambda - L_n \geq 0\} = Pr\{L_n \leq S_0 + \Pi_\lambda\} \geq (1-\varepsilon). \tag{7.2}$$

Inserting now $L_n = n \cdot x$ and dividing by x to isolate the random variable n, one obtains

$$Pr\left\{n \leq \frac{S_0 + \Pi_\lambda}{x}\right\} \geq (1-\varepsilon). \tag{7.3}$$

This states that the number of claims must not be larger than a certain value for the fund to remain solvent.

The premium income process is considered nonstochastic. At the end of the year ($t=1$), premium income is simply expected loss $EL_n = \Pi$ augmented by the safety loading,

$$\Pi_\lambda = (1+\lambda)EL_n. \tag{7.4}$$

7.1 Traditional Premium Calculation

Now the problem remains how to determine the distribution of the number of claims valid for one year. This requires a modeling of the stochastic claims process. A standard assumption is that the relevant time period t (the year) can be subdivided into τ subperiods (e.g., weeks) that are so short that the loss event either occurs with probability π or does not occur with probability $(1 - \pi)$. This is a binary distribution law; however, for small and constant π, it becomes the Poisson distribution. According to the Poisson distribution, the expected number of claims En during a year is given by π (the probability that the loss event occurs during a week) times τ (the number of weeks),

$$En = \pi \cdot \tau. \tag{7.5}$$

For instance, the probability of a member of a small fund dying during a given week is low, and more than one death per week can be excluded (almost with certainty). In expectation, the number of deaths therefore increases in linear proportion with the number of weeks the claims process continues.

Note that the standard Poisson distribution is fully characterized by En; therefore, the way the planning period is subdivided (in weeks, months, or days) is irrelevant. Knowledge of expected value En thus suffices to determine the value of n that is not exceeded with at least probability $(1 - \varepsilon)$ as defined by Eq. (7.3). Using the table of the Poisson distribution (see Table 7.A.2 of Appendix), three questions can be answered.

1. How large must initial surplus S_0 be for achieving a prescribed probability of solvency $(1 - \varepsilon)$ given the safety loading λ?
2. How large must the safety loading λ and hence the lower bound of the premium be such that given an initial surplus S_0, the prescribed solvency level $(1 - \varepsilon)$ can be achieved?
3. What is the consequence for required initial surplus S_0 if the prescribed solvency level $(1 - \varepsilon)$ is modified?

Example 7.1

A benefit fund with 1,000 members covers the expenses in case of death with a fixed amount of 2,000 MU (monetary units) paid to surviving dependents. The yearly mortality of the members is 0.01, and the premium contains a safety loading of 10%. The probability of ruin shall be 1% or less for a year.

1. *Determination of the necessary initial surplus S_0 (the equity capital).*
   According to the figures given, one has $En = 1000 \cdot 0.01 = 10$ deaths per year; $\varepsilon = 0.01$ and hence $1 - \varepsilon = 0.99$; $x = 2,000$.

   Table 7.A.2 indicates that a Poisson random variable with expected value of 10 has a probability of 0.993 for attaining values up to 18. Hence, $n = 18$ is the number of deaths that is exceeded with a probability of no more than 1% (which

would mean insolvency). Using Eqs. (7.1) and (7.4), one obtains for the premium income,

$$\Pi_\lambda = (1+\lambda) \cdot En \cdot x = (1+0.1) \cdot 10 \cdot 2{,}000 = 22{,}000. \qquad (7.6)$$

Inserting these figures into condition (7.3) results in

$$Pr\left\{18 \leq \frac{S_0 + 22{,}000}{2{,}000}\right\} = 0.993. \qquad (7.7)$$

Since the fund presumably has no interest in holding excess reserves, it is appropriate to determine the minimum value of S_0 satisfying condition (7.7). This is achieved by taking the case of equality to obtain

$$18 = \frac{S_0 + 22{,}000}{2{,}000}, \quad \text{and hence}$$
$$S_0 = 14{,}000 \text{ MU}. \qquad (7.8)$$

An initial surplus or equity capital of 14,000 MU or higher is sufficient to attain a probability of solvency of 99%.

2. *Determination of the safety loading λ.*

Assume that the benefit fund has succeeded in collecting 10,000 MU only as its initial capital S_0. How large must λ be to still guarantee a probability of solvency of 99%?

Since $(1-\varepsilon)$ has not changed, condition (7.3) continues to hold. However, this time S_0 is given while λ is to be determined. From Eq. (7.6), one has $\Pi_\lambda = (1+\lambda) \cdot 10 \cdot 2{,}000 = (1+\lambda) \cdot 20{,}000$ MU. Using this in the condition (7.3) yields

$$18 = \frac{10{,}000 + (1+\lambda) \cdot 20{,}000}{2{,}000}.$$

Therefore, $(1+\lambda) = 1.3$ and hence $\lambda = 30\%$. Since the fair premium income is 20,000 MU ($= 1{,}000 \cdot 0.01 \cdot 2{,}000$), the benefit fund needs to have a revenue of at least 26,000 MU per year for its service if it wants to maintain its solvency level at 99%.

However, note that the required increase of the safety loading (from 10% to 30%) means a hike in the price of insurance coverage. This could cause some members to leave the fund, especially those who estimate their mortality risk to be lower than average (this is an adverse selection effect; see Sect. 8.3).

3. *Changing the required probability of solvency $(1-\varepsilon)$.* Let the regulatory agency (or the members of the benefit fund) stipulate a higher probability of solvency (or a lower probability of ruin), e.g., $(1-\varepsilon) = 0.999$ rather than 0.99.

Table 7.A.2 shows that a Poisson random variable n with $En = 10$ takes on values between 0 and 21 with probability 0.9993. Therefore, the initial surplus together with the premium income must be sufficient to cover 21 deaths for the

7.1 Traditional Premium Calculation

Table 7.1 Convolution of two loss distributions (case of independence)

A	L_A	0	1	2	3			$EL_A = 1.0$
	π_A	0.3	0.5	0.1	0.1			
B	L_B	0	1	2				$EL_B = 0.8$
	π_B	0.4	0.4	0.2				
$A \cup B$	$L_{A \cup B}$	0	1	2	3	4	5	$\sum \pi_{A \cup B}$
	$\pi_{A \cup B}$	0.12	0.20	0.04	0.04	0.04	0.02	$=1.00$
			+0.12	+0.20	+0.04	+0.02		
			0.32	+0.06	+0,10	0.06		
				0.30	0,.18			$EL_{A \cup B}$
								$=1.8$

fund to obtain a solvency level of 99.9%. Condition (7.7) now reads (with the initial safety loading $\lambda = 0.1$ again),

$$21 = \frac{S_0 + 22{,}000}{2{,}000},$$

which, solving for S_0, results in $S_0 = 20{,}000$ MU.

As was to be expected, the increase in the required probability of solvency calls for a substantial increase in initial surplus. ∎

7.1.1.2 Both Number of Claims and Amount of Claim Uncertain

In most lines of insurance, the size of claims is not a fixed quantity but a random variable like the number of claims. A closer look at Fig. 7.1 shows that during the planning period, small claims occur frequently but large ones rather rarely. An example of a loss distribution is given by the probability distribution A of Table 7.1 (for simplicity, there are no claims L_A exceeding 3,000 MU).

The solvency requirement usually concerns the whole IC rather than just one portfolio of contracts. Besides A, let the IC considered have a portfolio B characterized by another loss distribution defined over losses L_B. To determine the overall loss distribution of the IC, the so-called *convolution* of the two densities must be formed.

The probability of zero total claims (denoted by $\pi_{A \cup B}$ pertaining to $L_{A \cup B}$ in Table 7.1) is derived from the product of the two component probabilities π_A and π_B, which equal $0.12 (= 0.3 \cdot 0.4)$ assuming independence. This means that π_B does not depend on whether or not a loss occurred in risk portfolio A. An aggregate claim of 1 thousand MU (TMU for short) can come about in two ways, not only $L_A = 0$ and $L_B = 1$ but also $L_A = 1$ and $L_B = 0$. Accordingly, for $L_{A \cup B} = 1$, the pertinent probability is the sum of $\pi(A = 1) \cdot \pi(B = 0) = 0.5 \cdot 0.4 = 0.20$ plus $\pi(A = 0) \cdot \pi(B = 1) = 0.3 \cdot 0.4 = 0.12$. Therefore, $\pi(L_A \cup L_B = 1) = 0.32$. Aggregate claim values of 2 and 3 TMU, respectively, can occur in three different ways. When all convolution possibilities are taken into account, their sum is 1. Obviously, the number of possible combinations rises rapidly with an increasing range of the two densities to be convoluted, making computer support necessary.

After convolution, it is possible to calculate the minimum premium which is compatible with a certain surplus (equity capital plus insurance reserves) as well as a required probability of solvency. For a required solvency level of 98% ($= 1 - 0.02$), the surplus at the end of the period must be 4 TMU (see Table 7.1). If initial capital and reserves amount to 1 TMU, then the IC must achieve a premium income of at least 3 TMU from its two products A and B in order to attain the required solvency level.

The example shows clearly that the asymmetry of the claims distribution is aggravated by the convolution. Aggregate claims can be as high as 4 and 5 TMU, occurring rarely but with probabilities that should not be neglected. It is quite possible that an aggregate claim of 5 TMU was never observed because of its low probability of 0.02. Thus, IC management would have a great interest in a tool enabling it to estimate the probability of an aggregate claim outside the observed range of value. The so-called normal power approximation to an arbitrary density function is one such tool, to be described below.

7.1.1.3 The Normal Power Approximation

The basic idea of the *normal power approximation* is simple. A probability density is a function, and any function can be approximated by a Taylor series. Using the approximation, one is able to calculate probabilities pertaining to outliers of total loss and hence the surplus necessary to ensure a desired probability of solvency.

Since the parameters of the Taylor series may be taken from the Normal distribution, the approximation amounts to mapping the unknown density $f(X)$ into $h(z)$, with $h(\cdot)$ denoting the standardized Normal distribution. Statements about probabilities with respect to X can then be read off from tables for the Normal distribution.

The only prerequisite is that the available information be sufficient to estimate the first three moments of the loss distribution, viz., the expected value, the variance, and the skewness of total claims given by $EX = \mu$, $E(X - \mu)^2$, and $E(X - \mu)^3$, respectively.

The (positive) third moment reflects the skewness of a typical loss distribution as depicted by $f(X)$ in Fig. 7.2. Small claims with $(X - \mu) < 0$ are frequent, but there are also very high claims with $(X - \mu) \gg 0$, causing the overall expected value $E(X - \mu)^3$ to be positive.

The first step is to normalize X to $\tilde{x} = (X - \mu)/\sigma$, where σ represents the standard deviation of X, equal to the square root of $Var(X)$. The resulting density is shown as $f(\tilde{x})$ in Fig. 7.2; it still exhibits skewness. Second, the value of the function can be approximated by a Taylor (also called McLaurin) series,

$$f[\tilde{x}] \cong \frac{f[0]}{0!} + \frac{f'[0]}{1!} \cdot \tilde{x} + \frac{f''[0]}{2!} \cdot \tilde{x}^2 + \frac{f'''[0]}{3!} \cdot \tilde{x}^3 + \ldots + \frac{f^{(n)}[0]}{n!} \cdot \tilde{x}^n. \quad (7.9)$$

By continuing up to the third power, one can take the skewness of the distribution into account. As for the derivatives, those of the Normal distribution are used because $f(\tilde{x})$ tends to the Normal distribution for large sample sizes according to the central

7.1 Traditional Premium Calculation

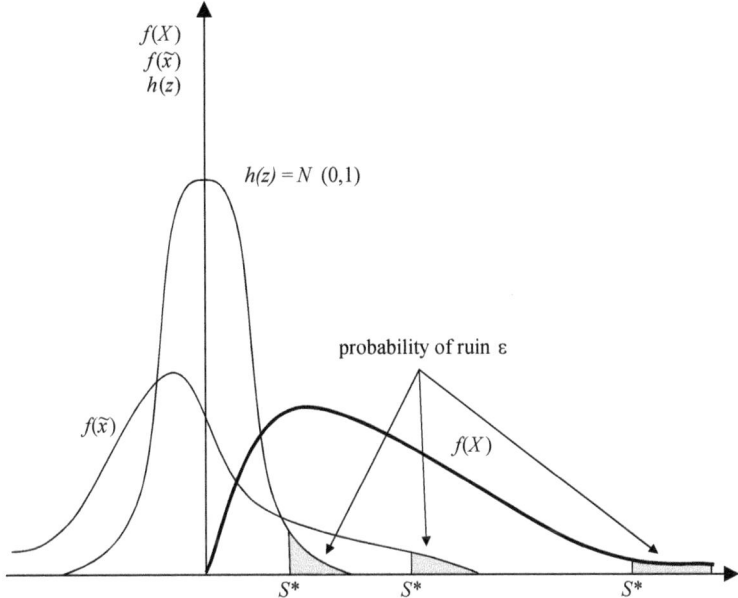

Fig. 7.2 Transformation of an arbitrary loss distribution, $f(X)$, into a standard Normal distribution, $h(z)$

limit theorem. These derivatives are given in Beard et al. (1984, 109f). They permit the transformation of $\tilde{x}$ with its unknown density $f(\tilde{x})$ into $h(z)$, with $h(\cdot) = N(0, 1)$ denoting the standard Normal distribution with mean 0 and standard deviation 1.

The third step is the transformation of $\tilde{x}$ values into corresponding z values. Here, Beard et al. (1984, 109f) show that for $\tilde{x} > 1$ (i.e., in the relevant domain of large claims), the corresponding value of z is given by

$$z = \sqrt{\left(1 + \frac{1}{4g^2} + \frac{\tilde{x}}{g}\right)} - \frac{1}{2g},$$

$$\text{with} \quad g = \frac{1}{3!}\gamma = \frac{1}{6}\gamma, \quad \gamma = E\tilde{x}^3 = \frac{E(X - \mu)^3}{(\sigma)^3}. \tag{7.10}$$

Example 7.2

An IC has equity capital and reserves amounting to 100 mn. MU. The expected value of total losses is 30 mn. MU, and from the convolution of component loss distributions, it is possible to estimate a standard deviation of $\sigma = 10$ as well as a skewness of $\gamma = 3$. Management wants to know whether the IC has a probability of solvency of 99.9%.

First, the standardization to $\tilde{x}$ is performed,

$$\tilde{x} = \frac{100 - 30}{10} = 7.$$

This indicates that total losses X can exceed their expected value of 30 by 7 standard deviations before using up equity capital and reserves amounting to 100. However, whether this is sufficient for 99.9% of possible values of X depends on the degree of skewness, indicated by the auxiliary variable g which is equal to $3/6 = 0.5$. Inserting $\tilde{x} = 7$ and $g = 0.5$ into Eq. (7.10), one obtains

$$z = \sqrt{\left(1 + \frac{1}{4 \cdot 0.5^2} + \frac{7}{0.5}\right)} - \frac{1}{2 \cdot 0.5} = 3.$$

Therefore, $\tilde{x} = 7$ ($X = 100$, respectively) is equivalent to $z = 3$ in the approximated standard Normal distribution. The probability of a deviation from the expected value by up to 3 standard deviations is $0.9987\ (= 0.5 + 0.49865)$ or 99.87%. This can be read off from Table 7.A.1 in Appendix to this chapter. Therefore, a surplus of 100 mn. MU is just about sufficient to achieve the desired solvency level of 99.9%. ∎

Calculating the minimum surplus and hence premium income compatible with a desired probability of solvency calls for the reverse transformation, from $h(z)$ to $f(X)$. Beard et al. (1984, 117) derive the following equation for values of $z \geq 1$ (i.e., again in the domain of large total losses),

$$\tilde{x} = z + g(z^2 - 1). \tag{7.11}$$

From the normalized value $\tilde{x} = (X - \mu)/\sigma_x$, one can then calculate the value of the surplus that suffices to match a total loss amounting to $X = \mu + \sigma \cdot \tilde{x}$.

Example 7.3

A very risk-averse management wants to attain a probability of solvency of 1/4 of one per thousand (0.00025). The expected value of total losses is $\mu = 30$ (mn. MU), the standard deviation $\sigma = 10$, and the skewness of the distribution $\gamma = 3$ as before. How large must surplus be to reach the desired solvency level?
The table of the standard Normal distribution shows that a $N(0, 1)$-random variable must be 3.48 standard deviations away from its expected value of zero to have passed through 99.975% ($=1 - 0.00025$) of its possible values. Hence, $z = 3.48$. By Eq. (7.11), the corresponding value of $\tilde{x}$ is given by

$$\tilde{x} = 3.48 + 0.5(3.48^2 - 1) \cong 9.$$

In view of the marked skewness of the loss distribution, surplus must cover total claims that are approximately 9 standard deviations away from their expected value. After retransformation, this results in

$$X = 30 + 9 \cdot 10 = 120.$$

The desired probability of insolvency of 0.00025 thus requires a surplus of 120 (mn. MU). Given, e.g., equity capital and insurance reserves of only 80, the lower bound of premium income must be set at 40. With an expected value of loss amount to $\mu = EX = 30$, this calls for a safety loading of $\lambda = 0.33$. What remains to be determined is the allocation of the required premium income to the lines of business and groups of insurance buyers (IBs). This allocation traditionally is performed on the basis of premium principles (see Sect. 7.1.2.2 below). ∎

▶ **Conclusion 7.1** To the extent that the loss distribution of an IC is positively skewed (many below-average claims, few very large outliers), the normal power approximation permits calculating the minimum premium income that is in accordance with a desired probability of solvency.

7.1.2 Basics of Probability Theory and Insurer's Risk

Up to this point, the exposition has focused on loss distributions for a given number of insurance contracts. In this section, the number of insurance contracts (also called risk units) denoted by n is varied. Important conclusions will be reached concerning underwriting risk and the probability of ruin when n increases beyond limits.

7.1.2.1 Basics of Probability Theory

The elements of probability theory to be examined below amount to statements about the convergence of the arithmetic mean of similar random variables to a common expected value when n goes to infinity. In an insurance context, the IC often underwrites risks that have the same expected loss (although realized losses differ during a given planning period). The average loss paid presumably approaches that common expected value when the portfolio comprising similar risks becomes very large.

However, it is important to note that realizations of a single random variable are subject to a limit. This follows from Chebyshev's inequality. Let X be a random variable with finite expected value μ and finite variance σ^2 (and hence finite standard deviation σ). Then, for every $k > 1$,

$$Pr\{|X - \mu| < k\sigma\} \geq 1 - \frac{1}{k^2}, \quad \text{or} \quad Pr\{|X - \mu| > k\sigma\} < \frac{1}{k^2}. \quad (7.12)$$

The second version of Chebyshev's inequality is obtained by inverting the inequalities. It states that a discrepancy between a single realization of X and its expected value amounting to more than k times the standard deviation has a probability of at most $1/k^2$. For an IC, this inequality is of great importance. In general, there is no reason to assume that the variance of a particular claim is infinite. Its standard deviation σ may be large, but a deviation from the mean exceeding, e.g., 3 standard deviations is already rare, having a probability of less than 1/9.

A stronger statement should be possible if the single realization X is substituted by $\bar{x}$, the arithmetic mean of many realizations. In fact, the *law of large numbers* holds. Let $x_1, x_2, \ldots, x_n$ denote realizations of a random variable with a common finite expected value μ and a common finite variance σ^2. Then, for the arithmetic mean $\bar{x}$ of these realizations, it is true for an arbitrarily small $\varepsilon > 0$ that

$$\lim_{n \to \infty} Pr\{|\bar{x} - \mu| > \varepsilon\} = 0. \tag{7.13}$$

In words, an absolute deviation of the arithmetic mean $\bar{x}$ from its (true) expected value μ by more than the arbitrarily small amount ε has a probability that approaches zero with an increasing number n.

7.1.2.2 Insurer's Risk

With the help of the law of large numbers, one can state properties of so-called insurer's risk. This means the risk that the IC has to pay higher claims than expected, potentially incurring a loss in its underwriting activity unless the premium contains a safety loading. The two dimensions of insurer's risk are as always the amount of loss and the probability with which it occurs (see Sect. 7.1.1). In the literature, three measures of insurer's risk can be found, (1) relative risk, (2) absolute risk, and (3) the probability of ruin [see, e.g., Cummins (1991a)].

1. *Insurer's relative risk* (IRR). Total losses of a portfolio consisting of n units (insured objects, persons, or firms) during a period are given by

$$L_n = \sum_{i=1}^{n} x_i = n \cdot \bar{x} \text{ with average loss equal to } \bar{x} = \frac{1}{n} \sum_{i=1}^{n} x_i. \tag{7.14}$$

   Provided claims are independent of each other (implying zero covariance between them), the variance of $\bar{x}$ amounts to

$$Var(\bar{x}) = \frac{1}{n^2} n \sigma^2 = \frac{\sigma^2}{n}. \tag{7.15}$$

   Therefore, the mean claim $\bar{x}$ has a standard deviation given by

$$\sigma_{\bar{x}} = \frac{\sigma}{\sqrt{n}}. \tag{7.16}$$

   Clearly, the law of large numbers can be applied to $\bar{x}$. Moreover, since ε in Eq. (7.13) is an arbitrary number, it may be substituted by $k \cdot \sigma_{\bar{x}}$ to obtain

$$\lim_{n \to \infty} Pr\{|\bar{x} - \mu| > \varepsilon\} = \lim_{n \to \infty} Pr\{|\bar{x} - \mu| > k \sigma_{\bar{x}}\}$$

$$= \lim_{n \to \infty} Pr\left\{|\bar{x} - \mu| > k \cdot \frac{\sigma}{\sqrt{n}}\right\} = 0. \tag{7.17}$$

   The statement of this equation leads to a first version of defining the insurer's relative risk, relating to mean loss rather than total loss.

7.1 Traditional Premium Calculation

Definition 7.1 The insurer's relative risk (IRR_1) consists of the possibility that the mean loss exceeds its expected value by more than the k-fold of its standard deviation. The probability of this discrepancy goes to zero with an increasing risk portfolio, however.

Division by μ in Eq. (7.17) and using (7.16) results in

$$\lim_{n\to\infty} Pr\left\{\left|\frac{\bar{x}-\mu}{\mu}\right| > \frac{\varepsilon}{\mu}\right\} = \lim_{n\to\infty} Pr\left\{\left|\frac{\bar{x}-\mu}{\mu}\right| > k\frac{\sigma_{\bar{x}}}{\mu}\right\}$$

$$= \lim_{n\to\infty} Pr\left\{\left|\frac{\bar{x}-\mu}{\mu}\right| > k\cdot\frac{\sigma}{\mu\sqrt{n}}\right\} = 0. \quad (7.18)$$

Note that this equation involves σ/μ, the so-called *coefficient of variation*. The statement of this equation leads to the second version of defining the insurer's relative risk.

Definition 7.2 The insurer's relative risk (IRR_2) consists of the possibility that the relative deviation of mean loss exceeds the k-fold of the coefficient of variation (σ/μ). The probability of this discrepancy goes to zero with an increasing risk portfolio, however.

2. *Insurer's absolute risk (IAR).* From Eq. (7.17), one obtains by multiplying through by n, and using Eq. (7.14),

$$\lim_{n\to\infty} Pr\left\{\left|\sum_{i=1}^{n} X_i - n\cdot\mu\right| > n\cdot\varepsilon\right\}$$

$$= \lim_{n\to\infty} Pr\left\{|L_n - n\mu| > k\sigma\cdot\sqrt{n}\right\} = 0. \quad (7.19)$$

The total loss of a portfolio composed of n similar units will repeatedly deviate from its expected value. However, the probability of a discrepancy exceeding the k-fold of the standard deviation of the average loss, $\sigma_{\bar{x}} = k\sigma/\sqrt{n}$, tends to zero for very large risk portfolios. Since k can be chosen arbitrarily, $k = 1$ is admissible. This makes the insurer's absolute risk equal to

$$\sigma\sqrt{n} = \frac{\sigma}{\sqrt{n}}\cdot n = \sigma_{\bar{x}}\cdot n.$$

Definition 7.3 The insurer's absolute risk (IAR) consists of the possibility that total loss exceeds its expected value by more than the k-fold of the standard deviation of the average loss. While the probability of such a discrepancy tends toward zero with an increasing risk portfolio, its amount $n\cdot\mu$ tends to infinity.

Table 7.2 Insurer's risk as a function of portfolio size n

n	Relative risk (Definition 7.1)	Relative risk (Definition 7.2)	Absolute risk (Definition 7.3)
1	800	1,600	800
10	253	0.506	2,530
100	80	0.160	8,000
1,000	25	0.051	25,298
10,000	8	0.016	80,000
∞	0	0	∞

Evidently, one has to be very precise when saying that an insurer's risk decreases or even disappears in large insurance portfolios. This holds true for the insurer's relative but not absolute risk.

Example 7.4

An IC planning to create an automobile insurance portfolio wants to estimate its insurer's risk. The portfolio comprises males above 25 years driving mainly in an urban area. The expected claim of this portfolio amounts to 500 MU, with a standard deviation of 800 MU (see Table 7.2).

The pooling of ever more risk units is advantageous for the IC because it reduces the relative risk per insurance contract and with it the amount of costly insurance reserves that must be held in order to attain a desired probability of solvency. However, note that absolute risk increases also with size, albeit at a rate $\sqrt{n}$ while premium income tends to increase with the number of contracts n. Ultimately, the law of large numbers therefore suggests that size could be beneficial for the insurance business. ∎

3. *The probability of ruin.* When the premium for an insurance contract is paid, it is credited to the reserve fund for unearned premiums (after deducting expenses especially for acquisition), which is used to pay the claims arising from that contract. At the end of the accounting period, the money may be put into other reserve funds (e.g., for losses incurred but not reported (IBNR). Furthermore, profits can either be retained as a contribution to surplus or paid out as dividends, depending on regulations and shareholder decisions. Hence, the IC can pay claims up to the value of (initial) equity capital plus accumulated surplus (also called buffer fund), denoted by S^*. The probability of ruin therefore is the probability with which total claims (L_n) exceed S^* (see Fig. 7.2). However, S^* is set in a way to cover the expected value of total claims [$n \cdot \mu$ in Eq. (7.19)] plus the relevant positive deviation $k\sigma\sqrt{n}$. Before and after taking the limit $n \to \infty$ in Eq. (7.19), one therefore has

$$\text{Probability of ruin} = Pr\{L_n > S^*\} = Pr\{L_n > n\mu + k\sigma\sqrt{n}\}$$
$$> 0 \text{ for } n < \infty \qquad (7.20)$$
$$\to 0 \text{ for } n \to \infty.$$

7.1 Traditional Premium Calculation

Definition 7.4 The insurer's risk in the sense of the probability of ruin tends to zero with an increasing risk portfolio. For a finite risk portfolio, however, it depends upon the properties of the loss distribution.

While in the limit, the probability of ruin goes to zero regardless of the loss distribution of $f(X)$, and for finite n it corresponds to the area below $f(X)$ beyond the threshold S^* (see Fig. 7.2 of Sect. 7.1.1.3). Obviously, this area depends on the shape of the loss distribution.

▶ **Conclusion 7.2** There exist three definitions of insurer's risk. (1) Relative risk decreases with an increasing number of insured units, while (2) absolute risk increases independently of the loss distribution. In contrast, for a finite risk portfolio, (3) the probability of ruin can be determined only when knowing the loss distribution.

7.1.3 Premium Principles

In underwriting, it is often necessary to quote a premium for an individual risk x that is characterized by an estimated loss distribution $f(x)$. In this situation, underwriters traditionally apply one of several premium principles (PP). In mathematical terms, a premium principle amounts to a functional Π which assigns a real number $\Pi(x)$, the premium for accepting the risk equivalent to the random variable x characterized by $f(x)$. Therefore, $\Pi(X)$ depends upon the loss distribution (the probability or density function of claims, respectively) of x, or conversely, the loss distribution determines the choice of a premium principle.

In the literature [see, e.g., Heilmann (1989)], the choice of PP is narrowed down by a set of desired mathematical properties.

1. The premium should not be less than expected loss, requiring $\Pi(x) \geq Ex$.
   Equivalently, the premium needs to contain a non-negative loading. Since realized claims sometimes exceed this expected value, a premium without a positive loading λ leads to ruin with a high probability.
2. The premium should be limited by the maximum loss, i.e., $\Pi(x) \leq max(x)$.
   This is the so-called no rip-off condition.
3. The premium should not contain an unjustified safety loading, i.e., $\Pi(x) = Ex$ for constant x.
4. A combined premium should increase by the same amount if the losses increase by a fixed amount, i.e., $\Pi(x+c) = \Pi(x) + c$.
   This condition is called consistency.
5. Total premium should not be affected by the pooling of independent risks, i.e., $\Pi(x+y) = \Pi(x) + \Pi(y)$.
   This property is called additivity.
6. Goovaerts and Laeven (2008, 121) also mention the property called iterativity. This means $\Pi(x) = \Pi[\Pi(x|y)]$ for all x, y. They state that the premium for x can be calculated in two steps. First apply $\Pi(\cdot)$ to the conditional distribution of

x, given y. The resulting premium is a function $h(y)$, say, of y. Then, apply the same premium principle to the random variable $\Pi(x|y) = h(y)$. However, the authors themselves qualify this criterion as rather artificial (ibid.).[2]

In risk theory, a multitude of premium principles are known of which only a few will be presented below. According to Heilmann (1988), they can be divided into those based on the net (fair) premium [(A) below] and those defined implicitly derived from a decision rule [(B) and (C) below].

(A) *PP based on the net premium*
An obvious starting point is the

- *Equivalence principle:* $\Pi_0(x) = Ex := \mu$.

The net premium is the (pure) risk premium equal to the expected loss (also called fair premium). It is the minimum premium, just sufficient to cover the expected loss of a risk-neutral IC. However, the expected value usually is close to the median which has the property that 50% of the area under the probability density lies below and 50% of the area above it. For an IC without equity capital or accumulated reserves, the ruin probability is therefore approximately 50% (see Fig. 7.2 of Sect. 7.1.1.3 again). Such a high frequency of insolvencies would be the end of private insurance. Hence, a safety loading in excess of the net premium is required. In the literature, this loading is called the "price of risk bearing".

A safety loading in proportion to expected loss leads to the

- *Expected value principle:* $\Pi_1(x) = (1 + \lambda)Ex$, $\lambda > 0$.

As before, only the expected loss associated with a risk to be underwritten needs to be known. However, the variability of losses is an important characteristic which should be used for determining the safety loading. This argument gives rise to the

- *Variance principle:* $\Pi_2(x) = Ex + aVar(x)$, $a > 0$.

Alternatively, one may use $\sigma = [Var(x)]^{1/2}$ to posit the

- *Standard deviation principle:* $\Pi_3(x) = Ex + b\sigma$, $b > 0$.

---

[2] As an example, Goovaerts and Laeven (2008, 121) cite a driver who causes a Poisson number X of accidents in one year, where the parameter λ [denoted by π in Eq. (7.5) of Sect. 7.1.1.1)] is drawn from Λ, the distribution of the structure variable. The number of accidents varies because of the Poisson deviation from the expectation λ and the variation in λ draws from Λ. If the premium is set to reflect the two sources of variation sequentially, iterativity requires that the resulting premium is identical to the premium determined for x directly.

Additionally, there is the

- *Exponential principle:* $\Pi_4(x) = (1/\alpha)ln[m(\alpha)]$.

The parameter $\alpha > 0$ reflects the degree of risk aversion on the part of the IC, and $m(\alpha)$ is the moment generating function with $m'(\alpha) > 0$. The exponential premium increases with α in spite of the multiplier $(1/\alpha)$ because the increase in the moment generating function dominates. For $\alpha \to 0$, one obtains the equivalence principle; for $\alpha \to \infty$, the resulting premium equals the maximum value of X.

Notwithstanding the arbitrariness in the choice of parameters λ, a, and b, principles $\Pi_1(x)$ to $\Pi_3(x)$ are used in actual practice. They all result in premiums that exceed expected losses. Because λ, a, and b can be arbitrarily large, they fail to satisfy the *no rip-off condition*, however.

(B) *PP defined implicitly*

In contrast to the premium principles cited above, the ones shown below are based on a decision-theoretic approach, with the respective parameter being replaced by a utility or value function.

Note that the principle of zero utility requires the existence of a strictly monotonically increasing, concave risk utility function $\upsilon(c)$, with $\upsilon[0] = 0$; $\upsilon'(c) > 0$; $\upsilon''(c) < 0$, implying that the IC is risk-averse. The premium has to be determined in such a way that the expected utility before ($c = 0$) and after accepting the risk x [$c = \Pi(x) - x$] is the same, resulting in zero excess utility as it were [see Bühlmann (1970), 86].

- *Principle of zero utility:* $E[\upsilon(\Pi_4(x) - x)] = \upsilon[0] = 0$.

For the usual types of risk utility function (quadratic, exponential), this equation has a unique solution $\Pi_4(x)$. For cases where a closed solution does not exist, one equates the IC to an individual who prefers the sure alternative to the financially equivalent risky prospect. As shown in Sects. 2.3.1 and 2.3.2, the maximum willingness to pay for certainty of such an individual facing a loss with expectation $Ex > 0$ is given by

$$\Pi_4(x) \approx Ex + \frac{R_A[0]}{2} Var(x), \qquad (7.21)$$

where the coefficient of absolute risk aversion is evaluated at $c = 0$. This equation is reminiscent of the variance principle $\Pi_2(x)$.

(C) *Loss function principles* start from the consideration that almost always the realized loss x deviates from the required premium Π. The larger the discrepancy $(x - \Pi)$, the higher the loss from underwriting. To the extent that large discrepancies evoke quick (and hence expensive) counter-measures, the loss function may increase progressively. A simple form is the quadratic, which, however, treats positive and negative discrepancies in the same way. In particular, minimization of the loss function $G(x, \Pi) = e^{hX}(x - \Pi)^2$ with $h > 0$ yields a PP that is called

- *Esscher principle:* $\Pi_5(x) = xe^{hx}/Ee^{hx}$.

The name of this PP is due to the fact that it can be derived using the so-called Esscher transformation [see Heilmann (1988)]. The parameter h is a measure of risk aversion on the part of the IC. The Esscher principle requires the absence of an unjustified safety loading, it exceeds expected loss, and satisfies the no rip-off condition. The numerator shows that the claims x are weighted in a way such that small claims contribute less to the premium than do the large ones.

This short overview is concluded with four critical remarks.

(i) The choice of PP remains *arbitrary*. It remains at the discretion of IC management whether and to what extent their risk aversion is reflected in the calculation of premiums. The parameters $\{\lambda, a, b\}$ of $\Pi_1(x)$ to $\Pi_3(x)$, reflecting also risk aversion of IC management, are not determined, no more than R_A and h in the two implicitly defined PP.
(ii) The PP are entirely *supply-oriented*. They implicitly assume that the IC has a monopoly that the IB cannot escape, not even by doing without insurance coverage. There is no reaction of IBs to differences in premiums. This makes PP adequate only for the determination of a lower limit price (which still varies with the PP applied) at which an IC should accept a risk portfolio of a given size rather than leaving it to a competitor.
(iii) The only source of risk is the *loss distribution*. Risks and returns of investment activity are not taken into account, although they are of great importance to the owners of the IC. It is therefore questionable whether PP of the type discussed above serves to maximize the market value of the IC.
(iv) The pooling of risks in the aim of benefiting from the diversification effect due to the law of large numbers is often considered the core mission of IC management. However, this creates the danger of limiting IC activity to aggregating a large number of IBs for forming risk pools, calculating risk-based premiums, building insurance reserves, and purchasing reinsurance. As a consequence other elements of insurance technology discussed in Sects. 6.3–6.8 tend to be neglected.

7.2 Financial Models of Insurance Pricing

In the wake of liberalization of insurance markets (see Sects. 9.1.5 and 9.3.2), the premium principles discussed in Sect. 7.1.3 have been losing importance. Deregulation has caused ICs to increasingly compete with other financial intermediaries.

Also, the insurance contract has become one investment alternative among many available in the capital market. Therefore, the optimization problem of an investor (who could be also an IC) is recapitulated on the basis of Sect. 5.1. Next, a first alternative to the traditional premium principles, deduced from the interests of a (potential) shareholder of the IC, is presented. This alternative ensures that pricing in underwriting is compatible with maximizing the market value of the IC. To this end, the Capital Asset Pricing Model is applied to the IC, taking into account both its underwriting and capital investment activities (see Sect. 5.1.3). The second alternative reflects the fact that under increasingly competitive conditions, it is market demand that determines price, i.e., the premium an IC can charge. In this situation, probability theory can only be used to determine a lower limit price at which offering a product becomes viable for the IC. In contrast, the market premium mirrors the certainty with which the IB can count on the performance of the service promised in the contract. This has to do with how claims against the IC are split between IBs and shareholders, calling for the application of the Black–Scholes option pricing model discussed in Sects. 5.2.2 and 5.3.3.

7.2.1 Portfolio Optimization by the IC

Risk-averse investors are interested in the expected value of returns and their riskiness. Riskiness is operationalized by the standard deviation of returns σ (for simplicity of notation, the same symbols $\{\mu, \sigma\}$ are used in Sect. 7.1, where they refer to losses rather than returns). A point on the efficient frontier in (μ, σ)-space is determined by the weights $\{w_1, \ldots, w_n\}$ making up the portfolio. These weights are chosen in a way such that for a given expected value of the returns $Er_P = \mu$ of the portfolio, the variance $\sigma^2(r_P)$ [and hence the standard deviation] of these returns is minimized [see Markowitz (1959)],

$$\sigma^2(r_p) = \sum_i^n w_i^2 \sigma_i^2 + 2 \sum_{i \neq j}^n \sum_j^n w_i w_j \sigma_{ij} \to \min.$$

$$\text{s.t. } Er_p = \sum_i^n w_i Er_i \geq \bar{r}_p, w_i \geq 0, \sum_i^n w_i = 1, i = 1, \ldots, n. \quad (7.22)$$

Here, $r_p :=$ rate of return of the portfolio, $r_i :=$ rate of return of the ith security, $E :=$ expected value operator, and $\sigma_{ij} :=$ covariance between the returns of securities i and j, all relating to an unspecified time period (usually a quarter or a year). The risk of the portfolio depends on the standard deviations (or, equivalently, variances) of the single securities, their weights in the portfolio, as well as the signs and amounts of correlations between the securities.

The portfolio theory of Markowitz (1959) is relevant for IC management in several ways.

- It *informs its investment policy* emphasizing possible diversification effects, which are acknowledged by the risk-based capital approach of the United States and Solvency II and (planned) Solvency III regulation in the European Union (see Sect. 9.4.4).
- It also informs underwriting policy (or liability management more generally), again by pointing out possibilities of *risk diversification* in the pooling of risks.
- It recalls the fact that an underwriting portfolio generates funds that can be used for capital investment which in turn permits the IC to lower premiums for improved competitiveness (so-called *cash flow underwriting*).

However, portfolio optimization only determines the efficient frontier in (μ, σ)-space. For a pricing of risk (and hence, an insurance product) that is compatible with conditions prevailing on the capital market, one needs to know the risk-adjusted equilibrium rate of return. This quantity is provided by the Capital Asset Pricing Model (CAPM).

7.2.2 Pricing According to the Insurance CAPM

The Capital Asset Pricing Model is presented in Sect. 5.1.3. Here, the equation for the security market line (SML) serves as the point of departure for applying the CAPM to insurance,

$$Er_i = r_f + \beta_i(Er_M - r_f) = (1 - \beta_i)r_f + \beta_i Er_M, \qquad (7.23)$$

with $r_i :=$ return of security i, $r_M :=$ return of the market portfolio, $r_f :=$ risk-free interest rate, and $\beta_i :=$ beta of security i with respect to the total market portfolio. The beta is equivalent to the slope coefficient of a regression equation with r_i as the dependent variable and r_M as the explanatory variable, $r_i = \kappa_i + \beta_i \cdot r_M + \varepsilon_i$. It is given by

$$\beta_i := \frac{Cov(r_i, r_M)}{Var(r_M)} = \frac{E[(r_i - Er_i)(r_M - Er_M)]}{E(r - Er_M)^2}. \qquad (7.24)$$

The application of the CAPM to the premium calculation is called *financial insurance pricing*. The connection results from the following reasoning. The calculation of premiums (i.e., pricing) is a genuine management activity. In the interest of the owners of the IC, it must be performed in a way that guarantees them a return on their investment that matches returns normally prevailing on the capital market after adjustment for risk. This argument calls for two steps. First, the expected return on equity capital (i.e., shareholders' investment) must be related to both underwriting and capital investment as the two core activities of IC management. Second, equality with the conditions prevailing in the capital market needs to be introduced as a requirement.

7.2 Financial Models of Insurance Pricing

7.2.2.1 Determination of the Expected Return on Equity Capital

The objective is to express the expected return on equity capital of an IC in terms of the expected returns to capital investment as well as risk underwriting. Indeed, expected profit ("gain") EG can be written as

$$EG = \underbrace{(K + k\Pi)Er_p}_{\text{contribution of investment}} + \underbrace{(\Pi - EL_n)}_{\text{contribution of underwriting}}. \quad (7.25)$$

Here, $K :=$ equity capital, $\Pi :=$ premium income, and $EL_n :=$ expected total loss. The so-called fund generating factor k indicates the multiple of per-period premium income that is available for investment. In the most simple case, where premiums are due at the beginning of the year and losses occur uniformly distributed over the year, half of the premium income can be invested, hence $k = 0.5$. However, many claims are paid only after a time-consuming process of control and assessment, making $k = 1.5$ a realistic estimate, i.e., losses are payable on average with a lime lag of 18 months after receipt of the premium.

Division by equity capital K and multiplication by $\Pi/\Pi = 1$ yields the following expression for Er_K, the expected return on equity capital of the IC:

$$Er_K := \frac{EG}{K} = \left(1 + k \cdot \frac{\Pi}{K}\right) Er_p + \left(\frac{\Pi - EL_n}{\Pi}\right) \cdot \frac{\Pi}{K}. \quad (7.26)$$

The percentage difference between premium and expected loss can be interpreted as the expected return on underwriting,

$$Er_u := \left(\frac{\Pi - EL_n}{\Pi}\right). \quad (7.27)$$

Hence, the expected return on equity capital becomes

$$Er_K = \left(1 + k \cdot \frac{\Pi}{K}\right) Er_p + \frac{\Pi}{K} \cdot Er_u. \quad (7.28)$$

The expected return on equity capital can therefore be expressed as a linear combination of the expected returns from investment and underwriting activity, respectively. The weights of this combination depend on the ratio of premium income and equity capital Π/K, often called leverage.

7.2.2.2 Equality with Conditions in the Capital Market

According to the CAPM, the stock issued by the IC must attain the same risk-adjusted expected return on capital as any other stock i. Therefore, applying Eq. (7.23) to the IC with $i = K$, one obtains

$$Er_K = r_f + \beta_K (Er_M - r_f). \quad (7.29)$$

However, Er_K is related to the expected returns of capital investment and risk underwriting as stated in Eq. (7.28). Equating (7.28) and (7.29) results in the equilibrium condition,

$$\left(1 + k \cdot \frac{\Pi}{K}\right) Er_p + \frac{\Pi}{K} \cdot Er_u = r_f + \beta_K (Er_M - r_f). \tag{7.30}$$

In addition, equality with the conditions in the capital market also holds for the investment activity of the IC. Since the IC has access to the same capital market as any other investor, the expected return on its investment portfolio satisfies the same Eq. (7.23) for the security market line,

$$Er_p = r_f + \beta_p (Er_M - r_f). \tag{7.31}$$

Given that Er_K can be expressed as a linear combination of Er_P and Er_u, it should be possible to express β_K as a linear combination of β_P and β_u, respectively, with β_u defined in Eq. (7.32) below as the slope coefficient of a regression relating r_u to r_M. As shown in Eq. (7.24), the beta is linear in the return of the security because r_i enters the $Cov(r, r_M)$ term in a linear way. Linear combinations of returns therefore result in linear combinations of betas. Equation (7.30) contains a linear combination of returns with weights $(1 + k \cdot \Pi/K)$ and (Π/K), respectively. Hence, β_K is given by

$$\beta_K = \left(1 + k \cdot \frac{\Pi}{K}\right) \beta_p + \frac{\Pi}{K} \cdot \beta_u, \text{ with } \beta_u := \frac{Cov(r_u, r_M)}{Var(r_M)}. \tag{7.32}$$

Substituting expression (7.31) for Er_p on the left-hand side of the equilibrium condition (7.30) and expression (7.32) for β_K on the right-hand side of condition (7.30) yields

$$\left(1 + k \cdot \frac{\Pi}{K}\right) [r_f + \beta_p (Er_M - r_f)] + \frac{\Pi}{K} \cdot Er_u$$
$$= r_f + \left[\left(1 + k \cdot \frac{\Pi}{K}\right) \beta_p + \frac{\Pi}{K} \cdot \beta_u\right] (Er_M - r_f). \tag{7.33}$$

Subtraction of $\left(1 + k \cdot \frac{\Pi}{K}\right) [\beta_p (Er_M - r_f)]$ and r_f on both sides gives

$$k \cdot \frac{\Pi}{K} \cdot r_f + \frac{\Pi}{K} \cdot Er_u = \frac{\Pi}{K} \cdot \beta_u (Er_M - r_f). \tag{7.34}$$

7.2.2.3 The Insurance CAPM

Dividing Eq. (7.34) by (Π/K), one immediately obtains the *equation for the insurance CAPM*,

$$Er_u = -kr_f + \beta_u (Er_M - r_f). \tag{7.35}$$

For investors, the holding of securities issued by the IC considered must be equivalent to the other investment alternatives available in the capital market. For this condition to be satisfied, premiums have to be set such that underwriting activity yields an expected return (Er_u) that corresponds to the usual risk-adjusted capital return. Hence, according to the insurance CAPM, premiums contain two components:

1. *A deduction for credit provided.* The IBs pay the premium in advance, while losses are paid k periods later on average. Therefore, they receive a deduction reflecting the risk-free interest the IC can earn during this time.
2. *A price for bearing systematic risk.* The owners of the IC bear the systematic risk to the extent that returns on underwriting are positively correlated with market returns. Accordingly, this price equals the underwriting beta times the market risk premium. However, note that even with a positive price for risk bearing, the return on underwriting activity may be negative in expectation without jeopardizing the economic survival of the IC as a whole. All it takes is a negative (or even positive but small) β_u indicating that the underwriting activity of the IC provides an excellent opportunity for risk diversification to investors.

▶ **Conclusion 7.3** The insurance CAPM states that premiums must be set such that underwriting yields a certain expected return, which can be negative. This benchmark value is given by the price for systematic risk bearing by the IC (according to the beta of its underwriting activity) minus the interest on the capital provided by the IB by paying the premium in advance.

The example below is designed to convey an impression of the parameters characterizing the insurance CAPM in the case of non-life business in the United States. However, it should be noted that estimates vary considerably with the period of observations [for some examples, see Cummins and Phillips (2005)].

Example 7.5

Estimating the CAPM parameters for U.S. non-life insurance

(Source: Lecture notes by N.A. Doherty)

β_p = 0.3 : If returns in the capital market vary by one percentage point, this typically goes along with a variation of 0.3 points in returns on capital investment by the IC, reflecting their conservative investment strategy.

Er_M = 0.1 : Returns in the capital market are approximately 10% p.a. nominal.

r_f = 0.03 : The risk-free interest rate is equated to that of government bonds, which is some 3% p.a. nominal.

Er_p = 0.051: According to equation (7.31), $0.051 = 0.03 + 0.3 (0.1 - 0.03)$. The IC achieves an average rate of return of 5.1% on its capital investment.

β_u = −0.1 : If returns in the capital market increase by one percentage point, this typically goes along with a reduction in the rate of return on underwriting by 0.1 points because building reserves through the capital market becomes the more attractive alternative compared to purchasing insurance.

k = 1.5 : Losses are paid 18 months after receipt of premium on average.

Π/K = 2 : Premium income is roughly twice that of equity capital.

Er_u = −0.052: According to equation (7.35), $-0.052 = -1.5 \cdot 0.03 + (-0.1)(1 - 0.03)$. In order to be able to offer investors conditions equivalent to those prevailing in the capital market in general, the IC needs to attain a negative rate of return of −5.2% p.a. on average in its underwriting activity, i.e. an average combined ratio of 1.052.

Er_K = 0.1: According to equation (7.28), $0.1 = [1 + (1.5 \cdot 2)] \cdot 0.051 + 2 \cdot (-0.052) \cdot 0.1$. The expected return on equity of an IC that is in line with the capital market amounts to 10% p.a.

β_K = 1: According to equation (7.32), $1 = [1 + (1.5 \cdot 2)] \cdot 0.3 + 2 \cdot (-0.1) \cdot 1$. Returns on IC equity vary in step with those on the capital market.

Another illustration of the insurance CAPM comes from Germany, which was characterized by very strict premium regulation prior to 1994 (see Sect. 9.3.3). German regulatory authorities stipulated an admissible rate of return on underwriting of 3% in auto insurance; beyond this threshold, profits were to be credited to insurance reserves. Using this threshold ($Er_u = 0.03$) in Eq. (7.28) and taking the remaining parameters from Example 7.5, one obtains

$$Er_K = (1 + 1.5 \cdot 2) \cdot 0.051 + 2 \cdot (0.03) \approx 0.264 = 26.4\%.$$

Hence, a seemingly modest profit margin of 3% in underwriting may admit a very high return on equity capital. Published rates of return were much lower because profits were transferred to reserves. The high implicit rates of return served to raise the price of insurance stock. ∎

However, the CAPM and hence its application to insurance are open to criticism in view of its simplifying assumptions. The CAPM would have to be rejected if all its assumptions were necessary, being part of the core of the model. This is not the case since essential restrictions of the CAPM were relaxed over the course of time. For instance, a risk-free rate of interest need not exist [see Black (1972)], expectations of investors may not be homogeneous [see Grossman (1976) and Williams (1977)], and the planning horizon can comprise more than one period [see Cummins (1991b)]. In addition, inflation, taxes, and insolvency risk have been taken into account as well [see Cummins (1991b)].

Still, the application of the CAPM to insurance pricing raises at least four issues.

(i) *Use of the fund generating factor k*. The fund generating factor represents the time lag between receipt of premium and payment of losses. In the finance literature, it is well known that flows of money have to be discounted to their present values. Therefore, k should not be a constant but a function of the interest rate used in discounting.

(ii) *The assumption that ruin of the IC does not occur*. Investing in an IC is put on par with investing in an enterprise that cannot go bankrupt. However, it is precisely the loss distribution of insurance that entails a significant probability of ruin. This means that the price of insurance coverage should be determined using a "risky claims" model rather than a "risk-free" model as the CAPM.

(iii) *Diverging average maturities*. In particular in life insurance, it is important for investments to have the same average maturity as loss payments. A divergence between the two may cause a lack of liquidity, which must be made up by selling securities at possibly quite unfavorable prices. As a consequence, the IC may be unable to hold the security market line.

(iv) *The neglect of regulation*. IC in many countries cannot choose their capital investment freely but are regulated more or less stringently in this respect. Hence, they need a higher return in their underwriting activity to compensate for the lower return in their investment activity.

Yet, the CAPM makes it clear that ICs can be looked upon as financial intermediaries since they hold an investment portfolio of primary securities while issuing the insurance product as secondary security. As Gurley and Shaw (1969, 192) state, the principal function of financial intermediaries is to purchase primary securities from ultimate borrowers and to issue indirect debt for the portfolio of ultimate lenders. And in a similar way, Pyle (1971, 737) sees "the essential characteristic [of a financial intermediary] that it issues claims on itself and uses the proceeds to purchase other financial assets". ICs therefore exist because they can perform this transformation of primary into secondary securities at a lower cost than others. Their cost advantage stems in particular from their underwriting know-how (information about probabilities and sizes of losses and the influence of IBs on them; see Sect. 5.3). If, on the other hand, perfect insurance and capital markets are assumed, then in equilibrium all efficient investment portfolios lie on the capital market line (CML), and market participants would not bear any risk unsuitable to them. A risk transformation by financial intermediaries would be unnecessary in such a world.

7.2.3 Pricing According to Option Pricing Theory

The (more recent) models of option pricing constitute an improvement over the CAPM because they can accommodate a positive probability of bankruptcy (see Sect. 5.2). The calculation of the option price is based on a risk-free hedging portfolio consisting of an option and the underlying security. Because of the efficient market assumption, the return on the hedging portfolio has to be equal to the return of the risk-free investment. Otherwise, there would be opportunities for arbitrage. Option prices

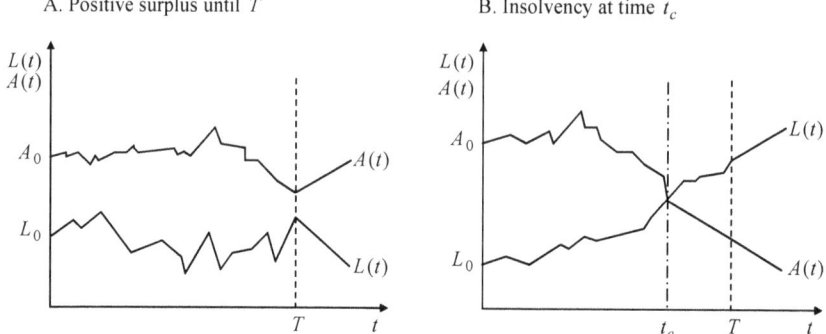

Fig. 7.3 Assets (A_t) and liabilities (L_t) as stochastic processes

are therefore determined independently of the risk preferences of market participants. It will be shown below (Sect. 7.2.3.2) that this feature permits the derivation of an underwriting premium reflecting the insolvency risk of an IC in an objective way. However, the first step is to interpret insurance stocks as options.

7.2.3.1 Insurance Stocks as Options

An insurance contract can be interpreted as a contingent claim, associated with payments that depend on other assets and liabilities. It therefore has the characteristics of an option. By aggregating all contracts to a total claim or obligation of the IC, the IC itself can be viewed as a security of the option type [see, in particular, Doherty and Garven (1986)]. Its net value at the end of the accounting period, time T, is given by the difference between assets A_T and the present value of its liabilities L_T (which amount to the obligation of the IC to pay future losses). Since A_T depends on risky rates of return on capital investment, both A_T and L_T of the IC need to be modeled as outcomes of stochastic processes. Two possible stochastic processes are shown in panels A and B of Fig. 7.3.

At time T, the question arises on how to divide the claims against the assets of the IC between owners and IBs. In principle, owners are entitled to the excess of assets over liabilities, while IBs are entitled to the liabilities, i.e., the benefits contracted. However, two states must be distinguished. In one state, the IC is solvent and can satisfy all claims; in the other state, the IC is insolvent, and IBs are entitled to the bankrupt IC's assets. The division of claims is therefore state-dependent. It can be represented by the formula,

$$A_T = \underbrace{max(0, A_T - L_T)}_{\text{Owners}} + \underbrace{L_T - max(L_T - A_T, 0)}_{\text{IBs}}$$

$$= \begin{cases} (A_T - L_T) + L_T = A_T \text{ if } A_T > L_T \\ 0 + L_T - (L_T - A_T) = A_T \text{ if } A_T < L_T. \end{cases} \quad (7.36)$$

If the inequality $A_T > L_T$ holds (as in panel A of Fig. 7.3), the first maximum of the equation is $A_T - L_T$, the second, zero. Claims are divided as follows. The

7.2 Financial Models of Insurance Pricing

IC pays IBs their losses L_T, while the surplus $S_T = A_T - L_T$ (consisting of equity capital plus insurer's reserves) is appropriated by its owners. If $L_T > A_T$ (as in panel B of Fig. 7.3), then the first maximum of Eq. (7.36) equals zero, the second, $L_T - A_T$. This time, the division of claims to the assets of the IC is as follows. The owners lose their entitlement to the surplus $A_T - L_T$ but are not held liable for the full amount of the net debt. Being shareholders, they are only liable up to the value of their funds invested. Conversely, IBs get only A_T instead of L_T, thus losing the amount $(L_T - A_T)$. This quantity represents the losses that are not paid to IBs in the case of insolvency.

Equation (7.36) provides the basis for describing the IC through options since the maxima are equivalent to the holding of a call and a put option, respectively,

$$A_T = C(A_T, L_T) + L_T - P(A_T, L_T). \tag{7.37}$$

The call option $C(A_T, L_T)$ corresponds to $max(0, A_T - L_T)$. It gives the owner the right to buy the security at a predetermined price. This right has value zero in the worst case (i.e., if liabilities exceed assets); in all other cases, it has a positive value which increases with the surplus $(A_T - L_T)$. The rights of IBs consist of two components. In the first place, they are entitled to loss payments L_T according to the contract. The second component is a deduction, equal to the put option held by the owners of the IC. Recall that the owners have the right to indemnify IBs using no more than the assets A_T remaining, permitting them to burden IBs with the difference between liabilities and assets $(L_T - A_T)$. This is equivalent to a put option that gives the owner (here, the shareholder of the IC) the right to sell the security at a predetermined price. In the case of solvency, this right has zero value since assets exceed liabilities; in the case of insolvency, it has the value $(L_T - A_T)$ because it allows the owners of the IC to burden IBs with the excess of liabilities over assets $(L_T - A_T)$. In sum, the put option has a non-negative value which is equivalent to the market value of the insolvency risk.

At the time of decision ($t = 0$), one needs to forecast the values $\{A_T, L_T\}$ at time T when the options will be exercised. In the case of an IC, this is the end of the accounting period. These forecasts are based on an extrapolation of the instantaneous rate of change in the values of assets and liabilities. One assumes a process that consists of a systematic and a stochastic part. In the case of assets, it has the form

$$dA = (\mu_A A + \delta N - \theta L)dt + A\sigma_A dZ_A, \tag{7.38}$$

with $\mu_A :=$ instantaneous expected return on assets per time unit; $\delta :=$ instantaneous rate of premium income per contract; $N :=$ number of contracts; and $\theta :=$ instantaneous rate of claims per monetary unit of liabilities (the time argument is dropped for simplicity of notation). These variables cause a systematic change of assets between two points in times (dt). Investment returns and premium income cause the rate of change to increase and loss payments to decrease assets.

In addition, there is the stochastic disturbance dZ_A which is assumed to be drawn from a standard Normal distribution (giving rise to a so-called standard Wiener process with respect to assets A). The resulting deviation is given by $A\sigma_A dZ_A$, with $\sigma_A :=$ instantaneous standard deviation of returns on assets, and $A :=$ value of the asset at the beginning of the period. Hence, the larger the volatility of the asset and the larger its value, the more marked is the influence of the normalized disturbance dZ_A on the process. In sum, $A\sigma_A dZ_A$ measures the deviation of the asset value from trend due to stochastic effects and thus amounts to a stochastic error term.

Similarly, one has for the liabilities,

$$dL = (\mu_L L + \eta N - \theta L)dt + L\sigma_L dZ_L, \qquad (7.39)$$

with $\mu_L :=$ instantaneous growth rate of liabilities (also due to inflation) and $\eta :=$ instantaneous rate of new losses presented. The larger the losses to be paid and the higher the number of contracts, the quicker is the increase in liabilities of the IC, *ceteris paribus*. A slowing of this increase is only possible by paying losses (θL). In sum, liabilities increase due to inflation and the presentation of new claims and decrease due to loss payments. In addition, there is the stochastic deviation $L\sigma_L dZ_L$, with $\sigma_L :=$ instantaneous standard deviation of liabilities per time unit, and $dZ_L :=$ a draw from the standard Normal distribution (giving rise to a standard Wiener process with respect to L).

Finally, the two stochastic components in Eqs. (7.38) and (7.39) may be correlated, resulting in an interdependence between the development of assets and liabilities. This can be represented by connecting the two standard Wiener processes,

$$dZ_A dZ_L = \rho_{AL} dt,$$

with $\rho_{AL} :=$ instantaneous coefficient of correlation. If ρ_{AL} is positive, then disturbances acting in the same direction are to be expected such that $dZ_A dZ_L > 0$. The higher the ρ_{AL}, the higher is the product of the two disturbances $dZ_A dZ_L$, reflecting their reinforcing effect.

In principle, one has to determine from normally distributed changes of the stochastic variable $(dA - dL)$ its level after T periods. This difficult task of integration of a stochastic variable was solved for the first time by Black and Scholes (1973) [see the Black–Scholes formula in Sect. 5.2.2, Eq. (5.29)].

Using the Black–Scholes formula to determine the asset value for the starting point $t = 0$ instead of $t = T$, one obtains

$$A_0 = \underbrace{C_0(A_T, L_T)}_{\text{Owners}} + \underbrace{L_T e^{-rT} - P_0(A_T, L_T)}_{\text{IBs, net}}. \qquad (7.40)$$

The assets of the IC must cover the entitlements of both shareholders and IBs. The call option reflects the right of the shareholders to the excess of assets over liabilities. The right of IBs in principle equals the present value of the losses to be paid; however, deduction must be made for the value of the insolvency risk reflected by the put option which hedges the owners in case of bankruptcy. The entitlement of IBs to loss payments is discounted using the market interest rate r, which is assumed constant during the period $[0, T]$, resulting in the present value factor e^{-rT}.

7.2.3.2 Underwriting Premium

The premium income that takes into account the call and put options held by IC shareholders can be derived from the equality of accumulated IC assets and claims to it. Let $t = 0$ denote the time when a first policy is issued by the IC. Its assets then consist of equity S_0 plus premium income Π (recall that premiums are paid in advance),

$$A_0 = S_0 + \Pi. \tag{7.41}$$

This expression can be equated to Eq. (7.40). Solving for Π results in

$$\Pi = L_T e^{-rT} - P_0(A_T, L_T) + C_0(A_T, L_T) - S_0. \tag{7.42}$$

However, at the start of underwriting, the value of the call option held by shareholders cannot differ yet from the equity of the IC. Therefore, $C_0(A_T, L_T) = S_0$, causing (7.42) to reduce to

$$\Pi^* = L_T e^{-rT} - P_0(A_T, L_t). \tag{7.43}$$

This is the insurance premium Π^* that is *compatible with the option pricing model*. Hence, premium income must cover the present value of losses with the insolvency put deducted [see Cummins (1991a, 291)]. The insolvency put equals the market value of the risk of insolvency of the IC. The higher this risk (represented by the standard deviation of losses σ_L), the higher the value of the put ($\partial P/\partial \sigma_L > 0$). Conversely, the higher the solvency level of the IC, the lower the value of the put option, and the higher the premium. This result is the *very opposite* of the prescription of risk theory, where a higher attained solvency level leads to a lower safety loading on the fair premium (see Sect. 7.1.3).

▶ **Conclusion 7.4** The premium income which is compatible with option pricing theory covers the present value of loss payments minus the put option contained in the stock of the IC.

Note the great simplification afforded by the assumption of an efficient capital market. While the risk theory approach calls for the difficult estimation of a loss distribution (which in addition may change over time) for determining the probability of ruin, this quantity becomes *irrelevant* here. The economic value of the risk of ruin is reflected by the market prices of options written on insurance stocks.

To summarize, one can argue as follows. The option pricing model operationalizes the total risk exposure of an IC by the volatility of its assets and liabilities. In this way, it captures the entire range of price and value fluctuations and hence both systematic and non-systematic risk. The risk concept of the option pricing model takes into account the capital structure and the shocks impinging on assets and liabilities as well as the correlation between them. Furthermore, the insolvency risk of the IC is not only measured by a probability distribution but is *valued* economically (and through the market). Hence, one may say that the option pricing model provides the connecting link between the mathematical and statistical approaches to insurance on the one hand and the economic, capital market-oriented approaches on the other hand. It integrates the risk concepts of risk theory and capital market theory.

However, the option pricing model hinges on the crucial assumption that the hedging portfolio always guarantees risklessness. This requires the ratio of hedging securities to underlying assets and liabilities to be adjusted continuously in response to the current market situation. Therefore, one must abstract from the many "discontinuities" that characterize real markets. The model is, however, consistent in itself and can also be used for calculating the surplus (i.e., the necessary solvency capital) as a function of several determinants.

The main criticism remaining is that problems of asymmetric information between IC management, capital owners, and IBs are not taken into account. Furthermore, the option pricing model reduces insurance products to streams of monetary payments. The interpretation of insurance products as pure debt results in a narrow perspective, neglecting important aspects of insurance markets. Some of them are taken up in Chap. 8.

7.2.4 Evidence on the Actual Behavior of the IC

7.2.4.1 Price Setting by the IC

The CAPM advises IC management on how to set premiums to ensure that holding shares of the IC is as attractive as holding other securities offered in the capital market, while the option pricing model (OPM) prescribes a premium that satisfies the claims of both shareholders and policyholders. Whether these prescriptions are reflected by actual premium setting behavior of the IC is a different matter. Research by Garven and D'Arcy (1991) provides information regarding this question. The authors measure the returns on underwriting activity [defined in Eq. (7.27)] of U.S. property-liability insurers during the time period 1926–1985, comparing them with the target values resulting from five premium setting rules. Besides (1) the CAPM and (2) OPM, these are (3) recommendations issued by the insurance commissioners, (4) the value resulting from a target return for the entire IC, and (5) a rule applied in Massachusetts requiring equality between the present value of premiums and of losses plus taxes to be paid by the IC (see Table 7.3).

The authors assess the five rules in terms of their mean square error (MSE), defined in the note to Table 7.3, which indicates how closely they correspond to observed returns in underwriting activity. First, as a general observation, realized underwriting returns are low with a mean value of 1.32% p.a., while their standard deviation amounts to 5.74 percentage points, indicating that years with negative returns were frequent. However, the CAPM rule (1) would have even called for a mean return of -1.11% p.a., which does not come as a surprise in view of the discussion of Eq. (7.35). Nevertheless, the CAPM rule is in close accordance with effective returns over the entire observation period 1926–1985, resulting in an MSE of 0.24. Toward the end of the observation period (1976–1985), it increases to 0.58, a possible indication that ICs were discarding this pricing rule.

The OPM (2) prescribes a rate of return on underwriting that on average coincides with the actual mean values of 1.32% p.a. However, some years are characterized by

7.2 Financial Models of Insurance Pricing

Table 7.3 Underwriting returns, U.S. property/liability insurers (1926–1985)

Actual and target values according to pricing rule	Mean	Standard deviation	Mean square error[a]		
			1926–1985	1966–1985	1976–1985
Actual	1.32	5.74	–	–	–
(1) CAPM	−1.11	4.45	0.24	0.35	0.58
(2) Option pricing	1.32	4.56	0.21	0.22	0.38
(3) Commissioners' recommendation	4.73	0.06	0.44	1.00	1.66
(4) Target rate for entire IC	0.57	5.43	0.16	0.25	0.39
(5) Equality of present values (Massachusetts)	−1.27	1.60	0.32	0.37	0.60

[a]The mean square error is given by MSE $= \frac{1}{N} \sum_{t=1}^{N} (r_{u,t} - \hat{r}_{u,t})^2$ with $r_{ut} :=$ actual underwriting return; $\hat{r}_{ut} :=$ target rate of return, calculated according to rules (1) through (5).
Source Garven and D'Arcy (1991), Tables 2 and 3

substantial discrepancies, resulting in an overall MSE that is comparable to that of the CAPM rule. Toward the end of the observation period, the OPM comes off best of all (MSE of 0.38).

Rates of return recommended by insurance commissioners (3) would have resulted in a much higher mean value (4.74% p.a.) than that actually observed. The high MSE toward at the end of the observation period suggests that ICs increasingly disregarded these recommendations. Conversely, they seem to have observed rule (4) where an underwriting return is derived from a target value calculated for the IC as a whole. Its MSE for 1976–1985 amounts to 0.39, comparable to that of the OPM (2). Finally, the discounted cash flow rule (5), developed for the Massachusetts automobile rate hearings as an alternative to the CAPM, would have resulted on average in an underwriting return of −1.27%, even below the CAPM value of −1.11% p.a. The MSE associated with this rule is similar to the MSE of the CAPM-based one.

▶ **Conclusion 7.5** Underwriting returns earned by U.S. property-liability insurers are in close accordance both in mean value and variation in time with prescriptions derived from the OPM, followed by those derived from the CAPM. In addition, the performance of OPM compares favorably with pricing rules that are theoretically less founded.

7.2.4.2 Risk Management of the IC

The OPM is suitable not only for deducing pricing rules but also for rules guiding risk management of the IC. The reason is that the pertinent decisions affect the value of the options held by the owners of the IC. These relationships are highlighted by Cummins and Sommer (1996), who take Eq. (7.43) as their point of departure. First, they rearrange Eq. (7.42) in order to focus on the net claim of policyholders on the right-hand side, resulting in

$$S_0 + \Pi^* - C_0(A_T, L_T) = L_T e^{-rT} - P_0(A_T, L_T). \tag{7.44}$$

Next, they note that the values of both the call and the put option generally depend on current values of assets and liabilities $\{A_0, L_0\}$, time to maturity T, market interest rate r, and volatility of the underlying, which is the surplus in this context (σ_S), in analogy to Sect. 5.2.2. Finally, they denote by L_0 the present value of claims $L_T e^{-rT}$ and factor L_0 out in order to state their hypotheses in terms of the asset–liability ratio $(a := A_0/L_0)$,

$$S_0 + \Pi^* - C_0(A_0, L_0; T, r, \sigma_s) = L_0 \left[1 - \frac{P_o}{L_0}(a, 1; T, r, \sigma_s) \right]. \tag{7.45}$$

The right-hand side of (7.45) shows that the net value of the claim held by IBs crucially depends on the value of the put option held by the owners which in turn is determined by the asset–liability ratio a, in addition to the other determinants cited above.

Statement of the Hypotheses

Maximization of expected profit entails a certain probability of insolvency (and hence expected costs of insolvency). The management of the IC considers two decision variables to maintain the probability of insolvency at its optimal value. First, through a high asset–liability ratio a, IC management can (almost always) avoid excessive indebtedness. Second, it can keep the volatility of surplus low, with a similar effect. This volatility is given by $Var(A - L)$, which, however, cannot be easily measured at the level of an individual company. As a substitute, Cummins and Sommer (1996) use the variance of surplus returns σ_S^2, for which market observations are available,

$$\sigma_S^2 := Var(r_A - r_L) = \sigma_{r_A}^2 + \sigma_{r_L}^2 - 2\rho_{AL} \cdot \sigma_{r_A} \cdot \sigma_{r_L}, \tag{7.46}$$

with $\sigma_{r_A}^2 :=$ variance of returns on assets, $\sigma_{r_L}^2 :=$ variance of returns on liabilities, and $\rho_{AL} :=$ coefficient of correlation between the returns on assets and liabilities. IC management has to take into account that risk-averse IBs react to an increase in the risk of insolvency by curtailing their demand for insurance. The (per-unit) market value of the insolvency risk is given by P_0/L_0, the value of the put option in the hands of shareholders per MU of claims to be presented to the IC, establishing the connection with the option pricing model.

Cummins and Sommer (1996) put forward two hypotheses describing the relationship between the two decision variables σ_S and a.

H1: $da/d\sigma_S > 0$. The desired asset–liability ratio is predicted to increase in response to a higher volatility of surplus returns σ_S. The higher the σ_S, the higher the value of both the call option C_0 and the put option P_0 of the shareholders (since they can get rid of their share in the worst case at a price of zero even though liabilities may exceed assets by far). But the latter increase entails a *redistribution of claims* in favor of the owners to the detriment of IBs, who respond by curtailing their demand for insurance. To counteract this effect, the IC can increase a, thus reducing the value of the put option (the density function of the surplus is shifted to the right, toward more positive values). In this way, IC management can mitigate or even neutralize the negative impact on demand and hence premium income.

H2: $d\sigma_S/da > 0$. The desired volatility of surplus returns is predicted to increase in response to a higher asset–liability ratio a. The reason is that a high asset–liability ratio keeps the probability of insolvency low. IBs honor this by increasing their demand for insurance. While this raises expected profit, it also entails a *redistribution of claims* at the expense of IC shareholders. In Eq. (7.45), the value of their put option decreases because the density function of the surplus is shifted to the right. To restore the value of the put option, IC management can act to increase the variance (and hence the standard deviation) of returns on surplus σ_S by opting for more volatility in its underwriting and investment activity.

The authors consider an alternative hypothesis designed to explain the relationship between risk-taking by the IC and its asset–liability ratio.

A1: $da/d\sigma_S > 0$ because of regulation. The insurance regulator could impose extremely high costs on the IC in case of insolvency, forcing IC management to keep the probability of insolvency at a very low level regardless of any response by IBs [see Shrieves and Dahl (1992)]. The consequence would again be $da/d\sigma_S > 0$ since an increase in the volatility of surplus returns has to be balanced by an increase in the asset–liability ratio (e.g., by non-renewal of insurance contracts) to prevent the probability of insolvency from increasing. If alternative hypothesis A1 is correct, a variable representing the stringency of regulation should contribute to the explanation of the relationship $da/d\sigma_S$ (and possibly $d\sigma_S/da$ to the extent that the regulator does not permit the IC to adjust volatility to the extent management deems optimal).

Furthermore, there are two additional hypotheses which do not address the two relationships *per se* but predict particular modifications of them because of the principal-agent problem between the owners and the management of the IC (see Sect. 5.3.3).

Z1: In an IC where owners are separate from management, managers are less diversified than owners. The economic success of the managers depends crucially on the performance of this particular IC, while the majority of owners have stocks in many firms. Therefore, in this type of IC, management is predicted to behave in a more risk-averse manner than would be expected on the basis of the option pricing model [see Mayers and Smith (1988)]. Conversely, a closely held IC (e.g., where management takes significant ownership) should take higher risks at a given asset–liability ratio, or operate with *a lower asset–liability ratio* given the same risk exposure, because the interests of management and owners are more closely aligned.

Z2: A closely held IC in the sense that its owners are engaged in management is predicted to take more risks, resulting in a lower asset–liability ratio. However, Fama and Jensen (1983) argue to the contrary, noting that owners of this type usually hold a large part of their wealth in the firm with which they are intimately connected.

Econometric Analysis of the Relation $da/d\sigma_S$ (H1)
First, the relation $da/d\sigma_S$ is analyzed. However, in Table 7.4 the dependent variable is not the asset–liability ratio but the ratio of equity capital to assets K/A. This choice is designed to facilitate comparison with earlier studies. It can be justified by the argument that a high value of K/A indicates that shareholders have much to lose in the event of insolvency, creating an incentive to keep its probability low. This incentive is also marked when assets exceed liabilities by far [see Eq. (7.36)]. Therefore, K/A may serve as an indicator of $a := A_0/L_0$.

Since the data are a time series, the residuals of the regression were tested for autocorrelation. Autocorrelation could not be excluded, causing the authors to purge the lagged dependent variable (a_{t-1}) of its stochastic error by estimating it using an auxiliary regression [two-stage least squares (2SLS); see, e.g., Greene (2003,

Table 7.4 Ratio of equity capital to assets, 142 American ICs (1979–1990)

Explanatory variable	Sign[a]	Hyp.[b]	Coefficient	t-value
Intercept			1.567	3.843
1. Equity capital-to-asset ratio (lagged 1 year), $\approx a_{t-1}$			0.096	0.584
2. Standard deviation of surplus returns σ_s	+	H1	1.891	2.567
3. ln(assets), size indicator			−0.073	−4.776
4. Closely held by management[c]	−	Z1	−0.099	−2.440
5. Closely held by another group of owners[c]	−	Z2	−0.102	−3.167
6. IC with nationwide activity[c]			0.068	1.920
7. Unaffiliated single company[c]			−0.127	−3.074
8. Intra-group Herfindahl index			−0.042	−1.702
9. Licensed in New York[c]	+	A1	0.019	0.702
10. Distribution through independent agencies[a]			0.005	0.135
11. Growth rate of industrial production			0.075	3.197
12. Bond yield			0.001	1.624
Coefficient of determination R^2			0.449	

[a] Predicted sign
[b] Pertinent hypothesis to be tested
[c] The variable = 1, if the characteristic applies, = 0 otherwise
Source Cummins and Sommer (1996)

7.2 Financial Models of Insurance Pricing

15)]. The same 2SLS procedure was applied to the standard error of surplus return σ_S, which according to H2 is a decision variable too, i.e., an endogenous variable. Finally, the values of the dependent and all explanatory variables of a given year were divided by the estimated residual pertaining to that year in order to neutralize the effect of a non-constant variance of the error term [correction for heteroskedasticity; see White (1980)]. After these several transformations, the data were finally used in a multiple regression which explains some 45% of the variance of the dependent variable ($R^2 = 0.449$).

The estimation results are interpreted below.

1. *Equity capital-to-asset ratio of the preceding year*. This variable serves to model lagged adjustment. The closer to 1 the regression coefficient, the more the value of the preceding year carries over to the value of the current year, and the slower therefore is the adjustment to exogenous changes. Since the coefficient pertaining to (a_{t-1}) is statistically non-significant, there is no evidence of such a lag in adjustment. Rather, IC management seems to react to changes in the economic environment during the same year.
2. *Standard deviation of surplus returns*. This variable is represented by σ_S as defined in Eq. (7.46). For its measurement, one needs data on the changes in the value of capital investment (r_A), the changes of loss payments including loss adjustment expenses (r_L), and the correlation coefficient between these two returns (ρ_{AL}) for every IC in the sample. The significantly positive regression coefficient of this variable *confirms hypothesis* H1. Thus, the evidence suggests that the IC responds to an increase in risk by financing their assets with additional equity capital (or with a higher asset–liability ratio, respectively), as predicted by the option pricing model.
3. *ln (assets)*. As expected, the pertinent coefficient is negative. Larger ICs achieve a better internal risk diversification and can therefore operate with less equity per MU of assets invested (see Sect. 7.1.2.2).
4. *IC closely held by management*. The negative coefficient indicates that, *ceteris paribus*, this type of IC operates with less equity capital, thus exhibiting less risk aversion than the comparison group (i.e., ICs with dispersed ownership), as postulated in the additional hypothesis Z1.
5. *IC closely held by another group of owners*. Hypothesis Z2 predicts a lessened influence of risk aversion than in the case of dispersed ownership. It is confirmed as well by the negative regression coefficient. Since the value of the coefficient is of comparable magnitude as the one of variable No. 4, it is apparently not important who forms the group of owners with a marked engagement in the IC.
6. *IC with nationwide activity*. A negative sign could have been expected for this variable, reflecting the benefit of regional diversification. The sign of the coefficient is positive (albeit not quite significant), tending to speak against this expectation.
7. *Unaffiliated single company*. Since an independent IC bears the costs of insolvency itself, it tends to be rated as less risky than other ICs by the capital market, permitting it to save on equity capital. The negative regression coefficient supports this notion.

8. *Intra-group Herfindahl index*. In this case, a high value of the Herfindahl index indicates a high contribution of the IC considered to the net premium income of the insurance group. The more marked this type of concentration is, the more the IC approaches the status of an independent company. On the basis of the arguments proffered with respect to variable No. 7, the regression coefficient is predicted to be negative (which it is, albeit without statistical significance).
9. *Licensed in New York*. The U.S. state of New York is known for its stringent insurance regulation. Hence, if the equity capital-to-asset ratio is governed by regulation (hypothesis A1), one would expect an especially marked adjustment to an increase in risk. The pertinent regression coefficient should therefore be positive. It is, but far away from statistical significance.
10. *Distribution through independent agencies*. Outsourcing distribution to agencies who can adjust their sales effort to the development of local markets contributes to internal risk diversification for the IC. Therefore, there is a reduced necessity to have a high K/A ratio for maintaining a low insolvency risk. The expectation is a negative sign for the regression coefficient, which however is not confirmed.
11. *Growth rate of industrial production*. This is a business cycle indicator. During an upswing, IC profits rise; since they are at least in part transferred to equity capital, K/A increases. This effect is confirmed by the positive regression coefficient.
12. *Bond yield*. A rise in the returns on bonds increases IC profit from investment activity and indirectly, its equity capital. While positive, the pertinent regression coefficient does not attain statistical significance, thus failing to confirm this effect.

Econometric Analysis of the Relation $d\sigma_S/da$ (H2)
Table 7.5 shows the regression result of the reverse relation $d\sigma_S/da$. The dependent variable is the volatility of surplus returns. The explanatory variables are almost the same as in Table 7.4, except for No. 12. The coefficient of determination (R^2) indicates that almost 90% of the variance of the dependent variable can be explained.

This time, interpretation is limited to a few key findings.

1. *Standard deviation of surplus returns of the preceding year*. The regression coefficient of $\sigma_{S,t-1}$ is small and insignificant. Apparently, decisions concerning risk-taking in underwriting and investment activities are made without a lag. Rather, IC management seems to act during the same year when a change in the economic environment occurs.
2. *Equity capital-to-asset ratio*. This variable has a positive influence on σ_S, confirming hypothesis H2. In combination with the evidence supporting H1 (see variable No. 2 of Table 7.4 in Sect. 7.2.4.2), this result suggests that the option pricing model can describe IC risk management behavior at least approximately.
4. *IC closely held by management*. ICs where the separation between management and owners is not very pronounced incur higher risks, as predicted by additional hypothesis Z1.

7.2 Financial Models of Insurance Pricing

Table 7.5 Standard deviation of surplus returns, 142 American IC (1979–1990)

Explanatory variable	Sign[a]	Hyp.[b]	Coefficient	t-value
Intercept			−0.156	−2.247
1. Standard deviation of surplus returns (lagged), = $\sigma_{S,t-1}$	+	H2	0.296	1.533
2. Equity capital-to-asset ratio	+	H2	0.135	2.161
3. ln(assets), size indicator			0.010	3.073
4. Closely held by management[c]	−	Z1	0.024	3.017
5. Closely held by another group of owners[c]	−	Z2	0.013	2.756
6. IC with nationwide activity[c]			−0.011	−2.755
7. Unaffiliated single company[c]			0.031	3.265
8. Intra-group Herfindahl index			0.015	2.616
9. Licensed in New York[c]	−	A1	−0.004	−1.124
10. Distribution through independent agencies[c]			−0.004	−1.498
11. Change in volatility of bond returns			0.017	2.208
12. Change in volatility of stock returns			0.010	2.172
Coefficient of determination R^2			0.896	

[a] Predicted sign
[b] Pertinent hypothesis to be tested
[c] The variable = 1 if characteristic applies, = 0 otherwise
Source Cummins and Sommer (1996)

5. *IC closely held by another group of owners.* According to hypothesis Z2, this type of IC should be characterized by marked risk aversion hence a lower value of σ_S. Again, Z2 is not confirmed.
9. *Licensed in New York.* While the negative sign of the regression coefficient seems to support hypothesis A1, its t-value is far from statistical significance. This suggests that stringent solvency regulation—at least in the case of the United States—does not really affect the risk behavior of ICs.

▶ **Conclusion 7.6** To a considerable degree, the risk behavior of American ICs is in accordance with the predictions of the Option Pricing Model (OPM). Management increases promptly equity capital in response to a higher surplus volatility, and it opts for a higher volatility of surplus returns when the equity capital-asset ratio is high.

Conclusion 7.6 gives rise to the presumption that the Option Pricing Model of the IC with its emphasis on the conflict of interest between owners and IBs can explain the risk management of ICs quite well.

7.3 Economies of Scope

Insurance supply is also described by the spectrum of products offered. An important consideration is the existence of so-called economies of scope.

7.3.1 Economies of Scope and Properties of the Cost Function

For many issues, modeling the enterprise as a one-product firm is sufficient. In the case of an IC, one can study the optimal size of the firm by singling out a line of business with a homogeneous product (see Sect. 7.4). The management of an entire IC, however, needs to also decide on the product spectrum. Some of these decisions are of the either-or type: Should an additional business line be taken over from another IC? Should capital investment continue to be delegated to a bank? Other decisions are of a gradual nature: Should additional risks be underwritten even though they differ from the existing portfolio? Should distribution be changed in favor of independent agents?

Economies of scope can be traced to several sources, which invariably exhibit characteristics of a local public good, i.e., the asset in question can be made available to additional users within the IC at almost no cost.

- *Management ability.* The management of an IC that takes over a line of business from a competitor may have a particular ability in solving problems pertinent to that line.
- *Information advantages.* In underwriting new types of risk, the IC may benefit from information gained through its current underwriting activity.
- *Know-how.* In changing the distribution in favor of independent agents, current employees may be able to transmit their specific knowledge about market conditions to new colleagues.

One criterion for determining whether economies of scope exist are the effects on cost (another view is presented in Sect. 7.3.3 below). A valid (although not sufficient) reason to enlarge the product spectrum is an increase in cost that is less than proportional, reflecting so-called subadditivity of costs [see Baumol et al. (1982); Panzar and Willig (1977)]. Given subadditivity, it is true that for all values y and z of two outputs Y and Z, the cost of joint production satisfies

$$C[y \cup z] \leq C(y) + C(z). \qquad (7.47)$$

This condition can be related to the shape of the cost function in the following way. Select an output level y^0 for the product Y, simultaneously setting the output level of Z at zero (see point P of Fig. 7.4). In point P, (7.47) holds as an equality because with $Z = 0$ there cannot exist economies of scope. For the sake of simplicity, let the sum of costs $C(y) + C(z)$ increase linearly up to point Q. By inequality (7.47), it is known that in the case of subadditivity, the combination of outputs $\{y, z\} > 0$ causes lower costs (represented by $C[y \cup z]$), than indicated by Q. Let this lower cost level be represented by point R. Yet, for the cost function to go through R its slope must decrease. By assumption, output Z increases from 0 to z in the transition from P to R. Obviously, the decreasing slope of the cost function reflects economies of scope

7.3 Economies of Scope

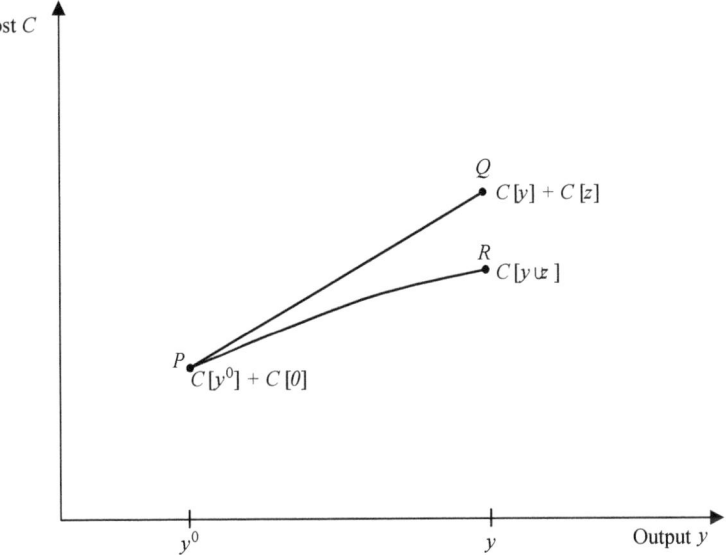

Fig. 7.4 Economies of scope and the cost function

which originate from the production of Z. This amounts to a decline in marginal cost,[3]

$$\frac{\partial}{\partial z}\left[\frac{\partial C}{\partial y}\right] = \frac{\partial^2 C}{\partial z \partial y} = \frac{\partial^2 C}{\partial y \partial z} < 0. \quad (7.48)$$

▶ **Conclusion 7.7** Economies of scope with regard to cost obtained if the cost function exhibits a decrease of marginal cost with respect to output Y when the output level of Z increases (or *vice versa*).

The "vice versa" in Conclusion 7.7 results from the fact that in Eq. (7.48) the order of differentiation does not matter.

7.3.2 Empirical Relevance of Economies of Scope

An empirical study that tests for economies of scope in the insurance industry using the cost function is by Suret (1991). The database comprises 50–70 small, medium, and large (above Can$ 80 mn. assets as of 1986) Canadian ICs covering the period 1986–1988. The author distinguishes four products (i.e., lines of business):

- Auto insurance (y_1);
- Property insurance (y_2);

---

[3] Note that the decline in marginal cost may also be due to an increase in y, i.e., $\partial^2 C / \partial y^2 < 0$.

- Liability insurance (y_3);
- Remaining lines (y_4, accounting for about 20% of premium income).

A test of condition (7.48) requires a cost function that allows for variable marginal costs that depend on the output levels of all four outputs considered. A popular variant satisfying this requirement is the so-called translog cost function, which is a second-order Taylor approximation to an arbitrary function in the logarithms of its arguments. In its basic form, it contains as explanatory variables only the logarithms of outputs and interaction terms, factor prices and their interaction terms, as well as interaction terms between outputs and factor prices. For n outputs y_i and m factor prices p_k, it reads

$$lnC = \alpha_0 + \sum_{i=1}^{n}\alpha_i lny_i + \sum_{k=1}^{m}\beta_k lnp_k + 1/2\sum_{i=1}^{n}\sum_{j=1}^{n}\gamma_{ij}lny_i lny_j$$

$$+1/2\sum_{k=1}^{m}\sum_{l=1}^{m}\delta_{kl}lnp_k lnp_l + 1/2\sum_{i=1}^{n}\sum_{k=1}^{m}\tau_{ik}lny_i lnp_k + \varepsilon \quad (7.49)$$

with $\{\alpha_0, \alpha_i, \beta_k, \gamma_{ij}, \delta_{kl}, \tau_{ik}\}$ denoting parameters to be estimated and ε an error term with zero expectation and constant variance.

The outputs $\{y_1, \ldots, y_4\}$ as defined above are measured as losses paid (see Sect. 6.2), and $p_1 :=$ average wage and $p_2 :=$ average rental cost of office space.

According to Murray and White (1983), economies of scope between outputs i and j exist if

$$\alpha_i \cdot \alpha_j + \gamma_{ij} < 0. \quad (7.50)$$

Note that a regular cost function has $\alpha_i > 0, \alpha_j > 0$; therefore, the interaction coefficients γ_{ij} must be strongly negative for the marginal cost of y_i to decrease with y_j (and vice versa), indicating economies of scope.

The estimation of this translog cost function yields regression coefficients $\{\hat{\alpha}_i, \hat{\alpha}_j, \hat{\gamma}_{ij}\}$ that can be compared with condition (7.50), with the following results.

- *Auto insurance* (y_1): In 2 out of 3 years and in all size categories, losses paid in property insurance (y_2) drives up rather than reduce marginal cost, i.e., condition (7.50) is violated. This also holds true of liability insurance (y_3) and the other lines (y_4), but in the category of small ICs.
- *Property insurance* (y_2): Here, only the influence of liability insurance (y_3) and other lines (y_4) on marginal cost must be tested because of the symmetry property (see Conclusion 7.7). There are no recurrent (in at least 2 out of 3 years) indications of economies of scope.
- *Liability insurance* (y_3): In 2 out of 3 years, a statistically significant cost-decreasing effect of the other lines (y_4) can be recognized, but only among medium-sized ICs.

7.3 Economies of Scope

In sum, there is no evidence of economies of scope across all size categories, at least if each line of insurance is examined individually. An IC with the intention of combining two lines of business could therefore hardly count on cost savings. However, the typical decision problem of an IC already in existence is different because it may want to add a line of business to its portfolio. For instance, could an IC operating in lines $\{y_2, y_3, y_4\}$ benefit from economies of scope by adding (or increasing the volume of, respectively) auto insurance (y_1)? This type of question is answered by the study of Suret (1991) as follows.

- *Added involvement in auto insurance* (y_1): ICs of all size categories repeatedly exhibit signs of scope economies with regard to $\{y_2, y_3, y_4\}$. They are statistically significant in 2 out of 3 years among the large ICs.
- *Added involvement in property insurance* (y_2): Complementing a portfolio $\{y_1, y_3, y_4\}$ with y_2 might also have a cost-reducing effect. However, it is again statistically significant only in 2 out of 3 years, and only among large ICs.
- *Added involvement in liability insurance* (y_3): Here, there are indications of diseconomies of scope with regard to the portfolio $\{y_1, y_2, y_4\}$, although the violations of condition (7.50) are never statistically significant.
- *Added involvement in other lines* (y_4): Neither economies nor diseconomies of scope can be recognized for the portfolio $\{y_1, y_2, y_3\}$.

On the whole, the study provides very limited support for the notion that non-life insurance might benefit from economies of scope. If at all, economies of scope may prevail among the large ICs. However, causation does not necessarily run from firm size to economies of scope. It may well be reversed because some ICs are better capable than others of benefiting from these economies, causing them to end up in the top size category.

More recently, Cummins et al. (2010) have tested for scope economies at a more aggregate level in the U.S. insurance industry. In the first step, the authors calculate efficiency scores for two types of ICs, one with activity mainly in the life and health lines, the other, with activity in the property-liability lines. Efficiency scores indicate the closeness of an IC to an efficiency frontier established by Data Envelopment Analysis (see Sect. 7.4.4.2 below). In a second step, these efficiency scores are related to a categorical (dummy) variable indicating whether the IC specializes in its main lines of business. In the life-health subgroup, this categorical variable consistently has a positive coefficient, indicating that specialization contributes to cost efficiency. By implication, there are diseconomies of scope in the traditional sense. However, the authors also measure the distance from an efficient frontier defined in terms of profit rather than cost. With inefficiency rather than efficiency scores constituting the dependent variable, the specialization dummy has a negative sign, pointing to advantages of specialization and hence again diseconomies of scope.

▶ **Conclusion 7.8** Empirical evidence from Canada and the United States suggests only limited economies of scope in non-life insurance and diseconomies of scope in life and health insurance.

7.3.3 Stochastic Economies of Scope

The management of an IC would be ill-advised to base decisions with respect to the product spectrum only on economies of scope with regard to cost. At most, this may be adequate for organizational measures, such as adding direct writers to the distribution network rather than relying on brokers. But even then, the risk exposure of the IC may be affected. It suffices for direct writers to attract risk types that differ from those attracted by brokers. Depending on the additional premium income on the one hand and the correlation of the additional losses to be paid with those of the existing portfolio on the other hand, the risk-return properties of the surplus will change.

The choice of product spectrum can therefore be analyzed in terms of a choice of portfolio structure (see Sect. 6.8). The decision variables are weights, i.e., the shares of the different lines of business in the premium volume. They are to be set in a way that for a given expected return on surplus, the volatility of this return is minimized. The return on surplus amounts to its percentage change over the previous period as defined in Eq. (6.34), with variance given by Eq. (6.37) in Sect. 6.8. If now, e.g., auto insurance (y_1) is to be added to the product spectrum of the IC, surplus is affected by not only additional premium income and investment returns, but also losses and expenses. The net contribution to surplus has an impact not only on the expected value of return on surplus but also on its volatility.

In Fig. 7.5, two efficiency frontiers are shown in (μ, σ)-space, with $\mu :=$ expected value of return on surplus and $\sigma :=$ its standard deviation. Let the initial efficiency frontier EE' be formed by the lines of business $\{y_2, y_3, y_4\}$. Three possible but fictitious allocations are marked with their weights $\{w_2, w_3, w_4\}$, reflecting the assumption that business line (y_4) is particularly lucrative but risky since its weight increases along EE'. Let IC management initially opt for $\{w_2^*, w_3^*, w_4^*\}$ represented by point Q^* on EE' as the optimal product spectrum.

Adding auto insurance (y_1) to the product spectrum causes the efficiency frontier to shift up (note the analogy with Fig. 5.3 of Sect. 5.1.2). If y_1 is lucrative but risky, it is not part of the minimum variance portfolio at point E. The new efficiency frontier therefore could be EE'', with optimum point R^*. The corresponding allocation $\{w_1^*, w_2^*, w_3^*, w_4^*\}$ will generally reflect changes in the efficient structure of underwriting activity. In the example, the optimal share of the newly added auto line (w_1) could be 0.10 (i.e., 10%), while the share of line y_2 decreases from 40 to 25%.

▶ **Conclusion 7.9** The choice of the product spectrum of an IC can be analyzed using an efficiency frontier defined over the expected value and the standard deviation of return on surplus.

7.4 Economies of Scale

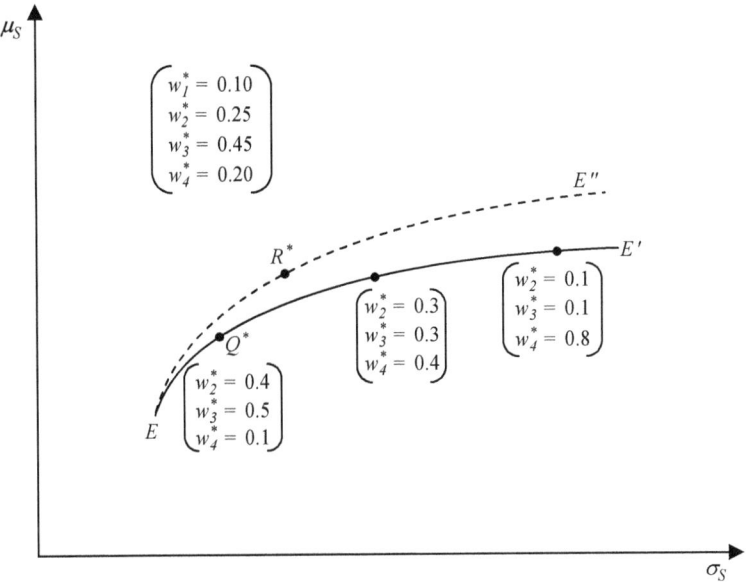

Fig. 7.5 Stochastic economies of scope due to adding a line of business

7.4 Economies of Scale

Economies of scale revolve around the question of whether expanding a particular line of business by scaling up all inputs in the same proportion goes along with falling unit cost. Alternatively, all lines of business could be scaled up by a common factor, leading to the question of whether the size of the IC as a whole confers a cost advantage. Note that economies of scope could become relevant in the course of expansion (see the connection between economies of scope and size among Canadian ICs found in Sect. 7.3.2). This makes the distinction between economies of scale and scope difficult in actual practice. To simplify the exposition below, the portfolio comprising the lines of business is assumed to always be the optimal one.

7.4.1 Definitional Issues

Positive economies of scale exist if, e.g., a doubling of all inputs leads to more than a doubling of outputs of the firm. Given cost minimization and predetermined input prices, production cost also doubles, while output increases more than proportionally. This implies that positive economies of scale cause (minimized) cost to decrease with a doubling (more generally: expansion) of outputs.

In insurance, economies of scale are thought to be grounded in theory at least as far as underwriting is concerned because they are implied by the law of large numbers. This law states that the arithmetic mean $\bar{x}$ of n stochastic variables with the same expected value μ and the same variance σ^2 approaches μ when n increases toward infinity (see Sect. 7.1.2.1). The arithmetic mean becomes an ever more reliable estimator of the expected value since its standard deviation decreases with n.

Therefore, an IC that succeeds in building a portfolio containing more and more risks with the same expected loss can estimate expected loss per unit with increasing precision based on recent experience. It needs less reserves per MU of premium for unforeseen deviations from the expected value that might jeopardize its solvency. The holding of reserves has an opportunity cost insofar as the underwriting of additional risks and investing the premium income usually results in a higher return than the low interest that can be earned on the risk-free investments typically prescribed for reserves [see, e.g., Zweifel (2015)]. Hence, the larger the portfolio of insurance contracts, the smaller is the cost of maintaining the desired probability of solvency.

Economies of scale are also relevant for IBs and regulatory authorities. Under competitive conditions, market entry by new competitors pushes down the sales price, eventually to the point where it just covers the minimum average cost. All firms operating in the market are forced by competition to produce at a minimum efficient scale, at least in the long run. From microeconomics, marginal cost is known to equal average cost at a minimum efficient scale, thus $dC/dy = C/y$. Division by (C/y) results in

$$\frac{dC}{dy} \cdot \frac{C}{y} := e(C, y) = 1. \tag{7.51}$$

Therefore, the elasticity of total cost with respect to output is equal to one at minimum efficient scale, meaning that an increase of output by 1% goes along with an increase of cost of just 1%. Accordingly, a value $e(C, y) < 1$ indicates economies of scale, while $e(C, y) > 1$ points to diseconomies of scale. With respect to insurance markets, it is of interest to know whether competition is in fact so vigorous as to make the IC adopt minimum efficient size, and if they fail to do so, whether they are characterized by economies or diseconomies of scale.

However, before applying condition (7.51) to ICs, two definitional issues must be clarified in advance.

- *How is "cost" to be measured?*

At first sight, cost can be simply measured using data taken from the profit and loss statement of an IC (in particular, acquisition and administration expense). Note that from an economic point of view, losses paid are not a component of cost because they reflect a redistribution from consumers without a loss to those who suffered a loss. On the other hand, the cost of providing insurance coverage goes beyond acquisition and administrative expense. First, one needs to add capital cost, i.e., interest on the capital tied up in the firm. Application of the CAPM (see Sects. 5.1.3 and 7.1.3) determines this quantity as the competitive risk-adjusted return on shareholders' investment. The second addition are transfers to reserves, which are designed to ensure the solvency (and hence continuing activity) of the IC. However, transfers to (and withdrawals from) insurance reserves importantly depend on expectations with respect to future claims and are therefore subject to considerable latitude on the part of IC management. In view of these difficulties, cost is simply measured as acquisition and administrative expense in the empirical studies presented in Sects. 7.4.2 and 7.4.3 below.

- *How is "output" to be defined?*

The debate about the definition of output presented in Sect. 6.3.1 need not be repeated here. The contingent commitment to pay benefits emerged as the most suitable definition. Aggregated over the risk portfolio of the IC, these commitments become total losses paid. However, premium income on its own is also used as an indicator of output. The two alternatives are juxtaposed below.

1. *Premium income as an indicator of output.* Premiums paid by IBs could reflect their willingness to pay for the several characteristics of the contract. Contractual characteristics (in particular, clauses limiting the obligation to pay on the part of the IC) were found to be reflected in premiums by Walden (1985) for the United States. In addition, evidence derived from so-called discrete choice experiments suggests that properties of health insurance policies influence willingness to pay in Germany, the Netherlands, and Switzerland [see Zweifel et al. (2006) and Vroomen and Zweifel (2011)]. Hence, in insurance markets with a sufficient amount of competition, using premium income as an indicator of output is not open to criticism.

   However, as soon as figures from different markets and time periods need to be made comparable, it is preferable to split premium income into a price and a quantity component with the help of the following identity (as in Sect. 1.5.2),

$$Premium income \equiv \frac{Premium}{Sum insured} \cdot Sum insured$$
$$\equiv Premium rate \cdot Sum insured. \qquad (7.52)$$

   For instance, let the premium for fire insurance be 1,000 MU per year; it consists of a premium rate amounting to 0.1% (price component) and a sum insured of 1 mn. MU (quantity component). However, this decomposition still fails to take into account differences in products, in particular, the exclusion (and inclusion, respectively) of certain risks or changes over time such as an acceleration (or slowing) in the settlement of claims.

2. *Losses paid as an output indicator.* The great advantage of this indicator is that it avoids any problem of aggregation. A MU paid for a claim is a MU, regardless of the competitive conditions and the business policy of the IC. By way of contrast, the premium is not independent of the costs of the IC in regulated markets, insofar as surcharges on the fair premium are approved by the supervisory authority. There is no dependence of output measured on cost if losses paid are used as the indicator of output [see Doherty (1981)].

7.4.2 Empirical Relevance of Economies of Scale in Life Insurance

Traditionally, economies of scale in life insurance have been at the center of interest because here the contractual partners of the IC are mostly individual consumers. For them, a failure of competition to force average cost down to its minimum has particularly grave consequences because of the long contract life in this type of business. The pioneering study is by Houston and Simon (1970). Their data relate to 327 life ICs with activity in the state of California as of 1962. As an indicator of output, the authors use premium income Π of that year. Rather than estimating the elasticity of total cost w.r.t. output, they test directly whether average cost (AC) decreases with output. Since a linear relation in Π would end up predicting negative AC values, its reciprocal value $(1/\Pi)$ serves as an explanatory variable to model

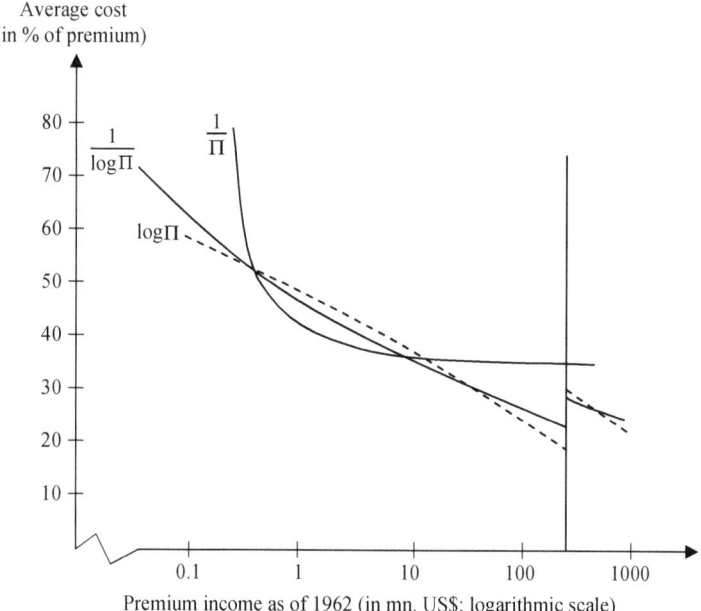

Fig. 7.6 Economies of scale in Californian Life Insurance (1962). *Note* The regressions including $1/\log \Pi$ and $\log \Pi$ contain a categorical variable which is $= 1$ if $\Pi > 200$ (million US\$), and $= 0$ otherwise. This explains the discontinuity at point $\Pi = 200$. (*Source* Houston and Simon 1970)

▶ **Conclusion 7.10** In testing for economies of scale, losses paid by the IC are best suited as an output indicator. Premium income (especially when split up into premium rate and sum insured) qualifies as well, in particular in markets with vigorous competition.

7.4 Economies of Scale

asymptotic convergence to zero (see Fig. 7.6). The OLS regression reads

$$AC = 0.248 + \underset{(0.063 \cdot 10^{-5})}{0.454 \cdot 10^{-5}} \frac{1}{\Pi} - \underset{(0.043)}{0,184} \frac{G\Pi}{\Pi} + \underset{(0.056)}{0.437} \frac{N\Pi}{\Pi} + \underset{(0.121)}{0.688} LR \quad (7.53)$$

$R^2 = 0.53$; $N = 327$; standard errors in parentheses.

AC	:	Average cost, operational expense including commissions for agents in distribution relative to premium income Π (often called expense ratio);
Π	:	Premium income (in mn. US$);
$G\Pi/\Pi$	:	Share of premiums originating from group insurance in premium income;
$N\Pi/\Pi$	:	Newly written premiums relative to premium income;
LR	:	Lapse rate, i.e., number of cancelations relative to portfolio of contracts.

This regression result can be interpreted as follows.

- *Constant:* With a value of 0.248, the constant indicates an absolute lower limit of the expense ratio. This limit would be reached if $\Pi \to \infty$, *ceteris paribus*, i.e., if there were no group and newly written contracts, and if the lapse rate were zero. However, the authors put the estimated minimum value of AC for $\Pi \to \infty$ at 0.37 (37% of premium; see Fig. 7.6), because they calculate it at the sample means of the other explanatory variables.
- *Reciprocal value of premium income* $1/\Pi$: The positive coefficient indicates economies of scale. The larger the premium income, the smaller is $1/\Pi$, and the lower is average cost AC. Beyond the value of 100 million US$, the AC curve runs practically horizontal (see Fig. 7.6). The largest IC of the sample with a premium income of 1,700 mn. US$ is still consistent with this curve. Specifications using $\log \Pi$ or $1/\log \Pi$ are also shown in Fig. 7.6. They contain an additional regressor in the guise of a categorical variable that is equal to one if premium income is in excess of 200 mn. US$ and zero otherwise. It serves to shift the average cost function back upward for a better fit with the observations relating to the very large IC. Therefore, these alternative specifications are likely to overstate economies of scale.
- *Share of group insurance contracts* $G\Pi/\Pi$: The larger the share of group contracts in premium income, the lower is the expense ratio *ceteris paribus*. This is intuitive since acquisition expenses per MU of premium are lower when, e.g., the entire staff of a firm can be enrolled compared to enrolling each employee individually. The estimated coefficient indicates that an IC with, e.g., a 50% rather than 40% share of group contracts has a cost advantage of 1.84 US cents $[0.1 \cdot (-0.184) = -0.0184]$ per US$ of premium (*ceteris paribus*).
- *Share of newly written premiums* $N\Pi/\Pi$: The greater the importance of new business, the higher the expense ratio of the IC. Indeed, not only acquisition expense but also much of administrative expense accrue at the beginning of the contract. A difference of 10 percentage points (e.g., 50% rather than 40% of premium income) is associated with an estimated increase of some 4.4 US cents per US$ of premium.

- *Lapse rate LR:* The more frequent are cancelations relative to the risk portfolio, the higher the expense ratio. Cancelations frequently occur after loss events that could not be settled without conflict, generating administrative expense. The effect of a 10 percentage point difference in the lapse rate is quite marked, amounting to an estimated 6.9 US cents per US$ of premium.

These and similar later results [see Pritchett (1973), Praetz (1980)] are subject to criticism [see Kellner and Matthewson (1983)], who point to observations that are not easily reconciled with economies of scale.

(i) *Long-run survival of small ICs in the life business.* In the United States at least, ICs that are one hundred times smaller than the market leader did not exit from the market over a period of over 10 years; they even remained independent. If economies of scale prevailed, exits and takeovers of small ICs should be observed during a period of this length.
(ii) *Decrease of concentration.* During the period 1961–1976, concentration (measured using premium income according to major lines of insurance) did not increase but decreased in Canada. Economies of scale should lead to an increase in concentration, however.
(iii) *Positive relationship between size and premium rates.* At least under competitive conditions, ICs benefiting from economies to scale pass on their cost advantage to consumers in the guise of lower premium rates. Kellner and Matthewson (1983) found to the contrary that large ICs charge higher premium rates than small ones.
(iv) *No outsourcing of activities.* Economies of scale occur in particular in activities that are typically assigned to headquarter offices (e.g., capital investment). They also foster the creation of enterprises specializing in these activities, to whom this type of task is outsourced. This does not seem to occur in the insurance industry, however.

One reason for the discrepancy between the estimation results in (7.53) and observed developments at the market level could be the use of a particular functional form. The translog function introduced in Sect. 7.3.2 provides a more flexible alternative. Fecher et al. (1991) use a translog cost function without imposing any restrictions (which could be derived from the assumption of cost minimization). This choice is advisable because 13 ICs in their French sample (out of 84 on average) are mutual companies, and four ICs are state-owned, making it risky to interpret observed total cost (which do not contain capital cost to begin with) as the outcome of cost minimization. In close analogy with Eq. (7.49), the cost function reads,

$$logC = \alpha_0 + \sum \beta_j s_j + \alpha_1 logy + \alpha_2 logz + \alpha_3 logr$$
$$+\alpha_{11}(logy)^2 + \alpha_{22}(logz)^2 + \alpha_{33}(logr)^2$$
$$+\alpha_{12}(logy)(logz) + \alpha_{13}(logy)(logr) + \alpha_{23}(logr)(logz) + \varepsilon \quad (7.54)$$

7.4 Economies of Scale

Table 7.6 Total cost and cost elasticities of French non-life ICs (1984–1989)

Explanatory variable		Output indicator			
		Gross premiums (1)		Loss payments (2)	
		Coefficient	t-value	Coefficient	t-value
Constant	α_0	2.681	1.6	8.446	12.9
Foreign IC[a]	β_1	0.431	4.9	0.128	1.6
Mutual[a]	β_2	−0.101	−1.1	−0.012	−0.2
Public[a]	β_3	0.196	0.9	−0.295	−1.7
Output (log y)	α_1	0.486	1.7	−0.046	−0.4
Distribution cost share (log z)	α_2	−0.233	−0.9	1.017	6.8
Reinsurance ratio (log r)	α_3	−0.214	−1.0	−0.084	−0.5
$(\log y)^2$	α_{11}	0.018	1.6	0.036	6.2
$(\log z)^2$	α_{22}	−0.120	−4.4	0.126	5.4
$(\log r)^2$	α_{33}	−0.066	−2.8	−0.139	−6.0
$(\log y)(\log z)$	α_{12}	−0.002	−0.1	0.007	−0.5
$(\log y)(\log r)$	α_{13}	0.029	1.7	0.007	0.7
$(\log z)(\log r)$	α_{23}	0.091	2.4	0.146	4.8
$e(C, y)$	foreign	0.811		0.692	
	mutual	0.857		0.740	
	public	0.957		1.031	
	stock	0.843		0.724	
	total	0.845		0.736	
R^2		0.863		0.887	
N		430		428	

OLS estimation
[a]This explanatory variable is = 1 if the IC is of this type and = 0 otherwise. Reference category is the domestic stock IC
Source Fecher et al. (1991)

s_j: a set of categorical variables with the domestic stock ICs serving as the reference group ($s_1 = 1$: foreign; $s_2 = 1$: mutual; $s_3 = 1$: public);
$log y$: Output, measured as premium income (in logs);
$log z$: Distribution expense as a share of total cost (in logs);
$log r$: Share of gross premium income ceded to reinsurance (in logs);
ε: Error term with zero expectation and constant variance.

Setting ε to its expected value of zero, the elasticity of cost with respect to output can be calculated as

$$e(C, y) = \frac{\partial \log C}{\partial \log y} = \alpha_1 + 2\alpha_{11} \log y + \alpha_{12} \log z + \alpha_{13} \log r. \quad (7.55)$$

Obviously, this elasticity varies with $log y$, $log z$, and $log r$. To obtain a representative value, it is evaluated at the respective sample means. Although many of the estimated coefficients $\{\alpha_1, \alpha_{11}, \alpha_{12}, \alpha_{13}\}$ are not statistically significant (see

Table 7.6), the authors estimate the cost elasticity with respect to gross premiums at $e(C, y) = 0.845$ for the total sample. This value is below 1, violating the condition for minimum efficient scale [see Eq. (7.51) of Sect. 7.4.1 again]. With an elasticity of 0.812, foreign ICs are particularly far away from minimum efficient size, while public ICs with 0.957 are so close to the minimum efficient size that they should not grow more at least on the grounds of cost.

In a second regression, loss payments are used as the output indicator. Here, among the coefficients entering (7.55), only α_{11} pertaining to $(\log y)^2$ is significant; therefore, the cost elasticities are very imprecise estimates. For the total sample, the cost elasticity is $e(C, y) = 0.736$, with a low value again in the group of foreign ICs (0.69) and a high one among the group of public insurers (1.03).

Despite reservations because of a lack of precision in estimates, one may draw Conclusion 7.11 below.

▶ **Conclusion 7.11** In the French life insurance market, there are indications suggesting that the private ICs have not reached their minimum efficient scale, while public insurers are probably at or beyond minimum efficient scale.

7.4.3 Empirical Relevance of Economies of Scale in Non-life Insurance

For ease of comparison between life and non-life insurance, the evidence presented again comes from Fecher et al. (1991). The authors estimate the same translog cost function (7.54) but using data from non-life ICs in France. The results are presented in Table 7.7.

The estimated coefficients of the first regression (with gross premiums as the output indicator) illustrate the importance of the flexible functional form. On the basis of $\alpha_1 = 1.028$ alone, reflecting just the linear relation between cost and output, one would conclude in favor of diseconomies of scale. However, the regressors $(\log y)^2$ and $(\log y)(\log r)$ have negative coefficients, resulting in a cost elasticity of $e(C, y) = 0.957$ evaluated at the respective sample means. This suggests economies of scale, with operations close to minimum efficient scale. The mutuals are likely too small with $e(C, y) = 0.940$ and the foreign firms just about at their efficient scale (0.979). If loss payments are used as an output indicator (see the right-hand side of Table 7.7), the estimate of the cost elasticity for the entire sample falls to $e(C, y) = 0.847$, suggesting substantial returns to scale. As in the life business, public insurers are likely beyond the minimum efficient scale with a value of 1.046. By way of contrast, both mutuals and foreign ICs could still grow before reaching the minimum average cost.

In sum, most of the estimated cost elasticities are clearly below the benchmark value of $e(C, y) = 1$. They are calculated on the basis of mostly highly significant parameter estimates $\{\alpha_1, \alpha_{11}, \alpha_{12}, \alpha_{13}\}$, justifying Conclusion 7.12 below.

7.4 Economies of Scale

Table 7.7 Total cost and cost elasticities of French non-life ICs (1984–1989)

Explanatory variable		Output indicator			
		Gross premiums (1)		Loss payments (2)	
		Coefficient	t-value	Coefficient	t-value
Constant	α_0	−0.568	−1.9	4.881	15.1
Foreign[a]	β_1	−0.119	−5.3	−0.367	−9.8
Mutual[a]	β_2	−0.182	−9.5	0.279	8.6
Public[a]	β_3	−0.025	−0.3	−0.283	−2.4
Output ($\log y$)	α_1	1.028	19.5	0.472	7.3
Distribution cost share ($\log z$)	α_2	−0.293	−4.6	−0.627	−7.0
Reinsurance ratio ($\log r$)	α_3	0.365	5.8	1.674	16.7
$(\log y)^2$	α_{11}	−0.003	−1.1	0.016	4.9
$(\log z)^2$	α_{22}	−0.055	−7.6	−0.089	−7.4
$(\log r)^2$	α_{33}	0.035	5.2	0.157	13.4
$(\log y)(\log z)$	α_{12}	0.031	5.9	0.065	8.1
$(\log y)(\log r)$	α_{13}	−0.011	−2.1	−0.054	−5.8
$(\log z)(\log r)$	α_{23}	0.108	9.8	0.189	9.3
$e(C, y)$	foreign	0.979		0.845	
	mutual	0.940		0.819	
	public	0.945		1.046	
	stock	0.956		0.872	
	total	0.957		0.847	
R^2		0.979		0.942	
N		1,284		1,284	

OLS estimation
[a] This explanatory variable is $= 1$ if the IC is of the this type and $= 0$ otherwise. Reference category is the stock IC
Source Fecher et al. (1991)

▶ **Conclusion 7.12** In French non-life insurance market, there are indicators that private ICs have not reached their minimum efficient scale, while public insurers are probably at or beyond minimum efficient scale.

One may note the similarity with Conclusion 7.11 in Sect. 7.4.2 relating to French life insurance.

7.4.4 Alternatives and Extensions

The studies presented up to this point are open to two types of criticism.

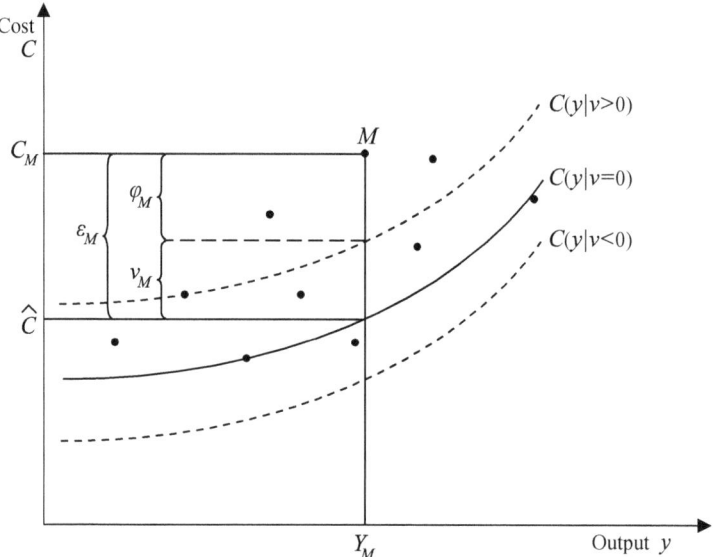

Fig. 7.7 Approximation of the efficiency frontier with the SFA method

(i) *Nature of the cost function:* Strictly speaking, a cost function traces a sequence of minimum cost allocations (tangency points between isoquants and budget lines). This implies that "excessive" cost given a certain output can result from two reasons. The IC may be truly inefficient, or it may have suffered an exogenous shock. Therefore, the error term ε of the regression should be split into a component with non-negative values only (reflecting inefficiency) and an unrestricted component (reflecting stochastic shifts in the cost function). This specification leads to so-called stochastic frontier analysis (SFA).

(ii) *Choice of functional form:* Since the production technology is not really known, the choice of functional form in a cost analysis is arbitrary. Rather, the production technology (and hence, best practice with regard to cost minimization) should be inferred from observed cost–output (or more generally, input-output) combinations. This approach is non-parametric, epitomized by data envelopment analysis (DEA).

7.4.4.1 Stochastic Frontier Analysis (SFA)

The point of departure of SFA is a cost function, frequently of the translog type as in (7.54) of Sect. 7.4.2. However, cost is now interpreted as the outcome of cost minimization. The cost function therefore constitutes an efficiency frontier. This frontier is considered stochastic, with shocks shifting it up ($\nu > 0$) and down ($\nu < 0$). For instance, an unexpected surge in office rental prices may cause the administrative expense of an IC to be higher than normal. In Fig. 7.7, this case is depicted by the dashed cost frontier $C(y|\nu > 0)$, whereas the "normal" frontier in the absence of such a shock is represented by $C(y|\nu = 0)$.

Now consider company M. At its output Y_M, the efficient cost level is indicated by $\hat{C}$. However, observed cost C_M exceeds that level by the amount of ε_M. Part of the excess (ν_M) is due to the stochastic shift of the cost frontier ($\nu > 0$). The remainder φ_M reflects the inefficiency of company M with $\varphi = 0$ indicating a fully efficient enterprise. The error term ε therefore needs to be split into a component ν that follows a symmetric distribution (usually the Normal) and a non-negative component φ that follows an asymmetric distribution (such as the half-normal, negative exponential, and gamma) [see, e.g., Kumbhakar and Lovell (2000)]. In contradistinction to a deterministic approach like Data Envelopment Analysis (described in Sect. 7.4.4.2), the distance from the efficient frontier is not wholly attributed to inefficiency.

Although the refinement of SFA is not needed to relate (minimum) cost to firm size, the method can be used to test for economies of scale. Several studies using SFA in the insurance industry have found evidence for scale economies [see Hardwick (1997) for the UK; Cummins and Weiss (1993) for the United States; Hirao and Inouse (2004) for Japan]. According to Cummins and Weiss (1993), scale economies are prevalent among ICs of small and medium sizes, suggesting potential for cost reductions from consolidation.

Finally, SFA can also be used to test for economies of scope. In a study of the Finnish insurance industry, Toivanen (1993) finds evidence suggesting that multi-line ICs are more efficient than single-line ones, contradicting the DEA-based findings of Cummins et al. (2010) cited in Sect. 7.3.2.

7.4.4.2 Data Envelopment Analysis (DEA)

In the non-parametric approach of DEA, output and input quantities are juxtaposed in an attempt to determine the production possibility frontier (the transformation curve, respectively) from the data. The *maximization of a distance function* between aggregated inputs and outputs (with relative values as weights) serves as the criterion. In the simplest case with only one output Y and one input X, one seeks to maximize Y for a given value of X, i.e., to make the distance between the two as large as possible.

In Fig. 7.8, let the (unknown) production frontier for a given period be approximated by the line $ABCDF$. Data envelopment analysis (DEA) determines whether or not an enterprise lies on the frontier [for a survey of the DEA method, see Seiford and Thrall (1990)]. To achieve this, convexity from above is a crucial assumption. Given convexity, one can select, e.g., firm C for a test. Since point C lies beyond the straight line connecting its neighbors B and D that would define the frontier in the absence of C, enterprise C belongs to the efficiency frontier. An analogous test can be performed for D, which also lies beyond the straight line connecting C and F (not shown). Firm M, to the contrary, is inefficient because it uses OS rather than OR units of input for the production of a given amount of output $\bar{y}$. The ratio $OS/OR > 1$ shows the degree of inefficiency. Conversely, an efficiency score of 1 indicates that the enterprise is on the efficient frontier. For more detail and an application to the U.S. insurance industry, see Cummins et al. (2010).

Using the DEA method, entire insurance markets can be ranked in terms of efficiency. In a study by Donni and Fecher (1997), the United States and Iceland determine the efficient frontier with an efficiency score of 1 in a sample of 15 OECD

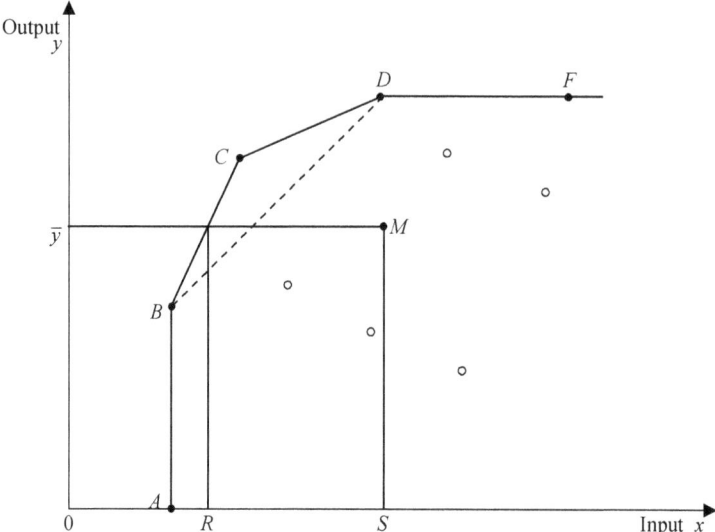

Fig. 7.8 Approximation of the efficiency frontier using DEA

countries (possibly because they constitute the extreme points, like firms A and F in Fig. 7.8). As outputs, the authors use net premiums of life and non-life insurance. As the only input, the number of employees including independent agents is used. Over the years 1983–1991, the Swiss ICs are near the frontier with an index of 0.98 followed by France, Great Britain, and Germany (0.91), Japan (0.59), Belgium (0.39), and Portugal (0.15) lagging behind at the time. However, the markets with low scores tended to have an above-average rate of increase in efficiency.

Evidently, SFA which treats inefficiency as part of a stochastic error term (see Sect. 7.4.4.1) and DEA which views it as a deterministic component of cost may lead to conflicting results. A study by Cummins and Zi (1998) compares the two approaches for the U.S. insurance industry. It finds agreement in that scale economies prevail among ICs with up to US$1 billion in assets. Only a few of the larger ones operate at minimum efficient scale; most of them exhibit diseconomies of scale. However, neither the parametric SFA nor the non-parametric DEA method are fully suited for tracking the efficiency of an IC in the course of its growth, which involves mergers and acquisitions or the enrollment of additional IBs. In either case, it is unlikely that these additional risks are comparable to the existing portfolio. This means that the risk exposure of the IC changes in the course of its growth. Analyzing cost as a function of premium income or identifying the production possibility frontier neglects this change in risk. As in the case of an expansion of the product spectrum, return on surplus changes both its expected value μ_S and standard deviation σ_S in the process of growth. The efficient frontier and its shifts need to be determined in (μ_S, σ_S)-space as in Fig. 7.5 of Sect. 7.3.3.

7.4 Economies of Scale

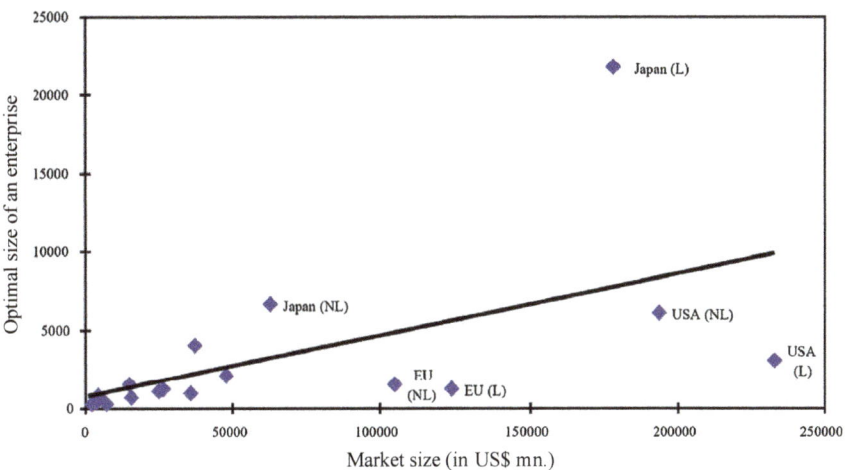

Fig. 7.9 Minimum efficient scale (MES3) and market size, life (L) and non-life (NL) insurance (1987). *Source* Eisen (1991)

7.4.5 Scale Economies and Size of Market

The available evidence points to scale economies in the insurance industry. Over time, the ICs in a given market approach their minimum efficient scale (MES). However, MES itself depends on the size of the market. The main reason is that an IC seeking to grow must accept risks with more unfavorable characteristics. In terms of the (μ_S, σ_S)-space alluded to at the end of the preceding section, the efficient frontier ceases to move up. However, this point is reached later in larger markets than in smaller ones, causing MES to have a higher value.

A rough test of this hypothesis is possible by comparing the European Union (EU) prior to its creation of the single market for insurance in 1994 with Japan on the one hand and the United States on the other. At that time, national boundaries caused MES still to be low in EU member states, whereas Japan was a large integrated insurance market permitting ICs to fully benefit from scale economies. The United States with its (partially harmonized) insurance regulation at the state level constitutes an intermediate case (see Sect. 9.4.1 for more details). Therefore, in a simple regression relating MES to the size of the respective market, Japan is predicted to present a positive outlier (higher MES than expected for its market size), the European Union, a negative one, and the United States in between.

This test was performed by Eisen (1991), using a sample of 11 OECD countries plus the European Union as of 1987. Since MES is not really known in these countries, the author proposes three indicators, each measured as net premiums of ICs,

MES1 := Average size (arithmetic mean);
MES2 := Median size (one-half of the IC lying above, the other half below this value);
MES3 := Average size of the leading IC accounting for 50% of premium volume.

The results do not depend much on the indicator chosen. Therefore, only the relationship between market size (measured by aggregate net premium income) and MES3 is displayed in Fig. 7.9. All countries as well as the EU are represented by two observations, one for life (L), the other for non-life (NL) insurance, respectively.

The lowest MES3 for life insurance is that of Austria with around 210 mn. US$ at the time (not visible in Fig. 7.9), the highest, that of Japan with almost 22,000 mn. US$. Market size varies between less than 2,000 and 233,000 mn. US$. The regression line suggests that a doubling of market size goes along with almost a doubling of MES3, confirming the hypothesized positive relation between minimum efficient scale and market size. In addition, the deviations from the regression line are of particular interest. In both life and non-life insurance, Japan is characterized by positive and the European Union, by negative deviations. Finally, in the case of non-life insurance, the United States is indeed located between Japan (NL) and the European Union (NL). With regard to life insurance, the observation pertaining to the United States is rather far away from the regression line, contrary to expectations. On the whole, the hypothesis that larger integrated markets induce higher MES receives a good measure of confirmation.

By implication, the transition to the single market for insurance by the EU should have spurred efficiency and resulted in higher MES. Performing a study utilizing DEA, Mahlberg and Url (1993) test this presumption for the case of Austria, an EU member country. Note that the MES condition $e(C, y) = 1$ of (7.51) requires ICs to be on a ray OC through the origin of Fig. 7.8 (x is proportional to cost). Therefore, part of the inefficiency characterizing firm M is due to its small scale. The authors do find that the scale efficiency score increases between 1994 and 1998 but then decreases in 1999, their last year of observation. An irregular development also characterizes the number of ICs operating at MES; it is 9 out of 65 in 1994 and drops to 5 out of 52 in 1999, down from 9 in 1998.

▶ **Conclusion 7.13** Minimum efficient scale is predicted to increase with the size of the market. While this prediction is confirmed across OECD countries, the evidence is mixed for Austria after the creation of the single market for insurance by the European Union.

Exercises

7.1

(a) Please summarize in no more than three sentences the core assumptions and conclusions of one of the premium principles, of the CAPM and the option pricing model in their application to insurance.

(b) As a representative of the agency that regulates the insurance industry, which alternative(s) would you use

 (b1) to assess the level of premiums charged?
 (b2) to assess the insurance reserves with a view to solvency?
 (b3) to examine whether an IC has an "acceptable" equity capital for its size?

7.2

(a) Please compare the pros and cons determining the efficiency frontier (transformation curve)

   (a1) by estimating a cost function;
   (a2) performing a data envelopment analysis (DEA).

(b) An IC is rated efficient according to (a1) but not according to (a2). How can this contradiction arise?
(c) An efficiency frontier can also be determined in a CAPM framework. Where does the difference between (a1) and (a2) lie?
(d) You find an IC to be efficient according to (3) but not according to (a1) or (a2). Why could this be the case?
(e) To the contrary, you ascertain that an IC is rated efficient according to (a1) or (a2), but not according to (3). Cite reasons for such a contradiction.
(f) You advise the management of IC "Star", contemplating a merger with IC "Top". According to research by the Statistics Department of "Star", the combined IC "Topstar" appears to be efficient according to (a1) and (a2), however, as inefficient according to (3). What is your advice concerning the merger?

7.A Appendix: Normal Distribution

Fig. 7.A.1 shows the area under normal distribution.

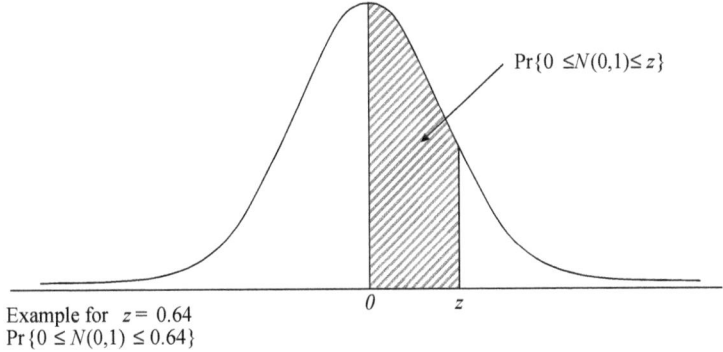

Example for $z = 0.64$
$\Pr\{0 \leq N(0,1) \leq 0.64\}$

Fig. 7.A.1 Area under the normal distribution

Table 7.A.1 Area below the standard normal density (from 0 up to z)

z	0.00	0.01	0.02	0.03	0.04	0.05	0.06	0.07	0.08	0.09
0.0	0.00000	0.00399	0.00798	0.01197	0.01595	0.01994	0.02392	0.02790	0.03188	0.03586
0.1	0.03983	0.04380	0.04776	0.05172	0.05567	0.05962	0.06356	0.06749	0.07142	0.07535
0.2	0.07926	0.08317	0.08706	0.09095	0.09483	0.09871	0.10257	0.10642	0.11026	0.11409
0.3	0.11791	0.12172	0.12552	0.12930	0.13307	0.13683	0.14058	0.14431	0.14803	0.15173
0.4	0.15542	0.15910	0.16276	0.16640	0.17003	0.17364	0.17724	0.18082	0.18439	0.18793
0.5	0.19146	0.19497	0.19847	0.20194	0.20540	0.20884	0.21226	0.21566	0.21904	0.22240
0.6	0.22575	0.22907	0.23237	0.23565	0.23891	0.24215	0.24537	0.24857	0.25175	0.25490
0.7	0.25804	0.26115	0.26424	0.26730	0.27035	0.27337	0.27637	0.27935	0.28230	0.28524
0.8	0.28814	0.29103	0.29389	0.29673	0.29955	0.30234	0.30511	0.30785	0.31057	0.31327
0.9	0.31594	0.31859	0.32121	0.32381	0.32639	0.32894	0.33147	0.33398	0.33646	0.33891
1.0	0.34134	0.34375	0.34614	0.34849	0.35083	0.35314	0.35543	0.35769	0.35993	0.36214
1.1	0.36433	0.36650	0.36864	0.37076	0.37286	0.37493	0.37698	0.37900	0.38100	0.38298
1.2	0.38493	0.38686	0.38877	0.39065	0.39251	0.39435	0.39617	0.39796	0.39973	0.40147
1.3	0.40320	0.40490	0.40658	0.40824	0.40988	0.41149	0.41308	0.41466	0.41621	0.41774
1.4	0.41924	0.42073	0.42220	0.42364	0.42507	0.42647	0.42785	0.42922	0.43056	0.43189
1.5	0.43319	0.43448	0.43574	0.43699	0.43822	0.43943	0.44062	0.44179	0.44295	0.44408
1.6	0.44520	0.44630	0.44738	0.44845	0.44950	0.45053	0.45154	0.45254	0.45352	0.45449
1.7	0.45543	0.45637	0.45728	0.45818	0.45907	0.45994	0.46080	0.46164	0.46246	0.46327
1.8	0.46407	0.46485	0.46562	0.46638	0.46712	0.46784	0.46856	0.46926	0.46995	0.47062
1.9	0.47128	0.47193	0.47257	0.47320	0.47381	0.47441	0.47500	0.47558	0.47615	0.47670
2.0	0.47725	0.47778	0.47831	0.47882	0.47932	0.47982	0.48030	0.48077	0.48124	0.48169
2.1	0.48214	0.48257	0.48300	0.48341	0.48382	0.48422	0.48461	0.48500	0.48537	0.48574
2.2	0.48610	0.48645	0.48679	0.48713	0.48745	0.48778	0.48809	0.48840	0.48870	0.48899
2.3	0.48928	0.48956	0.48983	0.49010	0.49036	0.49061	0.49086	0.49111	0.49134	0.49158
2.4	0.49180	0.49202	0.49224	0.49245	0.49266	0.49286	0.49305	0.49324	0.49343	0.49361
2.5	0.49379	0.49396	0.49413	0.49430	0.49446	0.49461	0.49477	0.49492	0.49506	0.49520
2.6	0.49534	0.49547	0.49560	0.49573	0.49585	0.49598	0.49609	0.49621	0.49632	0.49643
2.7	0.49653	0.49664	0.49674	0.49683	0.49693	0.49702	0.49711	0.49720	0.49728	0.49736
2.8	0.49744	0.49752	0.49760	0.49767	0.49774	0.49781	0.49788	0.49795	0.49801	0.49807
2.9	0.49813	0.49819	0.49825	0.49831	0.49836	0.49841	0.49846	0.49851	0.49856	0.49861
3.0	0.49865	0.49869	0.49874	0.49878	0.49882	0.49886	0.49889	0.49893	0.49896	0.49900
3.1	0.49903	0.49906	0.49910	0.49913	0.49916	0.49918	0.49921	0.49924	0.49926	0.49929
3.2	0.49931	0.49934	0.49936	0.49938	0.49940	0.49942	0.49944	0.49946	0.49948	0.49950
3.3	0.49952	0.49953	0.49955	0.49957	0.49958	0.49960	0.49961	0.49962	0.49964	0.49965
3.4	0.49966	0.49968	0.49969	0.49970	0.49971	0.49972	0.49973	0.49974	0.49975	0.49976
3.5	0.49977	0.49978	0.49978	0.49979	0.49980	0.49981	0.49981	0.49982	0.49983	0.49983
3.6	0.49984	0.49985	0.49985	0.49986	0.49986	0.49987	0.49987	0.49988	0.49988	0.49989
3.7	0.49989	0.49990	0.49990	0.49990	0.49991	0.49991	0.49992	0.49992	0.49992	0.49992
3.8	0.49993	0.49993	0.49993	0.49994	0.49994	0.49994	0.49994	0.49995	0.49995	0.49995
3.9	0.49995	0.49995	0.49996	0.49996	0.49996	0.49996	0.49996	0.49996	0.49997	0.49997
4.0	0.49997	0.49997	0.49997	0.49997	0.49997	0.49997	0.49998	0.49998	0.49998	0.49998

7.4 Economies of Scale

Table 7.A.2 Poisson distribution (cumulative)

n	\multicolumn{15}{c|}{E(n)}														
	0.5	1.0	1.5	2.0	2.5	3.0	3.5	4.0	4.5	5.0	6.0	7.0	8.0	9.0	10.0
0	0.60653	0.36788	0.22313	0.13534	0.08208	0.04979	0.03020	0.01832	0.01111	0.00674	0.00248	0.00091	0.00034	0.00012	0.00005
1	0.90980	0.73576	0.55783	0.40601	0.28730	0.19915	0.13589	0.09158	0.06110	0.04043	0.01735	0.00730	0.00302	0.00123	0.00050
2	0.98561	0.91970	0.80885	0.67668	0.54381	0.42319	0.32085	0.23810	0.17358	0.12465	0.06197	0.02964	0.01375	0.00623	0.00277
3	0.99825	0.98101	0.93436	0.85712	0.75758	0.64723	0.53663	0.43347	0.34230	0.26503	0.15120	0.08177	0.04238	0.02123	0.01034
4	0.99983	0.99634	0.98142	0.94735	0.89118	0.81526	0.72544	0.62884	0.53210	0.44049	0.28506	0.17299	0.09963	0.05496	0.02925
5	0.99999	0.99941	0.99554	0.98344	0.95798	0.91608	0.85761	0.78513	0.70293	0.61596	0.44568	0.30071	0.19124	0.11569	0.06709
6	1.00000	0.99992	0.99907	0.99547	0.98581	0.96649	0.93471	0.88933	0.83105	0.76218	0.60630	0.44971	0.31337	0.20678	0.13014
7	1.00000	0.99999	0.99983	0.99890	0.99575	0.98810	0.97326	0.94887	0.91341	0.86663	0.74398	0.59871	0.45296	0.32390	0.22022
8	1.00000	1.00000	0.99997	0.99976	0.99886	0.99620	0.99013	0.97864	0.95974	0.93191	0.84724	0.72909	0.59255	0.45565	0.33282
9	1.00000	1.00000	1.00000	0.99995	0.99972	0.99890	0.99669	0.99187	0.98291	0.96817	0.91608	0.83050	0.71662	0.58741	0.45793
10	1.00000	1.00000	1.00000	0.99999	0.99994	0.99971	0.99898	0.99716	0.99333	0.98630	0.95738	0.90148	0.81589	0.70599	0.58304
11	1.00000	1.00000	1.00000	1.00000	0.99999	0.99993	0.99971	0.99908	0.99760	0.99455	0.97991	0.94665	0.88808	0.80301	0.69678
12	1.00000	1.00000	1.00000	1.00000	1.00000	0.99998	0.99992	0.99973	0.99919	0.99798	0.99117	0.97300	0.93620	0.87577	0.79156
13	1.00000	1.00000	1.00000	1.00000	1.00000	1.00000	0.99998	0.99992	0.99975	0.99930	0.99637	0.98719	0.96582	0.92615	0.86446
14	1.00000	1.00000	1.00000	1.00000	1.00000	1.00000	1.00000	0.99998	0.99993	0.99977	0.99860	0.99428	0.98274	0.95853	0.91654
15	1.00000	1.00000	1.00000	1.00000	1.00000	1.00000	1.00000	1.00000	0.99998	0.99993	0.99949	0.99759	0.99177	0.97796	0.95126
16	1.00000	1.00000	1.00000	1.00000	1.00000	1.00000	1.00000	1.00000	0.99999	0.99998	0.99983	0.99904	0.99628	0.98889	0.97296
17	1.00000	1.00000	1.00000	1.00000	1.00000	1.00000	1.00000	1.00000	1.00000	0.99999	0.99994	0.99964	0.99841	0.99468	0.98572
18	1.00000	1.00000	1.00000	1.00000	1.00000	1.00000	1.00000	1.00000	1.00000	1.00000	0.99998	0.99987	0.99935	0.99757	0.99281
19	1.00000	1.00000	1.00000	1.00000	1.00000	1.00000	1.00000	1.00000	1.00000	1.00000	0.99999	0.99996	0.99975	0.99894	0.99655
20	1.00000	1.00000	1.00000	1.00000	1.00000	1.00000	1.00000	1.00000	1.00000	1.00000	1.00000	0.99999	0.99991	0.99956	0.99841
21	1.00000	1.00000	1.00000	1.00000	1.00000	1.00000	1.00000	1.00000	1.00000	1.00000	1.00000	1.00000	0.99997	0.99983	0.99930
22	1.00000	1.00000	1.00000	1.00000	1.00000	1.00000	1.00000	1.00000	1.00000	1.00000	1.00000	1.00000	0.99999	0.99993	0.99970
23	1.00000	1.00000	1.00000	1.00000	1.00000	1.00000	1.00000	1.00000	1.00000	1.00000	1.00000	1.00000	1.00000	0.99998	0.99988
24	1.00000	1.00000	1.00000	1.00000	1.00000	1.00000	1.00000	1.00000	1.00000	1.00000	1.00000	1.00000	1.00000	0.99999	0.99995
25	1.00000	1.00000	1.00000	1.00000	1.00000	1.00000	1.00000	1.00000	1.00000	1.00000	1.00000	1.00000	1.00000	1.00000	0.99998

References

Baumol, H.J., Panzar, J.C., & Willig, R.D. (1982). *Contestable Markets and the Theory of Industry Structure*. New York: Harcourt Brace Jovanowitsch.

Beard, R.E., Pentikäinen, T., & Pesonen, E. (1984). *Risk Theory. The Stochastic Basis of Insurance*. London: Chapman and Hall.

Black, F. (1972). Capital market equilibrium with restricted borrowing. *Journal of Business, 45*, 444–455.

Black, F., & Scholes, M. (1973). The pricing of options and corporate liabilities. *Journal of Political Economy, 81*, 637–659.

Bühlmann, H. (1970). *Mathematical Methods in Risk Theory*. Heidelberg: Springer.

Cummins, J.D. (1991b). Capital structure and fair profits in property-liability insurance. In J.D. Cummins, & R.A. Derrig, (Eds.), *Managing the Insolvency Risk of Insurance Companies*, 295–308. Dordrecht: Kluwer.

Cummins, J.D. (1991a). Statistical and financial models of insurance pricing and the insurance firm. *Journal of Risk and Insurance, 85*(2), 261–302.

Cummins, J.D., & Phillips, R.D. (2005). Estimating the cost of equity capital for property-liability insurers. *Journal of Risk and Insurance, 72*, 441–478.

Cummins, J.D., & Sommer, D.W. (1996). Capital and risk in property-liability insurance markets. *Journal of Banking and Finance, 20*, 1069–1092.

Cummins, J.D., & Weiss, M.A. (1993). Measuring cost efficiency in the property-liability insurance industry. *Journal of Banking and Finance, 17*(2–3), 463–481.

Cummins, J.D., & Zi, H. (1998). Comparison of frontier efficiency methods: an application to the US life insurance industry. *Journal of Productivity Analysis, 10*(2), 131–152.

Cummins, J.D., Weiss, M.A., Xiaoying, X., & Hongmin, Z. (2010). Economies of scope in financial services: a DEA efficiency analysis of the U.S. insurance industry. *Journal of Banking & Finance, 34*, 1525–1539.

Doherty, N.A. (1981). The measurement of output and economies of scale in property-liability insurance. *Journal of Risk and Insurance, 48*(3), 391–402.

Doherty, N.A., & Garven, J.R. (1986). Price regulation in property-liability insurance: a contingent claims approach. *Journal of Finance, 41*, 1031–1050.

Donni, O., & Fecher, F. (1997). Efficiency and productivity of the insurance industry in the OECD countries. *Geneva Papers on Risk and Insurance, 22*, 523–535.

Eisen, R. (1991). Market size and concentration: insurance and the European market 1992. *Geneva Papers on Risk and Insurance Issues and Practice, 60*, 263–281.

Fama, E., & Jensen, M.C. (1983). Separation of ownership and control. *Journal of Law and Economics, 26*, 301–325.

Fecher, E., Perelman, S.D., & Pestieau, P. (1991). Scale economies and performance in the French insurance industry. *Geneva Papers on Risk and Insurance Issues and Practice, 60*, 315–326.

Garven, J.R., & D'Arcy, S. (1991). A synthesis of property-liability insurance pricing techniques. In J.D. Cummins, & R.A. Derrig (Eds.), *Managing the Insolvency Risk of Insurance Companies*, Chap. 8. Boston: Kluwer.

Goavaerts, M.J., & Laeven, J.A. (2008). Actuarial risk measures for financial derivative pricing. *Insurance: Mathematics and Economics, 42*(2), 540–547.

Greene, W.H. (1997). *Econometric Analysis*, 5th ed. Upper Saddle River NY: Prentice Hall.

Grossman, S.J. (1976). On the efficiency of competitive stock markets when traders have diverse information. *Journal of Finance, 31*, 573–585.

Gurley, J.G., & Shaw, E.S. (1960). *Money in a Theory of Finance*. Washington DC: The Brookings Institution.

Hardwick, P. (1997). Measuring cost inefficiency in the UK life insurance industry. *Applied Financial Economics, 7*(1), 37–44.

Heilmann, W.R. (1988). *Fundamentals of Risk Theory*. Karlsruhe: Verlag für Versicherungswissenschaft.

References

Heilmann, W.R. (1989). Decision theoretic foundations of credibility theory. *Insurance: Mathematics and Economics, 8*(1), 77–95.

Hirao, Y., & Inoue, T. (2004). On the cost structure of the Japanese property-casualty insurance industry. *Journal of Risk and Insurance, 71*(3), 501–530.

Houston, D., & Simon, R. (1970). Economies of scale in financial institutions: a study in life insurance. *Econometrica, 38*, 856–864.

Kellner, S., & Matthewson, G.F. (1983). Entry, size, distribution, scale and scope economies in the life insurance industry. *Journal of Business, 56*(1), 25–44.

Kumbhakar, S., & Lovell, C.A. (2000). *Stochastic Frontier Analysis*. Cambridge: Cambridge University Press.

Mahlberg, B., & Url, T. (1993). Effects of the single market on the Austrian insurance industry. *Empirical Economics, 28*, 813–838.

Markowitz, H.M. (1959). *Portfolio Selection - Efficient Diversification of Investments*. New Haven and London: Yale University Press.

Mayers, D., & Smith, C.W. (1988). Ownership structure across lines of property-liability insurance. *Journal of Law and Economics, 31*, 351–378.

Murray, J.D., & White, R.W. (1983). Economics of scale and economics of scope in multiproduct financial institutions: a study of British Columbia credit unions. *Journal of Finance, 38*, 887–901.

Panzar, J.C., & Willig, R.D. (1977). Economies of scale in multi-output production. *Quarterly Journal of Economics, 81*, 481–493.

Praetz, P.D. (1980). Returns to scale in the U.S. life insurance industry. *Journal of Risk and Insurance, 47*(4), 525–532.

Pritchett, S.T. (1973). Operating expenses of life insurers, 1961–1970: implications for economies of size. *Journal of Risk and Insurance, 40*(2), 157–165.

Pyle, D. (1971). On the theory of financial intermediation. *Journal of Finance, 26*, 737–747.

Seiford, L.M., & Thrall, R.M. (1990). Recent developments in DEA. *Journal of Econometrics, 46*, 7–38.

Shrieves, R.E., & Dahl, D. (1992). The relationship between risk and capital in commercial banks. *Journal of Banking & Finance, 16*, 439–457.

Suret, M. (1991). Scale and scope economies in the Canadian property and casualty insurance industry. *Geneva Papers on Risk and Insurance Issues and Practice, 59*, 236–256.

Toivanen, A.M. (1993). Economies of scale and scope in the Finnish non-life insurance industry. *Journal of Banking and Finance, 21*(6), 759–779.

Vroomen, J., & Zweifel, P. (2011). Preferences for health insurance and health status: does it matter whether you are Dutch or German? *European Journal of Health Economics, 12*(1), 87–95.

Walden, M.L. (1985). The whole life insurance policy as an options package: an empirical investigation. *Journal of Risk and Insurance, 52*(4), 44–58.

White, H. (1980). A heteroscedasticity-consistent covariance estimator and a direct test for heteroscedasticity. *Econometrica, 48*, 817–838.

Williams, J.T. (1977). Capital asset prices with heterogeneous beliefs. *Journal of Financial Economics, 5*, 219–239.

Zweifel, P. (2015). Solvency regulation of banks and insurers: A two-pronged critique. *International Journal of Financial Research, 6*(3), 86-105.

Zweifel, P., Telser, H., & Vaterlaus, S. (2006). Consumer resistance against regulation: the case of health care. *Journal of Regulatory Economics, 29*(3), 319–332.

Insurance Markets and Asymmetric Information

8

This chapter deals with a property of insurance markets that has been repeatedly mentioned before (e.g., in Sects. 5.3, 6.5, and 6.6): Information may be distributed in an unequal way between the insurance company (IC) and the buyer of insurance (IB). Whereas in the markets for personal services, it is the consumer who is thought to suffer from a lack of information (patients vis-à-vis physicians, for example), it is usually the supplier in the case of financial services. For instance, the applicant for credit, as opposed to the bank, is better capable of judging the chances of success of the project to be financed. Likewise, it is the IB and not the IC who is better able to gauge the probability of a loss occurring in the future.

This assumption with regard to the asymmetry of information will be critically reviewed in Sect. 8.1. Next, asymmetric information is seen to give rise to two phenomena that have been known to insurers for a long time, which also characterize credit markets, labor markets, and the public sector, viz., moral hazard and adverse selection. Accordingly, Sect. 8.2 deals with moral hazard in its several variants, whereas Sect. 8.3 revolves around the problem of adverse selection. Section 8.4 contains an introduction to the analysis of the combined effects of moral hazard and adverse selection.

8.1 Asymmetric Information and Its Consequences

In classical microeconomics, perfect market transparency is assumed. Consumers as well as producers know prices and qualities of the goods and services they deal with. This assumption is not as restrictive as may seem at first sight. In many cases, the cost of acquiring information is sufficiently low for agents to attain transparency. For instance, prior to purchasing a computer, consumers can acquire information from distributors, neighbors, and colleagues at their workplace about prices and properties of the different models. However, rational agents gather information only up to the point where the marginal cost of acquiring it is equal to its expected return. For this reason, it is usually optimal to be only partially informed, resulting in some market participants knowing more than others. Therefore, information is often asymmetrically distributed.

In the insurance economics literature, the IBs are usually assumed to know more about their risk type and their future preventive behavior than the IC. Therefore, it is the IC who suffers from a lack of information.

This basic assumption is open to criticism. For example, one may question whether the potential purchasers of life insurance are capable of predicting their life expectancy. Yet, a survey of Americans aged 51–54 (at the time) revealed that their subjective estimates of remaining years of life were rather close to actual values [see Cawley and Philipson (1999)]. Compared to the IC, they likely enjoyed an informational advantage which presumably permitted those with a high mortality risk to buy term life insurance (i.e., life insurance without a savings component) with ample benefits at favorable rates. Interestingly however, this was not what the authors found. Rather, ample coverage was purchased by individuals with low mortality risk. This observation contradicts the conventional view of the adverse selection problem, where the IC enrolls "high" risks (those with high mortality in the present context) without recognizing them as such. Failing to charge these higher risks a higher premium, the IC runs into financial difficulties (see Sect. 8.3).

Quite generally, one would expect high risks to buy comprehensive coverage if the insurer lacks the information to adjust the premium accordingly. Since these risks cause losses more frequently and in higher amounts, a positive correlation between degree of coverage and losses paid is predicted. In the case of French auto insurance, this correlation is indeed observed by Chiappori et al. (2006), pointing to the presence of an asymmetry of information to the detriment of the IC.

It may still be that the IBs are less informed than the IC with regard to other features of the insurance contract. Surveys do show that many consumers do not compare insurance products. This may be due to the fact that the information to be gained is not very dependable. In the case of life insurance, e.g., consumers would have to estimate future surplus credited to policy holders. In property-liability insurance, they would have to predict the IC's customer accommodation policy (How stringently will the company interpret the small print of the contract?). Both assessments are subject to considerable margins of error.

In the commercial lines, market transparency could be achieved in principle because insurance is purchased by specialized managers, at least in larger firms. However, in these lines changes of contract and renegotiations are frequent, causing the marginal cost of collecting information to become high as well. Therefore, the asymmetry of information may again be to the detriment of the IB rather than the IC.

In spite of these reservations, the basic assumption in this chapter will be that it is the IC, not the IB who estimates the cost of acquiring information to be prohibitively high, resulting in an asymmetry of information to the disadvantage of the IC.[1] Therefore, the IC is assumed to know neither the risk type of nor the risk avoidance and risk prevention efforts of an individual IB. All it observes is choice of contract and occurrence of loss (both of which can be ascertained at no extra cost). This information asymmetry has two consequences, viz., moral hazard and adverse selection.

1. *Moral hazard (hidden action)*. This informational disadvantage of the IC takes effect after execution of the contract. Moral hazard effects result from the fact that both the probability of loss and its severity (amount or duration of damage) may depend on the behavior of the IB, which by assumption cannot be observed by the IC. Thus, all observation derives from an interaction between "Nature" (the chance element) and the IB, with no possibility of sorting out the respective influences. This ambiguity permits IBs to reduce their effort at preventing a loss or mitigating its consequences. Moral hazard therefore changes the basis for calculating the premium and drives up the cost of providing insurance protection [see Arrow (1963)]. Recall that an increase in premium reflecting a higher expected value of benefits does not amount to an increase in the cost of insurance; rather, it is the increased loading for administrative expense and risk bearing due to moral hazard that is referred to here.[2]
2. *Adverse selection (hidden information)*. This informational disadvantage of the IC takes effect at the time of the inception of the contract. Ignoring the true risk of an individual IB, the IC must resort to some averaging over risk types for its premium calculation. However, such a contract with a pooled premium (pooling contract for short) is too expensive for the low risks, who tend to migrate to a competing IC offering them a contract better tailored to their true risk type [see Pauly (1974); Rothschild and Stiglitz (1976)]. This causes the incumbent IC to increasingly lose low risks to competitors. It needs to adjust the premium of its pooling contract upward to counterbalance its rising loss payments. Yet, this renders the pooling contract even less attractive to the low risks. In the long run, the incumbent IC becomes insolvent, and its remaining high risks have to find another insurer. This makes one (or several) of the surviving ICs the new incumbent, with too many high risks on its books and a pooling premium that is excessive for the low risks. Therefore, the adverse selection spiral goes into its next turn. This possible "death spiral" property of adverse selection is considered to be a more serious problem than moral hazard effects by both insurers and policy-makers. For this reason, it will be of particular interest to examine the empirical evidence on this issue in Sect. 8.3.

---

[1] One exception is Sect. 6.6, where the contribution by Borch (1990) concerning optimal effort at controlling moral hazard is discussed.
[2] An important component of the increased expected benefit, however, may be higher prices in upstream markets, e.g., medical fees in the case of health insurance. Therefore, moral hazard may also affect contractual partners of the insured who suffer a loss [see, e.g., Zweifel et al. (2009, Ch. 10.2) and, in German, Eisen and Nell (1994)].

8.2 Moral Hazard

8.2.1 Definition and Importance of Moral Hazard

Moral hazard has little to do with morality but relates to human behavior (in French, "*les moeurs*" means the customs of a society). In economic theory, moral hazard stands for the change in unobservable behavior induced by the existence of a contract that provides protection against risk. This change is assumed unobservable for the IC, in keeping with the arguments of Sect. 8.1 above. Moral hazard effects are also common outside insurance. For example, after the end of a probation period, employers have to take into account that workers, now enjoying a degree of protection against being fired on short notice, tend to reduce their effort (which is difficult to monitor except for piecework). Of course, employers have developed incentive schemes designed to mitigate this type of moral hazard [this is the topic of the principal-agent relationship; see, e.g., Levinthal (1988)].

In the present context, moral hazard is triggered by the existence of an insurance contract. The reason for a change in IB behavior is that the returns to a risky choice of action continue to fully accrue to the IB, whereas its cost fall on the IC, who commits to cover possible losses. The following forms of moral hazard need to be distinguished.

1. *The probability of loss increases.* On the one hand, this occurs because the IB opt for more lucrative but also more risky courses of action. On the other hand, they can also skimp on costly efforts at loss prevention (often called self-protection). To the extent that these efforts are directed at the cause of the loss, they are called etiological (see Sect. 2.1). Since they necessarily occur prior to the occurrence of the loss, the associated modification of behavior is called *ex-ante* moral hazard.
2. *The amount of loss increases.* Three variants can be distinguished here.

   - The IB may undertake so-called palliative efforts prior to the occurrence of the loss, designed to limit its amount (often called self-insurance). Examples are fire extinguishers in buildings or airbags in cars. To the extent that such efforts are reduced due to insurance coverage, this variant constitutes another type of *ex-ante* moral hazard (see also Sect. 5.3).
   - After the occurrence of the loss, IBs can influence its amount in many cases. For instance, to repair a damaged car the IB may commission an expensive rather than a less costly body shop. With regard to timing, this is an *ex-ante* moral hazard effect. It is of the *static* type if the technology applied in the repair is predetermined.
   - However, the technology applied may be subject to choice rather than being predetermined. An important instance is medical care, where the insureds often have a choice between a new (typically more costly) technology and a conventional alternative. Insureds who opt for the new technology because health insurance covers most of its extra cost exhibit *ex-post* moral hazard of the *dynamic type*.

8.2 Moral Hazard

Table 8.1 Importance of insurance fraud, auto insurance in Florida (United States, 1988)

Aspect	Frequency, amount
Fraudulent claims	13% of total value of losses submitted
	10% of submitted losses
	8% of premium income
Fraudulently inflated claims	38% of repair invoices submitted by 18–34-year old insured are inflated by some 50%
	37% of invoices for medical treatment submitted by 18–34-year old insured are inflated by some 50%
Consumer opinion	33% of enrollees surveyed deem legitimate misrepresenting risk in order to benefit from low premiums
	50% of enrollees surveyed deem legitimate inflating repair invoices beyond the deductible in order to receive benefits
Sanctions expected	25% of enrollees surveyed think that the misrepresentation of information at conclusion of the contract will not be sanctioned
	50% know that misrepresentation can result in non-payment of claims and possibly legal prosecution

Source Mooney and Salvatore (1990)

In economic theory, moral hazard is not viewed as negatively as in insurance practice. Its demand-increasing effect can even be beneficial if it counteracts rationing caused by monopolistic pricing,, e.g., in the case of medical fees [Pauly and Held (1990)]. More generally, moral hazard amounts to a manifestation of the law of demand in that insurance coverage lowers the net relative cost of a risky action borne by the IB. If the risk materializes, only part of the loss falls on the IB, while the cost of effort directed at loss prevention and control have to be fully borne by the IB. Given that the insured cannot be monitored by assumption, standard microeconomic theory predicts that preventive effort is substituted by insurance in this situation (this does not always hold true, however; for more detail, see Sect. 8.2.2.1 below).

In its extreme form, moral hazard turns into criminal activity. Indeed, for a substantial subset of IBs, there is only a thin line separating moral hazard from insurance fraud. Table 8.1 contains estimates of insurance fraud. The data come from an audit of auto insurance records that was performed in Florida (USA) in the late 1980s. According to this study, fraudulent claims amount to no less than 13% of loss payments. In addition, more than one-third of all invoices for car repairs and medical treatment submitted by consumers below age 35 are inflated by a factor of some 50%.

Insurance fraud cannot be easily proved, however. For instance, according to German police statistics, only 0.3% of liability insurance claims were determined to be fraudulent in 1996 [Association of German Insurers GDV (1997, p. 122)]. The discrepancy between this figure and those of Table 8.1 does not necessarily imply that German IBs are more honest than their U.S. counterparts. First, the distinction between insurance fraud and moral hazard is difficult to draw. It also depends on the moral sentiments prevalent in a society. Second, the German data may reflect IC behavior. Note that Table 8.1 refers to an internal audit which tends to be comprehensive. Actual business operations differ from this in two important aspects. First, it would be suboptimal for the IC to check all claims for fraud (see the model of Sect. 6.6). Second, reporting a fraud to the police likely means losing the customer. Usually, a "slap on the wrist" in the guise of voicing concern and requiring additional evidence is the more efficient alternative because it preserves the flow of premiums.

In the exposition below, insurance fraud will be disregarded. Losses are assumed to occur by chance (influenced by preventive effort) rather than being deliberately caused by the IB.

8.2.2 *Ex-Ante* Moral Hazard

8.2.2.1 *Ex-Ante* Moral Hazard with Regard to the Probability of Loss

In this section, the amount of coverage provided by the insurance contract is assumed to be predetermined, i.e., negotiated previously.[3] Thus, the only decision variable of the IB is the amount of preventive effort V, the price of which is normalized to one for simplicity. In this way, V coincides with expenditure on prevention. Contrary to the theory discussed in Sect. 3.2, here the probability of loss π is not predetermined but depends negatively on V,[4]

$$0 < \pi(V) < 1, \quad \pi'(V) < 0, \quad \pi''(V) > 0, \quad \pi[0] = \bar{\pi}, \quad \pi[\infty] > 0, \quad V \geq 0. \quad (8.1)$$

Figure 8.1 displays the postulated relationship. For simplicity, only two possible states (loss and no loss) are distinguished. As before, the IB maximizes his or her expected utility as a function of V,

$$EU(V) = \pi(V) \cdot v[W_0 - V - P(I) - L + I] + \{1 - \pi(V)\} \cdot v[W_0 - V - P(I)]$$
$$= \pi(V)v(W_1) + \{1 - \pi(V)\}v(W_2). \quad (8.2)$$

- $v(\cdot)$: Risk utility function with the usual properties $v'(W) > 0$, $v''(W) < 0$;
- W: Final wealth of the decision maker;
- W_0: Predetermined initial wealth, exogenous;
- W_1: Final wealth in the case of loss L, $W_1 := W_0 - V - P(I) - L + I$;
- W_2: Final wealth in the case of no loss, $W_2 := W_0 - V - P(I)$;

---

[3] The model presented here is a simplified version of Zweifel et al. (2009), Ch. 6.4.1.4.
[4] Since V has nonnegative values only, $\pi(V)$ can never exceed $\bar{\pi}$. Hence, insurance fraud is excluded.

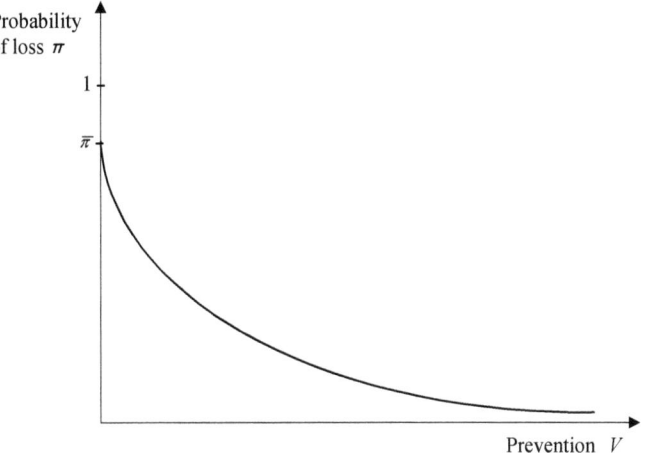

Fig. 8.1 Decreasing marginal effectiveness of prevention

V: Preventive effort, expenditure on prevention;
$P(I)$: Premium, increasing in the amount of insurance coverage (see below);
I: Insurance coverage (indemnity), $I \leq L$;
L: Loss, exogenous.

To save on notation, the following two symbols are used,

$$v[1] := v[W_1] = v[W_0 - V - P(I) - L + I], \text{ with}$$
$$\frac{dv[1]}{dV} = \frac{dv[1]}{dW} \cdot \frac{dW}{dV} = (-1)v'[1]; \quad (8.3a)$$
$$v[2] := v[W_2] = v[W_0 - V - P(I)], \text{ with}$$
$$\frac{dv[2]}{dV} = \frac{dv[2]}{dW} \cdot \frac{dW}{dV} = (-1)v'[2]; \text{ and} \quad (8.3b)$$
$$v[1] \leq v[2]. \quad (8.3c)$$

Taking the total derivative w.r.t. V and using Eqs. (8.3a) and (8.3b), one obtains (8.4). The fact that V must be nonnegative leads to the two Kuhn–Tucker conditions,

$$\frac{dEU}{dV} = \pi'(V)\{v[1] - v[2]\} - \pi(V) \cdot v'[1] - \{1 - \pi(V)\}v'[2] \quad (8.4)$$
$$< 0 \text{ if } V^* = 0;$$
$$= 0 \text{ if } V^* > 0.$$

Two cases can be distinguished.

1. $dEU/dV < 0$: This describes a *boundary solution*, with $V^* = 0$. Since expected utility evaluated at $V = 0$ is already decreasing in V, its maximum would be in

principle reached at a negative value of V. However, it coincides with $V^* = 0$ due to the nonnegativity constraint.
2. $dEU/dV = 0$: This describes an *interior solution*, indicating that V^* has a positive value.

The interior solution can be interpreted in the following way by rewriting the last two terms of condition (8.4) as $EU'(W) := \pi v'[1] + (1 - \pi)v'[2]$, i.e., as the expected value of the marginal utility of wealth. One then obtains

$$\frac{dEU}{dV} = \pi'(V)\{v[1] - v[2]\} - EU'(W) = 0, \quad \text{with}$$
$$EU'(W) := \pi v'[1] + (1 - \pi)v'[2]. \tag{8.5}$$

Therefore, prevention has a marginal return that consists of the reduced probability of incurring the utility loss given by $v[1] - v[2] < 0$ [see Eq. (8.3c) above]. It also has a marginal cost, which amounts to the wealth spent on it. Since the amount of prevention must be decided *ex ante*, its cost has to be borne in both states, to be valued using the pertinent marginal utility of wealth. At the interior optimum, the expected value of marginal return equals the certain marginal cost in utility terms.

The two cases distinguished above are of particular relevance in the case of full insurance coverage ($I = L$). Since the IB by assumption derives utility from wealth only, the utilities $v[1]$ and $v[2]$ coincide, causing the marginal return of prevention to become zero. Equation (8.5) then reduces to

$$\frac{dEU}{dV}\bigg|_{I=L} = -EU'(W) < 0. \tag{8.6}$$

According to condition (8.4), this implies $V^* = 0$; therefore, a total lack of prevention is predicted. Note that this result holds regardless of risk aversion since inequality (8.6) does not contain a second derivative of the risk utility function [recall Sect. 2.3.2 where the coefficient of absolute risk aversion is defined as $RA := -v''(W)/v'(W)$].

▶ **Conclusion 8.1** If the risk utility function has only wealth as its argument, full insurance coverage is predicted to cause zero prevention regardless of risk aversion.

Conclusion 8.1 does not hold, however, as soon as utility differs between the loss and the no-loss state even given full insurance coverage [see Eq. (8.5) again]. For example, at a given level of wealth, the IB typically values the "healthy" state more highly than the "ill" state [see Sect. 3.5 for more details]. Therefore, full coverage in health insurance does not necessarily entail zero prevention. In the context of fire insurance, even full coverage may not compensate for the loss of a home one has become familiar with over the years. More generally, whenever the utility difference in Eq. (8.5) depends not only on wealth but also on other determinants (health, familiarity with the asset, impact of the loss on others the IB cares for), a certain amount of preventive effort is predicted even under full insurance coverage.

8.2 Moral Hazard

Returning to the case of a risk utility function determined by wealth only, Conclusion 8.1 gives rise to the presumption that an increasing degree of insurance coverage undermines preventive effort, resulting in a negative relationship between I and V. This intuition can be checked using comparative static analysis. Let the necessary condition for an interior optimum $dEU/dV = 0$ be disturbed by additional insurance coverage $dI > 0$, contracted in a previous period. After the adjustment (which can only be a change dV since there is no other decision variable), the optimum condition must again be satisfied. Therefore, $dEU/dV = 0$ before and after the change, justifying the zero on the right-hand side of Eq. (8.7). However, both the disturbance dI and the adjustment dV do impinge on the expected marginal utility of prevention $\partial EU / \partial V$. Applying the implicit function theorem, one therefore has

$$\frac{\partial^2 EU}{dV^2} \cdot dV + \frac{\partial^2 EU}{\partial V \partial I} \cdot dI = 0. \tag{8.7}$$

This can be solved to yield

$$\frac{dV}{dI} = -\frac{\partial^2 EU/\partial V \partial I}{\partial^2 EU/\partial V^2} \lessgtr 0. \tag{8.8}$$

The sign of this expression indicates whether preventive effort increases or decreases in response to more insurance coverage. For simplicity, the sufficient condition for the maximization of expected utility is assumed to be satisfied in the neighborhood of the initial optimum, i.e., $\partial^2 EU/\partial V^2 < 0$. Therefore, the sign of expression (8.8) coincides with the sign of the mixed derivative in the numerator. If that sign is negative, ex-ante moral hazard is predicted, i.e., $dV/dI < 0$. Referring back to Eqs. (8.5), (8.3a), and (8.3b), and writing $\pi' := \pi'(V)$ as well as $P' := P'(I)$ to simplify notation, one obtains

$$\begin{aligned}\frac{\partial^2 EU}{\partial V \partial I} &= \pi'\{v'[1](-P'+1) - v'[2](-P')\} - \pi v''[1](-P'+1) \\ &\quad -(1-\pi)v''[2](-P') \\ &= \pi'\{\underbrace{v'[1](1-P')}_{(-)} + \underbrace{v'[2]P'}\} - \underbrace{\pi v''[1](1-P')}_{(+)} + \underbrace{(1-\pi)v''[2]P'}_{(-)} \\ &\lessgtr 0. \end{aligned} \tag{8.9}$$

Therefore, the sign of dV/dI is ambiguous. Closer inspection of the individual terms serves to explain this ambiguity and to reduce it. Note first that no one would rationally buy an insurance contract with $P'(I) > 1$, meaning that it costs more than 1 MU (monetary unit) of additional premium for 1 MU additional coverage (which is obtained only in the loss state). Therefore, $(1 - P') > 0$. Now with increasing insurance coverage, more wealth is transferred to the loss state, causing prevention ($\pi' < 0$) to be of less interest—a moral hazard effect. This is what the first term of Eq. (8.9) indicates. However, wealth in the loss state is valued at a decreased marginal utility due to risk aversion ($v''[1] < 0$ in the second term), causing the preceding effect to be counteracted and possibly reversed. Finally, the third term involves risk aversion in the no-loss state (where the IB only pays the additional premium without receiving a benefit) which serves to increase the marginal utility of wealth. This again works against costly prevention. If risk aversion is weak ($v''[1] \to 0$) or if $\pi \to 0$, the second term is dominated by the negative first and third, resulting in

$$\frac{dV}{dI} \begin{cases} \lessgtr 0 & \text{in general;} \\ < 0 & \text{if risk aversion is weak or the probability of loss small.} \end{cases} \quad (8.10)$$

▶ **Conclusion 8.2** While more comprehensive but still partial insurance coverage cannot be said to induce *ex-ante* moral hazard in general, it does so if risk aversion of the IB is weak or if the probability of loss is small.

Turning to the IC, premium calculation must take into account *ex-ante* moral hazard effects since they cause the frequency of payments to exceed the value estimated originally. Failure to do so would drive the IC into insolvency. Due to its assumed inability to verify moral hazard in the individual case, the IC lacks the justification for reducing benefits when suspicious claims are submitted. The only solution is to increase the premium. Note that this increase in principle should reflect the individual change in π of an IB, which while not observable initially becomes estimable as time goes by. For simplicity, however, $\pi'(V) = d\pi/dV$ will be written without a subscript i characterizing the (type of) consumer. In order to at least recover losses on average, the premium charged must satisfy the condition,

$$P(I) = \pi\{V(I)\} \cdot I. \quad (8.11)$$

This formula reflects the publicly available information that the probability of loss depends on preventive effort V. This in turn depends on the amount of coverage I contracted, which can be observed by the IC. Therefore, it is reasonable to structure the premium according to the amount of coverage purchased, knowing that additional coverage likely induces *ex-ante* moral hazard. The adjustment of premium in response to higher coverage is given by

$$P' := \frac{dP}{dI} = \underset{(+)}{\pi\{V(I)\}} + \underset{(-)}{\frac{d\pi\{V(I)\}}{dV}} \cdot \underset{(-)}{\frac{dV}{dI}} \cdot I > \pi. \quad (8.12)$$

In this formula, the first term reflects the marginally fair premium (see Sect. 3.3.3). However, there is a second term amounting to a surcharge for *ex-ante* moral hazard. This surcharge amounts to a loading and is greater,

- the more effective preventive effort would be ($d\pi/dV$ large in absolute value);
- the more marked is moral hazard if it exists (dV/dI negative and large in absolute value);
- the more comprehensive is coverage I (with $I \leq L$).

Since the premium now exceeds its actuarially fair value at the margin, coverage purchased will not be comprehensive anymore, in keeping with the theory developed in Sect. 3.5.

For the IC, it is important to know whether the surcharge for *ex-ante* moral hazard should be a progressively or a regressively increasing function of the amount of coverage purchased. This calls for examining the second derivative,

$$P'' := \frac{d^2 P}{dI^2} = \frac{d\pi}{dV} \cdot \frac{dV}{dI} + \left\{ \frac{d^2\pi}{dV^2} \cdot \frac{dV}{dI} \right\} \cdot \frac{dV}{dI} \cdot I + \frac{d\pi}{dV} \frac{d^2V}{dI^2} \cdot I + \frac{d\pi}{dV} \frac{dV}{dI}. \tag{8.13}$$

Factoring out terms in dV/dI and d^2V/dI^2 yields

$$P'' = 2\underbrace{\frac{d\pi}{dV} \cdot \frac{dV}{dI}}_{(+)} + \underbrace{\frac{d^2\pi}{dV^2} \left(\frac{dV}{dI}\right)^2}_{(+)} \cdot I + \underbrace{\frac{d\pi}{dV} \cdot \frac{d^2V}{dI^2}}_{(+/-)} \cdot I. \tag{8.14}$$

The first term is positive as soon as there is *ex-ante* moral hazard ($dV/dI < 0$) since $d\pi/dV < 0$ always. The second term is positive as well because the marginal productivity of prevention decreases with preventive effort [see Fig. 8.1 or assumptions (8.1) again]. The third term is also positive provided moral hazard becomes more acute with increasing insurance coverage ($d^2V/dI^2 < 0$). In sum, the result is

$$P'' > 0, \text{ if } \frac{d^2V}{dI^2} \begin{cases} < 0 \\ = 0 \\ > 0, \text{ but small.} \end{cases} \tag{8.15}$$

▶ **Conclusion 8.3** *Ex-ante* moral hazard calls for a surcharge to the fair premium that rises progressively with increased insurance coverage under rather general conditions. This loading causes a reduction of coverage purchased, thus strengthening the incentive for prevention and limiting *ex-ante* moral hazard.

It is worth noting that ICs in a competitive market without information exchange cannot impose a progressively increasing premium [see Pauly (1974), Eisen (1989)]. The reason is that by failing to disclose coverage purchased elsewhere, the IB can always benefit from the lowest premium category applicable to minimum coverage. Since this weakens incentives to limit *ex-ante* moral hazard, each consecutive IC suffers from a negative externality emanating from previously signed contracts. This can be prevented if ICs are permitted to exchange information about insurance coverage contracted with competition. More generally, pure price competition in insurance is not compatible with prevention (this will be shown in greater detail in Sect. 8.2.3).

8.2.2.2 *Ex-Ante* Moral Hazard with Regard to the Amount of Loss

In this section, the probability of loss π is a constant in order to focus on the relationship between prevention V and loss L. Thus, the IB is assumed capable of making effort designed to limit the amount of damage (sometimes called palliative intervention). The pertinent decision must be made *ex ante*, i.e., before the possible occurrence of the loss. Examples are fire doors that prevent fires from spreading through the building, or bulkheads that keep water from flooding the entire ship. As in the previous section, the cost of prevention thus has to be borne in both the loss and the no-loss state, making the IB's decision problem read,[5]

$$EU = \pi \cdot v[W_0 - L(V) - V - P(I) + I] + (1 - \pi) \cdot v[W_0 - V - P(I)],$$
$$\text{with } L' := L'(V) < 0. \qquad (8.16)$$

Let the amount of coverage I and hence premium $P(I)$ again be predetermined, permitting the IB to optimize preventive effort only. Using Eqs. (8.3a) to (8.3c) and in analogy with (8.4), one has for the first-order condition

$$\frac{dEU}{dV} = \pi \cdot v'[1](-L' - 1) - (1 - \pi)v'[2] \begin{cases} < 0 \text{ if } V^* = 0 \\ = 0 \text{ if } V^* > 0. \end{cases} \qquad (8.17)$$

For an interior solution with $V^* > 0$ to obtain, the first term must be positive, implying $L' < -1$. The marginal return to prevention therefore must be high enough to exceed its marginal cost (equal to one by assumption). At least for the case of comprehensive coverage ($I = L$), this requirement turns out to be rather stringent. One then has $v'[1] = v'[2]$, causing (8.17) in the equality case to simplify to

$$-L' = \frac{1}{\pi}. \qquad (8.18)$$

Since π is small in most lines of insurance, the marginal productivity of prevention given full coverage must obtain a value far above 1 to make an effort at loss reduction worthwhile. This means that under the influence of insurance, only the most productive variants of prevention are undertaken. Still, preventive effort is positive, in contradistinction to Conclusion 8.1 relating to *ex-ante* moral hazard affecting the probability of loss.

Now let the optimality condition (8.17) for $V^* > 0$ be disturbed by an increase in insurance coverage $dI > 0$. In full analogy with the comparative static analysis of Eqs. (8.7) and (8.8), one needs only to examine the sign of the mixed second derivative. From (8.17), one obtains

$$\frac{d^2 EU}{dV dI} = \pi \cdot \underset{(-)}{v''[1]} \underset{(+)}{(-L' - 1)} \underset{(+)}{(1 - P')} + (1 - \pi) \cdot \underset{(-)}{v''[2]} \cdot \underset{(+)}{P'} < 0 \qquad (8.19)$$

and therefore

$$\frac{dV}{dI} < 0. \qquad (8.20)$$

---

[5]Note the similarity with Sect. 3.5. However, prevention rather than the demand for insurance is the variable of interest here.

8.2 Moral Hazard

Ex-ante moral hazard with regard to the size of the loss is always to be expected. Preventive effort undertaken by consumers and insurance coverage are substitutes in this case [see, e.g., Ehrlich and Becker (1972)].

▶ **Conclusion 8.4** Preventive effort designed to reduce the amount of loss is predicted to always decrease in response to insurance coverage. However, full coverage does not necessarily wipe out preventive effort entirely.

8.2.3 Market Equilibrium with *Ex-Ante* Moral Hazard

As a benchmark, equilibrium in a competitive insurance market is analyzed first. The exposition follows Eisen (1990); for simplicity, the amount of loss is considered to be exogenous again. As stated below Conclusion 8.3, a progressively increasing premium cannot be imposed on a competitive insurance market without exchange of information between the IC. Excluding such an exchange, only a fixed premium rate $\bar{q}$ for a given indemnity I is sustainable. Since the individual IC cannot influence $\bar{q}$, one has

$$\bar{P} = \bar{q} \cdot I, \text{ implying } \frac{\partial \bar{P}}{\partial I} = \bar{q}. \tag{8.21}$$

Given costless market entry and exit, it is impossible to achieve a profit in expectation, denoted $E\Pi$. Neglecting administrative expense, the premium attainable in the market therefore cannot exceed the expected value of benefits paid,

$$E\Pi(\bar{P}, I) = \bar{P} - \pi\{V(\bar{P}, I)\} \cdot I = 0. \tag{8.22}$$

This notation makes clear that the probability of loss π depends on preventive effort V, which in turn varies with the amount of coverage I and potentially with the premium $\bar{P}$ (e.g., as the consequence of a substitution effect). For the IC who decides the amount of coverage, the first-order condition reads[6]

$$\begin{aligned} \frac{dE\Pi}{dI} &= \frac{\partial \bar{P}}{\partial I} - \left\{ \frac{\partial \pi}{\partial V} \cdot \frac{\partial V}{\partial \bar{P}} \cdot \frac{\partial \bar{P}}{\partial I} + \frac{\partial \pi}{\partial V} \cdot \frac{dV}{dI} \right\} I - \pi \\ &= \frac{\partial \bar{P}}{\partial I} - \frac{\partial \pi}{\partial V} \left\{ \frac{dV}{\partial \bar{P}} \cdot \frac{\partial \bar{P}}{\partial I} + \frac{dV}{dI} \right\} I - \pi = 0. \end{aligned} \tag{8.23}$$

Note that the expression in the bracket must be zero for the market equilibrium to be sustained. The increase in coverage induces a change in preventive behavior $dV/dI < 0$ that would throw the market out of equilibrium unless neutralized. For maintaining equilibrium, it takes an adjustment of the market premium $\partial \bar{P}/\partial I > 0$ that in turn causes a sufficient balancing increase of preventive effort $\partial V/\partial \bar{P} > 0$.

---

[6]The notation dV/dI is designed to recall that this moral hazard effect is the result of the comparative static analysis as performed in Eqs. (8.7) to (8.10).

The smaller this effect, the more marked the premium adjustment needs to be. Since $\partial \bar{P}/\partial I = \bar{q}$ from (8.21) holds for an arbitrary market premium, the premium function becomes

$$\bar{q} = \pi. \tag{8.24}$$

Under the pressure of competition, there is no scope for price differentiation according to the amount of insurance coverage demanded. Therefore, the premium function must be linear with slope $\bar{q}$, which is equal to the probability of loss (amounting to a marginally fair premium).

The resulting market equilibrium is depicted in Fig. 8.2, using (I, P)-space rather than the (W_1, W_2)-space of Sect. 3.2. In principle, the equation for the indifference curve, derived from $EU = \pi \cdot v[1] + (1 - \pi) \cdot v[2]$, also contains terms in V. However, behavior of consumers is given by the function $V(\bar{P}, I)$ in Eq. (8.22), while the total effect of an adjustment in V to a change in I is given by the expression in the bracket of (8.23). As argued above, the value of that expression is zero if market equilibrium is to be maintained, permitting to neglect the terms in V. Therefore, the equation of the indifference curve simply reads,

$$dEU = 0 = \pi \cdot v'[1]dI - \pi \cdot v'[1] \cdot d\bar{P} - (1 - \pi)v'[2] \cdot d\bar{P}. \tag{8.25}$$

In market equilibrium, the marginal rate of substitution must equal the marginal rate of transformation. The marginal rate of substitution is found by solving (8.25),

$$\frac{d\bar{P}}{dI} = \frac{\pi \cdot v'[1]}{\pi \cdot v'[1] + (1 - \pi)v'[2]} > 0. \tag{8.26}$$

The IB is therefore willing to pay more premium $d\bar{P}$ for additional indemnity dI. As to the marginal rate of transformation, it is given by π in Eq. (8.24) because this premium function guarantees that the IC can cover its loss payments in expectation. Market equilibrium calls for equating the two rates,

$$\frac{\pi \cdot v'[1]}{\pi \cdot v'[1] + (1 - \pi)v'[2]} = \pi. \tag{8.27}$$

Dividing the numerator and denominator on the left-hand side by $v'[1]$, one obtains the equilibrium condition,

$$\frac{\pi}{\pi + (1 - \pi)\frac{v'[2]}{v'[1]}} = \pi. \tag{8.28}$$

This condition is only satisfied if $v'[1] = v'[2]$. The equality of the marginal utilities of wealth implies equality of wealth in both states, and therefore, full insurance coverage. However, Conclusion 8.1 states that the optimal level of preventive effort is zero in this case.

8.2 Moral Hazard

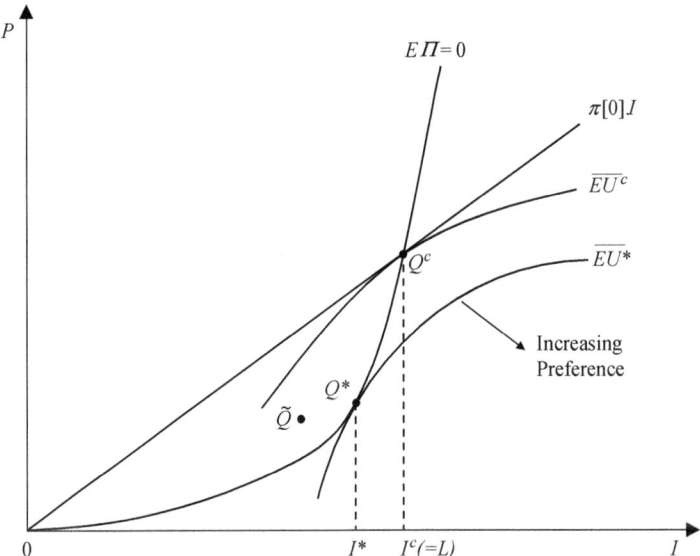

Fig. 8.2 Equilibrium in a competitive insurance market with *ex-ante* moral hazard

▶ **Conclusion 8.5** Given the conditions of perfect price competition without information exchange between the IC, the market equilibrium is characterized by full insurance and hence maximum *ex-ante* moral hazard (zero preventive effort).

Figure 8.2 illustrates the equilibrium condition (8.28). The indifference curves have positive slope in (I, P)-space, in keeping with (8.26). Any equilibrium must lie on the $E\Pi = 0$ locus, which as a rule calls for premiums that progressively increase with coverage I in order to balance moral hazard effects (see Conclusion 8.3). The IBs would find their optimum on this locus, e.g., at point Q^*. It indicates partial coverage in return for a relatively low premium rate.

However, given fully competitive conditions with no information exchange between the IC, the premium function is linear [see Eq. (8.24) again]. It has a relatively steep slope $\pi[0]$ because Conclusion 8.5 predicts full coverage and hence zero prevention under these conditions. Optimization by the IB requires the indifference curve to be tangent to the straight line representing the premium function with zero expected profit. Therefore, the equilibrium is at point Q^c. It has the following properties.

- The marginal rates of substitution and transformation are equal [see condition (8.27)] in order for the solution to qualify as an optimum for the IB;
- The indifference curve $\overline{EU^c}$ through Q^c is lower valued than the indifference curve $\overline{EU^*}$ through the Pareto-optimal point Q^*, indicating that the competitive equilibrium is Pareto-suboptimal;
- Points below the premium function $\pi[0] \cdot I$ and below Q^c are associated with positive expected profits. This causes the incumbent IC to offer additional coverage

and additional ICs to enter the market, resulting in an expansion of underwriting capacity. Therefore, these points cannot represent an equilibrium;
- Points above the premium function $\pi[0] \cdot I$ and above Q^c cannot represent an equilibrium either because the IC would suffer losses in expectation. Excess underwriting capacity would be wiped out, with some ICs exiting the market.

While point Q^c is an equilibrium, it is not a global optimum in most cases. A point like $\tilde{Q}$ indicates an improvement over Q^c for both the IB and the IC. However, it is associated with profits on the part of the IC which would be washed away by market entry. As to the IB, they would have to increase their preventive effort (causing the premium function to rotate down), for which they lack the incentive given asymmetric information. This failure to reach a Pareto-superior equilibrium amounts to a violation of the *two theorems of welfare economics*, stating that (1) a competitive equilibrium is Pareto-optimal, and (2) a Pareto-efficient allocation is a competitive equilibrium. Thus, a third party (specifically the government), disposing of additional information about the behavior of IB, might intervene to bring about a Pareto improvement [see Arnott and Stiglitz (1990)].

▶ **Conclusion 8.6** While *ex-ante* moral hazard in an insurance market with full price competition does not jeopardize the existence of an equilibrium, it does result in the equilibrium being Pareto-suboptimal due to insufficient prevention.

The welfare losses (expressed by the difference in expected utility associated with the pertinent indifference curves $\overline{EU^c}$ and $\overline{EU^*}$ of Fig. 8.2, respectively) are likely to be overstated by Conclusion 8.6, however. The IC operating in the market are aware of the fact that approaching comprehensive coverage boosts *ex-ante* moral hazard effects. Their risk of calculating too low a premium increases, and with it their interest in imposing a premium function progressively increasing in benefits I, like the locus $E\Pi = 0$ in Fig. 8.2. If an insurance association already exists, the extra cost of establishing an information exchange is low. The participating ICs only need to know the amount of coverage an IB has obtained from other ICs. The identity of the other ICs did not have to be disclosed, which preserves the privacy of insurers' business plans. However, this exchange must be made compulsory because it would be advantageous for each IC not to share the information while charging a marginally higher premium for its part of the coverage provided. This type of information sharing is mandated in major insurance markets because their laws prohibit excess insurance coverage (see Sect. 11.3.1). Nevertheless, excess insurance coverage does occur, e.g., in auto insurance, where vehicles can be insured for their purchase value rather than their (lower) current value, inducing *ex-ante* moral hazard effects [see Eisen (1990)].

To the extent that information exchange can be organized, charging a non-linear risk premium is therefore possible. The fact that an IC cannot observe preventive effort V of an IB does not stop it from charging for presumed moral hazard effects. However, the strength of these effects likely differs between individuals, causing ICs to charge premiums that are not fully scaled to risk, at least during the single period analyzed here (abstracting from so-called experience rating in a multi-period context, see Sect. 8.3.3). Thus, while gauging moral hazard effects enables ICs to write contracts that improve over the competitive equilibrium Q^c, the individual Pareto optimum Q^* of Fig. 8.2 typically is not attained.

8.2.4 Empirical Evidence on *Ex-Ante* Moral Hazard

Ex-ante moral hazard is not easily measured using insurance data. An important reason is the imperfect link between losses and claims. Indeed, the IC is informed about a loss only if the IB decides to present a claim. This decision is influenced by weighing costs and returns, which depend on the situation. For instance, a business strapped for liquidity may seek payment for a loss it would consider a petty claim not worth bothering about under normal conditions. Presenting fraudulent claims may be advantageous if the probability of being detected and sanctioned is low enough. However, major losses are recorded with sufficient precision to result in reliable frequency counts.

Specifically, losses in building and content insurance should be rather precisely measured in a country like Switzerland. Contents insurance covers furniture, durable consumer goods, and installations. Written by private ICs, it complements fire insurance. In the majority of the 26 cantons (member states) of Switzerland, fire insurance for buildings is mandatory. Claims regarding contents can be submitted only if there was fire in the building, an occurrence that is closely examined by fire police.

An analysis of claims data provided by a Swiss IC therefore should reflect π (the frequency of loss in the contents line) with great accuracy [see Bonato and Zweifel (2002)]. The IBs in this study are not households but businesses. The variable to be analyzed is D_LOSS, a dummy variable taking on the value of 1 if there was a claim relating to fire damage to contents during a year (and 0 otherwise). Therefore, D_LOSS can be interpreted as the realization of loss probability π, which approaches zero if the risk factors considered below are absent or at a low level. Conversely, π moves away from zero when these factors favor the occurrence of a loss.

For estimation, one typically uses the so-called Probit model [see, e.g., Greene (2003), Sect. 21.6]. The assumption is that the values of π describe a sigmoid function, reflecting the cumulative distribution function of a normal random variable. If the estimated probability given the values of the risk factors reaches $\pi \geq 0.5$, the indicator D_LOSS has the value one; if it is below 0.5, D_LOSS is set to zero. This dependent variable is related to almost 40 explanatory variables gleaned from the records of the insurance company. The Probit procedure determines the coefficients pertaining to the explanatory variables in a way similar to regression analysis such that the probability of observing the actual $\{0, 0, 0, 0, 1, 0, 0, ...\}$ values of D_LOSS is maximized for this sample (Maximum Likelihood Estimation).

In Table 8.2, selected results of two such Probit estimations are presented for the years 1993 to 1995. The left-hand side panel is based on information relating to the current year only. The right-hand side panel complements the set of explanatory variables with $BEFORE3$, the share of years with fire damage over the three pre-

Table 8.2 *Ex-ante* moral hazard in fire contents insurance, Switzerland (1993–1995)

| Explanatory variable | Dependent variable: $D\_LOSS = 1$ if claim, $= 0$ otherwise ||||||
| | Only information of the current year ||| Including information on losses of 3 previous years |||
	Coeff.	t-value	Marginal effect	Coeff.	t-value	Marginal effect
MAX_LOSS	0.03***	5.98	0.001	0.02***	3.57	0.001
SPECRISK	0.23**	2.17	0.006	0.12	0.94	0.005
D_COMP_B	−0.43***	−4.23	−0.010	−0.30***	−2.73	−0.014
NUMRISK	0.09**	4.05	0.002	0.07**	2.11	0.003
BEFORE3	–	–	–	1.28***	7.61	0.058
+31 more variables						
LR index, corrected			0.067			0.071
Observations			24,770			14,430
No. insured			12,855			7,092

*** (**): Statistically significant at the 1% (5%) level. *MAX_LOSS*: Sum insured against fire in 100,000 CHF (1CHF = 0.8 US$ at 2000 exchange rates); *SPECRISK*: Number of special risks covered; *D_COMP_B*: = 1 if fire insurance of buildings compulsory in the canton, = 0 otherwise; *NUMRISK*: Number of other risks [water, glass, theft, business interruption] voluntarily included; *BEFORE3*: Share of the previous 3 years with a loss
Source Bonato and Zweifel (2002), Table A.2

ceding years as an indicator of loss experience. Most of the 31 explanatory variables not shown are dummies serving to distinguish the three loss years, seven regions, six types of customers, and nine industries.

Starting with the left-hand side panel of Table 8.2, one notices that the statistical fit is not very good, as shown by the value of the corrected likelihood ratio (LR) index (which is similar to a coefficient of determination, R^2). This index is to be interpreted as follows. The value of the objective function obtained by maximizing the probability of having drawn the sample, without using the explanatory variables, is a mere 6.7% lower than the value obtained with the help of explanatory variables.[7]

Since Probit is a non-linear estimation procedure, its coefficients cannot be interpreted as marginal changes of loss probability π. Rather, a transformation is necessary for calculating these marginal effects (typically evaluated at the means of explanatory variables). Calculated marginal effects on π therefore appear in a separate column to the right of the t-values.

The four selected regressors relate to preventive effort and *ex-ante* moral hazard in the following way.

---

[7] The correction of the index relates to the number of so-called degrees of freedom. This is the number of independent observations not subject to a restriction. Since every estimated parameter to be estimated imposes a restriction, the number of degrees of freedom is reduced from 24,770 to 24,734.

- *Sum insured (MAX_LOSS)*: The higher the sum insured and hence maximum loss covered, the more comprehensive *ceteris paribus* insurance coverage I, and the stronger are moral hazard effects. This prediction derives from Conclusion 8.1 if one notes that in the case of commercial IBs, risk aversion is unlikely to be very marked because they have other possibilities to mitigate risks (see Sect. 5.3). Also, wealth can be assumed to indeed constitute the only argument of the risk utility function here. However, contrary to the theoretical development, loss L is not a fixed quantity. A high value of *MAX_LOSS* therefore could indicate a potential loss that is (much) higher than the insured value, resulting in a considerable amount of cost sharing by the IB. The partial relationship between *D_LOSS* and *MAX_LOSS* could therefore turn out negative rather than positive as predicted. Yet, note that the number of other risks covered (*NUMRISK*, see below) enters as a separate explanatory variable. A high number of other risks covered (such as theft or interruption of business operations) usually points to a high value of the building insured. *NUMRISK* therefore largely controls for the unknown potential loss, giving rise to the expectation of a positive partial relationship between *D_LOSS* and *MAX_LOSS*, reflecting *ex-ante* moral hazard.

  This interpretation is supported by the entries of Table 8.2 since *MAX_LOSS* displays a significantly positive relationship with the probability of loss π. The marginal effect of 0.001 appears to be minimal; however, it states that a building with the insured sum of an additional 1 million CHF ($=10 \cdot 10^5$ Swiss francs) has an increased probability of damage to contents caused by fire amounting to 1 percentage point ($0.01 = 10 \cdot 0.001$), i.e., of 0.028 rather than the sample mean of 0.018 (thus, $0.28 \cdot 10^5$ rather than $0.18 \cdot 10^5$ CHF). Aggregated over 12,855 contracts, this becomes an extra 12.9 million CHF ($= 12.855 \cdot 0.01 \cdot 10^5$ Swiss francs) per year of losses paid.

- *Number of special fire risks covered (SPECRISK)*: The greater the number of special fire risks covered, the greater the likelihood of an *ex-ante* moral hazard effect with respect to at least one of them. Therefore, *SPECRISK* has the same predicted impact as an increase in insurance coverage. Since the IB is a business, however, *SPECRISK* could also reflect the complexity of its equipment, resulting in a greater number of risk processes that are aggregated in the single dependent variable *D_LOSS*. In that case a positive relationship between *SPECRISK* and *D_LOSS* would have nothing to do with *ex-ante* moral hazard. The estimated effect is indeed positive and statistically significant. Since the maximum value of *SPECRISK* is 3, the increase in *D_LOSS* cannot exceed 1.8 percentage points ($0.018 = 3 \cdot 0.006$), however.

- *Compulsory fire insurance for buildings (D_COMP_B)*: The majority of Swiss cantons mandate fire insurance for buildings. This mandate likely goes along with an amount of fire prevention (e.g., through building codes) in excess of what the IBs would have opted for themselves. This excess might cause them to reduce their non-observable preventive effort (in the guise of everyday behavior), resulting in a probability of loss that is higher than in other cantons. On the other hand, there are close technological relationships between the building and its contents since prevention of fire in the building also benefits furniture and equipment. The

significantly negative coefficient of *D_COMP_B* supports this alternative view, suggesting that the probability of a fire damage to contents is one percentage point (0.01) lower in cantons with compulsory fire insurance of buildings. Apparently, there is a spillover of control of moral hazard from public mandatory insurance to private insurance here.
- *Number of other risks insured* (*NUMRISK*): Losses due to water, breakage of glass, and theft as well as business interruptions can be additionally insured on a voluntary basis. Given the sum insured, this variable therefore indicates a higher degree of complexity of operations and hence potential loss, relieving the interpretation of *MAX_LOSS* of some of its ambiguity (see above). Its estimated coefficient is positive, as expected.

The right-hand side panel of Table 8.2 contains an estimate (with reduced sample size) that also takes into account the loss experience of three preceding years. The explanatory variable *BEFORE3* indicates the share of years with at least one loss during those three years, causing it to take on the values [0; 0.33; 0.67; 1]. This information contributes significantly to the explanation of *D_LOSS*. More importantly, the estimated marginal effect of *MAX_LOSS* remains unchanged at 0.001. Since *BEFORE3* controls for unmeasured characteristics of the IB (who suffered losses earlier), this stability speaks in favor of the positive relationship between *D_LOSS* and *MAX_LOSS* reflecting a moral hazard effect.

▶ **Conclusion 8.7** There is evidence of *ex-ante* moral hazard in Swiss contents insurance, combined with a beneficial spillover effect from preventive effort imposed by mandatory fire insurance of buildings.

8.2.5 *Ex-Post* Moral Hazard in Short-Term Disability Insurance

8.2.5.1 The Model and Its Predictions

Ex-post moral hazard prevails when the IB has an unobservable influence on the size of a loss after its occurrence. This is possible in those lines of insurance where benefits are not fixed in advance but increase with the amount of the loss, especially in disability, health, and liability insurance. There, an IC often employs specialized agents for claims assessment. In health insurance, this is the task of physicians, who however also act on behalf of their patients.

In the case of disability, this *double agency of physicians* creates scope for workers to seek out a physician who will accommodate them to the greatest possible extent. This is the basic hypothesis advanced by Dionne and St-Michel (1991), in whose model a worker hopes to obtain prolonged disability payment by finding an accommodating physician.

Let e denote a worker's search effort, which is assumed for simplicity to vary in exact proportion with the length of paid leave for disability obtained in this way. If successful, search therefore conveys a benefit amounting to $\alpha l e$, because disability leave pays αl per day, with l denoting the wage rate and α ($0 < \alpha < 1$), the rate of income replacement. On the other hand, search has a cost $C(e)$ with marginal cost

8.2 Moral Hazard

$C'(e) > 0$. With probability ρ, however, the effort of the worker is in vain, i.e., it fails to result in a medical report certifying disability. Note that the cost of search has to be borne in this state as well. Assuming expected utility maximization in terms of final wealth, the worker's decision problem reads,

$$\max_{e} EU(W) = \rho \cdot v\{W_0 - C(e)\} + (1 - \rho) \cdot v\{W_0 + \alpha l e - C(e)\} \quad (8.29)$$

$C:$ Cost of search, depending positively on the duration of search e and hence paid short-term disability leave obtained, with $C'(e) > 0$;
$e:$ Duration of search as well as length of short-term disability leave;
$v(\cdot):$ Risk utility function, with $v' > 0$, $v'' < 0$;
$l:$ Wage rate of the worker;
$W_0:$ Initial wealth, predetermined;
$\alpha:$ Share of labor income that is paid as a benefit in the case of short-term disability (replacement rate, $0 < \alpha < 1$).
$\rho:$ Probability of not finding an accommodating physician, hence $(1-\rho)$: probability of success.

The first-order condition for an interior optimum is

$$\frac{dEU}{de} = 0 = -\rho \cdot v'[W_0 - C(e)]C'(e) + (1 - \rho)v'[W_0 + \alpha l e - C(e)]\left(\alpha l - C'(e)\right). \quad (8.30)$$

Here, $v'[\cdot]$ indicates that the marginal utility has to be evaluated at the respective value of final wealth. The first term reflects the (utility-weighted) expected value of the marginal cost of searching for an accommodating physician. The second term corresponds to the expected utility value of the marginal return to search, given by the short-term disability benefit minus the cost of extra search. For an interior solution, it needs to be positive; however, this implies that the marginal return to search effort in the guise of an additional day of paid leave (αl) exceeds the marginal cost C' of attaining it, i.e., $(\alpha l - C'(e)) > 0$.

An increase in the replacement rate ($d\alpha > 0$) may induce workers to step up their search effort ($de > 0$), which by assumption results in *a longer disability leave*. Since the work accident already happened, $de/d\alpha > 0$ would be an instance of *ex-post* moral hazard. Therefore, let $d\alpha$ be the exogenous impulse disturbing the first-order condition (8.30). This disturbance can only be neutralized by adjustment of effort by de since e is the only decision variable. Because the first-order condition (8.30) must again be satisfied after disturbance and adjustment, the effects of the two changes sum to zero. Application of the implicit function theorem therefore results in the comparative-static equation,

$$\frac{\partial^2 EU}{\partial e^2}de + \frac{\partial^2 EU}{\partial e \partial \alpha}d\alpha = 0. \quad (8.31)$$

Solving for $de/d\alpha$, one obtains

$$\frac{de}{d\alpha} = -\frac{\partial^2 EU}{\partial e \partial \alpha} \Big/ \frac{\partial^2 EU}{\partial e^2}. \tag{8.32}$$

The second-order condition for a maximum of the objective function is $\partial^2 EU/\partial e^2 < 0$. Assuming this condition to be satisfied for simplicity, Eq. (8.32) shows that the sign of $de/d\alpha$ coincides with the sign of $\partial^2 EU/\partial e \partial \alpha$. In view of (8.30) and writing C' for $C'(e)$, one obtains for this mixed derivative,

$$\frac{\partial^2 EU}{\partial e \partial \alpha} = (1-\rho)v''[W_0 + \alpha l e - C(e)] \cdot le \cdot (\alpha l - C')$$
$$+ (1-\rho)v'[W_0 + \alpha l e - C(e)]l. \tag{8.33}$$

This expression can be related to the measure of absolute risk aversion $R_A := -v''[\cdot]/v'[\cdot] > 0$, with the derivatives evaluated at the appropriate level of final wealth, i.e., $W = W_0 + \alpha l e - C(e)$ (see Sect. 3.2). Expanding the first term by v' and rearranging, one has,

$$\frac{\partial^2 EU}{\partial e \partial \alpha} = (1-\rho)v' \cdot \left\{ \frac{v''}{v'} \cdot le(\alpha l - C') + l \right\}$$
$$= (1-\rho)v' \cdot l \cdot \left\{ \frac{v''}{v'} e(\alpha l - C') + 1 \right\}$$
$$= (1-\rho)v' \cdot l \cdot \{1 - R_A \cdot e(\alpha l - C')\}. \tag{8.34}$$

This expression is positive if the worker is risk-neutral, i.e., if $R_A = 0$. In this case, the prediction is unambiguously $de/d\alpha > 0$, i.e., an *ex-post* moral hazard effect. If the worker is risk-averse ($R_A > 0$), the sign of (8.34) hinges on the degree of risk aversion. Provided R_A is sufficiently small, i.e., $0 < R_A < 1/e(\alpha l - C')$, the mixed derivative (8.34) is still positive and hence $de/d\alpha > 0$. Hence, one obtains

$$\frac{de}{d\alpha} > 0 \text{ if } \begin{cases} R_A = 0 \\ R_A < \frac{1}{e(\alpha l - C')}. \end{cases} \tag{8.35}$$

Therefore, *risk-neutral* and *weakly risk-averse* workers are predicted to exhibit *ex-post* moral hazard. Provided moral hazard obtains, Eq. (8.34) indicates an especially strong effect if the probability of success $(1 - \rho)$ is high. Arguably, the probability of finding an accommodating physician is high if the physician issuing the medical report does not fear being challenged. Being challenged is unlikely in the case of an ambiguous diagnosis such as "lower back pain". Ambiguous diagnoses are therefore predicted to induce marked moral hazard effects. Conversely, $(1 - \rho)$ goes to zero if the physician issuing the medical report could be easily challenged by a second opinion. For instance, this would be the case for a "broken arm" diagnosis, which can be checked at very low cost. A small *ex-post* moral hazard effect is predicted in this case.

Table 8.3 Categories of diagnoses and degrees of observability

Severity of accident	Degree of observability	
	High: Diagnosis of the physician easy to verify (E)	Low: Diagnosis of the physician difficult to verify (D)
minor: MI	MI_E (3 diagnoses: contusion, small amputation, rash)	MI_D (2 diagnoses: lower back pain with and without injury)
major: MA	MA_E (1 diagnosis: bone fracture)	MA_D (2 diagnoses: spinal column with and without injury)

▶ **Conclusion 8.8** In the context of insurance for short-term disability (including workers' compensation), *ex-post* moral hazard in response to a higher replacement rate is predicted to be marked among workers with limited risk aversion and in the case of ambiguous diagnoses.

8.2.5.2 *Ex-Post* Moral Hazard in Short-Term Disability Insurance: Empirical Evidence

Monopolistic insurance schemes are particularly well suited for investigating the influence of moral hazard. Their data are not confounded by risk selection effects that may characterize competitive insurers (see Sect. 8.3 below). One such monopolistic scheme is the Commission des accidents du travail Québec (Workplace Accidents Insurance of Quebec, Canada). Dionne and St-Michel (1991) analyzed the data of this public insurer covering the years 1978–1982. At the beginning of 1979, a new law brought about an increase in the income replacement ratio of most insured workers. This corresponds to the exogenous shock $d\alpha > 0$ introduced in Eqs. (8.31) to (8.35) of the preceding section.

To test the impact of asymmetric information, the authors distinguish two types of accident. In the case of category E, the degree of observability for the insurer is comparatively high because it is easy for any physician to verify the condition stated on the medical report. This causes the probability $(1 - \rho)$ of finding an accommodating physician to be low (see Sect. 8.2.5.1 above). By way of contrast, conditions of type D are difficult to diagnose and verify, resulting in a low degree of observability. Therefore, the probability of success $(1 - \rho)$ from the workers' point of view is considerable, which magnifies a given moral hazard effect [see Eq. (8.34)]. In addition, the authors distinguish between minor (MI) and major (MA) accidents, thus creating the four categories of Table 8.3. However, with regard to *ex-post* moral hazard, it is only the first distinction (D vs. E) that should be of relevance. This leads the authors to formulate the two following hypotheses.

H1: For a given severity of accident, the effect of additional coverage (i.e., the *ex-post* moral hazard effect) is especially marked if the diagnosis is difficult to perform and to verify (type E rather than D).

To test this hypothesis, denote by $b(\cdot)$ the absolute value of the pertinent regression coefficient that represents the effect of a variable on the duration of short-term disability. Then, the predicted pattern is $b(MI_D) > b(MI_E)$ and $b(MA_D) > b(MA_E)$, i.e., it holds regardless of whether the accident is minor or major.

H2: Given categories D and E, respectively, the severity of the accident does not influence the duration of disability.

Here, the predicted pattern is $b(MI_E) = b(MA_E)$ and $b(MI_D) = b(MA_D)$.

Table 8.4 contains the evidence. The dependent variable is $lnDUR$, with DUR denoting the duration of payment of benefits for short-term disability, measured in days. The logarithmic transformation serves to mitigate skewness caused by a few extremely high values of DUR. Therefore, coefficients b_2 and b_3 [pertaining to ln(age) and ln(annual income), respectively] can be interpreted as elasticities. The estimates say that 10% higher age goes along with roughly 5% longer duration of paid disability leave. A 10% higher wage income has a smaller estimated effect, of some 1.4% only. Note that there are two sets of estimates, one using the diagnosis "small amputation" (coefficients b_7 and b_8 set to zero) and the other, "rash" (coefficients b_9 and b_{10} set to zero) as the reference category. With few exceptions (notably b_{15}), the parameter estimates are comparable, indicating that the choice of benchmark condition does not matter.

The variable of crucial interest is *ALPHA*. It is of the binary type, reflecting a change of legislation in 1979 that increased the replacement ratio of many insured workers ($ALPHA = 1$ if the worker benefits from the change, $= 0$ otherwise). The following results are of particular interest.

- *Measuring the influence of insurance coverage:* The coefficient of *ALPHA* (b_1 in Table 8.4) is statistically insignificant. However, this does not mean that the change in coverage (increase of the replacement rate in the present context) is unimportant because *ALPHA* also appears in interaction with the dummy variables representing accident categories. Focusing, e.g., on the category MA_D (spinal column without injury, coefficients b_{19} and b_{20}), the pertinent segment of the regression equation reads,

$$lnDUR = b_0 + b_1 \cdot ALPHA + \cdots + b_{19} \cdot MA_D + b_{20} \cdot ALPHA \cdot MA_D + \epsilon, \quad (8.36)$$

with $\epsilon \sim N(0, \sigma^2)$, denoting an error term assumed to be normally distributed with zero mean and constant variance σ^2, regardless of the values assumed by the explanatory variables. Equation (8.36) reflects the hypothesis that the existence of an injury of type MA_D has an effect that depends on the degree of coverage *ALPHA*. This can be seen by setting ϵ to its expected value of zero and differentiating (8.36) with respect to MA_D to obtain,

$$\frac{\partial lnDUR}{\partial MA_D} = b_{19} + b_{20} \cdot ALPHA. \quad (8.37)$$

A positive value of b_{20} would be an indication of moral hazard, implying that more generous insurance coverage ($ALPHA = 1$) serves to increase the duration of

8.2 Moral Hazard

Table 8.4 Duration of payment for short-term disability (lnDUR), Québec (1978–1982)

Coeff.	Explanatory variable	Reference category: MI_E (small amputation)		Reference category: MI_E (rash)	
b_0	Constant	−1.114*	(−2.688)	−2.059*	(−4.914)
b_1	ALPHA (=1: higher coverage)	−0.012	(−0.142)	0.015	(0.165)
b_2	ln (age)	0.524*	(10.602)	0.524*	(10.602)
b_3	ln (annual wage income)	0.140*	(2.984)	0.140*	(2.984)
b_4	Male (=1)	−0.050	(−0.919)	−0.050	(−0.919)
b_5	MI_E (contusion)	−0.168*	(−2.092)	0.777*	(9.242)
b_6	ALPHA $\cdot MI_E$	0.045	(0.407)	0.017	(0.147)
b_7	MI_E (small amputation)	–		0.945*	(10.220)
b_8	ALPHA $\cdot MI_E$	–		−0.027	(−0.217)
b_9	MI_E (rash)	−0.945*	(−10.220)		–
b_{10}	ALPHA $\cdot MI_E$	0.027	(0.217)		–
b_{11}	MA_E (bone fracture)	1.246*	(11.405)	2.191*	(19.565)
b_{12}	ALPHA $\cdot MA_E$	−0.049	(−0.329)	−0.077	(−0.496)
b_{13}	MI_D (lower back pain with injury)	0.040	(0.431)	0.985*	(10.330)
b_{14}	ALPHA $\cdot MI_D$	0.363*	(2.883)	0.336*	(2.542)
b_{15}	MI_D (lower back pain without injury)	−0.473*	(−5.379)	0.472*	(5.178)
b_{16}	ALPHA $\cdot MI_D$	0.152	(1.258)	0.124	(0.981)
b_{17}	MA_D (spinal column with injury)	0.342*	(3.637)	1.287*	(13.251)
b_{18}	ALPHA $\cdot MA_D$	0.520*	(4.022)	0.492*	(3.643)
b_{19}	MA_D (spinal column without injury)	0.197*	(2.068)	1.142*	(11.618)
b_{20}	ALPHA $\cdot MA_D$	0.028	(0.217)	0.001	(0.005)

* Statistically significant with error probability of 5% or better in an OLS regression. t-values in parentheses. Not shown: 10 regression coefficients designed to reflect regional and industry differences. N=5,160

Source Dionne and St-Michel (1991), Table 4

payments for disability in this diagnostic category. However, while positive, the estimate of b_{20} cannot be distinguished from zero according to Table 8.4. There is no direct evidence of moral hazard in this particular case.

- *Test of hypothesis H1:* This hypothesis states that the influence of insurance coverage should be greater for conditions of type D than E for a given type (major or minor accident). As argued above, this calls for a comparison of regression coefficients pertaining to interaction terms, e.g., $b_{18} = 0.520^*$ (associated with $ALPHA \cdot MA_D$, statistically significant) with $b_{12} = -0.049$ (associated with $ALPHA \cdot MA_E$, not significantly different from zero). Clearly, the comparison of these two coefficients shows that the effect of increased coverage is greater if the condition is difficult to verify than when it is easy, confirming H1. Using the right-hand side estimates with a different reference category, one also obtains $b_{18} > b_{12}$, again as predicted by H1. Within the MI category, $b_{14} = 0.336^*$ (associated with $ALPHA \cdot MI_D$) is greater than b_6 (associated with $ALPHA \cdot MI_E$), indicating that the D-type diagnosis is subject to moral hazard more strongly than the E-type. Thus, there is evidence confirming hypothesis H1, i.e., $b(MA_D) > b(MA_E)$ and $b(MI_D) > b(MI_E)$.

  This means that *ex-post* moral hazard exists regardless of the severity of injury if the medical diagnosis is difficult to perform and to verify. In these cases, the treating physician does not contribute much to the mitigation of the information asymmetry, causing the chance of success $[(1 - \rho)$ in Eq. (8.34)] to be high for a worker who seeks to obtain a prolonged short-term disability leave.

- *Test of hypothesis H2:* This hypothesis states that within a given category of observability (diagnosis type of D, for example), the impact of increased insurance coverage should be the same regardless of whether a major (MA) or a minor (MI) injury occurred. This calls for a comparison between, e.g., the value $b_{18} = 0.520^*$ (associated with $ALPHA \cdot MA_D$) of Table 8.4 and $b_{14} = 0.363^*$ (associated with $ALPHA \cdot MI_D$). Both coefficients are statistically different from zero, pointing to moral hazard effects. More importantly, they are of the same magnitude in view of their standard errors of $0.126 \ (= 0.363/2.883)$ and 0.129 $(= 0.520/4.022)$. Indeed, 0.363 ± 0.126 overlaps with 0.520 ± 0.129, indicating that the two estimates cannot be distinguished statistically. Analogous results obtain for the other comparisons, confirming hypothesis H2, which predicts the equalities $b(MI_E) = b(MA_E)$ and $b(MI_D) = b(MA_D)$. Therefore, the extent of *ex-post* moral hazard has nothing to do with the severity of the condition; rather, it is the observability of the diagnosis that matters.

The results of hypotheses H1 and H2 combined imply that *ex-post* moral hazard does not depend on the type or extent of loss (here: major vs. minor injury) but on the degree of observability by the insurer (here: difficulty of performing and verifying the diagnosis), as predicted by theory.

▶ **Conclusion 8.9** There are clear indications of *ex-post* moral hazard in Quebec's short-term disability insurance scheme, suggesting that observability indeed is the crucial determinant.

8.2.6 Relational Moral Hazard in Long-Term Care Insurance

Public expenditure on long-term care (LTC) is increasing fast. As a share of GDP, however, recent developments do not appear quite so alarming (see Table 8.5; data before 2010 and after 2017 are unavailable). As a share of total government expenditure, public LTC expenditure (LTCE) has risen in all countries sampled with the exception of Canada, a country often hailed as an example of how to hold down healthcare expenditure as a share of GDP (at least compared to the United States). Canada has indeed managed to reduce the GDP share of LTCE from 1.4% to 1.3%, the GDP share of government expenditure from 42% to 40%, and to keep the LTCE component of government spending constant at 3.3%. By way of contrast, the LTCE share of the GDP of the United States has remained constant at a low 0.6%, but its share burdening the government (LTCE/G) has increased from 1.4% to 1.6% in the course of only eight years, while government expenditures as a share of GDP (G/GDP) shrank from 43% to 38%.

Not surprisingly, Norway as one of the Nordic welfare states exhibits both the highest (and rising) LTCE/GDP and G/GDP values in the sample, resulting in a share of 5.4% of government spending by 2017. In terms of growth of this share, however, Germany and Switzerland constitute the most extreme cases, with LTCE/G increasing from 2.3% to 3.4% and from 4.0% to 4.7%, respectively. It should be noted that the LTCE/GDP and LTCE/G values are likely underestimates because they measure only the health-related component of LTCE; recently estimates of the so-called social component have become available, which suggest that the total figures are substantially higher.

In view of Table 8.5, it comes as little surprise that governments are looking for ways to shift the burden of LTC to the private sector. However, the uptake of private LTC insurance has been very limited in almost all OECD countries, likely because the premiums contain a very high loading [recall that the loading is the true price of coverage; see Zweifel (2020) for an estimate in the case of the United States]. As argued in Sect. 8.1, the loading is driven up by moral hazard effects. Therefore,

Table 8.5 Public LTC expenditure in selected OECD countries, 2010 and 2017 (%)

	2010			2017		
	LTCE/GDP[a]	G/GDP[b]	LTCE/G[c]	LTCE/GDP[a]	G/GDP[b]	LTCE/G[c]
Canada	1.4	42	3.3	1.3	40	3.3
Germany[d]	1.1	48	2.3	1.5	44	3.4
Norway	2.3	45	5.1	2.7	50	5.4
Switzerland[d]	1.3	33	4.0	1.6	34	4.7
United States	0.6	43	1.4	0.6	38	1.6

[a]Public LTC expenditure as a share of GDP; [b]Government expenditure as a share of GDP; [c]Public LTC expenditure as a share of government expenditure, calculated as G/GDP=(LTC/GDP/(G/GDP))
[d]Figures include expenditure by (compulsory) social insurance
Sources stats.oecd.org/Index.aspx?DataSetCode=SHA; data.oecd.org/ggageneral-government-spending.htm

the question arises whether LTC insurance (provided by a private insurer, a social insurer, or the government) is subject to a particular type of moral hazard.

This indeed seems to be the case in the guise of "Intergenerational moral hazard" or more generally, "Relational moral hazard". Beyond a certain age, many individuals have to rely on help for a prolonged period of time to be able to run their household. Their need for LTC can often be satisfied informally by spouses, younger family members, and/or friends, whose motivation to provide care may however be undermined when there is LTC insurance covering the cost of formal care. In this way, LTC insurance runs the risk of inducing a particular moral hazard effect in the relationship between individuals.

8.2.6.1 Modeling the Interaction Between Parent and Potential Caregiver

As long as spouses age together, they tend to provide informal LTC to each other (see also the argument in Sect. 11.4.2). However, in the United States, e.g., almost 70% of women (and 30% of men) beyond the age of 75 were widowed, divorced, or never married in 2005 (Houser 2007). Especially for aged women, the choice boils down to one between informal care provided by a family member or friend and formal care provided by trained personnel (often in a nursing home) [for surveys of the issues involved, see Norton (2000) as well as Eisen and Sloan (1996)]. However, lay caregivers, while likely altruistic, will also consider opportunity costs and benefits. In an early contribution, Pauly (1990) noted that the prospect of receiving a bequest may motivate a potential caregiver (most often a daughter, "the child" henceforth) to make effort to delay or even avoid admission to a nursing home. Yet this incentive is weakened if the person facing the risk of needing LTC ("the parent" henceforth) purchases LTC insurance (which indirectly protects the bequest). This weakened incentive is a possible reason why the market for this type of insurance has been developing sluggishly (another reason is the widespread caps on benefits). Zweifel and Strüwe (1996, 1998) modeled the parent as the principal and the (one) child as the agent in a principal-agent framework, which allowed focus on a set of parameters governing what they called intergenerational moral hazard. However, this approach neglects the fact that as a rule the child is an adult in or near retirement.

This consideration led Zweifel and Courbage (2016) to view the parent and the child as two players in their own right who maximize expected utility. The parent is assumed to be rich enough to leave a bequest but poor enough to obtain some means-tested public support when needing formal LTC—a large group in those countries that have added an LTC component to their social security. Early in life, he or she decides to buy private LTC coverage I (usually as part of life insurance); after retirement, he or she has a propensity to save s (set earlier in life as an aspect of lifestyle) that determines the size of the bequest available to the caregiver. This reflects the fact that most people achieve savings no sooner than when their children have left the home.

The parent is characterized by a risk utility function $u(\cdot)$ defined over consumption in the two periods prior to and immediately after retirement, and over wealth as a

8.2 Moral Hazard

contingency reserve in a later third period, all of unit length. During this last period, he or she faces the risk of being dependent on formal LTC, which is equated to being admitted to a nursing home for concreteness.

Accordingly, the parent's utility function is conditioned on being in the nursing home $[v^i(\cdot)]$ or outside the nursing home $[v^o(\cdot)]$. His or her utility when out of the nursing home is higher for a given amount of wealth W than "when in", i.e., $v^o[W] > v^i[W]$, reflecting the greater degree of independence in the enjoyment of wealth. The parent is altruistic in the sense that final wealth w becomes a bequest for the child in its entirety. Since the parent is assumed to be retired, there is no labor income that could contribute to wealth. Therefore, final wealth in the nursing home is given by initial wealth saved $W_0 s$ accrued for interest $(1 + a)$ minus a share r of this amount multiplied by the relative price p_h of LTC; since time in the nursing home is normalized to one, p_h amounts to LTC expenditure. This relative price is greater than one ($p_h > 1$) because nursing home care is expensive compared to other goods and services. The parent has to pay a share $r(W)$ of this expense, which has the following property,

$$\frac{\partial r}{\partial W} := r'_W > 0 : \text{cost sharing increases with wealth defined by}$$

$$W = W_0 s(1 + a) + I(1 - \bar{\pi}) \text{ given admission} \quad (8.38)$$

to the nursing home,

with $\bar{\pi}$ denoting the premium rate. If the parent stays out of the nursing home, final wealth is simply given by $W_0 s(W + i) - \bar{\pi} I$. In sum, one has for expected utility (EU) of the parent defined over a total of three periods, assuming separability of utilities,

$$\begin{aligned} EU = &\, u_1[W_0 - \bar{\pi} I] + u_2[W_0(1 - s) - \bar{\pi} I] \\ &+ \pi(c) v^i [W_0 s(1 + a) + I(1 - \bar{\pi}) - r\{W_0 s(1 + a) + I(1 - \bar{\pi})\} \cdot p_h] \\ &+ (1 - \pi(c)) v^o [W_0 s(1 + a) - \bar{\pi} I]. \end{aligned} \quad (8.39)$$

The first component is (nonstochastic) utility early in the parent's adult life when saving is not possible yet, making the (present value of) consumption equal to wealth w_0 net of the premium $\bar{\pi} I$ for private LTC coverage. By assumption, the insurer charges an actuarially fair premium according to the average probability $\bar{\pi}$ of needing formal LTC services in the population, being unable to estimate an individual-specific probability so early in life. The second term is the utility of consumption after retirement when saving has become possible; but the LTC premium must still be paid as before.

The last two terms make up the risk utility function. With probability $\pi(c)$ which depends negatively on c, the child's effort directed at keeping the parent out of the nursing home, the parent nevertheless has to be admitted. The utility "when in" (the nursing home) depends on wealth as given in Eq. (8.39), consisting of wealth saved

$W_0 s$ accrued by the interest factor $(1+a)$, plus the LTC benefit net of the premium $I(1-\bar{\pi})$, minus the share r in the cost p_h of the nursing home stay. As stated this share increases with wealth $W_0 s(1+a)$ at the time of admission, augmented by the net LTC benefit $I(1-\bar{\pi})$ received (in some countries, this benefit may be exempted from cost sharing). Conversely, utility "when out" (of the nursing home) simply depends on accrued wealth minus the LTC premium.

From Eq. (8.39), one can derive the first-order condition (FOC) for an optimum with respect to the parent's purchase of LTC insurance, focusing on interior solutions with $I > 0$,

$$\begin{aligned}\frac{dEU}{dI} &= -\bar{\pi}\{u_1'(\cdot)+u_2'(\cdot)\}+\pi(c)\{(1-\bar{\pi})-r_W'(1-\bar{\pi})p_h\}\cdot v^{\prime i}(\cdot)\\ &\quad -(1-\pi(c))\bar{\pi}\cdot v^{\prime 0}(\cdot)\\ &= -\bar{\pi}\{u_1'(\cdot)+u_2'(\cdot)\}+\pi(c)\{(1-\bar{\pi})(1-r_W' p_h)\}\cdot v^{\prime i}(\cdot)\\ &\quad -(1-\pi(c))\bar{\pi}\cdot v^{\prime 0}(\cdot)=0. \end{aligned} \qquad (8.40)$$

Equation (8.40) can be interpreted as follows. In the two pre-retirement periods, an extra unit of coverage has the downside of incurring a higher premium at rate $\bar{\pi}$ which is valued according to the pertinent marginal utilities of consumption. In case of admission to the nursing home, however, it adds to wealth at the tune of the additional unit of LTC benefit net of the extra premium, with deduction made for the extra sharing in the cost of LTC. The net increase in wealth is valued using the marginal utility "when in". If admission can be avoided thanks to the effort exerted by the child, more LTC coverage merely costs the extra premium, which is valued using the marginal utility of wealth "when out".

Now let this optimum be disturbed by an increase in effort on the part of the child $dc > 0$, viewed as an exogenous change. Since the first-order condition (8.40) has to hold before and after the change, the right-hand side of the comparative-static Eq. (8.41) below is zero,

$$\frac{\partial^2 EU}{\partial I^2}dI+\frac{\partial^2 EU}{\partial I\partial c}dc=0. \qquad (8.41)$$

In analogy to Eqs. (8.31) and (8.32), this can be solved for dI/dc indicating how the parent optimally adjusts to the change in the child's behavior,

$$\frac{dI}{dc}=-\frac{\partial^2 EU/\partial I\partial c}{\partial^2 EU/\partial I^2}. \qquad (8.42)$$

Since the values of the parameters determining these second-order derivatives are unknown in general, all one can hope for is to be able to sign dI/dc. A considerable simplification is possible thanks to the assumption that the parent can distinguish a maximum from a minimum, hence $\partial^2 EU/\partial I^2 < 0$ for a maximum. One therefore obtains

8.2 Moral Hazard

$$\text{sgn}\left[\frac{dI}{dc}\right] = \text{sgn}\left[\frac{\partial^2 EU}{\partial I \partial c}\right], \tag{8.43}$$

a result also known as the implicit function theorem. Accordingly, the derivation of Eq. (8.40) w.r.t. c yields

$$\begin{aligned}\frac{\partial^2 EU}{\partial I \partial c} &= \pi'(c)(1-\bar{\pi})(1-r'_W p_h) \cdot v'^i(\cdot) + \pi'(c)\bar{\pi} \cdot v'^o(\cdot) \\ &= \pi'(c)\{(1-\bar{\pi})(1-r'_W p_h) \cdot v'^i(\cdot) + \bar{\pi} \cdot v'^o(\cdot)\}. \end{aligned} \tag{8.44}$$

In order to evaluate the sign of Eq. (8.44), one needs to consider two issues.

- Is $(1 - r'_W p_h)$ positive or negative? Note that $(1 - r'_W p_h) < 0$ can be rewritten as $r'_W > 1/p_h$; if, e.g., nursing home care is three times more expensive than all other goods and services ($p_h = 3$), a cost sharing rule that claims 34 percent of wealth is sufficient to make $(1 - r'_W p_h)$ negative. However, such a rule would make earlier savings result in lower net wealth in the event that the parent is admitted to the nursing home. Therefore, lenient cost sharing resulting in $(1 - r'_W p_h) > 0$ is assumed.
- Is the marginal utility of wealth high when in or out of the nursing home? At first blush, one is tempted to assume $v'^o(\cdot) > v'^i(\cdot)$ because extra wealth (and hence consumption) has more utility when one is out of a nursing home. However, at this point it is crucial to remember that parental wealth is risky, causing one to invoke Eeckhoudt and Schlesinger (2006) cited at the end of Sect. 3.2.2. Since panel A of Fig. 3.3 obtains where health and wealth are below expected value, marginal utility "when in" must be high, hence $v'^i(\cdot) > v'^o(\cdot)$.

In addition, $(1 - \bar{\pi}) > \bar{\pi}$ when the decision regarding LTC insurance is made. Therefore, one can justifiably argue that the bracketed expression in Eq. (8.44) is positive, thus

$$\frac{\partial^2 EU}{\partial I \partial c} < 0 \rightarrow \frac{dI}{dc} < 0, \text{ with } \left|\frac{dI^2}{dc^2}\right| < 0 \text{ because}$$
$$|\pi'(c)| \text{ decreases with } c. \tag{8.45}$$

The parent is predicted to reduce LTC insurance coverage in response to an increase in the child's effort designed to avoid admission to the nursing home, a behavior that constitutes a relational moral hazard effect on the part of the parent. The parent's reaction function is depicted in Fig. 8.3.

▶ **Conclusion 8.10** The provision of informal long-term care involves the interaction between a parent who decides the amount of insurance coverage and a caregiver. Caregiver effort has the effect of reducing the probability of needing formal care, in particular admission to a nursing home. If caregiver effort increases, the parent is predicted to scale back LTC insurance coverage, exhibiting relational moral hazard.

Turning to the child as the potential caregiver, she can expect a bequest amounting to the share $k(1-t)$ of the parent's final wealth, where k reflects the rules governing inheritance (usually, k decreases with the number of siblings and remoteness of kinship) and t denotes the rate of taxation (both assumed to be constant for simplicity). The child is affected by two parental decisions that determine the size of the bequest, the propensity to save s [which is viewed as predetermined here, contrary to Zweifel and Courbage (2016)] and the amount of LTC coverage I purchased. Providing care has an opportunity cost θ per unit; however, since the child typically is in retirement herself when the parent might enter a nursing home, $\theta = 0$ is assumed in that period for simplicity. With Z_0 denoting initial wealth and $\overline{EU}$ expected utility, one has for the child

$$\begin{aligned}\overline{EU} = &\ \bar{u}(Z_0 - \theta c) \\ &+ \pi(c)\bar{v}^i[Z_0 + k(1-t)\{W_0 s(1+a) + I(1-\bar{\pi})\} \\ &\quad - r\{W_0 s(1+a) + I(1-\bar{\pi}\} \cdot p_h] \\ &+ (1-\pi(c))\bar{v}^o[Z_0 + k(1-t)\{W_0 s(1+a) - \bar{\pi}I\}]. \end{aligned} \quad (8.46)$$

Here, $\bar{u}(\cdot)$ is the child's utility function pertaining to the pre-retirement period; further, the assumption is that she sets her level of effort c as a lifestyle variable. Next, $\bar{v}^i$ denotes her utility conditional on the parent being admitted to the nursing home, while $\bar{v}^o$ symbolizes her utility when the parent is "out". Altruism calls for $\bar{v}^o[Z] > \bar{v}^i[Z]$ meaning that the utility of wealth (and hence consumption) is greater for the child when the parent is out of the nursing home.

When the parent is in the nursing home, final wealth is bequeathed to the caregiver after deduction for cost sharing, given by $r[\{W_0 s(1+a) + I(1-\bar{\pi})\} \cdot p_h]$; in this way, the child also has a financial interest in avoiding admission to the nursing home. If her efforts are successful, her share $k(1-t)$ of final wealth becomes the bequest in full. In perfect analogy to Eqs. (8.40) to (8.44), one derives the FOC w.r.t. c,

$$\frac{d\overline{EU}}{dc} = -\theta \bar{u}'_1(\cdot) + \pi'(c)\{\bar{v}^i(\cdot) - \bar{v}^o(\cdot)\} = 0, \quad (8.47)$$

to be disturbed by an exogenous change dI (details not shown),

$$\frac{\partial^2 \overline{EU}}{\partial c \partial I} < 0 \rightarrow \frac{dc}{dI} < 0 \text{ with } |\frac{d^2 c}{dI^2} < 0| \text{ again}. \quad (8.48)$$

Therefore, there is a relational moral hazard effect also on the part of the child. Knowing that the parent has stepped up LTC coverage, the child has a reduced incentive to try avoiding the need for costly formal care, in particular admission to a nursing home.

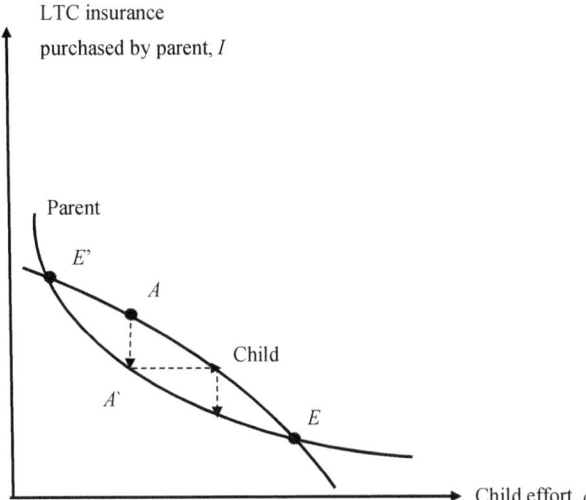

Fig. 8.3 Reaction functions of parent and child

Any intersection of the two reaction curves represents a so-called Nash equilibrium, from which neither player has an incentive to deviate (after all, each curve is a set of optimum solutions). Note that in Fig. 8.3, there might be no intersection (thus no Nash equilibrium) if the vertical distance between the two curves happens to be excessive. Excluding this possibility, there could be just one equilibrium where the two curves are tangent, a very unlikely outcome. This leaves the two equilibria E and E'; yet only E turns out to be stable. To see this, consider point A where at a given amount of child effort, the parent purchases too much LTC coverage. This triggers a reduction of coverage; but at point A', the child adjusts in turn to this scant coverage (which would result in a substantial reduction of the bequest in the event of a stay in the nursing home) by increasing her effort designed to avoid such a stay. Thus, the process of mutual adjustment ends at point E.

▶ **Conclusion 8.11** The interaction between a parent who decides the amount of insurance coverage and a caregiver who decides the level of effort results in a Nash equilibrium determining both of these quantities. It is characterized by bilateral relational moral hazard in spite of altruism.

8.2.6.2 Predicted Effects of Public Policy

Modeling the interaction between parent and child has the benefit of permitting one to predict whether a particular policy aggravates or mitigates relational moral hazard. In the case of aggravation, parents reduce their private LTC coverage (or abstain from purchasing it entirely), while their children provide less (or possibly no) informal care. This combination of responses would boost reliance on formal LTC services (use of nursing homes in particular), whose cost typically would fall on social insurance or taxpayers. Public interventions already performed or considered in future are

- Adding an LTC component to social health insurance (as in Germany and in the United States through Medicaid). In the case of Medicaid, this expansion has been found to crowd out private LTC insurance by Brown and Finkelstein (2008);
- Subsidizing nursing home care (as in Canada);
- Stepping up inheritance taxation (as proposed in Switzerland), the proceeds of which could be used to finance public LTC expenditure.

For concreteness, consider inheritance taxation in China ($dt > 0$), which currently does not exist but may be introduced in future. Since the variable t does not appear in the equations referring to the parent [Eqs. (8.39), (8.40), and (8.44)], the parental reaction function is not affected. As to that of the child, she can look forward to a positive bequest regardless of whether or not the parent requires formal LTC services because parental propensity to save s is very high in China, making both inner brackets positive in Eq. (8.49) below. Finally, while the government incurs LTC expenditure through public nursing homes, there is no subsidization of individuals, implying $r(\cdot) = 1$.

Using the FOC in Eq. (8.40) and recalling the wealth levels appearing in Eq. (8.39), one obtains for the crucial mixed derivative

$$\frac{\partial^2 \overline{EU}}{\partial c \partial t} = -\pi'(c)\{-k\{W_0 s(1+a) + I(1-\bar{\pi}) - r(\cdot)p_h\}\bar{v}'^i(\cdot)$$
$$+ (-k)\{W_0 s(1+a) - \bar{\pi}I)\}\bar{v}'^o(\cdot)\}$$
$$= \pi'(c)k\{\{W_0 s(1+a) + I(1-\bar{\pi} - p_h)\}\bar{v}'^i(\cdot) + \{W_0 s(1+a) - \bar{\pi}I\}\bar{v}'_o(\cdot)\}$$
$$< 0 \text{ since } \pi'(c) < 0 \text{ while wealth levels in brackets are } > 0. \qquad (8.49)$$

Therefore, the child's reaction curve is predicted to shift inward, toward a lower effort c for a given value of I. As shown in Fig. 8.4, this shift is comparatively small for high values of c since $|\pi'(c)| \to 0$ in Eq. (8.49), reflecting decreasing marginal effectiveness of effort provided by the child. Therefore, the transition from E to E' indicates a reduction in informal care provided by the child combined with a limited increase in LTC coverage bought by the parent. The reason is that by reducing the bequest the tax weakens the child's incentive to exert effort, which in turn induces the parent to step up LTC coverage.

8.2 Moral Hazard

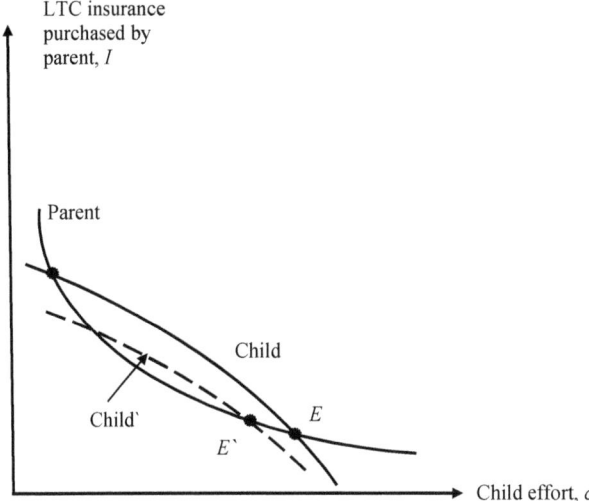

Fig. 8.4 Effect of a tax on inheritance

▶ **Conclusion 8.12** Public policy may affect bilateral relational moral hazard, depending on institutional detail. In the case of China, the possible introduction of an inheritance tax is predicted to decrease the amount of informal care provided but enhance parents' inclination to purchase private LTC insurance coverage.

In fact, this conclusion holds more generally than just for China. In Eq. (8.49), dropping the restriction $r(\cdot) = 1$ does not affect its sign as long as $r'_W > 0$ and wealth levels are positive both in and out of the nursing home, i.e., as long as LTC is means-tested in a lenient way and people cannot accumulate excess debt toward the end of their lives (these conditions apply to most industrial countries).

8.2.6.3 Empirical Evidence on Relational Moral Hazard

Empirical testing for relational moral hazard effect puts high demands on a database, which should contain information both on recipients and (potential) providers of LTC. Since China with its one-child policy (which has been rescinded only in 2016) fits the theoretical one-to-one model very well, an attempt was made by Xu and Zweifel (2014) to circumvent the lack of appropriate data. In a survey among 500 residents of Shanghai aged 30 to 60, respondents were asked to indicate whether acting as parents, they would have an increased, decreased, or unchanged inclination to purchase LTC insurance in response to an increase in their child's provision of informal care. While 68% said they would not be affected, 28% stated a diminished inclination and only 4%, an increased inclination to purchase, pointing to relational moral hazard on the part of the parent.

The same respondents were also asked whether acting as children, they would have an increased or decreased tendency to rely on formal LTC services in response to their parent having LTC insurance coverage. A full 98.4% opted for an increase and only 1.6%, for a decrease. However, when the question was framed directly in terms of providing informal care, results were not so clear. Likely out of filial piety (a norm deeply entrenched in Chinese culture), no fewer than 86.4% claimed that they would not adjust their effort, whereas a mere 6.4% would reduce it and 2.2%, even step it up. Therefore, while broadly supporting Conclusion 8.11, this evidence is preliminary because it does not derive from actual decisions but rather stated intentions [see Xu and Zweifel (2014) for more details].

Actual purchases of private LTC insurance and actual utilization of formal LTC are jointly analyzed by Konetzka et al. (2019). The emphasis of their work is on the moral hazard effect that may emanate from the extra insurance coverage in excess of the benefits provided by U.S. Medicare (for non-poor individuals of age 65 and higher) and U.S. Medicaid (for the poor, generally covering only nursing home care). To the extent that older people prefer to continue living at home rather than in a nursing home, the authors predict that moral hazard impacts the use of home care more strongly than nursing home stays. Accordingly, they define two dependent variables, one for any home care ($= 1$ if a medically trained person provided help or if there were personal care services that are typically covered by private LTC insurance, $= 0$ otherwise), the other, for nursing home care ($= 1$ if the stay was at least 100 days, $= 0$ otherwise).

For insurance coverage, only an indicator variable ($LTCI = 1$ if private LTC insurance is purchased, $= 0$ otherwise) is available. It does not fully correspond to the predictions in Figs. 8.3 and 8.4, which are presented in terms of quantities. More importantly, the decision to purchase LTC insurance depends on many factors (such as wealth and health status) which in turn may influence the use of formal LTC services. According to the theoretical model, this decision is made in view of the expected effort of a potential caregiver, which varies with LTC coverage contracted by the parent. Therefore, in order to discern the link between the presence of private LTC coverage and the use of formal LTC, a so-called instrumental variable is needed which only affects LTC coverage but not formal LTC services.

Konetzka et al. (2019) dispose of such an instrumental variable in the guise of the dummy variable TAX_{t-1}, which indicates whether the aged individual itemized his or her income tax deductions in the preceding year. Without itemizing their deductions, U.S. taxpayers cannot claim a tax deduction for the premiums paid for LTC insurance. Therefore, $TAX_{t-1} = 1$ shows that the decision to purchase LTC coverage was made based on a tax decision made in the past and was therefore unlikely to be correlated with the utilization of formal LTC during the current period.

In addition, the authors improve the comparability of the subsamples with $LTCI = 1$ and $LTCI = 0$, respectively. They note that individuals with $LTCI = 0$ are younger and less risk-averse, have lower income and less assets, and are more likely to be male, black, and Hispanic. They use TAX_{t-1} to derive a propensity score that predicts the probability of an individual having LTC insurance coverage. Next, individuals with equal scores are selected from the two subsamples. After this matching, the differences with regard to the personal characteristics cited above disappear, which suggests that TAX_{t-1} indeed is a powerful predictor of $LTCI$.

In this way, the authors arrive at two times 26,438 "comparable" observations drawn from the Health and Retirement Study covering the years 1996 to 2012. According to their estimates, individuals with private LTC coverage use home care services with a probability that is 16.5 percentage points higher than those without, an effect that amounts to more than a doubling of the mean probability of use (11.9%). However, their use of long nursing home stays cannot be said to be more likely.

It remains to show that parents buy less LTC insurance when they can count on informal care (see the reaction curve labeled "Parent" in Fig. 8.4). Indeed, Konetzka et al. (2019) also provide information supporting this hypothesis. In their subsample with $LTCI = 0$, a full 78.6% of the aged individuals have at least one daughter; in the subsample with $LTCI = 1$, this share is lower, at 76.8%. While this difference is small, it is statistically significant with an error probability of 0.001. The figures relating to the presence of a spouse or partner are 72.4% and 70.2%, respectively (again a significant difference). Therefore, the non-purchase of LTC insurance does correlate with the presence of a potential caregiver, which comes close to suggesting that the purchase of less LTC coverage goes along with more (expected) informal care provided by a potential caregiver.

▶ **Conclusion 8.13** The two reaction functions defined in the theoretical model can be inferred successfully from individual Chinese and U.S. data. While China is a country that accords with the one-to-one specification of the model, the data reflect intentions rather than actions. The U.S. data report actual purchases of long-term care insurance coverage and uses of formal long-term care services but do not measure the amount of informal care provided.

A third type of evidence reflects the influence of policy. The Netherlands (NL) and Germany (GER) are two (neighboring) countries that differ in their LTC policy. While Conclusion 8.12 is not directly applicable because neither country taxes inheritances to finance public LTC expenditure, public support is means-tested in NL ($r'_W > 0$) but not in GER ($r'_W = 0$). This means that in NL, parents cannot fully pass additional wealth to their children, which can be likened to a tax on inheritance which is absent in the case of GER. Therefore, the child's reaction curve runs closer to the origin in NL than in GER (see the dashed curve in Fig. 8.4). The interaction between parents and potential caregivers is predicted to result in point E in GER but E' in NL, with less effort on the part of potential caregivers.

Now Bakx et al. (2015) draw samples comprising 4,349 individuals aged 50 and older in NL and 4,390 in GER from the European SHARE database. They report a share of 10.2% receiving some type of formal LTC (defined here as ranging from home help to nursing home care) in the NL sample but only 3.6% in the GER sample, a difference suggesting that in NL less informal care is provided than in GER.

▶ **Conclusion 8.14** A comparison of policies in the Netherlands and Germany leads to the prediction that the reaction function of the child (generally, the potential caregiver) runs closer to the origin in the Netherlands, resulting in a Nash equilibrium with less informal care provided than in Germany. Available empirical evidence is in accordance with this prediction.

While the theoretical model of Sects. 8.2.6.1 and 8.2.6.2 can be said to have some predictive value, it still neglects the role of a third player, the provider of formal LTC services. In a recent empirical investigation, Hackmann and Pohl (2018) seek to close this gap by combining data on individual beneficiaries of U.S. Medicare and Medicaid with information about nursing homes, most of which are for profit. Their data set covers the years 2000–2005 and the states of California (CA), New Jersey (NJ), Ohio (OH), and Pennsylvania (PA). The authors find differences in the generosity of Medicaid between the three neighboring states NJ, OH, and PA to be associated with differences in probabilities of discharge by nursing homes (and hence length of stay and LTC expenditure). Thus, there may not only be relational moral hazard but also *ex-post* moral hazard on the part of providers of formal LTC services.

Finally, Zweifel and Courbage (2016) identify two exogenous changes that have the potential to aggravate relational moral hazard, namely higher parental wealth (depending on the type of model) and higher opportunity cost of the child (reflecting improved labor market prospects for women). Evidently, these changes constitute major challenges confronting both private and social LTC insurance.

8.3 Adverse Selection

Asymmetry of information not only gives rise to moral hazard effects after the inception of the insurance contract. Indeed, it may be even more important at the time of contracting. To the extent that insurers lack knowledge about the true risk of an IB, they cannot fend off high risks. This is called adverse selection; it may jeopardize the economic survival of an insurer who happens to be stuck with too many high risks. In the final analysis, adverse selection may undermine equilibrium in a private insurance market. However, this far-reaching conclusion holds only in a one-period model. In a multi-period framework, it becomes possible for the IC to learn about true risk from its loss experience. Such a model is presented in the second part of this section.

8.3.1 Adverse Selection in a Single-Period Framework

8.3.1.1 Unsustainability of a Pooling Equilibrium

The exposition follows the famous paper by Rothschild and Stiglitz (1976). The authors distinguish a "low" risk type with a low probability of loss π^L from a "high" risk type, with a higher probability $\pi^H > \pi^L$. To simplify the analysis, the two risk types do not differ with regard to the amount of loss L nor initial wealth W_0. The

8.3 Adverse Selection

first simplification is innocuous because the insurer can always write a policy with a limit on benefits I to deal with an IB who causes a high loss L. By way of contrast, it cannot put a limit on the probability of loss. The second assumption does entail a certain loss of generality because differences in initial wealth may go along with differences in risk aversion. However, the equality assumption permits focus on the impact of asymmetric information at the time of contracting.[8]

Expected utility for a low-risk type is defined as usual over two states,

$$EU^L = \pi^L \cdot v(W_0 - P^L - L + I^L) + (1 - \pi^L) \cdot v(W_0 - P^L), \quad (8.50)$$

where
- I^L: Insurance benefits paid according to a contract designed for a low-risk type;
- L: Amount of loss, independent of risk type;
- $v(\cdot)$: Risk utility function, with wealth as the only argument;
- W_0: Initial wealth, exogenous and independent of risk type;
- P^L: Premium of a contract designed for low risk;
- π^L: Probability of loss of a low risk.

Therefore, final wealth levels in the case of a low-risk type are given by (analogously for a high-risk type),

$$\begin{aligned} W_1^L &:= W_0 - P^L - L + I^L; \\ W_2^L &:= W_0 - P^L. \end{aligned} \quad (8.51)$$

For the subsequent graphical illustration, the indifference curve of such a low-risk individual is of interest. It is defined by holding expected utility constant, or equivalently, putting the change of expected utility to zero. Since Eq. (8.50) specifies a constant π^L and a risk utility function in terms of wealth only, such a change must be due to changed wealth levels,

$$dEU^L = \pi^L \cdot \frac{\partial v}{\partial W_1^L} \cdot dW_1^L + (1 - \pi^L) \cdot \frac{\partial v}{\partial W_2^L} \cdot dW_2^L = 0. \quad (8.52)$$

Therefore, the slope of the indifference curve (see $\overline{EU}^L$ in Fig. 8.5) is given by

$$\frac{dW_1^L}{dW_2^L} = -\frac{1 - \pi^L}{\pi^L} \cdot \frac{\partial v/\partial W_2^L}{\partial v/\partial W_1^L} = -\frac{1 - \pi^L}{\pi^L} \cdot \frac{v'[2]}{v'[1]}, \quad (8.53)$$

with $v'[2]$ ($v'[1]$) denoting the marginal utility of risky wealth in the no-loss (loss) state. As usual, this slope (i.e., the marginal rate of substitution between the two final wealth levels) is equal to the ratio of the marginal utilities, weighted with their probability of occurrence. Besides low-risk types, there are also high-risk ones. In

---

[8]The exposition utilizes the basic model of Sect. 3.2.1.

full analogy, the slope of their indifference curve (see, e.g., $\overline{EU}^H$ through point H^*) is given by

$$\frac{dW_1^H}{dW_2^H} = -\frac{1-\pi^H}{\pi^H} \cdot \frac{\partial v/\partial W_2^H}{\partial v/\partial W_1^H} = -\frac{1-\pi^H}{\pi^H} \cdot \frac{v'[2]}{v'[1]}. \tag{8.54}$$

Since a high risk's probability of loss exceeds that of a low risk ($\pi^H > \pi^L$), the indifference curve of the high risk must run flatter than the one pertaining to the low risk. The standard assumption is that the two indifference curves intersect only once in order to unambiguously define the risk type [so-called single crossing property, see, e.g., Kreps (1990), 638–645 and 661–674].

This difference in slope can be most conveniently read off at the certainty line ($W_1 = W_2$). There, final wealth is the same in both states, causing marginal utility of risky wealth to be the same. Equation (8.53) therefore simplifies to

$$\frac{dW_1^L}{dW_2^L} = -\frac{1-\pi^L}{\pi^L}, \quad \text{if } W_1^L = W_2^L. \tag{8.55}$$

An analogous result holds for the high-risk type.

The decision-making situation of the IC in the presence of asymmetric information can now be described as follows. By assumption, it does not know the probability of loss pertaining to the two types of risk. At best, it knows the average value $\bar{\pi}$ in the population, which reflects the shares of high-risk (h) and low-risk types ($1-h$), respectively,

$$\bar{\pi} = h \cdot \pi^H + (1-h) \cdot \pi^L. \tag{8.56}$$

Accordingly, the IC can only write a contract with a uniform average premium P and a uniform amount of insurance benefits I. It receives the premium P and pays the benefit I with probability $\bar{\pi}$, while with counter-probability $(1-\bar{\pi})$ it receives the premium P. With administrative expense amounting to C per risk insured, the break-even condition for the IC reads

$$E\Pi = \bar{\pi} \cdot (P - I) + (1 - \bar{\pi}) \cdot P - C \geq 0, \tag{8.57}$$

with $E\Pi$ denoting expected profit.

Now consider a variation of the contract in the guise of a marginally increased insurance benefit I in return for a marginally higher premium P such that expected profit remains constant. Justifiably, administrative expense is not affected by these changes ($dC = 0$), resulting in

$$dE\Pi = \bar{\pi} \cdot (dP - dI) + (1 - \bar{\pi}) \cdot (dP) = 0. \tag{8.58}$$

This condition is now projected into (W_1, W_2)-space to combine it with Eqs. (8.53) and (8.54) for the IB. Dropping superscripts for simplicity, one has

$$dP - dI = -dW_1; \quad dP = -dW_2. \tag{8.59}$$

8.3 Adverse Selection

In the loss state, the increase in the premium reduces consumer's final wealth while the increase in the insurance benefit adds to it. In the no-loss state, however, the extra premium simply amounts to a reduction of final wealth. Substitution of Eq. (8.59) into the condition (8.58) yields

$$dE\Pi = \bar{\pi}(-dW_1) + (1 - \bar{\pi})(-dW_2) = 0. \tag{8.60}$$

This results in the following slope of the insurance line in (W_1, W_2)-space,

$$\frac{dW_1}{dW_2} = -\frac{1 - \bar{\pi}}{\bar{\pi}} = -\left[\frac{1}{\bar{\pi}} - 1\right]. \tag{8.61}$$

Equation (8.61) can be used to show that the slope of the insurance line for a pooling contract must lie between the slopes pertaining to the two risks types. From $\pi^H > \bar{\pi} > \pi^L$, one obtains

$$\left|\frac{1}{\pi^H} - 1\right| \leq \left|\frac{1}{\bar{\pi}} - 1\right| \leq \left|\frac{1}{\pi^L} - 1\right|. \tag{8.62}$$

This is equivalent to the following ordering of slopes in absolute value,

$$\left|\frac{1 - \pi^H}{\pi^H}\right| \leq \left|\frac{1 - \bar{\pi}}{\bar{\pi}}\right| \leq \left|\frac{1 - \pi^L}{\pi^L}\right|. \tag{8.63}$$

Figure 8.5 illustrates the interaction of a competitive IC with an IB given conditions of adverse selection. The three insurance lines appear with slopes according to Eq. (8.63), symbolized as AH for the high-risk type, AL for the low-risk type, and AM for the pooling contract based on the average probability of loss $\bar{\pi}$. For a benchmark, consider first the case where information is publicly available. The probabilities of loss π^L and π^H would be known to the IC, enabling it to charge premiums according to π^H and π^L, respectively. Neglecting administrative expense for simplicity, $P^H = \pi^H L$ and $P^L = \pi^L L$ implement the condition, "price equal to marginal cost" because the (expected) marginal cost of enrolling a consumer amounts to the expected value of the claim to be paid. Therefore, in a world of symmetric information and competitive conditions, low risks would pay a low premium and high risks, a high one, reflecting the difference in marginal cost.

Since premiums are actuarially fair, both risk types would choose full coverage at points H^* and L^*, respectively. These contracts constitute a market equilibrium because all ICs calculate their premiums using the same (true) π-values. Moreover, this equilibrium is Pareto-optimal because both risk types attain their optimum. Note that if the high-risk premium is deemed excessive in comparison to a consumer's income and wealth, government could simply subsidize the purchase of insurance by high-risk types. By way of contrast, any cross-subsidization of premiums in favor of the high-risk types would amount to a loading of the premium charged to the low-risk types, making them worse off.

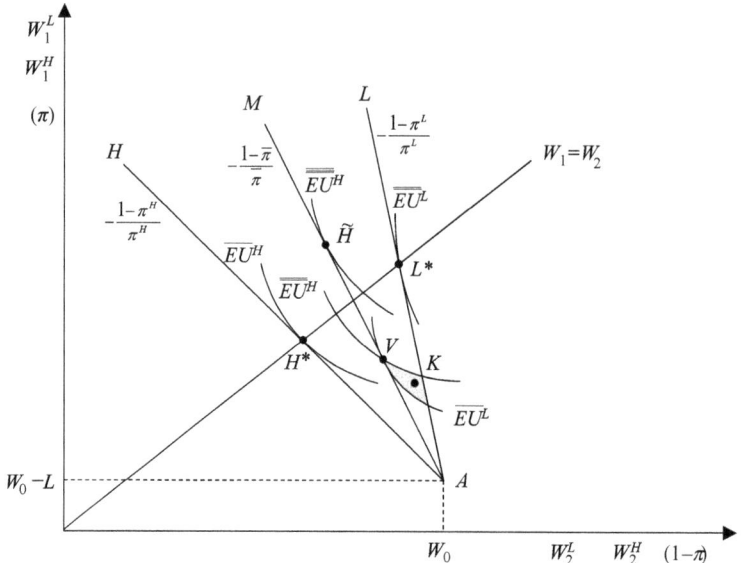

Fig. 8.5 Unsustainability of pooling equilibrium

However, information about the π-values is private by assumption, i.e., known only to the IB not the IC. For this reason, the IC has to fall back on the public information about the average value $\bar{\pi}$ of the population. It therefore writes a so-called pooling contract along the insurance line AM. Two consumer responses are predicted.

- The high-risk type seek to purchase excess coverage. In Fig. 8.5, the insurance line AM offers "too favorable" conditions, inducing an optimum at point $\tilde{H}$.
- The low-risk type opts for partial coverage because the contract is not actuarially fair anymore (see Sect. 3.3). The new optimum is indicated by point V in Fig. 8.5.

Since both contracts satisfy the zero profit condition in expectation, the combination $(\tilde{H}, V)$ is sustainable. However, excess coverage at $\tilde{H}$ means that high-risk types in fact benefit from the occurrence of loss, creating a powerful incentive to increase π^H. While moral hazard effects are otherwise neglected in this section, it would be highly unrealistic to assume that the IC remains inactive when the remedy is simple. In the absence of information about risk types and a uniform premium, a natural choice would be to ration the coverage offered to high-risk types by limiting it to the amount demanded by the others, i.e., the low-risk types. This amount is represented by point V of Fig. 8.5.[9] Since this point satisfies the zero profit condition on expectation, it constitutes a so-called *pooling equilibrium*.

However, such a pooling equilibrium is not sustainable because it can be broken up by a competitor. Consider another IC offering the contract represented by point K in the shaded area of Fig. 8.5. Contract K has the following properties.

- K has less comprehensive coverage in return for a lower premium than the pooling contract V;
- K is preferred by the low-risk type over V because it lies above the pertinent indifference curve $\overline{EU}^L$ through V;
- The high-risk type prefers the pooling contract V because K lies below the pertinent indifference curve $\overline{EU}^H$ through V.

By launching contract K, the competing IC can therefore attract the low-risk types while leaving the high-risk types with the incumbent IC. It has the incentive to do this in view of positive expected profit (note that point K lies below the insurance line AL for low risks). The incumbent IC now suffers from adverse selection because its pool of insureds increasingly consists of high-risk types only. Since this causes its average probability of loss to exceed the population average $\bar{\pi}$ that is compatible with breaking even, the incumbent IC incurs a loss in expectation.

▶ **Conclusion 8.15** In a single-period framework, a pooling equilibrium can always be broken up by a competitor and is therefore not sustainable.

However, it is doubtful whether a competitor whose planning horizon goes beyond one period will attack a pooling equilibrium. As argued by Wilson (1977), the incumbent IC can react by withdrawing contract V from the market as soon as it causes losses. This leaves high-risk types without coverage (at point A in Fig. 8.5). However, they certainly prefer contract K over point A. Provided the market consists of only two ICs, the challenger anticipates the high-risk types ending up with its contract K. Being exposed to the entire risk pool, it will face insurance line AM as the expected zero profit condition, causing contract K to generate losses. If however the incumbent's high-risk types have many other ICs to turn to, the challenger may get away with its attack on the incumbent's pooling contract.

The model of Rothschild and Stiglitz (1976) therefore provides an appropriate description of insurance markets with a low degree of concentration. However, given a high degree of concentration combined with competitive behavior, the pooling equilibrium is sustainable because the potential challenger anticipates the reaction of the incumbent and its consequences. This is the so-called *reaction equilibrium*

---

[9]The subsequent argument makes use of the "single crossing property" of indifference curves which implies that the high-risk types have flatter indifference curves not only at the certainty line but (almost) everywhere.

derived by Wilson (1977). For a generalization to a continuum of types and a survey of other instances of adverse selection, see Riley (2001).

▶ **Conclusion 8.16** Given a sufficiently long planning horizon on the part of insurers and a certain degree of concentration in the market for insurance, a reaction equilibrium is predicted that makes pooling contracts sustainable.

8.3.1.2 The Separating Equilibrium as a Possible Solution

Returning to a strictly single-period analysis, one has to accept Conclusion 8.15. Still, ICs may seek to protect themselves from an attack by a competitor by designing their contracts accordingly at the beginning of the period. Rather than writing a pooling contract, they could offer an actuarially fair contract to the low-risk types while preventing the (unrecognized) high-risk types from opting for it. This amounts to devising a mechanism for self-sorting which can be formulated as an optimization problem as follows [adapted from Eisen (1982)]:

$$\max_{P^L, I^L, P^H, I^H} EU^L = \pi^L \cdot v[W_0 - L - P^L + I^L] + (1 - \pi^L)v[W_0 - P^L] \tag{8.64}$$

$$\text{s.t. } \pi^H \cdot v[W_0 - L - P^H + I^H] + (1 - \pi^H)v[W_0 - P^H]$$
$$\geq \pi^H \cdot v[W_0 - L - P^L + I^L] + (1 - \pi^H)v[W_0 - P^L]; \tag{8.65}$$
$$\pi^H \cdot v[W_0 - L - P^H + I^H] + (1 - \pi^H)v[W_0 - P^H]$$
$$\geq \pi^H v[W_0 - L] + (1 - \pi^H)v[W_0]; \tag{8.66}$$
$$P^H = \pi^H L, \quad P^L = \pi^L L. \tag{8.67}$$

The objective function states that premiums and benefits for both risk types are to be set in a way that the expected utility of the low-risk type is maximized. This choice is dictated by the IC's need to retain its low risks. Constraint (8.65) is crucial because it establishes incentive compatibility. On the left-hand side, it contains the expected utility of a high-risk type (characterized by the unobserved probability of loss π^H) who obtains contract (P^H, I^H) reflecting that type. On the right-hand side, one has the expected utility of that high-risk type (characterized by unobserved π^H), who however could benefit from contract (P^L, I^L) tailored to the low-risk type. If the expected utility on the left-hand side exceeds that on the right-hand side, the high risks see an advantage in accepting the insurance contract designed for their risk type. This prevents them from migrating to the contract designed for the low risks. Condition (8.66) states that the high-risk type does buy the contract designed for it rather than opting for no insurance at all. Expected utility attainable by the high risks (the left-hand side) must thus be at least as high as the expected utility they would derive from bearing the loss occurring with probability π^H themselves (the right-hand side). Finally, Eq. (8.67) states that premiums are actuarially fair, permitting the two contracts to break even.

8.3 Adverse Selection

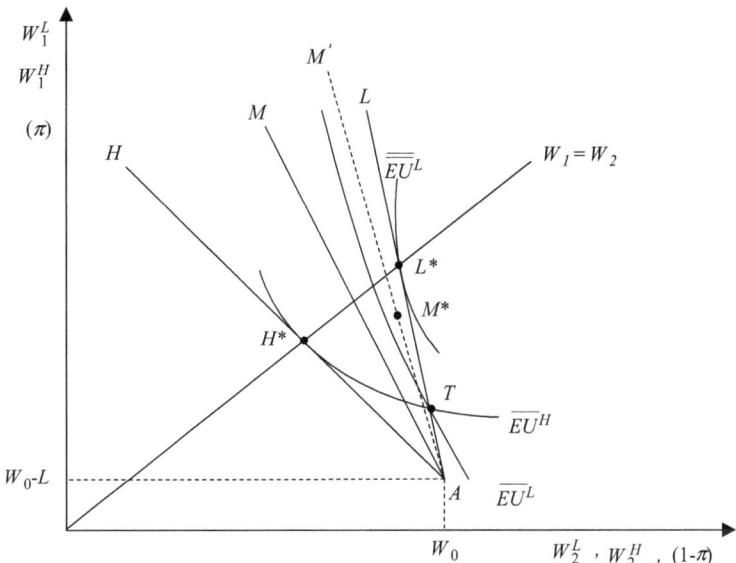

Fig. 8.6 Possibility of a separating equilibrium

The solution to this optimization problem gives rise to a so-called *separating equilibrium*. It is illustrated in Fig. 8.6, symbolized by the pair of points $\{H^*, T\}$. These points represent contracts having the following properties.

- The high-risk types obtain full coverage at H^*, paying the high premium that is actuarially fair to satisfy condition (8.66). Constraint (8.65) is satisfied as well since contract T for the low-risk types does not provide more expected utility than H^*, causing them to remain with H^*.
- The low-risk types obtain partial coverage at T, paying the low premium that is actuarially fair to them. However, they are rationed because at these actuarially fair terms, they would prefer to buy full coverage. But this is not possible because additional coverage along the insurance line AL beyond point T would attract the high-risk types (see their indifference curve $\overline{EU}^H$). The low-risk types therefore suffer a negative externality emanating from the IC's inability to recognize low-risk types. Still, contract T does maximize their expected utility as defined in (8.64), subject to the constraints imposed.
- Neither risk type has an incentive to deviate from their optimum while no IC has an incentive to enter or exit the market. Therefore, the contract pair $\{H^*, T\}$ constitutes a so-called *separating equilibrium*.

For an IC, devising a set of separating contracts is by no means an easy task. In the case of just two risk types however, implementation can be envisaged as follows. The IC first launches a contract offering full coverage at a high premium. It increases the premium to the point where some IBs begin to cancel it. This defines point H^* of

Fig. 8.6, which determines the probability of loss π^H of the high-risk types since H^* lies on the insurance line AH. Next, knowing the shares h and $(1-h)$ of high and low risks, respectively, the IC can infer the value of π^L for the low-risk types from Eq. (8.56). This enables it to launch a second contract designed for the low-risk types at the fair premium based on π^L but offering very limited coverage (at a point below T on the AL insurance line). Those opting for this contract must be low-risk types. Finally, the amount of coverage can be increased to the point where the purchasers of the contract with full coverage (i.e., the high-risk types) want to have it too (at point T).

This procedure can be pursued by all ICs in the market. It results in the same pair of separating contracts since the two risk types are characterized by the same probabilities of loss $\{\pi^L, \pi^H\}$ and indifference curves $\{\overline{EU}^L, \overline{EU}^H\}$. Competitive ICs therefore can implement the contract pair $\{H^*, T\}$ as the separating equilibrium.

However, the implementation of separating contracts becomes much more difficult when introducing just one additional risk type (e.g., a "medium risk"). The contract offered to the medium-risk type needs to be structured in a way as to keep the high-risk types out (which calls for somewhat limited coverage at a rather low premium). On the other hand, it must not provide an incentive for the medium risks to opt for the contract designed for the low-risk types (which calls for close to full coverage at a relatively high premium). Repeatedly modifying the three contracts poses quite a challenge to an IC (not least because consumers may be estranged by repeated contract changes). In addition, Riley (1979) proves sorting to be outright impossible if risk types form a continuum with a probability density over π-values lacking accumulation points (i.e., local maxima that clearly separate risk types).

Finally, note that separating contracts can be broken up by a challenger. Consider the insurance line AM' pertaining to a pooling contract. It runs so steeply as to contain points above the indifference curve $\overline{EU}^L$ through T. These points represent contracts that are able to attract low risks away from the separating contract (see point M^*). A steep insurance line is only possible if the share of low-risk types in the population is high. Intuitively, a high value of $(1-h)$ enables the challenger to offer substantially better benefits to them without attracting many high-risk types.

▶ **Conclusion 8.17** A separating equilibrium in the case of two risk types consists of a pair of contracts, one with comprehensive coverage but high premium for the high-risk type and the other with partial coverage but low premium for the low-risk type. However, it can be broken up by a pooling contract if the share of low-risk types is sufficiently high.

Note that the contract which may break up a separating equilibrium is a pooling contract. At the same time, Conclusion 8.15 states that any pooling equilibrium can be challenged by a competitor. Since this holds for the pooling contract M^* described in Conclusion 8.17 as well, private insurance markets appear to lack a stable equilibrium if the share of low-risk types is substantial, resulting in the dreaded "death spiral". However, if one admits a longer planning horizon on the part of IC, the question again arises of whether attempts at undermining a separating equilibrium are likely to occur. This is not the case for two reasons. First, the pooling contract of the challenger may

8.3 Adverse Selection

be broken up by yet another competitor. Second, Fig. 8.6 shows that contract M^* will attract the high risks of the incumbent, which is especially true once they are forced to migrate, because the incumbent IC withdraws the pair $\{H^*, T\}$ from the market. Their migration to contract M^* would cause the composition of the challenger's insured population to deteriorate; as a consequence, M^* would produce losses in expectation. For these reasons, a separating equilibrium may well be sustainable, with the IC using the procedure outlined above to implement it. In addition, if the conditions conducive to a Wilson (1977) reaction equilibrium hold, pooling contracts may survive as well.

8.3.2 Empirical Relevance of Adverse Selection

The core prediction of the preceding section (Conclusion 8.15) is that, in the presence of asymmetric information, a pooling equilibrium in the market for insurance may not be sustainable because it is profitable for a challenger to break it up. However, the empirical evidence presented below shows that it does not even take a challenger to trigger the process of adverse selection. It relates to the experience of health insurers who enroll employees of Harvard University [see Cutler and Reber (1998)].

Harvard University strikes group contracts with health insurers on behalf of its employees. It imposes pooling contracts by requiring that insurers admit all IBs at uniform terms (so-called open enrollment). Prior to 1995 (when Harvard University implemented a change) employees could choose between six variants A to F. Only the popular choices A, E, and F are detailed in Table 8.6. Up to 1994, more than 80% of individuals had subscribed to the most expensive policy A. It featured almost unlimited choice of physician and practically no utilization review.[10] This contract cost the employer US\$ 2,773 per year for an individual and about twice as much for a family (not shown here). The cheapest contracts E and F cost US\$ 1,945 and 1,957, respectively, or more than US\$ 800 less than A. In return, they were of the Health Maintenance Organization type, limiting provider choice to physicians participating in a network and requiring prior insurer approval for hospitalization. Before the 1995 change, employees contributed US\$ 555 to the most expensive contract and some US\$ 250 to the least expensive ones. Therefore, for about US\$ 300 more, Harvard employees could afford the most generous rather than a stingy low-cost policy.

In relation to healthcare services covered, contract A provides access to the entire universe but E and F only to a subset since they exclude some services, which must be paid for out of pocket by the IB. A transition from A to E or F can therefore be likened to a reduction of coverage, i.e., a movement away from the certainty line in (W_1, W_2)-space (see point K of Fig. 8.7).

Deeming the cost of health insurance unsustainable, Harvard University decided to make its employees contribute more, to the tune of US\$ 1,152 per year for the

---

[10] Utilization review involves checks of physician and hospital billings. It fosters a conservative practice style on the part of healthcare providers.

Table 8.6 Health insurance policies provided by Harvard University, before and after the 1995 change

	Gross premium US$		Net premium for the employee US$			
		Difference from (A)	Old	Difference from (A)	New	Difference from (A)
(A) Most expensive policy	2,773	–	555	–	1,152	–
(E) Second-cheapest policy	1,957	816	253	302	384	768
(F) Lowest-cost policy	1,945	828	235	320	396	756

Source Cutler and Reber (1998), Table II

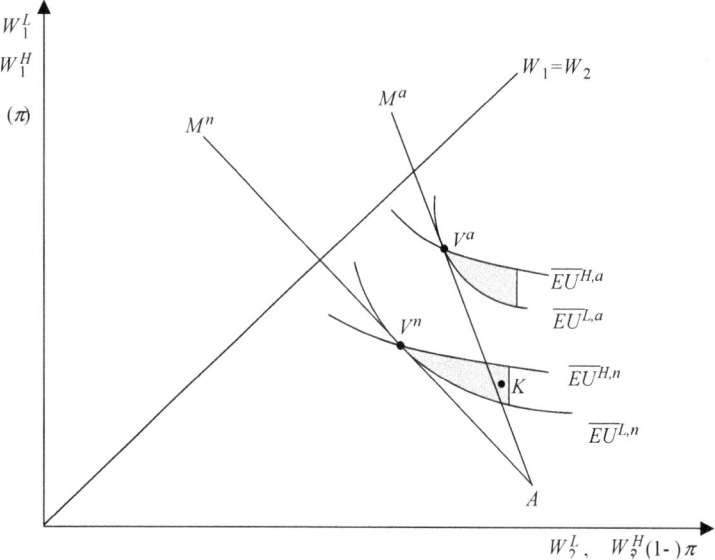

Fig. 8.7 The 1995 change as a trigger of adverse selection

most expensive contract A. Employee contributions were increased only slightly for contracts E, F. Opting for A now cost between US$ 756 and 768 extra rather than between US$ 302 and 320 as before (see Table 8.6). Thus, the extra benefits provided by contract A became more costly to employees.

Figure 8.7 shows the impact of this change on the pooling equilibrium associated with contract A. Prior to the change, it is represented by point V^a on the insurance line AM^a, close to the certainty line. Low-risk purchasers of contract A are characterized by a steeply sloped indifference curve $\overline{EU}^{L,a}$ through this point and high-risk ones, by a flat indifference curve $\overline{EU}^{H,a}$. The shaded area indicates that there had

8.3 Adverse Selection

Table 8.7 Migration between health insurance policies, Harvard University employees

	Structure of 1994 membership			
	A: Most expensive policy		(E, F): 2 least expensive policies	
Membership 1995	A	(E, F)	(E, F)	A
Stayers in 1995	85%		99%	
Movers in 1995		15%		1%
Average age in 1994[a]	50	46	41	46
Index of treatment cost[b]	1.16	1.09	0.96	1.0

Notes
[a] The differences in average age are statistically significant at the 5% level between members who change and those who do not change
[b] Average value over all individuals set to 1.00
Source Cutler and Reber (1998), Table IV

been scope for competing contracts to attract the low risks. Let point K be such a contract, representing both policies (E, F) for simplicity. However, the limited benefits associated with K outweighed its lower net premium, making it unattractive for the low risks, who stayed in the pool formed by contract A.

After the change, policy A, having become more costly to employees, now lies on the less favorable insurance line AM^n (for simplicity, the fact that net premiums for E and F also increased somewhat is neglected). As a consequence, alternative K is now attractive to low risks, without any activity on the part of health insurers. Thus, this analysis leads to two predictions:

- The most expensive policy A loses market share to contracts (E, F);
- Employees changing to (E, F) are low-risk types.

These predictions are fully confirmed by the migration patterns observed among employees of Harvard University (see Table 8.7). For simplicity, contracts B to D are neglected as before. Among those employees who had the most expensive policy A in 1994, 85% remained with A after the change. Their average age was a high 50 years, and their cost of treatment had an index value of 1.16, indicating an excess of 16% over the average of all employees. The other 15% migrated to the least costly alternatives (E, F); their average age was 46 years and their cost index, 1.09 only.

By way of contrast, migrations away from contracts (E, F) hardly occurred. Among those employees who had one of the two lowest-cost policies in 1994, 99% remained with them after the change. Their average age was 41 years, and their cost index a low 0.96. The 1% who did change in favor of the more expensive but more comprehensive alternative A were significantly older (46 years) and had higher treatment cost (index value 1.0) than those staying with (E, F). Evidently, policy A was turning into a high-risk pool.

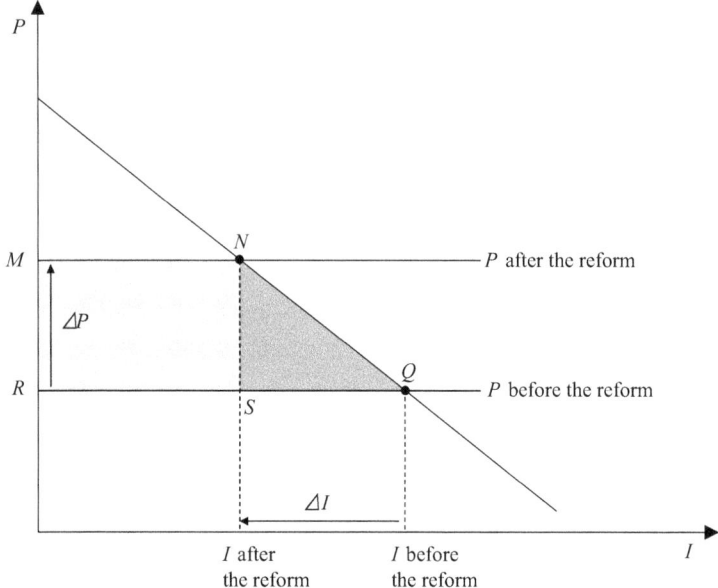

Fig. 8.8 Aggregate deadweight loss due to the increased net premium

In response to this development, the insurer in charge of A increased premiums by 16% in real terms. Low risks opted out of A even more, causing this policy to be withdrawn by 1997.

The change of 1995 generated savings to Harvard University amounting to 5–8% of premiums. According to Cutler and Reber (1998), however, these savings were in the main due to profit reductions suffered by participating health insurers rather than improved control of moral hazard effects since there were no signs of adjustment in the behavior of either insured patients or service providers. Therefore, the change resulted in a mere redistribution of income rather than a saving of resources.

Adverse selection caused contract A to be withdrawn although there certainly was demand for it. The associated loss of consumer surplus is illustrated in Fig. 8.8. Since there were no resource savings, true marginal cost of health insurance remains the same, leaving producer surplus unchanged (not shown). Therefore, the change in aggregate welfare reduces to the change in consumer surplus, which amounts to the deadweight loss given by

$$\text{Deadweight loss per IB} = \frac{1}{N}\left[\frac{1}{2} \cdot \triangle P \cdot \triangle I\right]. \tag{8.68}$$

$MNQR$: Loss of consumer surplus due to the increase in the price of coverage $\triangle P$;
$MNSR$: Additional premium revenue of insurer or savings accruing to Harvard University, respectively;
NQS: Deadweight loss.

According to the authors, the price elasticity of demand for contract A is so low that its net premium would have had to rise by US$ 2,000 annually to choke off the demand for it (thus, $\Delta P =$ US$ 2,000; this needs to be estimated because A was in fact withdrawn). Since the market share of A was 20% in 1994, the change in quantity amounts to $\Delta I = 0.2 \cdot N$, with N denoting the number of insured Harvard employees. The deadweight loss per IB can therefore be calculated as

$$\text{Deadweight loss per IB} = \frac{1}{N}\left[\frac{1}{2} \cdot 2000 \cdot 0.2 \cdot N\right]$$
$$= \text{US \$ 200 per year and enrollee.} \qquad (8.69)$$

This estimate equals almost 5% of average healthcare expenditure at the time. It quite likely is an overestimate since the IBs were not without coverage after withdrawal of contract A but still had a choice among alternatives B to E. On the other hand, some very sick Harvard employees may have been willing to pay even more than US$ 2,000 to be able to keep policy A. Note, however, that Cutler and Reber (1998) do not conclude there was a "death spiral"; they mention that the loss-making policy was withdrawn from the market, jeopardizing the economic survival only in the unlikely case of a single-product company.

More recently, Frech and Smith (2015) found evidence of another "death spiral", but again concerning a single plan rather than a company. Specifically, the authors note that Prudential began to cease selling its Coordinated Health Insurance Plan in California in 1981. Within the same year, the number of policies in force dropped from some 270,000 to fewer than 12,000. Yet it was only after another decade that Prudential started to sharply increase premiums affecting approximately 5,000 policyholders, leading the authors to refer to the process as "truly a slow-motion death spiral" (p. 68).

Even more importantly, Einav et al. (2010) arrive at an amazingly small welfare loss due to adverse selection. After a health insurance premium increase imposed by an employer (which precludes reverse causality running from healthcare expenditure to the premium), they do find that coverage decreased. Indeed the insurer needs to ration coverage to break even (where premium equals average cost). Since the authors are able to construct a demand curve, they can also estimate the efficiency loss as in Fig. 8.8, which equals the deadweight loss NQS. However, this area amounts to a mere US$ 10 per employee. Relative to total welfare attainable through the purchase of health insurance (i.e., the area below the demand curve up to the point on the I axis below Q), the welfare loss caused by adverse selection amounts to no more than 3% [for more details and additional evidence beyond health insurance, see Zweifel (2013)].

▶ **Conclusion 8.18** Experience with health insurance provided by Harvard University suggests that a pooling equilibrium can break up in the course of a few years even without an active challenger. The welfare loss caused by such a breakup may amount to several percent of benefits paid.

8.3.3 Adverse Selection in a Multi-period Context

When the analysis of adverse selection is extended to include several periods, the IC has the possibility of learning from consumers' loss experience, enabling it to infer the true probability of loss with increasing accuracy. The concomitant adjustment of premiums is called *experience rating*; it often takes the form of a *bonus-malus scheme*. A monopolistic IC, who does not stand to lose low-risk types to a competitor in the process, may even achieve a perfect risk categorization in the very long run. In the resulting equilibrium, each risk type pays the appropriate fair premium and obtains full coverage, indicating that a first-best solution is reached [see Dionne and Lasserre (1985)].

If this learning process occurs in a competitive market, an individual IC can perform experience rating only subject to certain restrictions. The presentation below follows Dionne and Doherty (1994).

First, note that the set of viable contracts is reduced as soon as more than one time period is considered. Indeed, contracts have to be *renegotiation-proof* in the following sense. The point of departure is a separating equilibrium as discussed in Sect. 8.3.1. In Fig. 8.6, the low risks are rationed at point T. However, their choice of contract serves as a signaling device; consumers who accept to be rationed at T during the first period are identified as low risks. At the same time, they have an interest in buying additional coverage at the favorable terms applicable to low-risk types. An IC satisfying this desire would offer a second-period contract that lies above point T on the insurance line AL. However, this creates an incentive for the high risks to first purchase the contract designed for the low-risk types (which makes them look like low risks) and then to add coverage at the favorable terms pertinent to low-risk types. Clearly, contracts that are not proof to this type of renegotiation cannot be part of a sustainable equilibrium. For this reason, an equilibrium on the insurance market must satisfy the additional constraint of being renegotiation-proof as soon as the analysis is extended to several periods.

On the other hand, the IC now can go beyond a mere differentiation of contracts in terms of premium and degree of coverage for controlling adverse selection. It observes whether or not a loss has occurred at the end of the first contractual period, providing it with the opportunity of restructuring contracts on the basis of this additional information. In the most simple version of the model, the IC applies experience rating while committing for two periods (years). This commitment is often called *guaranteed renewability*.

Figure 8.9 depicts the decision-making situation of an IB over two periods given commitment by the IC. In the first period, the IC offers two contracts. One is a conventional contract designed for the high-risk type, with benefits I^H and premium P^H regardless of loss experience. It therefore offers the same terms during the second period. In order to accommodate the low risks as much as possible (i.e., to avoid rationing their coverage to the greatest extent possible), this contract (I^H, P^H) must offer full coverage. The other contract is with experience rating. In the first period, this is a pooling contract since it will also be chosen by a share $(1 - x^H)$ of the high-risk types. Its terms are denoted by (I^M, P^M). In this way, consumers get a

8.3 Adverse Selection

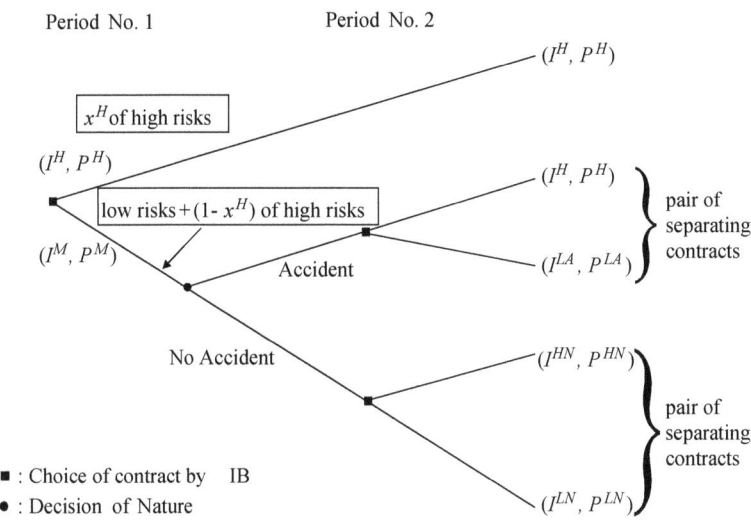

Fig. 8.9 Decision sequence in the context of experience rating

chance to benefit from better terms in period No. 2, provided they are without a loss. Note also that initially, a separating contract is not feasible because it would not be renegotiation-proof.

In the course of period No. 1, the IC can observe whether a loss ("accident", A) or no loss (N) has occurred, permitting it to propose two sets of separating contracts. One set applies to those insured who had an accident. It contains the contract (I^H, P^H) with full coverage and high premium that was designed for the high risks to begin with but also an alternative (I^{LA}, P^{LA}) designed for low risks who happened to have an accident. The other pair of separating contracts is offered to the insured who were without an accident. Below, only this second contract pair is described in detail. It is of particular interest because the IC must seek to retain the low-risk types while safeguarding its financial equilibrium (see also Table 8.8). Yet it still needs to set premiums and benefits in a way as to maximize the expected utility of low risks, who always have the option of migrating to a competitor. Therefore, the IC is forced to act as a perfect agent on their behalf, resulting in the objective function,

$$\max_{P^{HN}, P^{LN}, I^{LN}} EU^L = \pi^L \cdot v^L[W_0 - L + I^{LN}] + (1 - \pi^L)v^L[W_0 - P^{LN}]. \tag{8.70}$$

By setting the second-year premium P^{LN} and benefit I^{PN}, the IC seeks to attract low risks without a loss to its separating contract. These two quantities therefore appear as decision variables in Eq. (8.70). With regard to the high-risk types, the one remaining decision variable is P^{HN}; the pertinent benefit I^{HN} follows from the zero expected profit condition of the insurer. Contrary to the usual formulation, the notation here indicates that the premium must only be paid in the no-loss state. This simplification is without consequence, however, because premiums can be adjusted

Table 8.8 Optimization problem with experience rating (no-loss state in the first period)

$$\max_{P^{HN}, P^{LN}, I^{LN}} EU^L = \pi^L \cdot v[W_0 - L + I^{LN}] + (1 - \pi^L)v^L[W_0 - P^{LN}] \quad (8.70)$$

$$\text{s.t.} v^H[W_0 - P^{LN}] \geq \pi^H \cdot v^H[W_0 - L + I^{LN}] + (1 - \pi^H) \cdot v^H[W_0 - P^{LN}]; \quad (8.71)$$

$$\bar{T}^{LN}(x^H) \leq (1 - \pi^L)P^{LN} - \pi^L I^{LN}; \quad (8.72)$$

$$v^s[W_0 - P^{HN}] \geq v^H[W_0 - \bar{P}^{HN}(x^H)]. \quad (8.73)$$

I^{LN}:	Benefit of the second period, offered to a low risk without loss in the first period
P^{HN}:	Premium of the second period, paid by a high risk without loss in the first period
$\bar{P}^{HN}$:	Premium calculated for the second period given no loss in the first period, offered to high risks, $d\bar{P}^{HN}/dx^H < 0$
$\bar{T}^{LN}$:	Transfer in favor of low risks without loss, "premium rebate for no loss", depends on the share of high risks x^H sorted out during the first period; $d\bar{T}^{LN}/dx^H > 0$
x^H:	Share of high risks that choose the contract without experience rating
$1 - x^H$:	Share of high risks that choose the contract with experience rating

accordingly. In addition, the risk utility function $v^L(\cdot)$ has the same form as $v^H(\cdot)$; the superscripts are only added for easy identification of risk types.

A first constraint ensures that the high risks that happened to be without a loss during the first year stick to the contract designed for them. This is formulated in Eq. (8.71) below, whose left-hand side shows the (fixed) value of the risk utility function of a high-risk type who pays premium P^{HN} in return for full coverage I^{HN} in case of loss L; therefore the two quantities cancel. Its right-hand side contains the loss probability π^H and the risk utility function $v^H(\cdot)$, indicating that this continues to be a high risk having the option of contractual terms (I^{LN}, P^{LN}) with partial coverage, tailored to the low-risk type. However, as long as the inequality below is satisfied, the high risks will not infiltrate the contract designed for the low ones,

$$v^H[W_0 - P^{HN}] \geq \pi^H \cdot v^H[W_0 - L + I^{LN}] + (1 - \pi^H) \cdot v^H[W_0 - P^{LN}]. \quad (8.71)$$

Another constraint states that the IC must be able to finance the premium reduction awarded to consumers with a favorable loss experience, which comes from an excess of premium over benefits paid in the first period (in expectation). Note that such a transfer $\bar{T}^{LN}$ can only be due to the low risks; it therefore depends positively on x^H, the share of high risks who were made to choose the contract designed for them in the first period. The higher x^H, the more successful was the insurer in sorting the high risks out by offering the contract without experience rating. However, low risks without a claim during the first period can still suffer a loss during the second period. The transfer $\bar{T}^{LN}$ (which amounts to a rebate for no claims) therefore is bounded by the favorable second-period loss experience, i.e., an excess of expected premium

$(1-\pi^L)P^{LN}$ over expected loss $\pi^L I^{LN}$,

$$\bar{T}^{LN}(x^H) \leq (1-\pi^L)P^{LN} - \pi^L I^{LN}. \tag{8.72}$$

The last constraint excludes renegotiation. It reads

$$v^H[W_0 - P^{HN}] \geq v^H[W_0 - \bar{P}^{HN}(x^H)]. \tag{8.73}$$

This condition says that high risks who did not present a loss during the first year obtain a rebate for no claims as well. Their second-period premium P^{HN} must be lower than the premium $\bar{P}^{HN}$ the IC committed to at the beginning of the first period. This premium was calculated to be the lower, the more success the IC had with the sorting out of high risks during the first year (x^H). Note that condition (8.73) reflects the one-sided commitment on the part of the IC in the context of experience rating. Indeed, it states that the IC sets the premium in a way that high risks are willing to stay with the contract designed for them even though they were without a loss during the first period.

Rather than going through the mathematical details of this optimization problem, its solution is illustrated by Fig. 8.10. The IBs who incur a loss are offered relatively unfavorable separating contracts. The contract with full coverage and high premium (I^H, P^H) is designed for the high risks, who opt for point H_A^*. The low risks who were unfortunate to suffer a loss during the first period opt for a contract (I^{LA}, P^{LA}) with limited benefits but low premium (point T_A in Fig. 8.10) below.

All consumers without a loss receive a premium reduction in the second year. For the high risks, this is symbolized by the difference between H_N^* and H_A^*, which can be interpreted as a premium rebate for no claims. This rebate must not change the marginal price of additional coverage lest the IC would create an incentive for renegotiation. For the low risks, the transfer they receive is depicted by the transition from T_A to T_N. Note that the rebate paid to the high risks serves to relax the rationing constraint imposed on the low risks, who obtain additional coverage at the same premium rate.[11] Figure 8.10 also shows the pooling contract of the first period with insurance line $A^M M$. Depending on x^H, the share of the high risks that opt for the contract without experience rating, its slope lies between that of $A^E H$ for the high and $A^E L$ for the low risks. In addition, the pooling contract includes a fixed loading which serves to finance the second-period premium rebates to the insured without incurring a loss. Since this is charged in both states, it is reflected by the horizontal and vertical distance between endowment points A^E and A^M. For such a loading

---

[11] In actual practice, the IB obtains the same coverage at a lower *ex-post* premium. However, their additional coverage is usually charged at the full rather than a reduced premium rate. In Fig. 8.10, there accordingly is no rotation of the budget line $A^E L$. The most simple way to represent the benefit accruing to the low-risk type is therefore the movement from T_A to T_N on the unchanged insurance line.

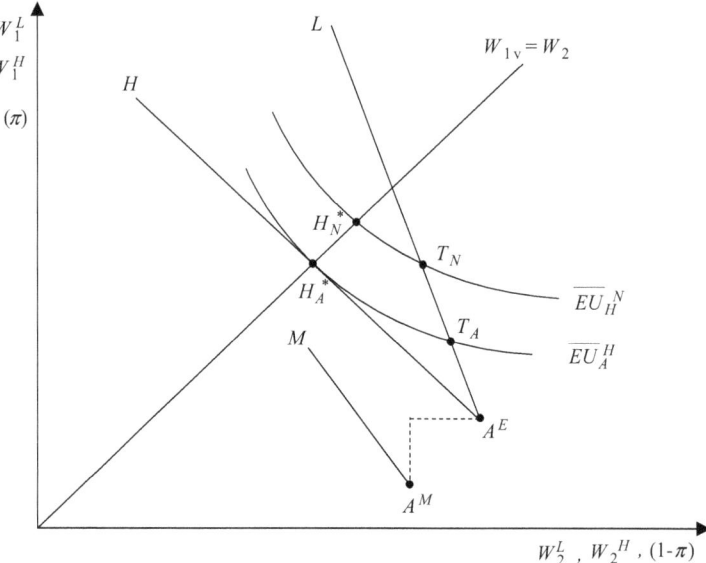

Fig. 8.10 Experience rating of premium and first-period pooling contract

to be accepted, the consumer's rate of time preference must not be very marked (a condition that is neglected here).[12]

In sum, Fig. 8.10 shows the configuration of contracts to have the following properties.

(1) During the first year, there is a pooling contract with partial coverage, combined with a separating contract for the second year;
(2) The separating contract of the second year offers full coverage to high and (still) partial coverage to the low risks;
(3) The second-year premiums are experience-rated, awarding rebates for no claims to both types of risk who are without a loss;
(4) The premium rebate for no claims paid to the high risks is financed by a loading contained in the premium of the first-year pooling contract;
(5) In equilibrium, the IC has positive expected first-year profits and negative second-year ones (summing to zero in present value).

By paying back the first-year surcharge through the second-year premium rebate, the IC can prevent the low risks from changing to a competitor [see condition (8.72)]. At the same time, the high risks enjoy guaranteed renewability on previously defined

---

[12] Individuals with a high rate of time preference discount heavily benefits (consumption, wealth) and costs that accrue in the future rather than in the present. In the two-goods model of Fig. 10.6 (Sect. 10.5.1.2), they are characterized by a steeply sloped indifference curve. For more detail, see, e.g., Hirshleifer et al. (2005), Ch. 15.7.

8.3 Adverse Selection
371

terms [condition (8.73)]. With all ICs finding their optimum on the basis of the same information and the same contractual configuration, there is no incentive for any one of them to deviate. This configuration therefore constitutes a Nash equilibrium.

▶ **Conclusion 8.19** in a private insurance market, an equilibrium is possible in which a first-period pooling contract is followed by an experience-rated second-period separating contract combined with guaranteed renewability. Both low and high risks are at least as well off as without experience rating; those without a loss during the first period are better off.

8.3.4 Empirical Evidence Regarding the Experience-Rating Model

Property (5) stated in the preceding section is not universally accepted. Notably, Kunreuther and Pauly (1985) had argued that the IB is short-sighted, permitting the IC not to commit with regard to the second-period premium [i.e., to neglect condition (8.73)]. At the same time, the IC continues to benefit from the information gain due to loss experience. In this way, it can reap an expected profit also during the second period.

If, however, experience rating typically combines with guaranteed renewability, the first-period premium would have to be "too expensive" since it contains a surcharge designed to finance the rebate for the IB without a loss during the second year. This is often called "highballing". Conversely, if the IC fails to commit to guaranteed renewability while "locking in" consumers using its information gain from loss experience, the premium would have to be excessive in the later years of the contract. This is called "lowballing".

A direct test of this set of predictions is not possible using publicly available data because ICs only report total premium volume rather than premiums earned according to the year of contract. However, Dionne and Doherty (1994) find a way around this difficulty by arguing that an IC exhibiting a rapidly increasing premium volume must have a high share of newly written contracts. If they indeed combine guaranteed renewability with experience rating, their ratio of losses paid relative to premiums written as (L/P) would have to be low.

On the other hand, there are other reasons why an IC with rapidly growing premium volume should exhibit a rather low value of the loss ratio L/P. One of them is that new business comes with a great deal of administrative expense (see Sect. 7.4.2), causing premiums to be high relative to loss payments. The strength of this connection depends on the IC's underwriting policy, however, permitting to test for its importance. According to Conclusion 8.19 of the preceding section, it is through experience rating combined with guaranteed renewability that the IC is able to attract low risks. Therefore, to the extent that the IC successfully attracts low risks using this policy, it is characterized by a low L/P value. In this case, the negative relationship between L/P and premium growth would have to be especially marked.

The hypothesis to be tested therefore states,

Table 8.9 Ratio of losses 1986–1988 to net premiums 1985–1987, auto insurance in California

Explanatory variable	Losses per vehicle[b]		
	Low	Medium	High
Constant	0.919**	1.2085**	0.833**
	(5.651)	(5.956)	(5.842)
Premium growth	−0.906**	0.0686	0.0209
	(−2.309)	(0.724)	(1.76)
Direct writer = 1	0.0689	−0.0911	0.1447
	(0.808)	(−1.306)	(1.016)
Rating by A.M. Best[a]	0.0430	−0.0726	0.0863
	(0.683)	(−1.098)	(1.617)
$\bar{R}^2$ (OLS)	0.12	0.01	0.19
N	20	32	30

[a]The ratings have the following numerical representation: A++ = 9, AA = 8, etc
The explanatory variable is the square root of this value
[b]t-ratios in parentheses
**Coefficient different from zero with 1% error probability
Source Dionne and Doherty (1994)

H_0 : There is a markedly negative relationship between the loss ratio (losses paid/premiums) and premium growth among those ICs who use experience rating.

In Table 8.9, the variable analyzed by Dionne and Doherty (1994) is L/P, the ratio of losses paid between 1986 and 1988 over premiums written between 1985 and 1987 (net of reinsurance) of a total of 82 ICs who had auto insurance business in California. Among them, 20 exhibit low losses per vehicle (the unit of risk insured). In each of three groups (low, medium, and high losses per vehicle), L/P is related to premium growth and two additional explanatory variables. In view of the very low coefficient of determination ($R^2 = 0.01$) in the regression for the medium group, results pertaining to this category are not discussed below.

- *Premium growth:* There is no clear relationship between L/P and premium growth in this sample, except for the ICs with low losses per vehicle. This result clearly supports hypothesis H_0. In addition, the effect is not negligible. A premium growth of 10 percentage points (0.1) above the average causes the L/P ratio to drop by an estimated 0.09 points, away from a sample average of 1.01.
- *Distribution through direct writers:* This variable has the value 1 if the IC sells its products through direct writers (i.e., employed personnel) and zero otherwise. At least in the United States, one ascribes a cost advantage to direct writers (especially in comparison to independent agents) that should be reflected in a lower premium and hence an increased value of L/P (see Sect. 6.4.3). In Table 8.9, this expectation is not confirmed, however.
- *Rating by A.M. Best:* Similar to, e.g., Moody's for banks, A.M. Best specializes in the collection and analysis of insurance data and publishes ratings. An "A++" rating indicates an insurer in a very sound financial condition with more than sufficient assets to match its liabilities. Such an IC achieves high incomes from

capital investment, enabling it to lower premiums in its underwriting business (performing so-called cash flow underwriting). A favorable rating by A.M. Best therefore is predicted to go along with a high L/P ratio. However, this expectation is not confirmed either.

In sum, the crucial prediction of H_0 with regard to experience rating combined with guaranteed renewability by the IC receives a measure of confirmation. However, since the (well-founded) predictions relating to mode of distribution and financial status of the IC fail to be confirmed, the findings do not amount to a full confirmation of the hypothesis.

▶ **Conclusion 8.20** Preliminary evidence tends to support the model "experience rating combined with guaranteed renewability" against the alternative "no experience rating, no guaranteed renewability".

8.4 Adverse Selection and Moral Hazard in Combination

The analysis of optimal structuring of contracts and equilibrium in insurance markets calls for considerable modeling effort even when only one of the two types of asymmetric information is presented. For this reason, it comes at no surprise that the analysis of the joint influence of adverse selection and moral hazard is not too well developed [see Eisen (1990)].

The exposition here is limited to the contribution by Steward (1994). The author introduces the extra assumption that a high risk has a high probability of loss π^H for two reasons:

(1) $\pi^H[V^H = 0]$ is high, i.e., the high-risk types have a high initial value of $\bar{\pi}^H$ before prevention effort V can have any impact;
(2) $C'(V^H)$ is high, i.e., the high-risk types must bear a high marginal cost if they seek to reduce π^H.

Figure 8.11 illustrates these assumptions. Its upper part shows that the high-risk type starts from a higher initial value of probability of loss $\bar{\pi}^H[0]$ than the low-risk type with $\bar{\pi}^L[0]$. However, the vertical distance between the functions $\pi^H(V^H)$ and $\pi^L(V^L)$ is constant, indicating that the marginal effectiveness of preventive effort does not depend on the risk type. Therefore, the high-risk type could in principle reduce his or her probability of loss below the level of the low-risk type by exerting a great deal of preventive effort.

The lower part of Fig. 8.11 explains why this does not happen. Due to the identical marginal effectiveness of prevention, marginal benefit MB [defined in Eq. (8.5) of Sect. 8.2.2.1] is identical between risk types as well. However, by assumption the marginal cost MC of prevention $C'(V^H)$ is high for the high risks but low $[C'(V^L)]$ for the low risks. Therefore, the high risks undertake less preventive effort

Fig. 8.11 Risk types and their marginal benefit and cost of prevention V

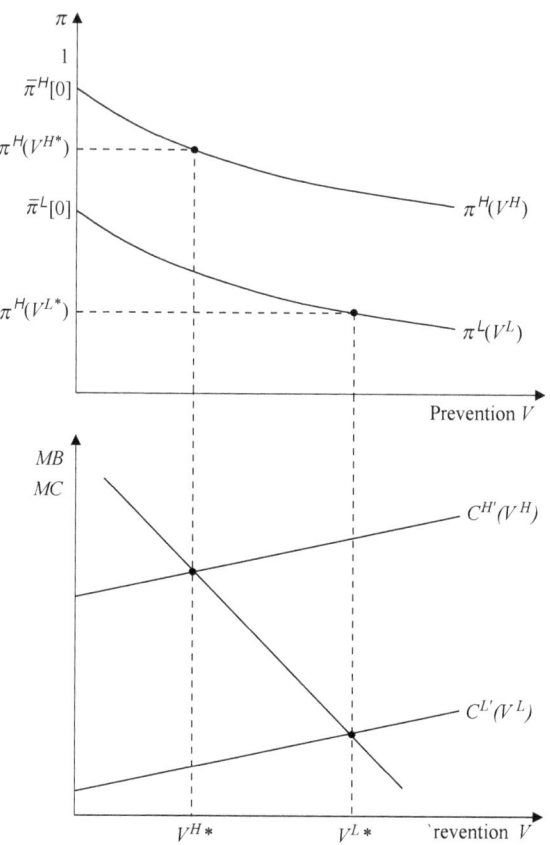

(V^{H*}) than the low ones (V^{L*}). These effort levels determine the effective probabilities of loss $\pi^H[V^{H*}]$ and $\pi^L[V^{L*}]$, respectively, which can be observed in the upper part of Fig. 8.11. Clearly, the difference between these two probabilities is greater than the difference between the two initial ones $\{\bar{\pi}^H[0], \bar{\pi}^L[0]\}$.

Therefore, moral hazard serves to accentuate the difference between the two risk types. This facilitates the design of separating contracts for the following reasons. The high-risk types must be offered full coverage, as before. This causes them to lose all (financial) incentive for preventive effort (maximum *ex-ante* moral hazard, $V^{H*} = 0$). Recall that the low-risk types cannot have full coverage for a separating equilibrium to be sustainable. However, this has the beneficial side effect of inducing prevention on their part. In view of assumption (2) above, this effect is comparatively marked. Therefore, separating contracts go along with a mitigation of *ex-ante* moral hazard on the part of the low-risk types. Considered from the other perspective, the necessity of dampening moral hazard effects among the low-risk types calls for a limitation of their coverage, which in turn helps to render separating contracts sustainable. Therefore, by attacking the problem of moral hazard, ICs can mitigate or even entirely solve the problem of adverse selection.

8.4 Adverse Selection and Moral Hazard in Combination

This argument shows that the efficiency losses caused by moral hazard and adverse selection are subadditive. Attributing to private insurance markets the negative effects of both moral hazard and of adverse selection without correcting for double counting is inappropriate.

Finally, one can also derive a statement about the efficiency gains social insurance may achieve over private insurance. Social insurance solves the problem of adverse selection through compulsory coverage combined with a public monopoly. Since separating contracts are not necessary anymore, uniform insurance conditions can be offered to all risk types. This permits low-risk types to obtain more comprehensive insurance coverage but undermines their incentive to limit moral hazard. In this case, the solution of the adverse selection problem does not contribute to the solution of the moral hazard problem—quite to the contrary. Therefore, the possible efficiency gain of social insurance over private insurance is limited if both types of asymmetric information are present.

This still leaves open the issue of how to empirically distinguish adverse selection from moral hazard. The key is to extend the analysis to more than one period, as in Dionne et al. (2013). In a first period, the IB determines the probability of future accidents by exerting preventive effort, which is unobservable. However, both IBs and ICs learn about risk over time. If the IB learns faster than the IC, the result is adverse selection; if they learn at the same pace, it ultimately is full information.

In the authors' theoretical model, contracts are renewed annually; they are of the comprehensive (CC) or limited liability (LL) type, with the amount of loss (L) fixed (this excludes *ex-post* moral hazard). Drivers with a CC contract pay a deductible $f < L$ when at fault. Both contract types are experience-rated,

$$P_t(b_t, d_t) = b_t exp(\rho_0 + \rho_1 d_t). \tag{8.74}$$

Here, P_t symbolizes the premium in period t, b_t ($0 < b_t < 1$) is the bonus-malus coefficient, and d_t the contract type (with $d_t = 0$ for an LL contract and $d_t = 1$ for a CC contract). The base premium is given by $exp(\rho_0) > 0$ and decreases over time depending on b_t and d_t as long as there is no loss reported.

Preventive effort has just two values such that $V_t = \{0, 1\}$. If $V_t = 1$, the probability of all accidents (i.e., regardless of whether the driver is at fault) is reduced during the contract year. The baseline accident probability also takes on two possible values only, $\alpha = \{0, \alpha_H\}$, where α_H denotes the high-risk type. A restrictive assumption is that α_H is uncorrelated over time. The share of high-risk types is h. The subjective probability of being a high risk π_t is given by

$$\pi_t(\alpha_H | n_1, \ldots, n_{t-1}; V_1, \ldots, V_{t-1}), \tag{8.75}$$

where $n_1, \ldots, n_{t-1}$ is a series of categorical variables reflecting the accident history ($n_i = 1$ if an accident occurred, $= 0$ otherwise) and $V_1, \ldots, V_{t-1}$ (again a series of categorical variables) reflects the effort history. Interestingly, preventive effort reinforces the effect of an accident on the subjective probability, and there is no learning by driving.

Decisions regarding preventive effort and contract choice are made simultaneously, with the aim of maximizing the objective function,

$$Z_t(s_t, n_{t-1}, y_t) = \max_{V_t, d_t} u(c_t, V_t) \tag{8.76}$$
$$+ \beta \int \sum_n Z_{t+1}(s_{t+1}, n, y_{t+1}) p(n_t | V_t, \pi_t) dF(y_{t+1} y_t),$$

where s_t denotes the driver's current state, n_{t-1} the anticipated occurrence of an accident (yes/no), y_t is current income, $u(\cdot)$ is utility defined over consumption c_t and effort V_t, β ($0 < \beta < 1$) is the subjective discount factor which transforms the sum of future values Z_{t+1} to present value, and $dF(y_t | y_{t-1})$ is the density function defined over incomes, assumed to be known. There are no assets, so consumption equals income minus the insurance premium and the deductible to be paid (CC policy).

Even with these simplifications, there is no closed solution to Eq. (8.76), forcing the authors to solve it numerically for a set of parameters (shown in their Table 1). They simulate 2,000 drivers over 40 periods until they reach 25 years of experience.

The results of these simulations are reported in Fig. 8.12. They exhibit a clear distinction between the two risk types (a solid line for low risk; a dashed line for high risk) which grows larger with experience as drivers learn their type. The first panel shows the subjective probability of being a high-risk type, reflecting a slow pace of learning in spite of relatively frequent accidents. Accordingly, their bonus-malus coefficient (and hence their premium) remains high, contrary to the low-risk types who rapidly reach a bonus-malus coefficient of 0.5, corresponding to one-half of the base premium. This is the consequence of few at-fault accidents (no-fault ones do not count).

While the share of CC contracts is similar for both risk types initially, it increases among the high-risk drivers who realize their type. The authors conclude, "Over time, a positive correlation emerges between risk and coverage, which leads to adverse selection, as predicted by asymmetric learning" (p. 907-8). They then formulate three hypotheses.

(1) MH1. There is no evidence of moral hazard if the bonus-malus coefficient b_t has no effect on the accident distribution. A negative effect is consistent with the presence of moral hazard (experience rating reins in moral hazard).

However, a negative effect of the bonus-malus coefficient on the accident distribution could also reflect learning about risk. This calls for testing a second hypothesis.

(2) MH2. There is no evidence of moral hazard if $d_{t-1} = 1$ (the driver has CC coverage in the preceding period) has no effect on the accident distribution. A positive coefficient is consistent with the presence of moral hazard (more comprehensive coverage induces more accidents).

8.4 Adverse Selection and Moral Hazard in Combination

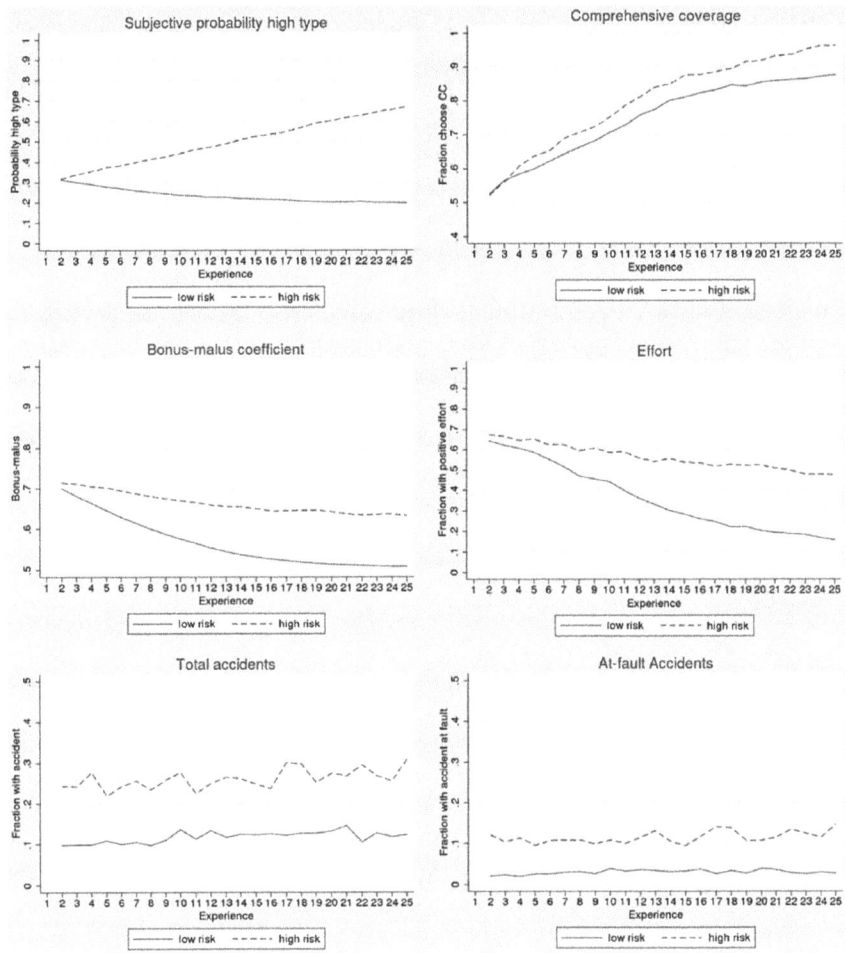

Fig. 8.12 Dionne et al. (2013), Figure 2 (p. 907)

Since in their model adverse selection is the consequence of "asymmetric learning" (i.e., IBs discover their risk type faster than do ICs), the authors formulate a third auxiliary hypothesis.

(3) AL. There is no evidence of asymmetric learning if n_{t-1} has no effect on $d_t = 1$ (the choice of a CC policy in the following period). A positive coefficient is consistent with asymmetric learning, hence adverse selection (note however that this does not show that IBs learn faster than do ICs).

The authors estimate an econometric model using a longitudinal survey of French car owners covering the years 1995–1997 complemented with data from the French Federation of Insurers, with the following results:

- MH1. Among drivers with less than 15 years' experience, the bonus-malus coefficient b_t is indeed negative, suggesting moral hazard.
- MH2. The fact that a driver has CC coverage ($d_{t-1} = 1$) is positively related to his or her probability of an accident later on, again suggesting moral hazard.
- AL. Drivers with less than five years' experience indeed are more likely to opt for the CC policy if they had an accident in the preceding year ($n_{t-1} = 1$). However, this effect vanishes quickly with increasing experience, suggesting that ICs catch up as it were by becoming increasingly successful in their experience rating.

With an observation period of only three years, these findings are unlikely to reveal the full dynamic process of leaning and contract adjustment. For instance, drivers may be slow to modify their choice of policy in response to their accident experience because this involves an investment of time (studying the fine print of a contract) and money (the comprehensive coverage contract is more expensive than the limited liability alternative). Still, this research shows how repeated observations can be used to sort out moral hazard and selection effects.

Exercises

8.1

(a) The investigation by Dionne and St-Michel (1991) is cited as one of the most convincing pieces of evidence in favor of the existence of *ex-post* moral hazard. Why could this be so? Why is this not *ex-ante* moral hazard?

(b) In the text, the hypothesis test is carried out using coefficients b_{18} and b_{12} as well as b_{14} and b_{18} of Table 8.4. Are you capable of performing equivalent tests with two other pairs of coefficients? Do you arrive at the same result? Why (not)?

(c) The empirical evidence relates to a public monopoly insurer in Canada. Is the result of the investigation transferable

   (c1) to a public monopoly insurer for occupational disease in another industrial country?
   (c2) to public insurers for occupational disease that are in competition with each other?
   (c3) to a private monopoly insurer?
   (c4) to private insurers that are in competition with each other?
   Justify your assessment.

(d) On the basis of your answers with regard to (c), what is more important for the limitation of *ex-post* moral hazard,

   (d1) the difference between private and public insurers;
   (d2) the difference between monopoly and competition?

8.2

(a) Describe in no more than five sentences the type of asymmetric information that can cause an insurance market to be unstable.
(b) Cite at least two measures of public insurance regulation that may be appropriate for avoiding instability. Justify your suggestion.
(c) Should the regulation (and with it usually uniformity) of premiums be among the measures cited in (b)? Why (not)?
(d) Cite at least two measures that can be taken by private parties in insurance markets in order to avoid the problem of instability as a consequence of adverse selection. How do you rate their effectiveness and also their efficiency, comparing them to the alternatives cited in (b) of public regulation?

References

Arnott, R., & Stiglitz, J.E. (1990). The welfare economics of moral hazard. In H. Loubergé, (Ed.), *Risk, information and insurance* (pp. 91–122). Boston, Kluwer.

Arrow, K.J. (1963). Uncertainty and the welfare economics of medical care. *American Economic Review, 53*, 941–973.

Association of German Insurers GDV (1997). *Jahrbuch 1997 (Yearbook 1997)*. Karlsruhe: Verlag Versicherungswirtschaft.

Bakx, P., de Meijer, C., Schut, F., & van Doorslaer, E. (2014). Going formal or information, who cares? The influence of public long-term care insurance. *Health Economics, 24*: 631–643.

Bonato, D., & Zweifel, P. (2002). Information about multiple risks: the case of building and content insurance. *Journal of Risk and Insurance, 69*(4), 469–487.

Borch, K.H. (1990). The price of moral hazard. *Scandinavian Actuarial Journal,* 173–176 (1980); reprinted. In K.H. Borch (Ed.), *Economics of Insurance* (pp. 346–362). Amsterdam: North-Holland.

Brown, J.R. & Finkelstein, A. (2008). "The interaction of public and private insurance: Medicaid and the long-term care insurance market." *American Economic Review 98*(3), 1083–1102.

Cawley, J., & Philipson, T.J. (1999). An empirical examination of information barriers to trade in insurance. *American Economic Review, 89*(4), 827–846.

Chiappori, P.A., Jullien, B., Salanié, B., & Salanié, F. (2006). Asymmetric information in insurance: general testable implications. *RAND Journal of Economics, 37*(4), 738–789.

Cutler, D.M., & Reber, S.J. (1998). Paying for health insurance: the trade-off between competition and adverse selection. *Quarterly Journal of Economics, 113*, 433–466.

Dionne, G., & Doherty, N.A. (1994). Adverse selection, commitment, and renegotiation: extension and evidence from insurance markets. *Journal of Political Economy, 102*(2), 209–235.

Dionne, G., & Lasserre, P. (1985). Adverse selection, repeated insurance contracts and announcement strategy. *Review of Economic Studies, 50*(7), 719–723.

Dionne, G., & St-Michel, P. (1991). Workers' compensation and moral hazard. *Review of Economics and Statistics, LXXXIII*(2), 236–244.

Dionne, G., Michaud, P-C. & Dahchour, M. (2013). Separating moral hazard from adverse selection and learning in automobile insurance: longitudinal evidence from France. *Journal of the European Economic Association, 11*(4), 897–917.

Eeckhoudt, L., & Schlesinger, H. (2006). Putting risk in its proper place. *American Economic Review, 96*(1), 280–289.

Ehrlich, J., & Becker, G.S. (1972). Market insurance, self-insurance and self-protection. *Journal of Political Economy, 80*, 623–648.

Einav, L., Finkelstein, A., & Cullen, M.R. (2010). Estimating welfare in insurance markets using variation in prices. *The Quarterly Journal of Economics, 125* (3), 877–921.

Eisen, R. & Nell, M. (1994). Die Wirkung von Versicherungsschutz auf Drittmärkte: Das externe moralische Risiko, in: Hesberg, D. et al. (Hrsg.): Risiko, Versicherung, Markt: Festschrift für Walter Karten zur Vollendung des 60. Lebensjahres, pp. 221–241, Karlsruhe: Verlag Versicherungswirtschaft.

Eisen, R. & Sloan, F.A. (1996). *Long-Term Care: Economic Issues and Policy Solutions*. Boston MA: Kluwer.

Eisen, R. (1982). Competition and regulation in insurance – informational equilibria and rate regulation. *Discussion Paper* IIM/IP 82-3, International Institute of Management, Berlin.

Eisen, R. (1989). Regulierung und Deregulierung in der Deutschen Versicherungswirtschaft (Regulation and deregulation in the German insurance industry). *Zeitschrift für die gesamte Versicherungswirtschaft, 78*(2), 157–175.

Eisen, R. (1990). Problems of equilibria in insurance markets with asymmetric information. In H. Loubergé (Ed.), *Risk, Information and Insurance* (pp. 123–141). Boston: Kluwer.

Frech, H.E. III, & Smith, M.P. (2015). Anatomy of a slow-motion health insurance death spiral. *North American Actuarial Journal, 19*, 60–72.

Greene, W.H. (1997). *Econometric Analysis*, 5th ed. Upper Saddle River NY: Prentice Hall.

Hackmann, M.B. & Pohl, R. (2018). Patient vs. provider incentives in long-term care. *Working Paper*, Loss Angeles: UCLA Dept. of Economics.

Hirshleifer, J., Glazer, A., & Hirshleifer, D. (2005). *Price Theory and Applications*. Cambridge University Press: Cambridge Books.

Houser, A. (2007). Women and long-term care. *Fact Sheet* 77R, Washington DC: AARP Public Policy Institute.

Konetzka, R.T., He, D., Dong, J., & Nyman, J.A. (2019). Moral hazard and long-term care insurance. *Geneva Papers on Risk and Insurance Issues and Practice, 44*, 231–251.

Kreps, D.M. (1990). *A Course in Microeconomic Theory*. Princeton: Princeton University Press.

Kunreuther, H., & Pauly, M.V. (1985). Market equilibrium with private knowledge: an insurance example. *Journal of Public Economics, 26*(3), 269–288.

Levinthal, D. (1988). A survey of agency models of organizations. *Journal of Economic Behavior and Organization, 9*, 153–185.

Mooney, S.F., & Salvatore, J.M. (1990). Insurance fraud project: report on research. In G. Dionne, A. Gibbens, & P. Saint-Michel (Eds.) (1993), *Analyse Économique de la Fraude. Risques, 16*, 9–34.

Norton, E.C. (2000). Long-term care. In Culyer, A.S. & Newhouse, J.P. (Eds.), *Handbook of Health Economics* (pp. 955–994), New York: North Holland.

Pauly, M.V. (1974). Overinsurance and public provision of insurance: the role of moral hazard and adverse selection. *Quarterly Journal of Economics, 88*(1), 44–62.

Pauly, M.V. (1990). The rational non-purchase of long-term care insurance. *Journal of Political Economy* 98(1), 153–168.

Pauly, M.V., & Held, P.J. (1990). Benign moral hazard and the cost-effectiveness analysis of insurance coverage. *Journal of Health Economics, 9*(4), 447–461.

Riley, J.G. (1979). Informational equilibrium. *Econometrica, 47*(2), 331–359.

Riley, J.G. (2001). Silver signals: twenty-five years of screening and signaling. *Journal of Economic Literature, 39*, 432–478.

Rothschild, M., & Stiglitz, J.E. (1976). Equilibrium in competitive insurance markets: an essay on the economics of imperfect information. *Quarterly Journal of Economics, 90*(4), 629–650.

Stewart, J. (1994). The welfare implications of moral hazard and adverse selection in competitive insurance markets. *Economic Inquiry, 32*, 193–208.

Wilson, C.A. (1977). A model of insurance markets with incomplete information. *Journal of Economic Theory, 16*(2), 167–207.

Xu, X. & Zweifel, P. (2014). Bilateral intergenerational moral hazard: empirical evidence from China. *Geneva Papers on Risk and Insurance: Issues and Practice, 39*(4), 651–667.

Zweifel, P. & Courbage, C. (2016). Long-term care: is there crowding-out of informal car, private insurance as well as saving? *Asia Pacific Journal of Risk and Insurance* 10(1), 107–132.

Zweifel, P. (2013). The division of labor between private and social insurance. In G. Dionne (Ed.), *Handbook of Insurance* (pp. 1097–1118). Boston: Kluwer.

Zweifel, P. (2020). Innovation in long-term care insurance: Joint contracts for mitigating relational moral hazard. Working paper.

Zweifel, P., & Strüwe. W. (1996). Long-term care insurance and bequests as instruments for shaping intergenerational relationships. *Journal of Risk and Uncertainty* 12(1), 65–76.

Zweifel, P., & Strüwe. W. (1998). Long-term care insurance in a two-generation model. *Journal of Risk and Insurance* 65(1), 33–56.

Zweifel, P., Breyer, F., & Kifmann, M. (2009). *Health Economics*, 2nd ed. New York: Springer.

Regulation of Insurance

9

This chapter deals with a fact that has been largely neglected up to this point: the insurance industry is one of the most tightly regulated. The arguments proffered for justifying this regulation are reviewed in Sect. 9.1, which also introduces the distinction between two types of insurance regulation. Section 9.2 outlines three theories of regulation designed to explain the changing intensity of insurance regulation and some of its consequences. Empirical evidence regarding the effects of regulation on the industry and consumers is presented in Sect. 9.3. Finally, Sect. 9.4 contains a discussion of recent trends in insurance regulation.

9.1 Objectives and Types of Insurance Regulation

9.1.1 Objectives of Insurance Regulation

Regulation of the insurance industry is justified in the main by consumer protection. Claims held by buyers of insurance (IBs) against an insurance company (IC) are at the center of attention. An insolvent IC is unable to honor its commitment to pay losses; as a rule, it goes bankrupt. Yet removal from the market (also called liquidation) has always been the final sanction of a market economy for an enterprise whose products have an unfavorable performance–price ratio. In principle, the cost of liquidation is borne by the owners of the enterprise. This is not the situation in the case of an IC, which is financed predominantly not by its owners (shareholders in the case of a stock company) but by its policyholders (see the balance sheet of Sect. 6.1.1 again). In the event of bankruptcy, policyholders lose their claims to promised benefits. These claims can be substantial particularly in the case of life insurance. Many IBs and their families may be without an income due to the insolvency of the IC, causing

---

© The Author(s), under exclusive license to Springer Nature Switzerland AG 2021
P. Zweifel et al., *Insurance Economics*, Classroom Companion: Economics,
https://doi.org/10.1007/978-3-030-80390-2_9

them to fall back on public support. Therefore, the insolvent IC burdens the rest of society with external costs.

The argument that the insolvency of an IC gives rise to a negative externality is even more convincing in the case of liability insurance. When an accident happens, the injured party can be said to suffer from a negative externality (if found innocent), which liability law seeks to internalize. However, since full internalization of the externality through payment of the damage would wipe out the economic existence of the injurer in many cases, the law permits the transfer of liability risks to an IC. Yet, an IC that fails to honor its commitment jeopardizes the functioning of the entire internalization mechanism.

9.1.2 Avoidance of Negative Externalities

Evidently, society is confronted with a risky externality that may be dealt with through regulation of insurance designed to prevent insolvency. In analogy with private preventive effort (see Sect. 2.5), there are two basic approaches to insurance regulation:

(A) Prevent the insolvency of ICs, seeking to avoid the externality altogether (etiological measures);
(B) Limit the external costs associated with insolvencies while abstaining from minimizing their probability of occurrence (palliative measures).

Approach (A) has the benefit of reducing the expected cost of insolvency to zero provided regulation is successful. However, it causes efficiency losses by making bankruptcy impossible. The original motivation of consumer protection turns into the protection of ICs, including inefficient ones. In addition, an IC that must not become insolvent needs to be continuously monitored. Approach (A), therefore, calls for so-called material regulation, governing the activity of insurers in detail. Variant (B) typically requires ICs to make provisions for the case of insolvency at the time when they enter the market. It usually goes along with regulation that is limited to formal requirements, without intervening in day-to-day insurance operations.

9.1.3 Material Regulation

9.1.3.1 Theoretical Justification of Material Regulation

Variant (A) of insurance regulation became the dominant model in German-speaking and Scandinavian countries around 1900 after the bankruptcy of several ICs. It went along with the development of a whole theory of its own emphasizing the peculiarities of insurance and predicting market failure of non-regulated insurance markets. This theory never became part of the international literature. Nevertheless, its major arguments are sketched below because they are still proffered at times [see Eisen et al. (1993)].

- *Ruinous competition due to decreasing marginal cost:* In underwriting, required reserves per unit risk indeed decrease with volume because of the law of large numbers (see Sect. 7.1.2.2). However, this does not imply that the marginal cost of enrolling an additional risk unit is decreasing because expenses for marketing and claims settlement may well increase at the margin. If the overall marginal cost of business were declining, a tendency toward natural monopoly would be predicted. This is not observed in view of hundreds of small ICs in major markets (see Sect. 1.3).
- *Uncertain premium calculation:* Admittedly, an unexpected series of extremely high losses can exceed premium income by so much as to cause the insolvency of an IC. But this is true of any type of premium calculation principle that stops short of making the premium equal to the amount of actual (rather than predicted) loss. This in turn would mean the end of market insurance which relies on risk pooling.
- *Lack of transparency:* Assessing and comparing insurance policies are difficult for consumers. However, the lack of transparency can be remedied by brokers and wealth management consultants. In addition, some ICs specialize in developing simple policies that are suitable for marketing through the media.
- *Need for co-operation:* Large risks often are underwritten by several ICs (so-called shared and layered programs), and their settlement may be organized by a reinsurer. This type of co-operation is similar to consortia in the construction industry and does not call for encouragement by regulation but rather scrutiny by competitors.
- *Need to avoid excessive insolvencies:* The argument here is that, without regulation, the probability of insolvency is too high. The validity of this argument depends on the theoretical reference point. If IC management is assumed to simply maximize the expected value of the firm, it is predicted to hold a sufficient amount of reserves to preclude insolvency, under the important proviso that the loss distribution "thins out" for high losses, reaching zero frequency at a finite amount of loss [see Rees et al. (1999)]. By way of contrast, Plantin and Rochet (2007, Ch. 5) argue that IC management, driven by shareholders recognizing the call option characteristic of their stock, tends to take excessive risk, resulting in insolvencies (see Sect. 5.3.1.3). Outside the insurance industry, this tendency is counteracted by a powerful creditor (usually a bank). However, in the case of an IC, the creditors are the policyholders, who typically face exceedingly high costs of the organization to form a powerful pressure group.

Note, however, that while this argument may justify some type of solvency regulation, it does not justify the material regulation of variant (A). In particular, variant (B) is still an option (see Sect. 9.1.4 for more detail), seeking to secure the claims of policyholders in the event of bankruptcy. In contradistinction to banks, where a loss of confidence may spill over from one bank to the next, the possibility of a "run on insurers" is rather remote. Contrary to the great majority of bank creditors, policyholders often face high costs when canceling their contracts. They can be charged with the cost of acquisition, which easily offsets three years' worth of premiums. Moreover, premium payments sometimes continue for several months after the cancelation of the policy. All of this gives the IC time for liquidating and restructuring assets in an attempt to avert insolvency.

▶ **Conclusion 9.1** The arguments designed to justify minimizing insolvency risk and the concomitant material regulation for avoiding market failure in insurance are not fully convincing.

Interestingly, these arguments fail to relate to core theoretical results that point to the possibility of market failure due to asymmetric information. These are moral hazard and adverse selection.

1. *Moral hazard:* As shown in Sect. 8.2.3, moral hazard effects do not jeopardize the existence of an equilibrium in competitive insurance markets. However, they typically result in an equilibrium that is not Pareto-optimal. The reason is the lacking observability of preventive effort, which results in an increased expected loss of the IC. Consequently, increased premiums are paid by all consumers, constituting a negative externality.
   However, the concomitant efficiency loss must be balanced against inefficiencies likely caused by regulation itself (see Sect. 9.3). Moreover, market participants will themselves make attempts at limiting efficiency losses. For instance, the IC can offer premium rebates for no claims to encourage preventive effort. Admittedly, experience rating of this type honors preventive effort only to an approximation since the occurrence of a loss depends on chance as well. In insurance markets with product regulation, it is also seen as a means to attract favorable risks, exposing competitors to adverse selection effects.
2. *Adverse selection:* Asymmetry of information of this type may render a pooling equilibrium unsustainable (see Sect. 8.3), which has to be interpreted as a market failure. However, ICs may solve the problem by launching separating contracts, making high-risk types pay high premiums for a contract with (more) comprehensive coverage (see Sect. 8.3.1.2 for a qualification) and for low-risk types, low premiums for a contract with limited coverage.
   If the high risks cannot afford the high premium because they are poor while there is a consensus in society that they should have access to insurance, there is still the possibility of a means-tested subsidy. For instance, health insurance is mandatory in Switzerland; however, low-income individuals receive a personal subsidy permitting them to purchase health insurance coverage [see Kreier and Zweifel (2010)].
   Rather than paying subsidies, countries often revert to premium regulation to make insurance coverage affordable to everyone. Yet note that this likely triggers risk selection effort on the part of ICs (see Sect. 6.5.2). In the case of health insurance, risk adjustment schemes have been introduced notably in Germany, the Netherlands, Switzerland, and the United States in an attempt to neutralize the "cream-skimming" incentives of insurers [see Van de Ven and Ellis (2000); for a theoretically founded "impossibility theorem", see Frech and Zweifel (2017)]. If an IC is found to have a higher than average share of low risks in its portfolio, it

9.1 Objectives and Types of Insurance Regulation

must pay into the risk adjustment scheme. Conversely, ICs with a prevalence of high risks obtain a subsidy from the scheme. However, a risk adjustment scheme would have to balance the present value of expected future returns to risk selection effort, which proves impossible in actual practice [see Zweifel and Breuer (2006)].

On the whole, adverse selection gives rise to a regulatory dilemma. The private solution would be separating contracts, which often (e.g., in the case of health insurance) are deemed incompatible with equity because they imply high premiums for high risks. On the other hand, attempts at enforcing a pooling contract through imposing uniform premiums induce risk selection effort on the part of competing ICs, which are deemed undesirable as well.

▶ **Conclusion 9.2** Contrary to moral hazard, adverse selection can cause market failure. The private solution is separating contracts, while enforcing a pooling equilibrium by uniform premium regulation gives rise to efficiency losses of its own.

9.1.3.2 Instruments of Material Regulation

To serve the stated goal of consumer protection, material oversight must govern not only the underwriting activity but also the capital investment of an IC. Accordingly, its instruments fall into three main categories.

1. *Price regulation.* The regulation of premiums is designed not only to improve transparency for consumers, but also payment of claims under all circumstances, in keeping with variant (A) of Sect. 9.1.1. Therefore, the premium income of the IC should be high enough to cover the value of claims with a very high probability. Provided the price elasticity of demand is low, high premium rates are conducive to high premium income. However, in the presence of asymmetric information, high premiums (at least when combined with rather comprehensive coverage) attract high risks in competitive insurance markets (see Sect. 8.3). To avoid this side effect, premium regulation often ends up imposing a uniform premium for a given type of contract, stifling price competition between insurers.
2. *Product regulation.* As soon as ICs cannot compete with price, they try to attract consumers through other product attributes. Insurance contracts are easily amenable to product differentiation because there is the need to define insured events and consequences, sums insured, fixed and variable copayments, premium rebates for no claims, the level of consumer accommodation, and rights to future surplus especially in life insurance. However, differentiation along these lines undermines price regulation because a given premium does not relate to a given product anymore. For this reason, price regulation almost always goes along with product regulation in insurance, with the side effect of hampering product innovation. Once uniform product regulation is in place, supervisory authorities can allow a product innovation only if adopted by all ICs. This simultaneity goes against the very idea of innovation, which is to gain an advantage over competitors.

Therefore, product regulation not only increases the cost of product innovation but also reduces its expected return, causing it not to be undertaken in many cases.

3. *Regulation of capital investment.* An IC whose premium income satisfies the regulator's solvency standard may still suffer losses on its capital investment jeopardizing its solvency. Therefore, material oversight prohibits types of investment that are deemed risky while prescribing other forms (in particular, the holding of government bonds). However, this limits the scope of diversification, with the result that the same expected return on surplus $\mu = Er$ cannot be achieved anymore given a certain variance $\sigma^2 = Var(r)$ (see Sect. 6.8). The efficiency frontier in (μ, σ)-space, therefore, shifts down and away from the origin. Since a low value of Er makes negative returns on surplus more probable, regulation of capital investment even runs the risk of increasing rather than reducing the risk of insolvency [see Zweifel and Auckenthaler (2008)].

▶ **Conclusion 9.3** Regulation of price, product, and capital investment are the core elements of material regulation of insurance. They have the side effect of stifling price competition, hampering product innovation, and making efficient portfolio allocations unattainable.

9.1.4 Regulation Limited to Formal Requirements

Insurance regulation limited to formal requirements corresponds to the variant (B) described in Sect. 9.1.1, which admits the insolvency of an IC as a possibility but seeks to mitigate the concomitant externalities. Typical elements of this type of regulation are as follows:

- A minimum of equity capital for founding the IC;
- Insurance reserves sufficient for a major part of loss payments;
- At least one member of the management of the IC that is responsible for sound underwriting (often an actuary); and
- Provision of information proving that the conditions imposed at market entry continue to be satisfied.

This type of regulation is characteristic of the United Kingdom and Ireland. Countries such as Australia, New Zealand, and traditionally South Africa also belong to this type. U.S. insurance regulation differs between states but has been moving toward variant (B) as well with the introduction of Risk-Based Capital (see Sects. 9.3.1 and 9.3.2).

There may be the desire to internalize the cost of insolvency by shifting it back to the IC. This can be achieved without having recourse to material regulation and continuous monitoring of insurance activities. There are several instruments for implementing this type of regulation:

9.1 Objectives and Types of Insurance Regulation

1. *Reinsurance.* The IC can be mandated to purchase reinsurance (RI) for a certain share of its expected loss payments. This serves to secure at least part of policyholders' claims. While this internalization measure has its cost (in the guise of the loading contained in the RI premium), it does reflect the expected external cost of insolvency because ICs with a high insolvency risk must pay a higher RI premium than those with a low insolvency risk. Still, there will be debate about the precise amount of RI in the interest of consumer protection. In essence, this issue revolves around the value of the put option held by the owners of the IC that is in fact financed by the IB (see Sect. 7.2.3). Also, note that the reinsurer might become insolvent itself (a rare occurrence).
2. *Individual guarantee fund.* The IC can be required to create a guarantee fund. Since the funds accumulated are lost to the IC in the event of insolvency, there is a strengthened interest in avoiding it (see Sect. 7.2.3). The downside is that the regulator must gauge the marginal benefit in terms of avoided external cost, to be compared to the marginal cost incurred which consists of underwriting activity forgone and hence less provision of insurance coverage. This is of special importance to a newly formed IC, causing such a fund to become a barrier to market entry.
3. *Transfer of policyholders to a competing IC.* Insolvency means that an entire portfolio of risks becomes available. To a competing IC that seeks to grow, taking over a portfolio provides an alternative to acquiring additional risks through an expensive sales force. A transfer of policyholders to another IC can also be the solution. Let there be a particular IC that is left with the high risks, causing it to end up in bankruptcy. Although its risk portfolio is unfavorable, the decision to acquire it need not run counter to shareholders' interests in the acquiring IC. The expected surplus (assets A minus liabilities L) given by $E(A - L)$ of the IC may be lowered; however, the losses of the new portfolio could correlate negatively with those of the existing one, causing $Var(L)$ to decrease and with it, variance of the surplus $\sigma^2 = Var(A - L)$, *ceteris paribus*. Now shareholders are not *per se* interested in a reduction of σ^2 because it serves to lower the value of their call and their put options (see Sect. 7.2.4). However, the IBs benefit from a reduced value of the put option, and their increase in demand could boost $E(A - L)$ sufficiently to result in an increased total value of the call and put options for shareholders. For a risk-averse management, reduction of σ^2 is an additional reason for acquiring the risk portfolio of the insolvent competitor. Therefore, the transfer of policyholders can be in the interest of both management and owners of the surviving IC.
4. *Joint guarantee fund.* There may be a mandate to take part in a guarantee fund financed by all ICs in the market. Since a simultaneous insolvency of several ICs is improbable, the claims of the IBs are fully guaranteed [see Nektarios (2010) for a detailed description]. However, this benefit must be balanced against the problems of moral hazard and adverse selection. The management of an IC is susceptible to moral hazard effects knowing that in the case of insolvency the claims of its IBs would be paid using contributions from other ICs. This causes the probability of insolvency to be higher than otherwise. This in turn would make insolvencies more frequent, possibly driving the entire fund into insolvency. The

problem of adverse selection is also of relevance. Consumers may be tempted to purchase low-cost policies from high-risk ICs, knowing that their claims will be paid by the joint guarantee fund regardless.

For dealing with these effects, the instruments of insurance technology discussed in Sect. 6.2 are available. In particular, a joint guarantee fund could estimate the risk of insolvency of a participating IC to differentiate contributions accordingly. ICs with a high risk of insolvency would have to factor these contributions into their premiums, which would again reflect the expected value of external cost. Provided they are not commissioned by the ICs themselves, rating agencies can help the fund to recognize ICs with a high risk of insolvency.

5. *Guaranteeing claims with tax money.* A final possibility of mitigating the impact of insolvency is to simply have the government provide relief. For instance, nuclear liability insurance typically provides only very partial coverage in the event of a major accident. Evidently, this alternative fails to internalize external costs. Also, it gives rise to moral hazard effects in that, e.g., nuclear operators may invest too little in safety—a tendency that is counteracted by specialized regulatory agencies [see also Eckles and Hilliard (2015) for a discussion of moral hazard around the Troubled Asset Relief Program in the United States]. Still, these inefficiencies must be balanced against those that are caused by material regulation interfering with day-to-day operations of insurers (and sometimes, insured enterprises).

▶ **Conclusion 9.4** Insurance regulation limited to formal requirements can rely on several instruments, some of which serve to internalize the external costs of insolvency (in particular, mandatory reinsurance). Problems of asymmetric information are to be expected especially in the case of the joint guarantee fund and payment of claims by the government.

9.1.5 Historical Differences in Insurance Regulation Between Countries

Historically, there have been marked differences between countries with regard to the regulation of insurance. One popular case is New Zealand, where starting in 1985 only an annual financial statement with a commentary must be submitted, while decisions regarding premiums, products, and capital investment are left to the ICs to a very high degree [see Adams and Tower (1994)]. The other popular case up to the mid-1990s was the material variant of insurance regulation typical of Germany, Austria, Switzerland, and the Scandinavian countries [see Zweifel (1994)].

The United States occupies an intermediate position. Through the McCarran–Ferguson Act of 1945, the authority for regulation of insurance was assigned to the individual states [see Abraham (1995), 97]. However, a considerable degree of harmonization has been achieved by the National Association of Insurance Commissioners. The main objective of regulation is a low probability of insolvency, with premium income relative to surplus in underwriting (the so-called premium-to-surplus ratio) serving as an indicator. During the 1970s, there was a shift of emphasis toward protecting consumers from excessive premiums, creating a conflict of objectives (see Sect. 9.3.1).

As to the European Union (EU), its objective is to create an internal market for all goods and services. This calls for an opening of national markets for insurance to international competition. Accordingly, the European Commission issued four generations of insurance guidelines to this effect. The first generation (resulting from changes in 1973 and 1979) in the main equalized the conditions governing the local representation of ICs with origin in another EU member country. The second generation (1978–1990) introduced free trade in non-life insurance for large enterprises as IBs (250 employees and more). With regard to life insurance, however, this freedom was only granted if the IB took the initiative for purchasing insurance from an IC of another EU country. The third generation (1991–1994) dropped the distinction between large commercial and small IBs both for life and non-life contracts [see Merkin and Rodger (1997), Ch. 1]. The fourth generation (2007–2009) deals with distribution, permitting brokers to be active in all countries of the EU (see Sect. 9.4 for most recent developments).

9.2 Three Competing Theories of Regulation

In the preceding section, important differences between countries with regard to the regulation of insurance are noted. If the internalization of the external cost was the primary objective of regulation, such marked differences are not to be expected. These observed differences call for a theory of regulation that can predict the intensity of regulation in different countries and its development over time.

Since the pioneering work of Peltzman (1976), one distinguishes three theories of regulation: public interest theory, capture theory, and market for regulation theory. These theories are discussed below and tested for their empirical content.

9.2.1 Public Interest Theory

The point of departure of the public interest theory of regulation is a market failure that needs to be remedied. In the case of insurance, an instance of market failure would be the insolvency of a life insurer. The IBs, who rely on insurance benefits for their old age, may be insufficiently informed to know that insurance reserves of an IC fall short of its future liabilities. The insolvency of the IC confronts the country with a group of mainly older citizens who had accumulated savings through life insurance but are now without an income. To prevent this from happening, the government intervenes, acting in the public interest. However, this theory raises at least three issues.

- *Unclear definition of market failure:* As pointed out in Sect. 9.1.3, the insured population of an insolvent IC can be taken over by a competitor, possibly at less favorable conditions for the insured. The question now arises whether such a

change of conditions still constitutes a market failure. Frequently, the definition of market failure is simply left to the government, who, however, may pursue its own interests (see Sect. 9.2.3 below).
- *Lacking explanation of the choice of instruments:* The public interest theory of regulation is not capable of predicting the choice of measures that are used for achieving stated objectives. Specifically, in some countries, the government merely imposes conditions on market access while in others, it proceeds to regulate premiums and products as well.
- *Lack of incentive to act as hypothesized:* The motivation for members of government and of public administration to act in the public interest is simply postulated. However, since the public interest is difficult to define, these decision-makers have leeway to pursue their own interests, which need not coincide with that of the majority of citizens.

9.2.2 Capture Theory

The capture theory of regulation was proposed by Posner (1974). It starts from the notion that the investors of an industry have an interest in maximizing risk-adjusted returns. However, achievable returns often depend on public regulation. Investors, therefore, have an incentive to capture the regulatory authority to their advantage, typically resulting in excessive prices. Note that the benefits reaped by investors in one industry are to the detriment of investors in other industries who have to pay these high prices. This raises the question of which group of investors can impose their interest on the regulator. One answer has been provided by Olson (1965), who argues that the winners are the owners of dominant enterprises who can be organized at low cost to form a pressure group. Such a pressure group can also offer valuable industry-specific knowledge to the regulatory authority.

However, this theory also has an important shortcoming. It assumes that the regulatory authority is willing to be captured. The possibility that these decision-makers have interests of their own that differ from those of the regulated industry is neglected. Given such a difference, they would have to be compensated for accepting capture, with the amount of compensation depending on the country and historical circumstances. Conversely, assuming capture, it is difficult to explain why there are countries with a low intensity of insurance regulation, while others have a high intensity. In addition, why would there ever be a trend toward deregulation rather than maintaining a given intensity of regulation?

9.2.3 Market for Regulation Theory

This theory can be seen as a generalization of the two preceding alternatives, combining elements of public interest and capture. It is due to Peltzman (1976) who posits a market for regulation.

9.2 Three Competing Theories of Regulation

Supply of regulation is provided jointly by the government and public administration. The government weighs the advantages and disadvantages of additional regulation in view of its chance to be re-elected. The regulated industry can contribute to this chance especially by campaign donations. As to public administration, the benefits of additional regulation are power, prestige, and pay [the three P's emphasized by Niskanen (1971)]. The marginal cost of regulation falls on the government because voters become aware of the efficiency losses engendered, causing them to favor a challenger in the next election. In the case of insurance, high minimum premiums in the interest of solvency make some consumers do without insurance coverage. This constitutes a loss of efficiency that typically increases progressively when regulation becomes more intensive.

In sum, the ratio between marginal cost and marginal benefit for government and public administration increases as a function of the intensity of regulation. It, therefore, takes a higher price of regulation (in the sense of advantages associated with regulation) to motivate government and public administration to supply more regulation (see the positively sloped supply function S of Fig. 9.1, with marginal benefit normalized to one).

As to the demand for regulation, it originates with two groups. On the one hand, there are IBs who have to pay high premium rates as long as they reflect true risk. Some IBs may fail to receive promised benefits due to the insolvency of an IC. On the other hand, there are ICs who expect to be protected by regulation against competition from newcomers both domestically and from abroad. Regulatory norms provide protection because they usually favor incumbents; at the very least, they first have to be learned, which is far easier for incumbents who are more familiar with regulatory traditions than foreigners. The usual assumption is that these beneficiaries are willing to pay a high price for the first steps toward an increasing intensity of regulation. However, their marginal willingness to pay for additional regulatory effort typically decreases. This is equivalent to a downward-sloping demand function such as D_0 in Fig. 9.1. The intersection of the supply and the demand functions determines the equilibrium intensity of regulation, associated with a price (which is not directly observable).

Adams and Tower (1994) note that in New Zealand, where the five leading ICs traditionally account for up to 70% of premium volume, demand for regulation is expected to be low, like D_0 in Fig. 9.1. The reason is that the dominant ICs do not have to rely on public government and administration for protection. With so few players, it is easy to reach collusive agreements regarding the treatment of outsiders. Accordingly, equilibrium in the case of New Zealand (indicated by point Q^*) is characterized by a low intensity of regulation R^*. Also, the price of regulation is low, at p^*. It can be interpreted as the extra cost government and public administration are willing to bear in view of the extra benefits in terms of re-election support. The marginal costs to be covered are still low and with them, the total cost of regulation (amounting to the area below the supply schedule).

In another country, the demand for regulation of the insurance industry may be much more marked, e.g., D_1 in Fig. 9.1. In keeping with the argument above, this could be a less concentrated market, where many small ICs find it more difficult to

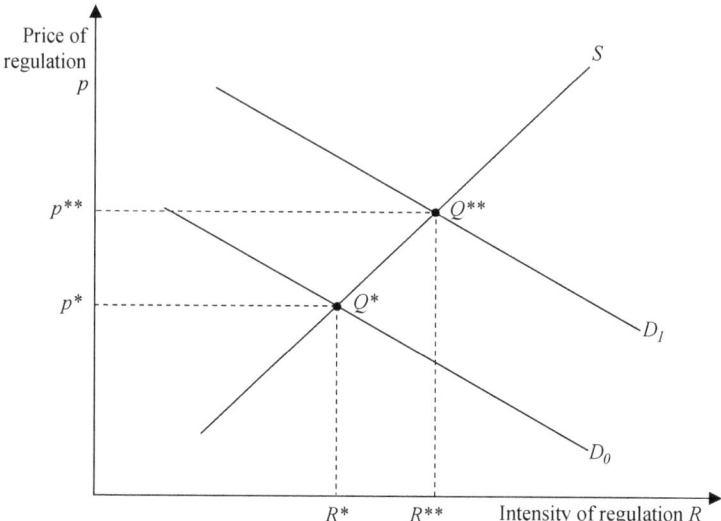

Fig. 9.1 Market for regulation

influence market conditions without the support of the government. However, this may also be a market shaken by a series of insolvencies, leaving many IBs without the promised benefits. In either case, the equilibrium would be at point Q^{**}, with higher intensity of regulation R^{**} and a higher price p^{**}.

In contradistinction to the theory of capture of Sect. 9.2.2, the market model can also explain why there may be waves of deregulation and regulation. Focusing first on the demand side as before, willingness to pay for regulation may decrease from D_1 to D_0, making point Q^* with a lower intensity of regulation R^* the new equilibrium. One reason could be an increase in the degree of concentration in the industry; another, the introduction of cost-reducing innovations that permit incumbent ICs to defend their market shares against newcomers without the protection of public regulation. However, a trend toward deregulation can also originate on the supply side due to an upward shift of the supply function S (not shown in Fig. 9.1) that moves the equilibrium intensity of regulation toward a lower value. The reason could be that government estimates the (marginal) cost of regulation to be higher than previously. Notably, the trend toward deregulation of the EU insurance industry between 1970 and 2010 (see Sect. 9.1.5) can be explained by the European Commission adopting the view of a central government. In contrast to national authorities who may be able to shift part of the cost of regulation to foreign ICs and consumers, the European Commission considers these agents domestic as long as they have an EU domicile. From its point of view, the supply function of Fig. 9.1 thus runs higher than for a national regulator, reflecting the fact that the cost of regulation accrues internally.

9.2 Three Competing Theories of Regulation

Table 9.1 Predictions of the three competing theories of regulation

Hypothesis (concerning insurance)		Predicted by		
		Public interest theory	Capture theory	Market[a] theory
H1.	Crises (especially insolvencies) result in a higher intensity of regulation	yes	no	yes
H2.	Highly regulated markets are characterized by highly active groups on the demand and the supply side	no	no	yes
H3.	Small interest groups dominate large groups (especially consumers)	no	yes	yes
H4.	Highly regulated markets are characterized by many small IC	no	no	yes
H5.	Highly regulated markets have a large regulatory bureaucracy	no	no	yes
H6.	Highly regulated markets are characterized by high lobbying expenditure	no	no	yes

[a] Short for: Market for regulation theory

9.2.4 Empirically Testable Implications for Insurance

The market for regulation model gives rise to six hypotheses that can be contrasted with the public interest and capture models [see Adams and Tower (1994); for an application to health insurance, see Zweifel (2007)]. Yet the focus on the insurance industry has the downside that the hypotheses do not explain why the intensity of regulation is higher than in other industries. The six hypotheses are listed in Table 9.1.

H1. *Financial crises (especially insolvencies) lead to an increased demand for insurance regulation.* The source of this increase is consumers who fail to obtain insurance benefits as promised. *Ceteris paribus*, this results in increased regulatory intensity (see the shift from D_0 to D_1 in Fig. 9.1). According to the public interest theory, increased externalities caused by crises motivate the government to become active, resulting in the same prediction. By way of contrast, capture theory does not predict a change because the regulator is assumed to serve the interests of investors who are sufficiently diversified to be unaffected by the insolvency of a particular IC.

H2. *In highly regulated insurance markets, both sides of the market for regulation are characterized by highly active and organized associations and bureaucracies.* According to the market model, high demand (such as D_1 in Fig. 9.1) is the result of activities by groups that benefit from a high degree of organization. To deal with them, a developed bureaucracy is necessary. The public interest theory does not make any prediction in this regard because the government simply seeks to improve efficiency through its effort at internalization. The same is true of capture theory since regardless of the degree of regulation, the government acts in the interest of investors in the insurance industry.

H3. *Small interest groups dominate the market for regulation.* For instance, let private households as a large group have little interest in regulation (D_0 in Fig. 9.1), while a small group with homogeneous interests (typically comprising insurance companies) is represented by D_1. Since it can form a lobby at a much lower cost, it is their higher demand D_1 that determines the market outcome. This difference is predicted also by the capture model, where the government always serves the interests of the investors controlling the IC. By way of contrast, according to the public interest model, the government simply acts to remedy a market failure.

H4. *Highly regulated markets are characterized by many small ICs.* To the extent that regulation provides protection from newcomers, it ensures the economic survival of the small domestic ICs. Conversely, if there are just a few dominant ICs (as, e.g., in New Zealand), they do not need the government for closing the market. These considerations are not of importance in both the capture theory and the public interest theory.

H5. *Highly regulated markets have a large regulatory bureaucracy.* This hypothesis can again be derived from Fig. 9.1. Let the high intensity of regulation be the result of a market demand D_1. At the equilibrium point Q^{**}, the marginal cost of regulation (reflected by the supply schedule) is high. This also is true of its total cost, which can be read off as the area below the marginal cost schedule. A high cost of regulation reflects a large bureaucracy. According to capture theory, this is not necessary because investors control the regulatory activity of the government anyway, resulting in little bureaucracy. According to the public interest theory, the government achieves a high intensity of regulation by issuing very stringent norms without necessarily creating a large bureaucracy.

H6. *Highly regulated markets are characterized by high lobbying expenditure.* This hypothesis also derives from the market model of regulation. Intensive regulation usually affects many assets, and it affects them strongly. Therefore, those potentially exposed to regulation have a great interest in influencing it through lobbying. This prediction does not follow from either the public interest or the capture theory.

While a systematic test of these six hypotheses has not been performed yet, they do seem to contribute to an understanding of differences between countries as well as changes over time.

▶ **Conclusion 9.5** Both the public interest theory and the capture theory of regulation fail to predict phenomena that seem to characterize insurance markets. The market for regulation theory seems to have highest explanatory power.

9.3 Effects of Insurance Regulation

In the preceding section, the intensity of regulation is considered an endogenous variable. However, when studying the effects of insurance regulation, authors typically take the intensity of regulation as exogenously given. This can be justified by noting that regional differences in regulatory intensity persist as a result of influences that lie in the past. Changes over time usually constitute adjustments to the past as well.

However, in view of the quick regulatory response to the financial crisis of the years 2007–2009 (see Sect. 9.4.1), this may not hold true anymore.

9.3.1 Evidence from the United States

The United States provides an excellent basis for testing the effects of insurance regulation because regulatory authority is vested with the 50 states, resulting in marked differences. Frech and Samprone (1980) took advantage of this fact in their early study of the insurance lines "automobile liability" and "automobile physical damage". Both the capture theory and the market theory of regulation predict that the higher intensity of regulation in some states is the result of stronger demand for regulation (typically exerted by the local insurance industry in the interest of limiting price competition) rather than of a quest for improving efficiency.

With price competition limited or even eliminated, ICs who seek to gain market share turn to non-price competition. This means offering IBs things like a dense network of agencies, expediency in claims settlement, and consumer accommodation through a generous interpretation of contract clauses. However, these extras drive up the marginal cost of providing insurance coverage.

As a consequence, more stringently regulated states should be characterized not only by a higher premium level but also by higher marginal costs. In Fig. 9.2, price is assumed to be equal to (constant) marginal cost for simplicity. Both are higher in regulated states (p^r and MC^r, respectively) than in more competitive states (p^c and MC^c). Two cases can be distinguished. In panel A, non-price competition fails to result in an improvement of the product from the point of view of consumers, who, therefore, do not display an increased marginal willingness to pay. Accordingly, their demand function is given by D regardless of whether regulation is more or less stringent. This case results in a maximum loss of welfare due to a reduction in consumer surplus (see the shaded triangle). Also, one would observe that aggregate insurance coverage in the market is I^r rather than I^c under competitive conditions, with $I^r < I^c$.

However, non-price competition induced by regulation could result in product improvements that are indeed valued by the IB. This would result in a shift of the demand function from D^c to D^r (see Panel B of Fig. 9.2). Let this shift be marked enough to avoid the loss in consumer surplus shown in panel A. The condition for this outcome to obtain is that the increase in marginal willingness to pay by IBs exactly offsets the increase in marginal cost. The observable implication is that the amount of insurance coverage would be the same regardless of regulatory conditions ($I^r = I^c$), ceteris paribus.

The fact that in 1973 some jurisdictions of the United States regulated automobile insurance while some did not permit an amazingly simple test of the effect of regulation. If in the regulated states the demand function lies farther out than in the unregulated ones, this would suggest that regulation induced product improvements due to non-price competition. Such a shift could balance (or even over-compensate) the loss in consumer surplus caused by the increase in marginal cost. If, however,

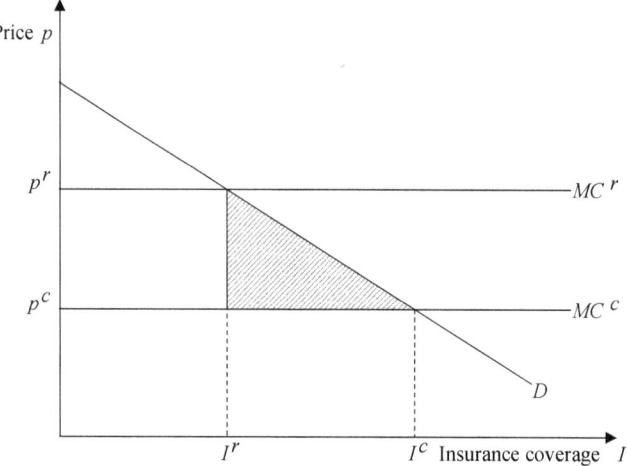

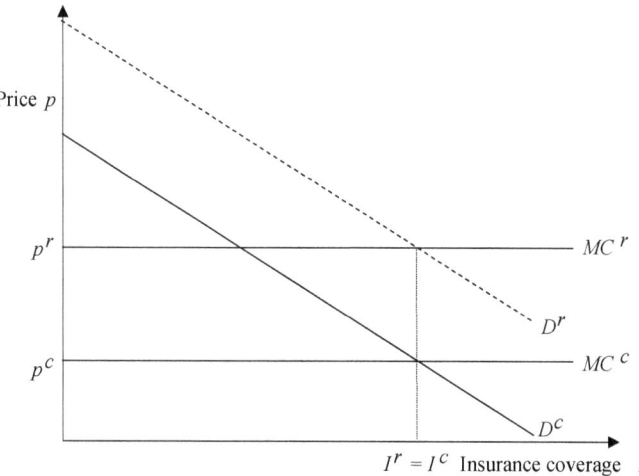

Fig. 9.2 Effects of insurance regulation

there is no such shift of the demand function in the regulated jurisdictions, then regulating this line of insurance would be wasteful.

As the dependent variable for the analysis, Frech and Samprone use paid losses L per capita in 1973. This choice is in accordance with the definition of output considered in Sect. 7.4.1. Their price variable is the ratio of premium volume over claims paid PV/L. As argued in Sect. 1.5.2, the cost of insurance is not reflected by the premium but by the excess of the premium over the expected value of losses $(\pi \cdot L)$ because in the aggregate, premiums are in part channeled back to IBs in the

9.3 Effects of Insurance Regulation

guise of losses paid. Denoting the surcharge over the fair premium by λ, one obtains

$$\frac{PV}{L} = \frac{\pi \cdot L \cdot (1+\lambda)}{L} = \pi(1+\lambda). \tag{9.1}$$

This is indeed the true price per unit insurance coverage. The higher PV/L, the higher the loading λ and the higher the price of insurance coverage.

However, using PV/L as the explanatory variable gives rise to a simultaneity problem because the amount of coverage is also measured by L. The demand function, therefore, reads $L = f(..., PV/L)$, causing price (given by PV/L) to be necessarily low when L is high. For instance, L may be higher than expected in a particular state; this would push PV/L below the expected value. In this way, the response of demand to price could be easily overestimated. In order to avoid this bias, the authors use the value of price in the preceding year, making it a predetermined quantity. The additional explanatory variable is the income per capita in the state.

For *auto liability insurance*, the regression equation reads

$$I = \underset{(6.34)}{50.34^{***}} - \underset{(-6.59)}{24.03^{***}} P_{-1} + \underset{(3.74)}{0.0033^{**}} Y - \underset{(-1.72)}{2.25} REG \tag{9.2}$$

$N = 51$; $R^2 = 0.63$; $F = 27.2$; $^*(^{**},^{***})$: Statistical significance at the 0.05 (0.01, 0.001) level; t-values in parentheses.

For the *auto physical damage*, the regression reads

$$I = \underset{(7.44)}{33.38^{***}} - \underset{(-5.34)}{14.52^{***}} P_{-1} + \underset{(2.09)}{0.0014^{**}} Y + \underset{(0.18)}{0.17} REG \tag{9.3}$$

$N = 51$; $R^2 = 0.39$; $F = 10.1$; $^*(^{**},^{***})$: Statistical significance at the 0.05 (0.01, 0.001) level; t-values in parentheses.

I: Amount of insurance coverage per capita, measured by claims paid per capita, in 1973;
P_{-1}: Price of insurance in the preceding year, measured by premium volume relative to claims paid;
Y: Income per capita;
REG: Dummy variable, $= 1$ if the state regulates the pertinent line of insurance, $= 0$ otherwise.

The two estimates are plausible. The price elasticity of demand (calculated at the means of I and P_{-1}) amounts to 1.7 for liability insurance and 1.6 for auto physical damage coverage. Also, higher average income Y is estimated to be related to more insurance coverage.

However, the main variable of interest is REG, symbolizing the presence of regulation. In the case of auto liability insurance, this coefficient is negative, giving rise to the suspicion that the demand function D^r of Fig. 9.2 runs closer to the origin than the function D^c, pertaining to the states without regulation. It is a suspicion only because the coefficient of *REG* in Eq. (9.2) fails to attain statistical significance. In the case of auto physical damage insurance, the coefficient of *REG* suggests that the demand function is not moved out by regulation and the induced non-price competition. Equation (9.3), therefore, points to the case shown in panel A of Fig. 9.2, where regulation simply increases the marginal cost of writing insurance.

▶ **Conclusion 9.6** Available evidence suggests that the regulation of U.S. automobile insurance (liability and physical damage) may have caused welfare losses in the early 1970s by merely inducing costly non-price competition.

A later study by Pauly et al. (1986) uses data covering the years 1975–1980 to test whether the loss ratios L/PV are lower in the regulated states of the United States than in the non-regulated ones. The authors conclude that regulation serves to increase the loss ratio (decrease PV/L), contradicting Frech and Samprone (1980). This result was confirmed by Harrington (1987) who associated regulation with an increase in the loss ratio by 3–5 percentage points.

These contradictions led Grabowski et al. (1989) to hypothesize that the market for insurance regulation in the United States changed in the course of the 1970s. During that decade, the premiums of automobile insurance increased markedly not only in nominal but also in real terms (i.e., compared to other goods and services). This triggered consumer protests that apparently forced a change in regulatory philosophy. Recall that demand for regulation emanates in part from consumers and in part from the insurance industry, with the latter component constituting the crucial component as a rule. However, in the United States of the later 1970s, the interests of IBs in lower premiums (for a given expected value of benefits) seemed to have become decisive in the wake of "consumerism" under the leadership of Ralph Nader. Accordingly, some states turned to deregulation (predicted by a shift in demand for D_1 to D_0 in Fig. 9.1), while in others, regulatory authorities started to put pressure on premiums (a change of regulation that goes beyond the "intensity" variable of the simple market for regulation model).

The presence or absence of regulation in a jurisdiction is a rough measure, however. Grabowski et al. (1989), therefore, use surveys among insurance managers to measure the perceived stringency of regulation. Their sample is limited to the 30 major U.S. insurance markets, covering the period 1975–1981. The authors again estimate a demand function but use price rather than the amount of coverage as the dependent variable. Their result reads

$$P = \underset{(0.085)}{1.954} - \underset{(1.5\cdot 10^{-5})}{4.9 \cdot 10^{-5***}} Y - \underset{(0.023)}{0.029} \, WAGE$$
$$- \underset{(0.021)}{0.121^{***}} \, NOFAULT - \underset{(0.021)}{0.064^{**}} \, REG - \underset{(0.037)}{0.105^{**}} \, STRING \quad (9.4)$$

$N = 180$; $R^2 = 0.39$; *(**,***): Statistical significance at the 0.05 (0.01, 0.001) level; standard errors in parentheses. Their variables are further defined as follows:

9.3 Effects of Insurance Regulation

P: Price of auto liability insurance (premium volume/claims paid);
Y: Income per capita;
$WAGE$: Average wage rate of production workers;
$NOFAULT$: $= 1$ if the jurisdiction is characterized by a no-fault rule, meaning that the responsible party in an accident does not need to be identified, $= 0$ otherwise;
REG: $= 1$ if the jurisdiction regulates the insurance line (may change between years), $= 0$ otherwise;
$STRING$: $= 1$ if the jurisdiction is characterized by extremely stringent regulation according to IC managers surveyed, $= 0$ otherwise.

In addition, the estimated equation contains dummy variables (not shown) that characterize the year of observation. The results can be interpreted as follows.

- *Income and marginal cost:* Interestingly, auto liability insurance coverage is not more expensive in states with high average income (Y) but rather cheaper. The authors relate this to a higher degree of information of a well-educated and hence higher income population. Better information goes along with a more marked price elasticity of demand, which in turn limits any monopolistic markup over marginal cost. Marginal cost is represented by the average wage rate ($WAGE$); however, the pertinent coefficient is negative rather than positive and lacks statistical significance.
- *Conditions governing insurance benefits:* In some U.S. jurisdictions, claims payment is separated from the identification of the party responsible for the accident ($NOFAULT$). On the one hand, this saves the IC a great deal of legal effort, resulting in lower administrative expenses. On the other hand, there is a moral hazard effect to be expected because IBs can count on payment without testing for negligence (i.e., lack of preventive effort). If the latter effect prevails, aggregate payments increase, and with them premiums, resulting in an ambiguous effect on price defined as premium volume relative to losses paid. As shown in Sect. 8.2.2.1, however, moral hazard effects induce a loading in excess of the fair premium, causing an increase in the price of insurance. Yet the coefficient of *NOFAULT* is negative, pointing to cost savings of some 12 percentage points of premium. For instance, given an average price (i.e., loading) of insurance coverage amounting to 20% of premium, it would be reduced to a mere 8% due to the no-fault rule, *ceteris paribus*.
- *Effect of regulation:* The impact of regulation is clear according to Eq. (9.4). In states with regulation ($REG = 1$), coverage in the price of auto liability insurance is about 6 percentage points lower than in other jurisdictions. This finding is in accordance with another estimate, for property-liability insurance (not shown here). In addition, the price of insurance is another 10 percentage points lower in those states (Massachusetts, New Jersey, and North Carolina) that stand out for their stringent regulation ($STRING = 1$).

The changed approach to insurance regulation in the United States with its focus on low premiums, therefore, has a recognizable effect according to this study. However, note that insurance markets are not pictured as competitive anymore [contrary to Frech and Samprone (1980)]. Rather, they are seen as closed markets permitting monopolistic pricing by the IC at least as a group. This capacity to set prices is limited by regulatory authorities.

However, there is still the alternative hypothesis that the observed prices in non-regulated jurisdictions at least approximately are competitive prices, whereas they might be below equilibrium values in regulated jurisdictions. To test this hypothesis, Grabowski et al. (1989) note that some drivers cannot find coverage at prevailing conditions since there is no open enrollment for automobile liability in the United States. These drivers are assigned to a so-called "assigned risk pool". Now if regulation were to push the average price of insurance coverage below its equilibrium value, the share of risks assigned to such a pool would have to be especially large in regulated jurisdictions. The regression equation below shows the result of this test:

$$Z = 0.062 - 0.892^* \underset{(0.358)}{YMALE} + 0.36 \underset{(0.031)}{URBAN} + 1.6 \cdot 10^{-4**} \underset{(5.5 \cdot 10^{-5})}{INJURY}$$
$$+ 0.013^* \underset{(0.006)}{COMPLIAB} - 0.002 \underset{(0.009)}{LIABONLY} + 0.026^{**} \underset{(0.010)}{JOINT}$$
$$+ 0.043^* \underset{(0.019)}{REINSURE} + 0.010 \underset{(0.006)}{REG} + 0.17^{***} \underset{(0.017)}{STRING} \qquad (9.5)$$

$N = 240$; $R^2 = 0.75$; $^*(^{**},^{***})$: Statistical significance at the 0.05 (0.01, 0.001) level; standard errors in parentheses.

Z:	Share of risks that are assigned to the pool;
$YMALE$:	Share of young male drivers (below 25 years) in the insured population;
$URBAN$:	Share of automobile traffic occurring in urban regions;
$INJURY$:	Number of car accidents with injuries;
$COMPLIAB$:	= 1 if the state mandates auto liability insurance, = 0 otherwise;
$LIABONLY$:	=1 if the assigned risk pool is limited to auto liability insurance only, = 0 otherwise;
$JOINT$:	= 1 if the ICs of the state jointly finance the deficit of the assigned risk pool, = 0 otherwise (i.e., each IC has to contribute according to its market share);
$REINSURE$:	= 1 if the ICs are mandated to accept open enrollment but are paid for this by reinsurance, = 0 otherwise;
REG:	= 1 if the state regulates auto liability insurance, = 0 otherwise;
$STRING$:	= 1 if the state is characterized by extremely stringent regulation, = 0 otherwise.

Once again, a series of annual dummies is included but not reported here. The results suggest the following interpretation.

- *Characteristics of the insured population:* The explanatory variables *YMALE*, *URBAN*, and *INJURY* describe the insured population in the respective state. One would expect high-risk groups to lead to an increased share of drivers assigned to the risk pool [the ICs also seem to learn the true risk rather quickly, see (Dionne et al., 2013) referenced in Sect. 8.4]. However, this effect is recognizable only for *INJURY*, the number of accidents with injuries. This expectation is even contradicted in the case of *YMALE*, possibly because the ICs prefer to accept high-risk

young males for a few years rather than losing them to the assigned risk pool, from where they would have to be retrieved later at high acquisition expense.
- *Scope of insurance mandate:* The pertinent indicators are *COMPLIAB* and *LIABONLY*. In states where auto liability insurance is compulsory ($COMPLIAB = 1$), the share of assigned risks is about one percentage point higher than otherwise (for instance 6% rather than 5%). By way of contrast, the fact that the assigned risk pool applies to auto liability coverage only ($LIABONLY = 1$) does not have a recognizable effect.
- *Organization of the assignment:* In some states, the ICs jointly operate a special purpose company for risks they decline to accept ($JOINT = 1$). The positive coefficient of this variable points to an increased incentive to transfer risks to this special company. In other states, a reinsurance scheme is in charge ($REINSURE = 1$). This variant has an even stronger effect on the propensity to assign risks to the pool.
- *Influence of regulation:* Contrary to expectations, the mere fact that a state regulates auto liability insurance ($REG = 1$) cannot be said to increase the share of risks assigned to the pool; the pertinent coefficient is positive but fails to be significant. However, the three states characterized by very stringent regulation ($STRING = 1$; Massachusetts, New Jersey, and North Carolina) have an especially high share of assigned risks. The estimate is in excess of 17 percentage points compared to the other states.

In sum, Eq. (9.5) suggests that at least in U.S. jurisdictions with extremely stringent regulation, premiums for auto liability insurance may be pushed below their equilibrium level, causing a particularly high share of drivers to be transferred to the assigned risk pool.

▶ **Conclusion 9.7** Starting in the 1970s, the regulation of insurance in the United States has increasingly aimed at keeping premiums low. In a few states, the point seems to be reached where ICs exhibit an increased propensity to transfer high-risk drivers to the pool of assigned risks.

9.3.2 Risk-Based Capital as the U.S. Regulatory Response

The evidence presented in the previous section suggests that in the United States, the conventional premium regulation increasingly became subject to a conflict of objectives between solvency and consumer interest in low premiums. In terms of Fig. 9.1, the conventional approach did not meet with as much interest as originally estimated, reflected by an inward shift of the demand schedule (e.g., from D_1 to D_0). The transition to a lower level of regulatory intensity occurred in 1992 when the National Association of Insurance Commissioners (NAIC) instituted the risk-based capital (RBC) approach. The RBC approach reflects the insight that the total of equity and insurance reserves can be set in a way to attain a target probability (usually 99.5%) with which it can cover claims of policyholders but also losses from other activities of the IC. In this way, detailed regulation of premiums, products, and capital investment can be obviated.

In return, a broad spectrum of risks is contemplated [see NAIC Capital Adequacy Task Force 2009].

1. Asset Risk—Affiliates: The IC may have subsidiaries, and the investment in them may lose value;
2. Asset Risk—Other: This category contains three components. Fixed-income assets (mainly bonds), equity assets (mainly stocks and real estate), and loans provided may all fluctuate in value;
3. Underwriting Risk: There may be more and higher claims than expected, or pricing for a future level of claims may have been inaccurate; and
4. Business Risk: In life insurance, a variation of the interest rate can cause losses because cash flows relating to assets and liabilities have different maturities.

Since exposure to these risks differs between life, property/casualty, and health insurers, the applicable RBC formula varies. However, all formulas adjust for the estimated amount of covariance between the elements making up the four risk categories to take diversification effects into account. For example, let category (1) consist of just two components with standard errors of 3 and 4 monetary units (MU), respectively, and a correlation coefficient of 0.46. Applying formula (5.4) of Sect. 5.1.1, the portfolio variance is given by 36 MU ($= 3^2 + 4^2 + 2 \cdot 0.46 \cdot 3 \cdot 4$) and a standard error of 6 MU (which determines the RBC necessary for attaining the target probability of solvency). Neglecting the fact that the two risks are less than perfectly correlated, one would have put the standard error of the portfolio at 7 MU [$(3^2 + 4^2 + 2 \cdot 1 \cdot 3 \cdot 4)^{\frac{1}{2}}$], resulting in an excess capital requirement.

RBC calculated using the formulas prescribed by the NAIC is compared to the equity and reserves reported in the financial statement of the IC. For instance, a ratio of 150 to 200% (the so-called Company Action Level) triggers regulatory intervention in that the IC must prepare a proposal to again reach a ratio of 200% or more. A ratio of less than 70% requires the regulator to seize control of the IC.

9.3.3 Evidence from Europe

Member states of the European Union (EU) differ in terms of their insurance regulation. Finsinger and Schmidt (1994) took advantage of these differences to investigate the influence of regulation on insurance premiums. They compare premiums charged for a homogeneous product, namely term life insurance (which does not contain a savings component, contrary to whole life insurance). They collected quotations for men and women aged 25, 35, and 45 and for a contract life of five and ten years, obtaining 12 homogeneous contract types. In this way, the authors are able to exclude differences in quality that would occur if an international comparison were performed at a more aggregated level. Premiums quoted relate to 1988.

9.3 Effects of Insurance Regulation

The authors cite evidence suggesting that among the five countries considered, three (Germany, France, and Italy) were stringently regulated toward the end of the 1980s ($REG = 1$). By way of contrast, the Netherlands and the United Kingdom were already characterized by largely open insurance markets, combined with a low intensity of regulation ($REG = 0$). The twelve contract types and five countries result in 60 observations for the OLS estimate (t-ratios in parentheses),

$$\begin{aligned} ln(PREM) = {}& \underset{(27.02)}{13,86^{***}} + \underset{(8.96)}{0.48^{***}} \, REG + \underset{(7.65)}{0.39^{***}} \, HERF \\ & - \underset{(-16.09)}{2.07^{***}} \, ln(LIFEEXP) \end{aligned} \qquad (9.6)$$

PREM: Premium quoted for a term life insurance contract in ECU (the precursor of the Euro at the time);
REG: Indicator of regulation; = 1 for Germany, France, and Italy; = 0 for the Netherlands and the United Kingdom;
HERF: Herfindahl measure of concentration defined by the sum of squared market shares of the ICs in the country considered;
LIFEEXP: Life expectancy at birth in the country considered.

The results of this regression equation can be interpreted as follows.

- *Influence of regulation:* The positive coefficient of *REG* suggests that at least in the EU member countries considered, regulation causes an increase in the premiums of life insurance. The difference amounts to about 61% *ceteris paribus* [calculated as $exp(0.48)/exp(0) = 1.61)/1$].
- *Influence of concentration:* From a theoretical perspective, it is not quite clear whether a higher degree of concentration (*HERF*) should go along with higher or lower premiums. A high degree of concentration could indicate an oligopolistic market structure characterized by vigorous price competition among the leading ICs. On the other hand, it also facilitates collusion among the few. Apparently, in the EU countries considered, this second effect dominated. For example, let the market structure change from 5 to 4 companies of equal size, causing *HERF* to increase from 0.20 to 0.25 points. This change is associated with an estimated increase in *ln(PREM)* by 0.0195 ($= 0.39 \cdot 0.05$) or some 2%. This is a small effect compared to the influence of regulation. Therefore, although deregulation usually goes along with an increase in concentration, its net effect likely is lowered premiums for a given expected value of benefits.
- *Influence of life expectancy:* With contract life limited to 5 and 10 years, respectively, a high life expectancy implies a low probability of loss since term life insurance pays benefits in the event of death only. Therefore, high life expectancy should be associated with low premiums. The estimated elasticity of -2.07 is in accordance with this expectation. A market with a 10% higher life expectancy would be characterized by premiums that are an estimated 21% lower *ceteris paribus*.

▶ **Conclusion 9.8** The regulation of important European insurance markets had the effect of a marked increase in premiums for a given expected loss, at least up to the end of the 1980s.

Another study concerns Germany, which until the mid-1990s clearly adhered to variant (A) of insurance regulation, aiming at the prevention of insolvency (see Sect. 9.1.2). Employing premium regulation as part of material regulation, the federal agency in charge faced a dilemma. On the one hand, it favored high premiums for ensuring solvency; on the other hand, it sought to protect consumers from excessive premiums. It saw the solution in a mandate to share profits with IBs.

Allowable premium volume PV had to cover expected losses, which, however, were not estimated at the level of the individual IC but corresponded to the average of the industry for a given line of business. In calculating its cost (C), the IC could add expenses for acquisition and administration as well as a safety loading. The difference $(PV - C)$ relative to PV, amounting to a rate of return on premiums, was regulated. Beyond some benchmark, profits were to be shared with IBs according to complicated rules. For instance, in auto liability insurance, the regulation stated that if the actual rate of return was below 3% of PV, the IC was allowed to keep profits entirely. For rates of return between 3 and 6% of PV, it had to distribute profit entirely to policyholders. For rates of return between 6 and 15%, the IC could retain one-third of profits; and above 15%, all profits again. In spite of this regulation, application of the CAPM suggests that ICs enjoyed a high level of profitability (see Example 7.5 in Sect. 7.2.2.2)

Finsinger (1983a) simplifies by introducing one threshold value α of the rate of return on PV that must not be exceeded. For most ICs, $\alpha = 0.03$ presumably was relevant at the time. The regulatory constraint can then be written

$$\frac{PV - C}{PV} \leq \alpha \qquad (9.7)$$

PV: Premium volume;
C: Admissible expenses (loss payments, administrative expense, and safety loading).

For a profit-seeking ICs, the optimization problem amounts to (neglecting risk)

$$\max_{e} PV - C - rK, \quad \text{s.t.} \quad PV - C \leq \alpha PV \qquad (9.8)$$

e: Set of decision variables;
r: User cost of capital, per unit;
K: Capital employed, including insurance reserves.

9.3 Effects of Insurance Regulation

Among the possible decision variables e, the author considers two in greater detail:

1. *Marketing and advertisement expenses.* They have the effect of increasing both premium volume PV as well as admissible expenses C. However, an increase in PV serves to relax the regulatory constraint [see αPV in (9.8)], permitting the contribution to profit ($PV - C$) before the cost of capital to increase as well. By spending on marketing and advertisement, the German ICs were, therefore, able to improve their degree of goal attainment.
2. *Lavish use of capital.* This could in principle be excess reserves. Indicating that the IC has a very low probability of insolvency, they serve to increase demand for its products and hence premium volume PV. Additional reserves, therefore, relax constraint (9.7) in a similar way as marketing and advertisement expenses. However, using extra capital indirectly through the purchase of reinsurance (RI) is more effective than the accumulation of insurance reserves because RI premiums count as admissible cost C. The prediction is, therefore, that regulated ICs purchase a high degree of RI coverage.

These side effects of the rate of return regulation are predicted for stock companies, who can be assumed to be profit-seeking. However, in Germany, there are two additional types of insurers, mutuals and public insurers. They are both not for profit, which means that the regulation defined by (9.7) and (9.8) usually does not bind. Therefore, one can derive the two following hypotheses.

H1: *German stock companies exhibit a higher expense ratio than mutuals and public insurers, in particular, because they spend more on advertisement and marketing.*

This hypothesis is confirmed by a regression analysis based on 75 ICs as of 1980. Estimates point to an expense ratio of 22% for stock corporations, higher than the average value of 21% (see Table 9.2). By way of contrast, mutuals exhibit an expense ratio of only 18%. These differences are significant with an error probability of 1% or less. Moreover, they can be traced in the main to administrative expense (which in turn consists importantly of marketing expenses). The type of distribution system also influences the expense ratio to a comparable degree (see Table 9.2 again). This points to the importance of the choice of the distribution system, which is discussed as an element of insurance technology in Sect. 6.4. However, in this context, the comparison is not between independent agents and employed agents but between a centralized and decentralized system of agents because, at the time, independent brokers hardly existed in German auto liability insurance.

H2: *German stock companies cede a greater share of their premium volume to reinsurance compared to mutuals and public insurers.*

This hypothesis is also confirmed based on data from 81 German ICs. The dependent variable is the retained (rather than ceded) share of premium volume, which is a high 87% *ceteris paribus* among public insurers, exceeding the average value

Table 9.2 Expense ratios of 75 German ICs with auto liability business (1980)

Explanatory variable	Estimated deviation from mean value of 21%[a]	Error probability
Type of insurer:		
Stock company	+1 percentage point	
Mutual	−3 percentage points	$p < 0.01$
Public	−1 percentage point	
Distribution system:		
Centralized network	−4 percentage points	$p < 0.001$
Network with many agents	+1 percentage point	
$R^2 = 0.32$		

[a]Based on the estimated regression coefficient pertaining to a categorical variable ($= 1$ if characteristic is present, $= 0$ otherwise)
Source Finsinger (1983b)

of 64% (see Table 9.3). This difference could, however, also be due to the fact that public insurers can use the government as a reinsurer, permitting them to rely less on RI coverage. Note that this alternative explanation does not apply to mutuals. As can be seen from Table 9.3, mutuals also retain a greater share of their premium volume than do stock companies. This fact lends support to the hypothesis that the marked use of reinsurance by stock companies is induced by the incentive effects of insurance regulation.

The type of distribution system again is among the explanatory variables. Reliance on centralized agencies may entail a risk of geographical accumulation of risks, with the possibility of a positive correlation between them. Therefore, the underwriting result may be characterized by particularly high variance in the case of centralized distribution agencies. In the light of the option pricing model for the demand of reinsurance, high variance is in the interest of shareholders (see the negative relationship between the Herfindahl index of geographical concentration and the demand for reinsurance in Sect. 6.7.4). Here, the estimation results weakly point to a positive partial relationship between geographical concentration and demand for reinsurance in that the rate of retention is lower for ICs with centralized agencies. However, the difference is not statistically significant.

The two effects predicted by hypotheses H1 and H2 amount to a cost disadvantage of stock companies that *ceteris paribus* should be reflected in higher premiums for comparable products. Indeed, the author finds that premiums for automobile liability insurance in the compact class are 2.5% higher than average among stock companies but 14.6% lower than average among public insurers.

▶ **Conclusion 9.9** There is evidence suggesting that German rate of return regulation in the 1980s thwarted incentives of for-profit ICs in a way as to increase their expense ratios, amounts ceded to reinsurance, and premiums above the level of non-profit ICs.

This conclusion points to the possibility that regulation designed to protect consumers has counter-productive effects, reminiscent of the study by Frech and Sam-

Table 9.3 Retained share of premium volume, 81 German ICs (1980)

Explanatory variable	Estimated deviation from mean value of 64%[a]	Error probability
Type of insurer:		
Stock company	−4 percentage points	
Mutual	+12 percentage points	$p < 0.03$
Public	+23 percentage points	
Distribution system:		
Centralized network	−5 percentage points	$p < 0.29$
Network with many agents	+1 percentage points	
$R^2 = 0.14$		

[a]Based on the estimated regression coefficient pertaining to a categorical variable ($= 1$ if characteristic is present, $= 0$ otherwise)
Source Finsinger (1983a)

prone (1980) discussed in Sect. 9.3.1. The usual expectation is that the profit motive induces enterprises to keep their costs low. Under the pressure of competition, any cost advantage must be passed on to consumers, possibly with a delay. In the present case, however, a rule governing the use of profits to the advantage of IBs seems to have influenced the for-profit ICs in a way that consumers end up paying more for the product than in the case of public ICs, who are suspected of inefficiency as a rule.

9.4 Recent Trends in Insurance Regulation

9.4.1 The Financial Crisis of 2007–2009

Observers agree that the 2007–2009 financial crisis had its origins in the Californian real estate market. The U.S. government had sought to encourage housing and property ownership also among low-income households who would not have been considered eligible when applying conventional criteria for mortgage financing. At the same time, prices of real estate had been soaring especially in California. Consumers had reason to believe that even a debt amounting to 100% of the value of the house would be reduced to a much lower share of property value within a few years. Banks did not want to keep these "subprime mortgages" in their books; rather they combined them with higher rated assets, to be offered as securities in capital markets. This so-called securitization allows agents to structure their portfolios in a way to benefit from improved risk diversification effects. However, this time securitization continued into second and third rounds, with the undesired side effect that many banks ended up with having such amalgamated products among their assets [see OECD (2010a)].

When the economic development in California and other areas of the United States turned out less favorable than envisaged, property values began to fall. Homeowners saw their net housing wealth drop and even become negative. Since affected properties could not be sold anywhere near the purchase price, this caused mortgage-based securities to quickly lose value. Banks worldwide became concerned about these losses, which negatively affected their asset position. They also began to doubt the creditworthiness of their (mostly anonymous) partners in the interbank credit business (particularly after the collapse of Lehman Brothers). With interbank credit not easily available anymore, banks had to cancel credits provided to businesses outside the banking sector. For this reason, the financial crisis spilled over to the real economy, causing one of the worst recessions, second only to the great depression in some countries [see OECD (2010a)].

With the exception of the American International Group (AIG), insurers had not invested a great deal in these mortgage-based securities. Therefore, they were not directly involved. However, they suffered indirectly because of their capital investment activity. Figure 9.3 makes clear that average returns on the U.S. capital market were as low in 2008 as in 1931 only, far lower than in 2001/02 when the so-called dot-com bubble burst. The year 2007, marking the beginning of the crisis, was still in the 0–10% category of returns, and 2006, in the 10–20% category.

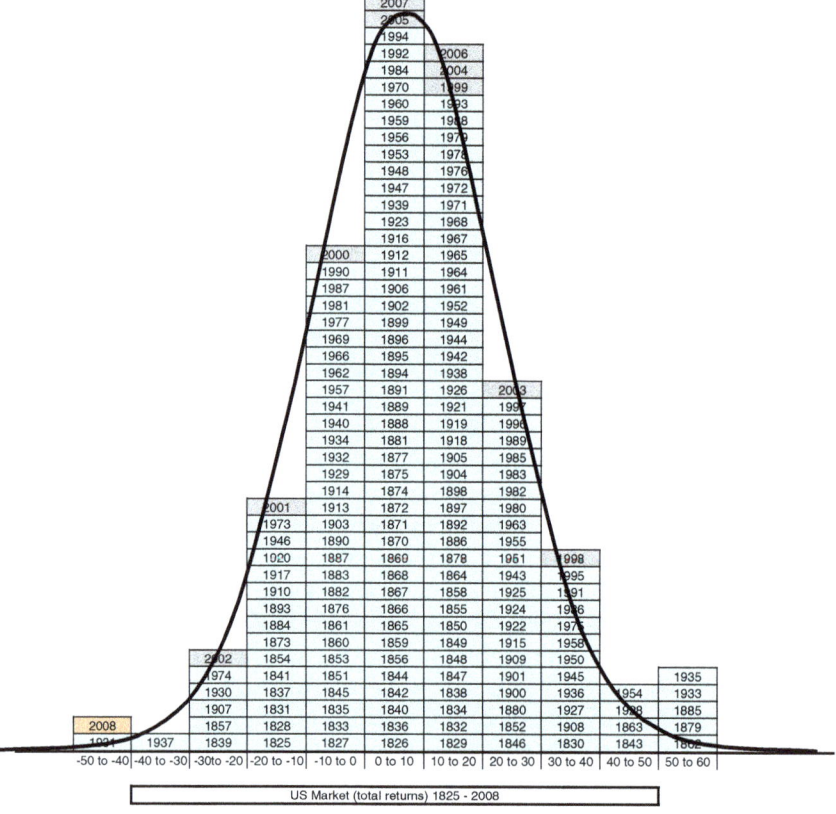

Fig. 9.3 Annual returns on investment in the United States, 1825–2008. *Source* Courtesy of Swiss Re (originally: AXA Insurance)

Referring back to the market for regulation model of Fig. 9.1 of Sect. 9.3.2, the predicted effect of this shock is clear. It amounts to an outward shift of the demand curve, resulting in a higher regulatory intensity, more future lobbying activities, and expansion of regulatory bureaucracy (see hypotheses H1, H2, and H5 of Table 9.1). Also, a renewed emphasis on preventing insolvency [variant (A) of regulation, see Sect. 9.1.1] is to be expected, which likely will go along with a return to the material supervision. The fact that the origin of the crisis was in the banking rather than in insurance is largely ignored, based on two main arguments. First, banks and insurers have been converging, and second, like banks, insurers may pose a systemic risk calling for so-called prudential regulation [see a report by the International Association of Insurance Supervisors IAIS (2010)].

9.4.2 The Convergence of Banking and Insurance and Their Regulation

In the wake of the "bancassurance" movement in the 1980s, banking and insurance were increasingly seen as two types of financial services that were bound to converge. More recently, this convergence has been related to the increased use of capital markets [in the guise of alternative risk transfer (ART)] by insurers and reinsurers. However, note that Cummins and Weiss (2009) entitle their survey of ART solutions with "Convergence of insurance and financial markets" rather than "Convergence of insurance and banking".

This is an important distinction when it comes to the regulatory implications of convergence. It is certainly true that banks have been using securitization for transferring liabilities to capital markets and that insurers have been catching up with them through their increased use of ART. It is also true that both banks and insurers are exposed to risks associated with these capital market products, causing even Baltensperger et al. (2008) in their careful survey to argue in favor of an integrated supervision of banks and insurers.

But it is not at all clear that this type of convergence implies that the two should be regulated in the same way. To put things in an (admittedly stark) perspective, one could replace banks by the hotel industry, insurers by the chemical industry, and the capital market by the market for information technology (IT) services. Both the hotel and the chemical industry have been increasingly relying on IT services. Any failure in IT markets would expose both industries to great risk. However, this "convergence" would hardly justify the use of the same regulatory approach to the two industries [for another critique of uniform regulation of banks and insurers, see Zweifel (2015)]. Returning to banking and insurance, the regulation would have to reflect the role of capital market products in their respective business models, which differ (see Sect. 9.4.3 below). Conversely, given that the capital market is the source of risk, it is there where the focus of regulation would have to be.

However, due to the global nature of capital markets, a global regulatory agency would have to be created. From the point of view of national agencies, the associated loss of local support causes the marginal cost of regulation to be extremely high. In terms of Fig. 9.1 above, their supply function runs so high that the predicted degree of regulatory intensity is zero. Rather, regulatory authorities in particular in the European Union propose Solvency II, which is modeled after the Basel II agreement for banks (see Sect. 9.4.4).

9.4.3 Is there Systemic Risk in Insurance Markets?

Cummins and Weiss (2014) distinguish three main criteria that must be cumulatively satisfied for systemic risk to exist (see panel A of Table 9.4). These are size, connectedness, and lack of substitutes in the case of crisis. Already when it comes to size (measured in terms of the industry's share in GDP, say), banking is typically far more important than insurance in an industrial country (see Table 1.6 of Sect. 1.3). This is due to their payment services, i.e., organizing current financial transactions between debtors and creditors. The same function also causes a more marked connectedness of banks compared to insurers. Connectedness is at the root of "domino effects", which can be largely excluded in the case of insurance except for credit insurance, where insolvency of a company might cause a liquidity squeeze for an enterprise that experiences a credit default.

Another possible source of connectedness comes from reinsurance transactions. To the extent that many insurers reinsure with fewer reinsurers, a loss to a single reinsurer may create a "domino effect" of insolvencies. Chen et al. (2020) note that some "[c]ore insurers usually are highly connected" and "make the market susceptible to a too-interconnected-to-fail problem" [p. 255]. However, they ultimately conclude that "even the failure of [a highly connected reinsurer] would not lead to widespread insolvencies in the U.S." [p. 255].

Moreover, banks are not easily substituted when it comes to their payment services. Failure of an insurer to perform its function, resulting in the non-availability of life and non-life products, likely would have small short-run effects but would hurt the economy in the longer run.

Cummins and Weiss (2014) also identify several contributing factors that serve to magnify any systemic risk. The first is leverage (see panel B of Table 9.4). Whereas the asset side of the balance sheet shrinks in the event of a crisis, liabilities remain the same, causing equity to strongly decrease if leverage (i.e., the liabilities/equity ratio) is high. However, insurers typically are much less leveraged than banks. Another contributing factor is a discrepancy in the maturity of assets and liabilities. In the case of banks, this discrepancy is part of their business model, which consists of transforming liabilities of short duration into assets of much longer duration. By way of contrast, insurers have a long tradition of maturity matching as part of their asset–liability management.

Next, the complexity of the business is also a factor. Here, the use of complex structured capital market products is typical of banks but has become increasingly popular with insurers as well. According to Plantin and Rochet (2007, Ch. 5.1), another contributing factor is governance (i.e., the tendency of management to invest in risky products on behalf of shareholders). This is a common feature of both banks

Table 9.4 Systemic risks in banking and insurance

Aspect	Banks	Insurers
A. Main criteria (cumulative):		
A1. Size	√ (GDP share ≈ 3%)	? (GDP share ≈ 1.5%)
A2. Connectedness	√ (payment services)	? (credit insurance/reinsurance)
A3. Lack of substitutes	√ (payment services)	? (long-run effects)
B. Contributing factors:		
B1. Leverage	√ (liabilities/equity ≥ 9/1)	? (liabilities/equity ≈ 5/1)
B2. Discrepancy in duration	√ (duration high for assets, low for liabilities)	- (duration matching)
B3. Complexity	√ (use of structured products)	√ (increasing use of structured products)
B4. Governance problems	√ (dispersed creditors)	√ (dispersed policyholders)
B5. Regulation	√ (phase of marked deregulation)	- (little deregulation)

Source Cummins and Weiss (2014)

(who usually have many dispersed creditors) and insurers (who have many dispersed policyholders). In both instances, there is no major creditor that could monitor and influence risk exposure. A final difference between the two industries is regulation. Deregulation of banking was more marked compared to insurance.

On the whole, Table 9.4 points to several important differences both with regard to the three main criteria for systemic risk and contributing factors. If the public interest model of regulation were to apply (see Sect. 9.2), banks and insurers would be predicted to be subject to different intensities (and types) of regulation. In view of the market for the regulation model, however, the supply and demand schedules are relevant. As long as the regulatory authority conceives the extra cost of extending regulation to be similar for banks and insurers (same supply schedule), an increase in demand, e.g., driven by a financial crisis is likely to result in an increase in regulatory intensity for both industries and in a rather undifferentiated way.

9.4.4 Characterization of Recent Regulatory Initiatives

This overview focuses on three recent regulatory initiatives. The first is Solvency II of the European Union, which predates the 2007–2009 crisis and was adopted in 2009. It comprises three pillars; (1) quantitative requirements regarding solvency capital, (2) supervisory review, and (3) disclosure requirements [see Pricewaterhouse Coopers (2008–2011)]. The quantitative requirements are of particular interest because they seek to reflect the particular risk profile of an IC, in contrast to the previous Solvency I standard. Required solvency capital (the sum of equity and insurance reserves) consists of three layers.

- Layer A is the best estimate of liabilities augmented by a risk margin. It should be market based, reflecting the rate of return on capital a potential buyer of the liabilities would require.
- Layer B is the additional solvency ("add-on") capital necessary to reach the so-called minimum required capital (MRC) threshold. Here, national regulatory authorities enjoy a measure of discretion, and the EC Directive (Article 27) qualifies add-ons as an exceptional measure. This is important because whenever solvency capital falls short of the MRC threshold, the authority has the right to intervene.
- Layer C amounts to a risk-sensitive additional requirement reflecting operational, underwriting, investment, and other financial risks (e.g., arising from counterparty default). Correlations between these risks are taken into account in the formulas making up the "standardized approach". Insurers can opt for an internal model reflecting their specific risk profile, which, however, must be approved by the national regulatory authority. This innovation was copied from the Basel II accord for banks.

In the spirit of the U.S. approach (see Sect. 9.3.2), Plantin and Rochet (2007, Ch. 6.2) propose a "double trigger" for governing regulatory intervention. As long as the top layer C is not breached, the authority makes sure that the reports submitted by the IC are correct. If it is breached, it must investigate more thoroughly and work with the IC to establish a plan to restore the situation. In the event that solvency capital falls short of the limit defined by layer A, the regulator would manage the case jointly with the guarantee fund (rather than simply seizing control as in the United States). In this way, the authors seek to ensure incentive compatibility because the other IC will seek to avoid the reputation loss caused by bankruptcy, e.g., by taking over the insured population.

Two additional regulatory initiatives are in the planning stage.

The International Association of Insurance Supervisors IAIS (2010) conducted a survey of planned regulatory initiatives among its members. It may be of interest to compare its findings with the specific proposal of the European Commission (EC) in the guise of a White Paper [see European Commission (2010)] proposing mandatory insurance guarantee schemes. In both documents, solvency (more precisely, the assurance of a low probability of insolvency) is the stated objective.

One can characterize the two documents along the dimensions listed in Table 9.5. The basic theory underlying the planned initiatives of IAIS members and the EC is the public interest model. There is no reflection on the marginal cost to members for providing more regulation nor the demand factors that could influence its intensity. However, the IAIS survey finds very different intensities of regulation in member countries, as predicted by the market for the regulation model. In the same vein, neither IAIS (2010) nor European Commission (2010) contain an evaluation of past regulatory performance. Rather, the burden of proof for or against additional regulatory intervention is never on the regulator but on the insurance industry. And in spite of the differences found with regard to systemic risks between banking and insurance (see Table 9.4), the concept of so-called macro-prudential

9.4 Recent Trends in Insurance Regulation

Table 9.5 Recent regulatory initiatives

	Aspect	Members of IAIS	EC (Guarantee schemes)
1.	Basic theory	Public interest model	Public interest model
2.	Evaluation of past regulatory performance?	No	No
3.	Burden of proof for/against regulatory intervention	On insurance industry	On insurance industry
4.	Adoption of macro-prudential regulation also for insurance?	Yes (planned by majority of members)	Yes
5.	Details of implementation	Not specified	Creation of EBA, EIOPA, ESMA; ESRB
6.	Specific treatment of reinsurance	Yes	No

IAIS: International Association of Insurance Supervisors; EC: European Commission; EBA: European Banking Authority; EIOPA: European Insurance and Occupational Pensions Authority; ESMA: European Securities and Markets Authority; ESRB: European Systemic Risk Board

regulation is carried over from banking to insurance in both reports. Neither report contains details of implementation that would permit to assess the potential for efficiency enhancement. In particular, the European Commission simply proposes to create four new supervisory authorities. Finally, it is interesting to see that reinsurance is mentioned in the IAIS report only, while the European Commission does not address the purchase of reinsurance as a possible alternative to a mandated guarantee fund.

Although Table 9.5 points to several deficiencies in the planned intensification of regulation at the EU level, there may be the advantage of a common future regulatory standard. These advantages are emphasized by Von Bomhard (2010), who notes in particular that the Risk-Based Capital approach adopted by the U.S. National Association of Insurance Commissioners (NAIC) differs from the Solvency II standard adopted by the European Union. Of course, it would be of great benefit especially for (re)insurers with international business activity to have to comply with one regulatory standard only. However, this standard would have to be the efficient one, which is somewhat doubtful in view of the predictive power of the market for the regulation model. The so-called regulatory arbitrage could make some of the more extreme jurisdictions realize that their supply function in Fig. 9.1 runs higher than envisaged, inducing them to reduce regulatory requirements. This option is lost once a global standard is implemented [this is also emphasized by Baltensperger et al. (2008)]. The trade-off between these costs and benefits has not been addressed in a systematic way yet, however. This continues to be true of the most recent regulatory initiative, Basel III, which is likely to become Solvency III for insurers (Basel Committee on Banking Supervision, 2019). Compared to Solvency I and II, which may well have led ICs to take on more rather than less risk (Zweifel, 2015), Solvency III would require ICs to consider shocks in the capital market which modifies the relationship between their solvency level and their attractiveness to investors. Therefore, Solvency III (if modeled after Basel III) holds the promise of inducing at least under-capitalized ICs to adopt less risky strategies (Zweifel, 2019).

Exercises

9.1

(a) For explaining the regulation of insurance markets, one can draw on efficiency arguments emphasizing the public interest as well as on arguments of political economy emphasizing capture and the market for regulation. Describe the core elements of these alternatives in no more than three sentences each.
(b) You are a member of the management of the IC "Star" in your own country. You consider a merger with the U.S. company "Top". Explain the differences with regard to supervision of insurance in the two countries. Give two reasons why knowing these differences is of importance for the merger project.
(c) For your merger project, is it also important to know (e.g., for your lobbying effort) whether in one or both of the countries, regulation conforms to the public interest, capture, and market for the regulation model, respectively?

9.2

(a) In their study of some European insurance markets, Finsinger and Schmidt (1994) find that $ln(PREM)$ can be explained to a high degree by $REG, HERF$, and $ln(LIFEEXP)$ (see Sect. 9.4.2). How is the dependent variable defined? What could be a reason for its logarithmic transformation?
(b) How are the explanatory variables defined? Why do they appear in the equation?
(c) What is the hypothesis to be tested when one assesses the sign and statistical significance of REG? What is the implication of the finding that the coefficient of REG amounts to $+0.48$ and is significant statistically?
(d) Germany is part of the sample. In which category of REG does it fall? And which category would apply to the Scandinavian countries, the United Kingdom, and the United States? Using (c), how should the premium level of these countries compare with the EU average?
(e) This study is based on data from 1988. Would a similar study using present-day observations yield similar results? Justify your answer.

References

Abraham, K.S. (1995). *Insurance Law and Regulation: Cases and Materials*, 2nd ed. Westbury NY: The Foundation Press.

References

Adams, M.B., & Tower, G.D. (1994). Theories of regulation: some reflections on the statutory supervision of insurance companies in Anglo-American countries. *Geneva Papers on Risk and Insurance Issues and Practice, 71*, 156–177.

Baltensperger, E., Buomberger, P., Iuppa, A., Wicki, A., & Keller, B. (2008). Regulation and intervention in the insurance industry – fundamental issues. *The Geneva Reports, Risk and Insurance Research 1*.

Basel Committee on Banking Supervision (2019). Basel framework: Scope and definitions. Basel: Bank for International Settlements.

Chen, H., Cummins, J.D., Sun, T., & Weiss, M.A. (2020). Property-casualty insurers: Microstructure, insolvency risk, and contagion. *Journal of Risk and Insurance. 87*(2), 253–284.

Cummins, J.D., & Weiss, M.A. (2014). Systemic risk and the U.S. insurance sector. *Journal of Risk and Insurance, 81*(3), 489–528.

Cummins, J.D., & Weiss, M.A. (2009). Convergence of insurance and financial markets: hybrid and securitized risk-transfer solutions. *Journal of Risk and Insurance, 76*(3), 493–545.

Dionne, G., & Michaud, P-C. & Dahchour, M. (2013). Separating moral hazard from adverse selection and learning in automobile insurance: longitudinal evidence from France. *Journal of the European Economic Association, 11*(4), 897–917.

Eckles, D.L. & Hilliard, J.I. (2015). Government intervention through an implicit federal backstop: is there a link to market power? *The Geneva Papers on Risk and Insurance–Issues and Practice, 40*, 538–555.

Eisen, R., Müller, W., & Zweifel, P. (1993). Entrepreneurial insurance: a new paradigm for deregulated markets. *Geneva Papers on Risk and Insurance, 66*, 3–56.

European Commission (2010). *White Paper on Insurance Guarantee Schemes*.

Finsinger, J. (1983a). *Versicherungsmärkte (Insurance Markets)*. Frankfurt: Campus Verlag.

Finsinger, J. (1983b). The performance of property line insurance firms under the German regulatory system. *Journal of Theoretical and Institutional Economics/Zeitschrift für die gesamte Staatswissenschaft, 139*, 473–489.

Finsinger, J., & Schmidt, F.A. (1994). Prices, distribution channels, and regulatory intervention in European insurance markets. *Geneva Papers on Risk and Insurance Issues and Practice, 70*, 22–36.

Frech, H.E. III, & Zweifel, P. (2017). Market Socialism and community rating in health insurance. *Comparative Economic Studies, 59*, 3, 405–427.

Frech, H.E. III, & Samprone, J.C. (1980). The welfare loss of excess nonprice competition: the case of property-liability insurance regulation. *Journal of Law and Economics, XXI*, 429–440.

Grabowski, H.G., Viscusi, W.K., & Evans, W.N. (1989). Price and availability tradeoffs of automobile insurance regulation. *Journal of Risk and Insurance, 56*, 275–299.

Harrington, S.E. (1987). A note on the impact of auto insurance rate regulations. *Review of Economics and Statistics, 69*, 166–170.

International Association of Insurance Supervisors (IAIS) (2010). Macroprudential surveillance and (re)insurance, *Global Reinsurance Market Report*, Mid-year ed.

Kreier, R., & Zweifel, P. (2010). Health insurance in Switzerland: a closer look at a system often offered as a model for the United States. *Hofstra Law Review, 39*(1), 89–111.

Merkin, R., & Rodger, A. (1997). *EC Insurance Law*. London: Longman.

National Association of Insurance Commissioners (NAIC) Capital Adequacy Task Force (2009). *Risk-Based Capital. General Overview*, www.naic.org/documents/committees_e_capad_RBC.

Nektarios, M. (2010). Deregulation, insurance supervision and guaranty funds. *The Geneva Papers on Risk and Insurance-Issues and Practice, 35*, 452–468.

Niskanen, W.A. (1971). *Bureaucracy and Representative Government*. Chicago: Aldine.

OECD (2010a). *The Impact of the Financial Crises on the Insurance Sector and Policy Respones*. Paris: OECD Publishing.

Olson, M. (1965). *The Logic of Collective Action: Public Goods and the Theory of Groups*. Boston: Harvard University Press.

Pauly, M.V., Kunreuther, H., & Kleindorfer, P. (1986). Regulation and quality competition in the U.S. insurance industry. In J. Finsinger, & M.V. Pauly (Eds.), *The Economics of Insurance Regulation* (pp. 65–107). London: Macmillan.

Peltzman, S. (1976). Towards a more general theory of regulation. *Journal of Law and Economics, 19*, 211–240.

Plantin, G., & Rochet, J. (2007). *When Insurers Go Bust: an Economic Analysis of the Role and Design of Prudential Regulation*. Princeton: Princeton University Press.

Posner, R.A. (1974). Theories of economic regulation. *Bell Journal of Economics and Management Science, 5*(2), 335–358.

Pricewaterhouse Coopers (2008–2011). *Countdown to Solvency II*, several editions, http://www.pwc.com/gx/en/insurance/solvency-ii/.

Rees, R., Gravelle, H., & Wambach, A. (1999). Regulation of insurance markets. *Geneva Papers on Risk and Insurance Theory, 24*(1), 55–68.

Van de Ven, W.P., & Ellis, R.P. (2000). Risk adjustment in competitive health plan markets. In A.J. Culyer, & J.P. Newhouse (Eds.), *Handbook of Health Economics* (pp. 155–845), Chap. 14. London: Elsevier.

Von Bomhard, N. (2010). The advantages of a global solvency standard. *The Geneva Papers, 35*, 79–91.

Zweifel, P. (1994). EEA (European Economic Area) and insurance in the EFTA countries. In Sveriges Riksbank (Ed.), *Financial Integration in Western Europe, Occasional Paper, 10*, 93–99.

Zweifel, P. (2015). Solvency regulation of banks and insurers: A two-pronged critique. *International Journal of Financial Research, 6*(3), 86–105.

Zweifel, P. (2019). Planned Solvency III regulation: Should it be adopted outside the European Union? *Asia-Pacific Journal of Risk and Insurance, 13*(1), 1–12.

Zweifel, P. (2007). The theory of social health insurance. *Foundations and Trends in Microeconomics, 3*(3), 183–273.

Zweifel, P., & Auckenthaler, C. (2008). On the feasibility of insurers' investment policies. *Journal of Risk and Insurance, 75*(1), 193–206.

Zweifel, P., & Breuer, M. (2006). The case for risk-based premiums in public health insurance. *Health Economics, Policy and Law, 1*(2), 171–188.

Social Insurance   10

This chapter deals with social insurance (also called social security especially in the United States) and its interaction with private insurance (PI). After a short survey of the importance of social insurance (SI) in Sect. 10.1, the question of why there should be SI is raised in Sect. 10.2 (after all, there is no social banking!). In view of the fact that in the domain of personal insurance, SI has a volume of contributions that is several times as high as that of PI (see Table 10.1), this question may appear to be a moot point. Very often, the market failure of PI is cited as a possible reason for the existence of SI. The problem is only that the rapid growth of SI would have to be explained with reference to market failures becoming more acute over time. This points to another explanation of SI, namely as a tool in the hands of political decision-makers.

Section 10.3 contains a comparison of social expenditure (which importantly consists of SI benefits) in a few industrial countries. It is found that the structure of expenditure differs substantially, likely reflecting idiosyncrasies of a country's political process determining SI.

In view of the stepwise and not very systematic expansion of SI in the countries considered, there is a reason for concern that the interaction between the different branches of SI on the one hand and between PI and SI on the other hand may diverge from an optimum from the perspective of citizens and consumers. For this reason, a simple criterion is developed in Sect. 10.4 in order to assess the performance of an entire system of social protection. This criterion is derived from portfolio theory (see Sect. 5.1.2); it points to substantial scope for efficiency improvement.

The impacts of social insurance on the economy as a whole are examined in Sect. 10.5. Focus is on influences on markets for factors of production (supply and demand for labor and capital) rather than product markets because most branches of SI leave it up to recipients how to use their benefits received. One exception is health insurance, which makes its benefits conditional on the use of medical services; in the

Table 10.1 Social expenditure of selected OECD countries as percent of GDP

Country	1980	1990	2000	2010	2017	PI (life)[a]
France	20.8	24.3	27.5	30.7	32.0	6.7
Germany	22.1	21.4	25.4	26.7	25.5	2.7
Italy	18.0	20.7	22.0	27.6	27.3	6.1
Japan	10.4	11.1	16.3	22.1	21.8[b]	6.0
Mexico	n.a.	3.2	4.8	7.5	7.3[c]	1.0
Sweden	27.0	27.2	26.8	26.3	25.9	5.7
United Kingdom	10.5	15.2	17.7	22.8	20.9	5.5
United States	13.2	13.2	14.3	19.3	18.6	4.5

[a]Life insurance premiums; [b]2015; [c]2016
Sources Ladaique (2011); OECD Social Expenditure Database (SOCX) 2018; OECD.Stat Insurance industry indicators: Penetration 2018

extreme case, it directly pays for services rendered (benefits in kind). In principle, the impacts on the economy attributed to SI are to be expected for PI as well. However, in PI, only relatively few consumers opt for a given type of contract, the conditions of which are tailored to their specific circumstances. By way of contrast, SI subjects major parts of the population to uniform regulation per decree. Such homogeneous impulses are likely to trigger far-reaching adjustments in the economy. Since these adjustments in part give rise to external effects, they may amount to efficiency losses.

10.1 Importance of Social Insurance

The term "social insurance" is understood to comprise all lines of personal insurance that are organized by the government. Still, it has to be insurance in the sense that it takes the payment of contributions to create a right to uncertain future benefits. If this condition is not satisfied, one speaks of public welfare, comprising, e.g., housing benefits, child benefits, and educational subsidies but also exemption from income taxation. Typically, public welfare is means-tested. Since the dividing line between SI and public welfare depends on the country considered, international statistics aggregate the two into social expenditure.

Table 10.1 exhibits the total social expenditure as a share of gross domestic product (GDP) in selected OECD member countries.

Since the 1980s, Sweden, Germany, and France have been at the top of the sample; in the case of Sweden, already in 1980, the GDP share of social expenditure was as high as 27%. Meanwhile, France even exceeds this value with 32%. However, even in the rich OECD countries with comparatively little social expenditure such as Japan, the United Kingdom, and the United States, this share has risen through 2010, reaching around 20%. Mexico, a recent addition to the OECD, seemed to be on a similar path of expansion, albeit starting from a far lower level. During the last

10.1 Importance of Social Insurance

few years, however, there has been a slight decline in the GDP share of the countries listed in Table 10.1, with the exception of France.

Table 10.1 also reveals that even in countries known for their sizable insurance industries (see Fig. 1.1), SI dwarfs its counterpart, the personal lines of private insurance (PI). Since combined data for these lines (health, life, and personal liability) are not easily accessible, premiums of life insurance relative to GDP may serve as an approximate indicator (note that premium revenue differs from expenditure). Its value presently is highest in France (6.7%); the United Kingdom follows at some distance, with 5.5%. Yet even in these countries, SI is four times more important than its private counterpart; this also holds true of the United States, where social health insurance is confined to Medicare (for pensioners) and Medicaid (for the poor). The predominance of SI in countries that are otherwise market economies calls for an explanation, to be proffered in Sect. 10.2.

At this point, it is important to note that there are two conceptions of social policy or welfare state underlying expenditure on social programs. One is associated with the German Chancellor Bismarck (1815–1898); it emphasizes the financing of social insurance through contributions by workers, possibly in combination with employers. The other conception (based on a 1944 report to the government) is associated with the British Prime Minister Beveridge (1879–1963); it relies on financing by taxation. This distinction is particularly clear in the case of social health insurance. Whereas the health funds of Germany charge contributions as a percent of labor income, the British National Health Service obtains its revenue from the government.

Public welfare or social assistance is not analyzed any further because it does not constitute insurance and is also less important than SI. The main objective here is to throw light on the parallels and differences as well as the interactions between PI and SI. An important parallel is given by the fact that both types of insurance relate payments in the case of loss to contributions (or premiums) paid, with the link between the two frequently much attenuated in the case of SI.

On the whole, there may be a certain degree of substitution between PI and SI. Indeed, it is theoretically plausible that an excessive coverage of risks (from the point of view of consumers) by SI induces the insured to accept more risk outside SI, causing demand for private coverage to be reduced [see Schulenburg 1986]. In addition, it may well be that social insurance crowds out PI, for example, if it guarantees a minimum income.

A reverse causation running from PI to SI can be neglected as far as demand is concerned. Consumers who prefer to have more private coverage cannot substitute this coverage by contracting for less SI. For reasons that are discussed in Sect. 10.2 below, SI as a rule imposes a uniform product that does not permit citizens to express their differentiated preferences.

Yet, there are interactions between private and social insurance that have to do with market failures in private insurance. Markets for private insurance may have problematic characteristics due to asymmetric information, especially in the guise of adverse selection (see Sect. 8.3). Adverse selection can be avoided by uniform mandatory insurance. However, this does not imply that PI should be fully replaced by SI; rather, a certain complementarity can be shown to be efficiency-enhancing (see Sect. 10.2.1.3 below).

10.2 Why Social Insurance?

In the economics literature, there are two explanations for the existence of SI.

1. *SI as an efficiency-enhancing institution.* SI constitutes a far-reaching regulation of insurance markets. For such an intervention to be efficiency-enhancing, it has to correct a market failure. The most important reasons for market failure in the present context are excessive time preference of consumers, altruistic motivation, and adverse selection combined with transaction costs.
2. *SI as an instrument in the hands of politicians and bureaucrats.* Insurance is perfectly suited for redistribution of incomes and redistribution in kind (e.g., medical services through social health insurance). This fact makes SI attractive for politicians who seek (re)election as members of government or of parliament because it helps them channel benefits to their constituencies. In addition, public administration can also benefit from SI through enhanced pay, power, and prestige [see Buchanan and Tullock 1976; Niskanen 1971].

These two explanations are presented below and tested to see whether they are compatible not only with the existence but also with the growth of SI over time.

10.2.1 Social Insurance as an Efficiency-Enhancing Institution

10.2.1.1 Excessive Time Preference as a Reason for Market Failure

A popular argument states that most people have an "excessive" time preference, implying that they discount "too heavily" returns and costs that occur in the future. For this reason, they fail to purchase a life insurance policy that would guarantee them a sufficient income in old age. Likewise, they might underestimate the risk of high healthcare expenditure in old age, making them go without health insurance coverage.

These arguments suppose that politicians who decide about the implementation and expansion of SI in parliament know the correct time preference. The difficulty with it is that in a democracy the same individuals with their excessive time preference vote for politicians who impose on them a lower time preference. Therefore, individuals must for some reason have a lower time preference acting as voters than acting as consumers. Such a discrepancy is puzzling.

On the other hand, the two modes of behavior on the part of citizens could be the consequence of the fact that using SI, politicians offer a product that cannot be offered by PI. This is the promise in favor of offspring not yet born to provide benefits regardless of inherited capabilities and incapacities that ensure a minimum standard of living. Parents, therefore, have the guarantee that their unborn offspring will not end up at the very bottom of an unequal income distribution. In principle, they could

10.2 Why Social Insurance?

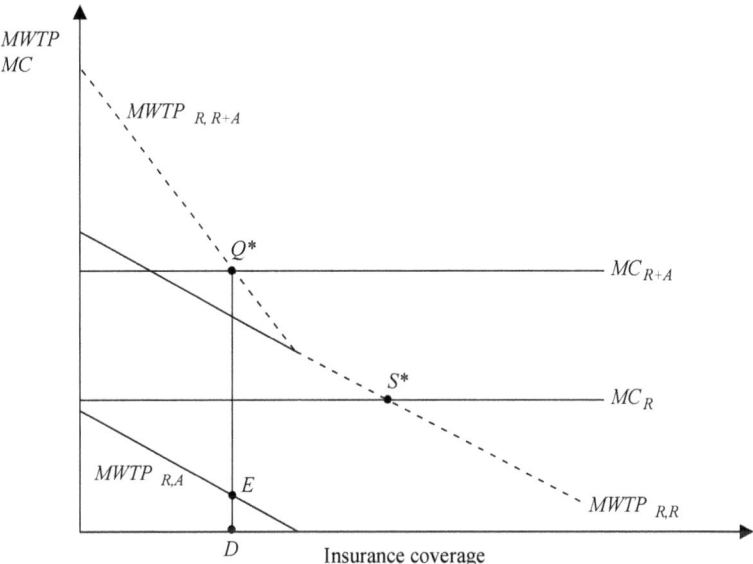

Fig. 10.1 Insurance of a poor individual as a public good

contract for their offspring to finance this insurance coverage themselves. However, an arrangement of this type would be close to bondage and unlikely to be enforceable in the context of PI [see Sinn 1996]. Such a "contract between generations" in the presence of low time preference can only be implemented through the government.

Still, this argument in favor of SI is not quite convincing since there is the alternative of subsidization. Parents who cannot afford to pay an insurance contribution to PI for their unborn children could be subsidized. The crucial problem seems to be the insurer's difficulty with categorizing unborn individuals according to the risk and of experience rating them after birth. This lack of information encourages adverse selection (see Sect. 8.3). In the final analysis, it is not so much excessive time preference but rather asymmetric information that is at the root of a possible market failure (for more detail, see Sect. 10.2.1.3). However, note that this type of market failure would have to become more common or more acute in the course of time in order to explain the growth of SI.

10.2.1.2 Altruistic Motivation Combined with Free Riding as a Reason for Market Failure

Social insurance may be seen as a means to alleviate a market failure in the context of altruistic motivation. This argument is due to Culyer (1980). Let a rich individual R be affected negatively by the poverty of individual A, with the cause of poverty constituting an insurable event (if the poverty of A were permanent, the alternative for R would be a gift). In order to mitigate this negative externality, R is prepared to pay the premium for providing A with insurance coverage. The (decreasing) marginal willingness to pay of R in favor of A is shown in Fig. 10.1 as $MWTP_{R,A}$.

In addition, there is also a marginal willingness to pay of R for himself or herself, $MWTP_{R,R}$ (which is higher than the one in favor of A). This raises the question of how to aggregate the two components of willingness to pay. Note that insurance coverage for A is a public good because other rich individuals benefit from externality relief even if they do not contribute to it. In the case of a private good, horizontal aggregation indicates the quantity each consumer wants to buy at a given price. In the present context, vertical aggregation is appropriate to determine the total willingness to pay for a predetermined quantity of the public good. Vertical addition of $MWTP_{R,R}$ and $MWTP_{R,A}$ results in the kinked line $MWTP_{R,R+A}$ of Fig. 10.1; for simplicity, let A have zero willingness to pay for insurance on his or her own.

The marginal cost of additional insurance MC is assumed constant and the same for R and A, ignoring moral hazard effects and differences in risk type. Point Q^* then indicates the social optimum, where aggregate marginal willingness to pay is just sufficient to cover the aggregate marginal cost. This corresponds to an amount of coverage D for R as well as A, financed by a contribution of R on behalf of A (DE) as well as for himself of herself (EQ^*).

However, this solution is not likely to be attained on a voluntary basis. Other rich individuals will hesitate to reveal their willingness to pay, hoping that R fully finances the public good in question, namely "relief from the negative externality caused by poverty". However, R will anticipate this free-riding behavior by not disclosing his or her willingness to pay. In order to reap the benefit achieved by insuring A, the rich members of society can agree to introduce mandatory insurance for the poor. It is financed by an obligation of the rich to pay a contribution presumably reflecting their estimated marginal willingness to pay, using income and wealth as an indicator. On the other hand, the poor must commit to accept the insurance mandate because otherwise the negative externality would continue to exist.

This model predicts mandatory insurance for everyone with a uniform basic coverage amounting to D in Fig. 10.1. The rich members of society additionally buy private coverage to reach their own optimum at point S^*.

▶ **Conclusion 10.1** Altruism combined with free riding can be used to explain the existence of mandatory insurance with uniform coverage, with the well-off purchasing additional coverage privately.

This model, therefore, can explain the existence of SI as well as some of its properties. However, it cannot explain why SI has expanded so much over time unless one would want to argue that altruistic motivations have become stronger with rising incomes.

10.2.1.3 Adverse Selection as Market Failure

As shown in Sect. 8.3, even the launch of separating contracts by the incumbent IC does not guarantee its economic survival. Provided the share of low risks is sufficiently high, a competitor can still offer them more favorable terms than the incumbent through a pooling contract which in turn is not sustainable. Therefore,

10.2 Why Social Insurance?

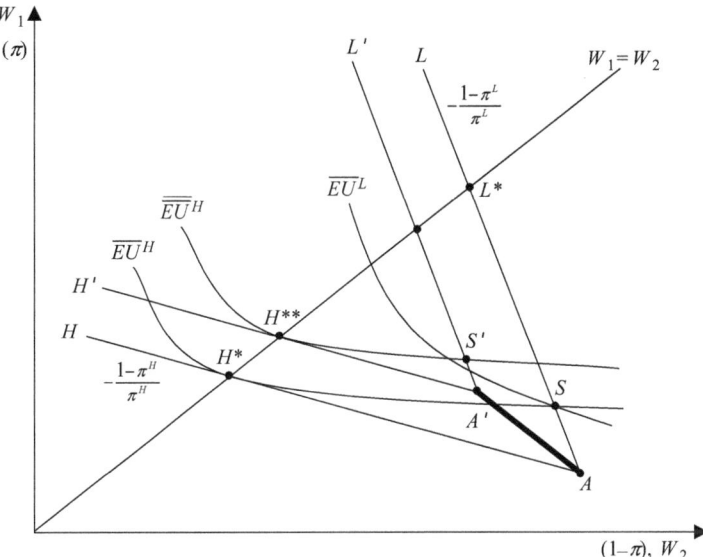

Fig. 10.2 Pareto improvement through SI given adverse selection

an equilibrium may fail to exist in private insurance markets. There is an externality again, this time related to the impossibility of identifying risk types (rather than the existence of poor individuals as in Sect. 10.2.1.2). It makes the unrecognized low risks either to pay a high premium (pooling contracts) or else to accept rationing of their insurance coverage (separating contracts). A partial insurance mandate can be shown to alleviate this externality and hence improve efficiency.

Figure 10.2 follows Dahlby (1981).[1] Let the two risk types L (low) and H (high) have the same endowment point A in the space of contingent claims (W_1, W_2). Thus, they do not differ in terms of their wealth in order to emphasize the difference with regard to risk. A risk-neutral IC offers insurance buyers (IBs) a transfer of wealth from the no-loss state (W_1) to the loss state (W_2). In the case of low risk, the so-called insurance line of Fig. 10.2 has slope $(1 - \pi^L)/\pi^L$ in absolute value, with π^L symbolizing the probability of loss (see Sect. 8.3.1). Since the high-risk types have a higher probability of loss ($\pi^H > \pi^L$), their insurance line runs flatter. Let the IC come up with a pair of separating contracts $\{H^*, S\}$ preventing high-risk types (who obtain full coverage at H^*, albeit at a high premium) from migrating to contracts designed for the low-risk types. Note that the indifference curve $\overline{EU}^H$ passes through H^* as well as S. Given certain conditions as cited in Sect. 8.3.1, the pair of contracts, therefore, constitutes a separating equilibrium. The low risks who by assumption cannot signal their risk type are disadvantaged because they cannot have full coverage (point L^*) but are rationed at S.

Let there now be partial mandatory coverage with a contribution calculated on the basis of the shares of low and high risks in the population (50% each for sim-

---

[1] See also Eisen (2006).

plicity). In this way, SI can offer coverage amounting to, e.g., AA' in Fig. 10.2 at a uniform contribution without jeopardizing its financial equilibrium. This results in a new endowment point A' from where a private IC can offer complementary coverage. There is no reason for them not to establish a separating equilibrium again. In Fig. 10.2, the corresponding pair of contracts is marked as $\{H^{**}, S'\}$. Compared to $\{H^*, S\}$, it amounts to a Pareto improvement thanks to SI:

- The high risks benefit because they are offered again full coverage at H^{**}, however, at more favorable terms. Coverage provided by SI is available to them at a contribution reflecting the average of low and high risks in the population.
- The low risks may also benefit. In the case illustrated by Fig. 10.2, they actually do because point S' lies above the indifference curve $\overline{EU}^L$ passing through S, indicating higher expected utility. The fact that SI and PI in combination serve to relax the rationing of the low risks (because the high risks have a more attractive alternative thanks to SI) may constitute a sufficient advantage for the low risks to compensate them for the disadvantage of having to pay the contribution to SI, which is calculated based on the average rather than on their own risk type.

▶ **Conclusion 10.2** In the presence of adverse selection, partial coverage imposed by SI can result in increased efficiency and hence Pareto improvement for both low and high risks and hence increased efficiency.

Note that Conclusion 10.2 mentions partial rather than comprehensive coverage by SI. Indeed, full coverage mandated by SI is unlikely to be in the interest of the low risks because two losses of expected utility reach their maximum. The first is that the contribution is now entirely calculated on the average risk. The second derives from the fact that in response to this unfair premium, the low risks prefer less than full coverage (see Sect. 3.3.1), an option denied by comprehensive SI. The partial relaxation of their rationing at point S may be important enough to offset these disadvantages. One condition for this to occur is a very marked convexity of the indifference curve and hence risk aversion (see Sect. 3.2.1).

10.2.1.4 Transaction Costs as Reason for Market Failure

Insurance coverage is never written at fair premiums since the premium must cover not only expected loss but also administrative and acquisition expenses. In addition, the IC typically charges a loading for risk bearing (see Sect. 7.1.3 for details). These surcharges amount to transaction costs that cause the IB to only demand partial coverage or even go without insurance coverage altogether.

This result may be interpreted as a market failure if SI can offer a less costly alternative. Indeed, being a monopolistic, uniform mandated scheme, SI does not have any acquisition expense; moreover, it benefits from the law of large numbers thanks to its size (see Sect. 7.1.2). Finally, a solvency guarantee by the government usually relieves it from any insurance risk, obviating a safety loading. Therefore, the contribution charged by SI comes close to the fair premium, resulting in a competitive advantage over private insurance.

10.2 Why Social Insurance?

One must also take into account that SI forces individuals to buy insurance coverage who otherwise would have not purchased it even given a low loading. Moreover, it prescribes a minimum amount of coverage that may be excessive for low risks. Also, the uniformity of SI neglects differences in the risk preferences of individuals and makes it difficult to adjust to their changes over time. Finally, the uniformity of contributions to social health insurance ultimately makes payment of service providers in the interest of patients impossible. Zweifel and Frech (2016) present ample evidence of preference heterogeneity with regard to the provision of healthcare services which calls for structuring insurance plans in a way as to make physicians act in accordance with the preferences of their respective clientele. This in turn implies that their payment and hence contributions paid by the insured must differ, something the solidarity principle of SI does not permit.

However, SI could indeed alleviate a market failure to the extent that transaction costs occur in a risk-specific way. Note that the political debate usually revolves around the problem that high risks have difficulty obtaining sufficient coverage from PI.[2] One reason could be that ICs, while recognizing high risks as such, are not able to correctly estimate their probability of loss and hence their risk-based premium.[3] Indeed, the fixed amount of loss is a simplification, abstracting from a loss distribution that usually is heavily skewed (meaning that small losses are very frequent, while extremely high losses do occur but rarely). Typically, high risks not only cause losses more frequently but also high losses are so rare that their probability of occurrence cannot easily be estimated. Therefore, there is uncertainty regarding their probability of loss π^H. This uncertainty contributes to the risk of insolvency, causing the IC to increase its safety loading.

Figure 10.3 illustrates. For simplicity, the low risks are offered a contract without any loading, whereas the high risks are confronted with a substantial loading (indicated by the flat insurance line AH''). On these conditions, the high-risk type is predicted to opt for partial coverage or for no insurance coverage at all. In Fig. 10.3, the indifference curve $\overline{EU}^H$ is drawn in a way that the second alternative obtains; the high risks could be said to be denied (affordable) insurance coverage. Note in passing that the private IC cannot launch a separating contract in this situation because the high risks would migrate to the contract designed for the low risks even if it offers very partial coverage (point S dominates point A in Fig. 10.3 since $\overline{\overline{EU}}^H$ indicates a higher expected utility for the high risks than does $\overline{EU}^H$). This means that in order to break even on its underwriting activity, the IC needs to maintain a high level of risk selection effort to implement contract S designed for the low risks.

By way of contrast, SI benefits from not having to charge a safety loading to control its risk of insolvency. This makes the insurance line AH (reflecting the fair premium) relevant for SI. If the low risks are assumed to constitute the majority of the population, SI can offer a pooling contract along the line AA'. However, this

---

[2] This concern contradicts the one described in Sect. 10.2.1.3, stating that due to the asymmetry of information, a negative externality emanates from the high risks, burdening the low ones.
[3] This argument is a modification of the model by Newhouse (1996).

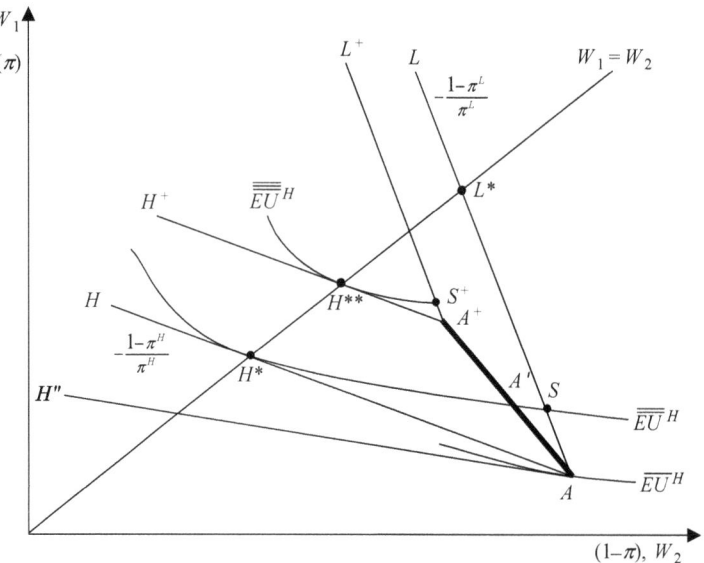

Fig. 10.3 Risk type-specific transaction costs and Pareto improvement through SI

limited amount of coverage may still confront private insurers writing complementary coverage with high losses from high risks. Therefore, let there be a mandated amount of coverage AA^+ that permits PI to suppress the safety loading for high risks. Starting from A^+ as the new endowment point, PI can now offer a contract along the insurance line A^+H^+. Given the favorable conditions of a fair premium, the high risks now purchase the private complementary insurance, resulting in full coverage at point H^{**}. Therefore, they reach the expected utility $\overline{\overline{EU^H}}$. The low risks can sign the separating contract S^+, which may result in an improvement for them compared to S (not shown).

▶ **Conclusion 10.3** The inability of PI to estimate the probability of loss of high risks with sufficient precision gives rise to risk-specific transaction costs. These can be avoided by SI, creating a possibility for Pareto improvement.

However, it remains questionable whether risk type-specific transaction costs can also explain the growth of SI over time. This growth is importantly due to the implementation of collective old age provision. Contrary to health insurance and liability insurance, there has been no technological change resulting in increasingly higher losses. Indeed, due to the rectangularization of the survival curve (see Sect. 2.1.2.1), predicting expected loss in old age provision has become easier over time. Therefore, there is no apparent reason for an increase in transaction cost that would motivate an expansion of coverage provided by SI.

10.2 Why Social Insurance?

10.2.1.5 Moral Hazard as a Limitation of SI

While mandatory SI provides a solution to the problem of adverse selection, it exacerbates the other problem associated with asymmetric information, i.e., moral hazard. As shown in Sect. 8.2.2.1, one way to counteract moral hazard is to have the premium progressively increasing with the amount of coverage. Moreover, while the probability of loss may not be known initially, insurers can observe the occurrence of losses. Therefore, insurance lines with a certain frequency of loss often use experience rating of premiums to control moral hazard effects.

However, progressive and experience-rated premiums induce differences in the price of insurance coverage. The theory laid out in Sect. 8.2.2.1 predicts differentiation in contributions that are not compatible with the basic idea of SI. Recall that Sects. 10.2.1.2 and 10.2.1.3 emphasized that SI offers a pooling contract with the same amount of benefits to high and low risks at a uniform contribution. This uniformity of SI does not square with differentiation of contributions with the aim of controlling moral hazard effects.

Conversely, the absence of differentiation of contributions according to risk in SI encourages moral hazard effects in two ways. The direct effect is that skimping on prevention is not sanctioned financially. The indirect one is that contributions approaching the fair premium create demand for full coverage by SI, which in turn strengthens moral hazard effects as a rule (see Sect. 8.2).

10.2.2 Social Insurance as an Instrument Wielded by Political Decision-Makers

10.2.2.1 Theoretical Background

Beginning in the 1950s, a few pioneers started applying economic theory to the behavior of political decision-makers [see Downs 1957; Olson 1965; and Mueller 1989]. Their main hypothesis states that members of governments, parliaments, administrations, and lobby groups pursue their own interest in the same way as do consumers and producers in markets. For example, members of a government would like to stay in power, which in a democracy requires winning the next election. An important way to win an election is to redistribute income and wealth in favor of one's own voters (and to the detriment of voters at large). For two reasons, SI suits this purpose especially well:

- Insurance of any type is a mechanism for redistributing wealth. Wealth is transferred from the many who pay premiums without incurring a loss to the few who suffer a loss. However, in PI, this redistribution is governed by chance and therefore non-systematic. In SI, it is easy for politicians to add systematic elements to the redistributive scheme without those bearing the burden being able to recognize this in the specific case. By way of contrast, redistribution through taxation is more easily recognized, triggering resistance.
- To the extent that there is a degree of altruism (see Sect. 10.2.1.2), politicians can even cite a desire for systematic redistribution as an argument in favor of SI. In contrast again, taxation first of all serves to finance public administration; there is

no guarantee for taxpayers that the funds are channeled to citizens who otherwise could not afford insurance coverage.

More generally, SI can be interpreted as the (extreme) outcome of demand for and supply of public regulation (see Sect. 9.3). Since the supply of public regulation comes from political decision-makers, both the existence and structure of SI likely reflect their interests to an important extent.

▶ **Conclusion 10.4** The creation of SI, but also its extension can be interpreted as the outcome of the supply of public regulation. Therefore, their existence and structure can be expected to be determined by the interests of political decision-makers to an important degree.

10.2.2.2 The Interest of Government in Social Insurance

There are hardly any empirical investigations designed to test the hypothesis that governments use SI as an instrument for ensuring their (re)election. An early study by Schneider (1986) found that the popularity of government in Australia, Germany, and the United States increased with a rising share of SI in total public expenditure. He estimated an increase in the share of SI benefits by 10% (for instance, from 30 to 33%) to have the same favorable impact on popularity as a reduction in the rate of unemployment by 10% (for instance, from 5 to 4.5%). Van Dalen and Swank (1996) conducted an event study for the Netherlands, covering the years 1954 to 1993. According to the authors, in the year of an expected election (sometimes, elections were deferred due to political crisis) as well as around those years, the share of SI in GDP increased by 13%, e.g., from 20 to 22.6%. The estimated increase in the public good, "defense spending" was only about half that size.

More recently, Bove et al. (2017) analyzed public spending by 22 OECD countries between 1988 and 2008. While the authors' pit government outlay on defense against social expenditure rather than SI proper, they also present results for expenditures on old age, family support, and incapacity support, which are important components of SI (see Table 10.2). Their dependent variables are the change in the log of the two shares in GDP [$\Delta ln(SOC/GDP)$ and $\Delta ln(DEF/GDP)$, amounting to percentage changes which filter out trends in the social and defense share, respectively]. In fact, average SOC/GDP increased from 19.2% in 1988 to 21.6% in 2008, while average DEF/GDP fell from 2.4 to 1.6%. Since the data set is of the panel type, the authors distinguish between countries by using dummy variables that equal one if the observation pertains to the United States (say) and zero otherwise, as in col. (2) of Table 10.2. However, the error terms might be correlated across countries and over time, reflecting spillovers not accounted for by the explanatory variables. As a rule, accounting for this causes an increase in standard errors and hence a reduction in the value of t−statistics, as in col. (3). According to the authors, statistical tests suggested this specification to be preferable for $\Delta ln(DEF/GDP)$ in col. (4).

Most of the regressors are self-explanatory. The categorical variables *Volunteer army*, *Involvement in wars*, *NATO*, and *Election* take on the value of one if the country has an army that is not composed of conscripts, has been taking part in wars

10.2 Why Social Insurance?

Table 10.2 Social expenditure and (re)election in OECD countries (1981–2008)

Explanatory variables (1)	Dependent variables		
	$\Delta ln(SOC/GDP)$		$\Delta ln(DEF/GDP)$
	OLS with country dummies (2)	Corrected std. err. for panel data (3)	Corrected std. err. for panel data (4)
Lagged dependent variable	0.228*** (5.30)	–	–
Δln GDP per capita	–0.800*** (7.93)	–0.812*** (6.34)	–0.564*** (3.43)
Δln unemployment rate	0.030*** (2.59)	0.048*** (3.83)	–
Δln globalization	–0.004** (2.48)	–0.003 (1.22)	0.001 (0.24)
Δln dependency ratio	0.064 (0.26)	0.006 (0.02)	–
Δln(Armed forces/workforce)	–	–	0.022** (2.46)
Δln(Rivals' armed f./workf.)	–	–	0.134*** (2.96)
Volunteer army	–	–	0.002 (0.23)
Involvement in wars	–	–	0.017*** (2.92)
Political constraints	–0.013 (0.41)	–0.032 (1.27)	–0.024 (0.84)
NATO	–	–	–0.004 (0.65)
Election	0.009*** (2.92)	0.008*** (5.13)	–
Pre-election*NATO	–	–	–0.011** (2.42)
Election*NATO	–	–	–0.014*** (2.59)
Left-wing ideology	0.002 (1.43)	0.001 (0.73)	–0.004* (1.68)
No. obs / R^2	460 / –	460 / 0.56	462 / 0.23

Not shown are the following explanatory variables: Income, relative prices of the expenditure categories, size of population, ideological preferences of the cabinet, and time trend. OLS estimation, with random effects (i.e., the error term is split up into a time-dependent and a country-dependent component); *(**)[***] indicating statistical significance at the 0.10 (0.05) [0.01] level of significance
R^2: corrected coefficient of determination
Source Bove et al. (2017), Tables 1 and 2

recently, is a member of the defense alliance NATO in the year(s) of observation, and had an election in the year(s) of observation. The remaining regressors are continuous: *Dependency ratio* measures the share of < 14 and > 64-year olds in the population; *Globalization* is an index composed of trade flows, foreign investment, and regulation; *Armed forces* and *Rivals' armed forces* symbolize the share of the respective workforce enlisted in the country's own army and of the armies of countries in (potential) conflict; *Political constraints* is again an index reflecting, e.g., the presence of a two-chamber system which makes a government's pursuit of its goals more difficult; and *Left-wing ideology* is another index with values 1 to 5, the latter for a Parliament with at least a 2/3 left-wing majority.

Starting with col. (2) of Table 10.2, one notices a quick adjustment of social spending: The coefficient of the lagged dependent variable [i.e., $\Delta ln(SOC/GDP)_{-1}$] of

0.228 means that, e.g., an acceleration leading to 1% change in its GDP share takes only a bit more than one year [$1.3 = 1/(1 - 0.228)$] before a return to normal growth is achieved [i.e., for the condition, $\Delta ln(SOC/GDP) = \Delta ln(SOC/GDP)_{-1}$ to hold, *ceteris paribus*]. Growth of per-capita GDP by 10% is estimated to slow the growth in social expenditure by 8%, whereas an increase in the unemployment rate does not seem to make a difference. Crucially, the evidence suggests that in the year of an election, OECD country governments let social expenditure grow at a rate of 5.56% ($= 0.055 + 0.009$) rather than the average rate of 5.5% during the observation period. This is a much smaller effect than the one estimated by Van Dalen and Swank (1996) but exceeds that of a government's more marked left-wing orientation with a coefficient (lacking statistical significance) of 0.002.

Alternatively, OECD country governments can be seen as being subject to common unobserved influences causing (auto)correlation in the error terms, which takes account of the lag in adjustment [as in col. (3) of Table 10.2]. Most of the coefficients of interest turn out to be stable, with the important exception of *Globalization* (which cannot be said to slow the growth in social expenditure anymore).

The contrast between social and defense spending is striking. In col. (4) of Table 10.2, an accelerated increase in the size of the domestic defense group (*Armed forces /workforce*) seems to matter, at least as much as *Rivals' armed forces/workforces* and *Involvement in wars*, indicators of external pressure. In particular, membership in NATO before and in election years (Pre-Election*NATO and Election*NATO, respectively) is associated with an acceleration rather than deceleration in the decline of *DEF/GDP*, likely reflecting free riding on military spending by the United States in particular. Finally, more left-wing governments are found to accelerate the rate of reduction in *DEF/GDP*, from 1.9% to almost 2% p.a. ($-0.0194 = -0.019 - 0.0004$).

10.2.2.3 Interests of Other Political Decision-Makers in Social Insurance

Since empirical research on this topic is rare, the presentation is limited to theoretical arguments.

- *Delegates to parliament and SI.* Similar to governments, delegates must win elections in democracies. For this purpose, they are predicted to use their time in parliament in a way so as to maximize political support [Crain 1979]. With regard to support from PI, it may not be lost by voting in favor of mandatory insurance coverage as long as it can be provided by a private IC. In that event, insurers still have to balance the advantage of the expansion of demand due to the mandate against the disadvantages resulting from almost inevitable restrictions in the guise of regulated premium levels, limited premium differentiation, and the prohibition of risk selection.

    Very often, however, the creation of SI goes along with the implementation of a monopolistic public supplier. Even such a proposal need not make delegates lose support from PI provided risk-specific transaction costs play an important role (see Sect. 10.2.1.4). By covering large losses, SI may enable PI to reduce its safety loading, making insurance coverage attractive even to high-risk types. Delegates

who lose the support of PI for voting in favor of SI still have the possibility to replace it by that of workers in the public insurance administration, especially when the outcome is an expansion of SI.
- *Public administration and SI.* In contradistinction to the government, public administration does not face the hurdle of re-election. Ever since the pathbreaking work by Niskanen (1971), pay, prestige, and power are believed to motivate chiefs of public agencies. All of these objectives are served by the creation but also by the extension of SI, especially if the implementation of an insurance mandate is not delegated to private providers but remains within the public sector. Under these circumstances, there are employment possibilities for many subordinates and the creation of a multilevel hierarchy, serving the advancement of pay, prestige, and power.

▶ **Conclusion 10.5** There are clear indications suggesting that social insurance serves as an instrument wielded by political decision-makers. This view may explain not only the existence but also the expansion of social insurance.

10.3 Overview of the Branches of Social Insurance

Insurance is one possible way to protect one's assets against losses. These losses in turn are caused by perils (e.g., theft, fire, or illness). In private insurance (PI), the main division is between the types of assets affected, i.e., life and non-life. By way of contrast, social insurance (SI) focuses on the protection of non-marketable assets, health, and human capital (ability to work, skills). Therefore, the branches of SI are typically structured according to the perils that result in losses to these assets.

- *Old age*: Old age is associated with a reduction in labor income. In France, provision for old age is the most important mission of SI; in all other countries listed in Table 10.3, it ranks second. Still, even among this group, the GDP share of SI devoted to this risk varies between 6.1% in the United Kingdom and 16.3% in Italy, although with 22%, Italy's share of those older than 64 years is not that much higher than the 18% of the UK.
- *Incapacity to work (disability)*: This is a peril whose incidence seems to be comparable between Germany and Sweden (as an example). Since incapacity statistics are not easily accessible, the estimated prevalence of Parkinson's disease may serve as a rough indicator (its diagnosis and treatment may differ between countries). The values are 190 per 100,000 in Germany but 150 in Sweden (Von Camphausen et al. 2005); yet, the share of SI benefits for incapacity in Sweden is twice that of Germany.
- *Benefits paid to households*: This is a category importantly related to the peril of "health loss". However, it also comprises housing subsidies, which hardly amount to the protection of a risky asset.

Table 10.3 Structure of social expenditure of selected OECD countries, as percent of GDP (2013)

Country	Old age	Incapacity	Households[a] (health, housing)	Family[a]	Labor market programs	Unemployment
France	23.8	1.7	19.5	2.8	2.8	1.6
Germany	10.1	2.1	15.6	2.1	1.6	1
Italy	16.3	1.7	19.3	1.9	1.9	1.7
Japan	10.2	1	13.2	0.4	0.4	0.2
Mexico	n.a.	n.a.	2.2	0.4	0	n.a.
Sweden	7.7[b]	4.3	14	3.5	1.9	0.5
United Kingdom	6.1	2	14.5	3.9	0.5[b]	0.3
United States	6.9	1.4	14.4	0.6	0.5	0.4

[a]2012; [b]2011
Source OECD.Stat, Indicators on social expenditures 2018

- *Family*: Additions to the family occur due to pure chance only rarely nowadays. Child benefits should, therefore, be qualified as transfers or subsidies rather than an insurance payment.
- *Labor market programs*: Here, the peril is job loss, and SI benefits are designed to replenish workers' human capital by occupational retraining.
- *Unemployment*: This is a peril resulting in a loss of labor income and potentially human capital and health. Private insurers shy away from covering unemployment because, in an economic downturn, it may simultaneously affect millions of people, resulting in an accumulation of risks.

The multitude of relationships between perils and branches of SI suggests that the possibilities of efficiency enhancement discussed in Sect. 10.2.1 may apply to the existence of SI but not necessarily its structure, which is likely to reflect the peculiarities of a country's political process.

This view is supported by col. (1) of Table 10.3 in that Sweden, the United Kingdom, and the United States stand out for their low GDP share of SI devoted to old age SI programs. Swedish workers can top up their occupational pension by contributing to one of 500 competing pension funds (which do not count as SI). In the United Kingdom, participants have the option of signing up with a SIPP (Self-Invested Private Pension) scheme. In the United States, participants currently can contribute roughly USD 20,000 annually to a pension plan of their choice (where available), benefiting from an exemption from the income tax under subsection 401(k) of the Internal Revenue Code. In all three cases, these schemes permitting choice were introduced (or substantially extended) by conservative governments, confirming the role of ideology evidenced in Table 10.2 of Sect. 10.2.2.

▶ **Conclusion 10.6** In industrial countries, the two most important missions of social insurance are provision for old age and health insurance. The other branches of social insurance differ widely in their importance, likely due to differences in national political processes.

10.4 Requirements for Efficient Social Insurance

Attempts at assessing the efficiency of social insurance (SI) have a long tradition in economics. Usually, the focus has been on provision for old age because it claims the greatest share of contributions in industrial countries. Moreover, it can be looked upon as a capital investment at the microeconomic level in that current contributions lead to future benefits. For this reason, the first objective of this section is to show how the efficiency of social provision for old age can be compared to the private alternative (PI). Since this amounts to a comparison of rates of return, one can generalize the analysis in a second step. Contributions to PI and SI can be viewed as claims against the respective insurance schemes with their expected returns and volatilities. Individuals, therefore, have a portfolio of claims, and they may be interested in the total volatility of their insured assets. The issue of efficiency then concerns the entire system of social protection that is characterized by a certain division of labor between PI and SI. A system that puts individuals on the efficient frontier defined over expected returns and volatility can be deemed efficient.

10.4.1 Comparing the Efficiency of Provision for Old Age

The contributions paid to SI for old age provision could also be used for provision through PI. For a risk-neutral individual who is only interested in consumption during old age, a comparison of performance between SI and PI reduces to a comparison in terms of rates of return. The preferred alternative is the one that achieves a higher rate of return on contributions paid. It also is Pareto-superior since in principle all individuals reach a higher level of welfare.

One might think that provision for old age through SI has a rate of return equal to zero because it is typically of the pay-as-you-go type. This means that contributions paid are used to finance the benefits for the retired during the same period. By way of contrast, in a funded scheme, contributions are invested in the capital market, generating benefits augmented by accrued interest. In a pioneering contribution, Samuelson (1958) demonstrated that a pay-as-you-go scheme can have a positive rate of return.

The exposition below follows Breyer (1990, Chap. 2). It considers a representative individual that derives utility from consumption during the phase of activity (c_t) as well as consumption possibilities during the retirement phase (z_{t+1}),

$$U_t = U(c_t, z_{t+1}). \qquad (10.1)$$

For simplicity, both phases are normalized to a period of length one, and all parameters (wage rate, rate of interest, wealth of active population, and retirement age) are fixed and in real terms. Consumption during the active phase is given by the difference between net labor income $w_t \cdot (1 - \tau_t)$ and savings s_t,

$$c_t = w_t \cdot (1 - \tau_t) - s_t. \tag{10.2}$$

In this equation, τ_t denotes the contribution rate defined as share of wage income given by $w_t \cdot 1 = w_t$. It is a payroll tax that is charged to the worker regardless of the choice between pay-as-you-go and capital-based schemes. Other taxes as well as property income are neglected for simplicity.

Possibilities for consumption during the retirement phase z_{t+1} consist of two components. On the one hand, individuals hold claims against the old age scheme amounting to x_{t+1}; on the other hand, they have savings that are augmented by the rate of interest r_{t+1} (i.e., the rate of interest that will prevail at the end of the first period),

$$z_{t+1} = x_{t+1} + (1 + r_{t+1}) \cdot s_t. \tag{10.3}$$

It is the benefit x_{t+1} that depends on the type of old age scheme. The *funded alternative* (K) is characteristic of PI; however, it may also constitute an employment-related component of SI. In either case, contributions $\tau_t \cdot w_t$ are invested in the capital market. For simplicity, it is assumed that these investments achieve the same rate of interest as private savings [see Eq. (10.3)],

$$x_{t+1}^K = (1 + r_{t+1}) \cdot \tau_t \cdot w_t. \tag{10.4}$$

Turning now to the *pay-as-you-go alternative*, the relationship between contributions and benefits is entirely different. Additional benefits can be financed if the number of contributors or the wage rate has increased in the meantime. The number of contributors is equated to the number of workers since the two quantities develop in parallel at least in the long run. Accordingly, m_t symbolizes the rate of growth of the working population. For example, in initial period t, let there be 100 workers paying contributions covering the livelihood of 100 retired people. The same per-capita contribution paid by 105 active workers in period $t + 1$ ($m_t = 0.05$) allows for an increase of 5% in benefits to the 100 individuals who have retired in the meantime. An increase in the wage rate by 5% (symbolized by g_t) has the same effect. For this reason, benefits of the representative individual in the pay-as-you-go (P) alternative can be written,

$$x_{t+1}^P = \tau_t \cdot w_t \cdot (1 + m_t)(1 + g_t). \tag{10.5}$$

The rate of return of either alternative is given by the excess of benefits over contributions paid,

$$1 + i_{t+1} = x_{t+1}/(\tau_t \cdot w_t), \text{ and therefore, } i_{t+1} = x_{t+1}/(\tau_t \cdot w_t) - 1. \tag{10.6}$$

10.4 Requirements for Efficient Social Insurance

For the comparison of rates of return, let the rate of interest, the growth rates of workers, and of the wage rate be constant for simplicity and denoted by r, m, and g respectively. Division of Eq. (10.4) by $(\tau_t \cdot w_t)$ and solving yield for alternative K

$$i_{t+1}^K = r. \tag{10.7}$$

This states that the rate of return of the funded scheme equals the rate of return on the capital invested. In the case of the pay-as-you-go alternative P, division of Eq. (10.5) by $(\tau_t \cdot w_t)$ and substitution into Eq. (10.6) result in

$$i_{t+1}^P = (1+g)(1+m) - 1. \tag{10.8}$$

Since both m and g usually have low values (e.g., 0.03), their product can be neglected, resulting in the simplification,

$$i_{t+1}^P = 1 + g + m + mg - 1 \approx g + m. \tag{10.9}$$

Therefore, although the pay-as-you-go alternative does not create a capital stock, contributions to SI still achieve a rate of return in the guise of a "biological rate of interest". This insight is due to Aaron (1966), who calls it the "paradox of social security".

▶ **Conclusion 10.7** Contributions to the pay-as-you-go alternative of old age provision have a "biological rate of interest" that amounts to the sum of the growth rate of workers and of the wage rate (paradox of social security).

If individuals are indeed interested only in consumption during their period of activity and consumption possibilities during their period of retirement [see Eq. (10.1) again], then for a given contribution, it is the value of x_{t+1} which determines the preferred alternative. This value depends on the rate of return achieved.

▶ **Conclusion 10.8** If $m + g > r$ (the rates of growth of the working population and of wages together exceed the rate of interest), the pay-as-you-go alternative dominates the funded alternative characteristic of PI. If $m + g < r$, the pay-as-you-go alternative may be dominated by the funded alternative; however, the loss of utility of the initial generation that must build up the capital stock in a transition from the pay-as-you-go to the funded alternative would have to be accounted for in the comparison.

This simple model needs to be complemented in several aspects.

1. *The longevity of individuals is uncertain.* Utility as defined by Eq. (10.1) needs to be replaced by the expected utility. However, this does not modify Conclusion 10.8.
2. *Individuals may have a bequest motive.* This does not modify Conclusion 10.8 either because bequests must be financed through savings; therefore, individuals will continue to compare alternatives for old age provision on the basis of rates of

return as before. A higher rate of return enables individuals to bequeath a greater amount of wealth.

3. *Retirement age depends on provision for old age.* The prospect of a higher pension causes the retirement age to fall, thereby reducing aggregate income (see Sect. 10.5.1.1). While this leaves the rate of return on the capital market unaffected at least to a first approximation, it does cause wage income available for financing pensions to grow more slowly. Provided there was an efficiency advantage of the pay-as-you-go alternative ($m + g > r$), it is reduced.

4. *The wage rate depends on provision for old age.* This is to be expected because a higher pension reduces the supply of labor through its effect on retirement age, causing the wage rate to be higher than otherwise. If the short-run elasticity of demand for labor w.r.t. the wage rate is less than one in absolute value, the wage bill rises faster than otherwise ($m + g$ is higher). In the more relevant long term, this elasticity increases due to enhanced substitution possibilities. This makes a slower growth of the wage bill likely, again undermining a given efficiency advantage of the pay-as-you-go alternative (as long as the rate of return on the capital market is unaffected by these changes).

Therefore, Conclusion 10.8 turns out to be robust with regard to modifications that are of a more actuarial nature, such as uncertain longevity. However, as soon as the relationship between retirement income and labor supply is taken into account, a possible efficiency advantage of the pay-as-you-go alternative is lessened and may change to a disadvantage arguing against provision for old age through SI. A final consideration is the fact that both the rate of return on the capital market r and its "biological" counterpart in the pay-as-you-go alternative (in particular, the rate of growth of the wage rate g) are volatile. If these two parameters are negatively correlated, a combination of the two alternatives can achieve a diversification effect, serving to enhance the efficiency of the system as a whole. This argument is developed in Sect. 10.4.2 below.

10.4.2 Efficiency Assessment from a Portfolio Theory Perspective

10.4.2.1 Insurance Claims as Components of a Portfolio

In Sect. 7.2, the performance of an IC is evaluated from the point of view of an investor who has to decide whether to hold shares of that IC rather than some other securities. Here, the point of view of a risk-averse IB is adopted, giving rise to the question of whether an insurance policy is part of his or her efficient portfolio defined in terms of expected return and volatility. Premiums paid can be regarded as deterministic; however, insurance payments by necessity are stochastic, being triggered by chance events. This makes the rate of return of an insurance policy a random variable.

Conclusion 10.7 of Sect. 10.4.1 also shows that not only claims against PI but also against SI have a rate of return. What is more, not only claims against PI but also SI have a certain volatility. There are at least four reasons for this.

10.4 Requirements for Efficient Social Insurance

1. Payments of SI are triggered by a chance event precisely as in PI, for instance by a premature death in the case of old age provision.
2. The benefits of SI are defined in nominal terms in principle. There may be adjustments to compensate for future inflation, which, however, are subject to the political process with its own risks. This makes the real value of benefits uncertain.
3. The benefits of SI can quite generally be adjusted upwards or downwards by political decisions which are difficult to foresee.
4. As in PI, payments of SI may not always match the expectations of the insured. The benefits of SI often depend on certain conditions such as marital status, number of dependent children, and continuity of employment and residence in the home country—parameters that can change in unexpected ways.

In sum, all lines of PI and branches of SI are characterized by expected returns and volatilities. Indeed, holding a claim against them amounts to a risky capital investment. For instance, in the case of accident insurance, a stream of SI contributions gives rise to a stream of future payments, which, however, occur only with a certain probability. In health insurance, future illnesses induce healthcare services that have a monetary value even if they are purchased by the health insurer on behalf of the patient.

Figure 10.4 illustrates the impact of SI on the efficient frontier in (μ, σ)-space (in analogy to Sect. 5.1.2). For simplicity, other assets are neglected and no risk-free alternative is admitted. Fully informed individuals free to combine insurance coverage offered by both PI and SI would reach the efficient frontier labeled EE' by assumption. Depending on their degree of risk aversion, they would choose a point such as C^{**}, indicating a certain set of policies according to lines of insurance written by PI only.

Now let SI levy contributions whose amount cannot be influenced by the individual. This causes the frontier EE' to be replaced by two specific frontiers, namely $E_p E'_p$ for PI and $E_s E'_s$ for SI, respectively. Compared to EE', these frontiers are modified in the following ways.

- Since the structure of the claims portfolio is predetermined by SI, several efficient allocations cannot be reached anymore. For instance, for total funds amounting to 100 MU (monetary units), assume the combination "80% old age, 20% health, both through PI" to be efficient. However, with 70% of funds claimed by SI, a maximum of 30 MU can be allocated to a private pension, likely causing inefficiency. More generally, since SI is compulsory, PI can only offer complementary coverage. This limits the diversification effect of contracts written by the several lines of PI. In sum, the efficient frontier $E_p E'_p$ pertaining to the subset of PI contracts runs not only lower but also more concave than EE'.
- As to SI, it is characterized by its own efficient frontier $E_s E'_s$. For some subgroups that are subsidized by SI (high risks, well-organized voter groups; see Sect. 10.2), this frontier may run higher than EE' as offered by PI alone since the efficient frontier pertaining to SI is subject to a number of limitations. There are three

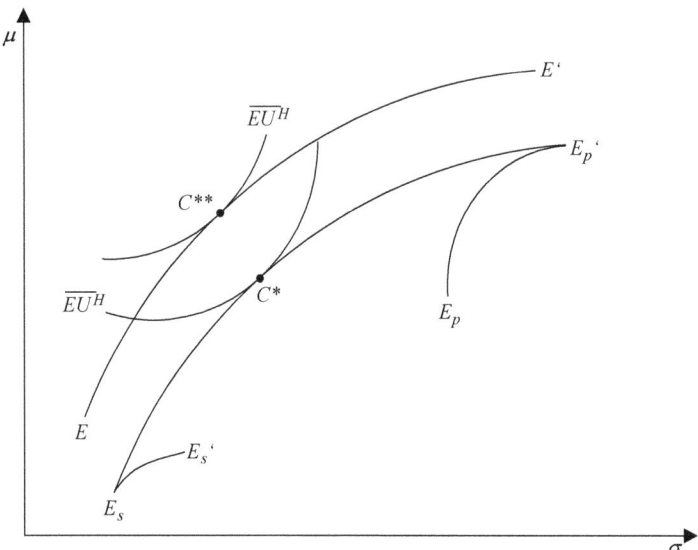

Fig. 10.4 PI and SI as components of an individual portfolio of claims

arguments as to why, for the population as a whole, $E_s E'_s$ runs lower and closer to the origin compared to EE', as shown in Fig. 10.4.

1. *Partial pooling of risks only.* The branches of SI were not created as part of a comprehensive system but constitute the outcome of the political process most of the time. This means that their consolidation into one organization in analogy to the lines of an IC is hardly possible even if appropriate in the light of portfolio theory. This limits the possibilities for hedging across branches of SI (so-called internal risk diversification, see Sect. 5.1.1). As a consequence, the efficient frontier $E_s E'_s$ pertaining to SI runs lower and more concave than EE'. At the same time, SI can offer a minimum variance portfolio with a lower volatility than PI because of its unique "biological rate of return" $g + m$, which typically varies less than rates of return r on the capital market. Finally, the frontier $E_s E'_s$ may even degenerate to a single point in case the amount of coverage is mandated for each risk covered by SI, precluding any choice of portfolio.
2. *Limited scope of investment opportunities.* In a pay-as-you-go scheme, SI disposes of a small amount of reserves for capital investment, which typically is of short-term nature. In addition, the class of admissible investments is more narrowly circumscribed than in PI, consisting mainly of government bonds. These restrictions cause the efficient frontier $E_s E'_s$ to run lower and closer to the origin than EE'.
3. *Absence of currency diversification.* Contributions as well as SI benefits are always in the national currency. This not only limits the efficiency of SI as an investor but also that of consumers. For example, a U.S. citizen could be interested in having paid her retirement income in Canadian Dollars because she

plans to live there after retirement. Once more, certain possibilities of diversification are excluded, causing $E_s E'_s$ to run lower and to be more concave than EE' (see Fig. 5.2 in Sect. 5.1.2).

The combination of two portfolios that are composed separately results in the combined efficient frontier $E_s E'_p$ (in analogy to Fig. 5.3 of Sect. 5.1.2). Let the constrained optimum be represented by point C^*, which is associated with a lower level of expected utility than C^{**} attainable with complete freedom of choice.[4]

▶ **Conclusion 10.9** From the point of a representative individual, both components of a portfolio composed of claims against PI and SI have expected return and volatility. They combine to form an efficient frontier in (μ, σ)-space.

10.4.2.2 A Simple Efficiency Test

Conclusion 10.9 gives rise to the question of whether payments by PI and SI permit individuals to reach the combined efficient frontier $E_s E'_p$ of Fig. 10.4. This section is devoted to a test under simplified conditions. In particular, differences in the rate of return between PI and SI are neglected. Indeed, Eisen and Zweifel (1997) as well as Schoder et al. (2013) find that in the long run there are no discrepancies between rates of return in the capital market (pertinent to PI) and "the biological rate of return" (pertinent to SI). However, if expected returns are the same, then the mission of a system of social protection combining PI and SI boils down to minimizing the volatility of the portfolio of claims. In terms of Fig. 10.4, minimization of σ for a given value of μ defines the efficient frontier $E_s E'_p$.

While this efficient frontier is not known, a portfolio of claims with a lower volatility σ must lie closer to it. However, as shown in Sect. 5.1.2, the volatility of a portfolio crucially depends on the covariances of its components. Treating premiums and contributions paid as nonstochastic, covariance (correlation) between the claims against SI and PI can only originate from deviations of actual benefits from their expected value. This argument motivates the following simple efficiency test:

Efficiency test *If deviations from the expected value of benefits are negatively correlated, then PI and/or SI contribute to the efficiency of the system. If, however, these deviations are positively correlated, then PI and/or SI contribute to volatility in the portfolio of claims, detracting from efficiency.*

This test was performed by Eisen and Zweifel (1997) and Schoder et al. (2013) using aggregate data from Germany, the United States, and Switzerland. In both studies, the unexpected components of insurance payments are constructed as annual deviations from their trend value for each line of PI and branch of SI. These deviations are used to calculate pairwise correlation coefficients. There are three sets of correlation coefficients, between the lines of PI, between the branches of SI, and between the lines of SI and PI. The third set of coefficients is displayed in Table 10.4.

---

[4]Contrary to Fig. 10.4, a very risk-averse individual may opt for the minimum variance portfolio of SI, which may dominate the minimum variance portfolio of PI.

Table 10.4 Correlations of trend deviations in U.S. private and social insurance, 1980–2004

	PLID	PLIDI	PLAI	PHI	PHI2
SDCB	0.1436	0.2790	−0.1358	0.2016	0.5015*
	(0.5239)	(0.2086)	(0.5467)	(0.3682)	(0.0174)
SWCB	−0.0211	0.3357	−0.1055	0.1521	0.5314*
	(0.9256)	(0.1266)	(0.6403)	(0.4993)	(0.0109)
SOACB	−0.0268	0.0604	−0.2457	−0.2593	0.2701
	(0.9058)	(0.7895)	(0.2704)	(0.2440)	(0.2240)
SPSB	−0.3367	−0.1628	−0.3591	−0.4135	0.4683*
	(0.1255)	(0.4690)	(0.1007)	(0.0558)	(0.0280)
SSB	0.2115	0.3337	0.0304	0.3112	0.2641
	(0.3447)	(0.1290)	(0.8931)	(0.1586)	(0.2349)
SFCB	0.1113	0.3175	0.0619	0.3054	0.2840
	(0.6218)	(0.1500)	(0.7845)	(0.1670)	(0.2003)
SUB	−0.3437	0.0431	−0.3343	−0.3070	0.5746**
	(0.1173)	(0.8488)	(0.1284)	(0.1647)	(0.0052)
SHB	0.2407	0.3455	−0.1405	0.3005	0.3582
	(0.2806)	(0.1152)	(0.5330)	(0.1743)	(0.1017)

Social insurance benefits: disability (SDCB), worker's compensation (SWCB), old age (SOACB), paid sick leave (SPSB), survivor's (SSB), family cash (SFCB), unemployment (SUB), and health (SHB)
Private insurance benefits: life in case of death (PLID), life in case of disability (PLIDI), life annuity (PLAI), health (PHI), and health according to U.S. Dept. of Health and Human Services (PHI2)
*, ** coefficient significant at the five and one percent level, respectively.
Source Schoder et al. (2013)

These correlations are of particular interest because on the one hand, SI claims to provide a safety net regardless of what happens in the economy (and hence PI), while PI often claims to fill gaps left by SI. Both arguments lead one to expect negative correlations.

First, it should be noted that results differ importantly according to the source of data in the United States. As to private health insurance (PHI) as recorded in the Life Insurers Fact Book, there is not a single significant positive correlation with deviations from trend values in SI benefits and a few albeit insignificantly negative ones. The impression prevails that private health insurance at least does not magnify the volatility originating from payments of social insurance. However, when PHI benefits are measured according to the U.S. Department of Health and Human Services (PHI2 in Table 10.4), four out of eight correlations are significantly positive. Therefore, one-half of the branches of SI tend to magnify volatility imparted by private health insurance according to that source. As to the other lines of PI, no SI branch displays a significantly positive correlation and vice versa. On the other hand, there are no negative correlations of statistical significance either. In sum, the interplay of U.S. private and social insurance does not systematically cause an increase in the risk exposure of citizens, contrary to Germany and Switzerland for instance [see Zweifel 2013].

10.4 Requirements for Efficient Social Insurance

▶ **Conclusion 10.10** In the United States, most of the lines of PI and branches of SI combine in a way as to avoid magnification of existing volatility of individuals' insurance claims.

An important criticism of this conclusion is that aggregate data may not reflect the situation at the individual level [see Schlesinger 1997]. However, as shown in Schoder et al. (2013, Appendix A), the data at hand do not lend support to this criticism. Still, observations relating to individuals, the losses occurred by them, and benefits received from SI and PI under several titles would be of great interest for future research.

10.5 Impacts of Social Insurance Beyond Insurance Markets

Social insurance has impacts both on the business cycle and the longer-run development of the economy. With regard to the business cycle, it serves as a so-called automatic stabilizer because, in a downswing, individuals who lose their jobs or are released into early retirement retain at least part of their income [for more detail, see, e.g., Blanchard and Perotti (2002)]. In keeping with the microeconomic orientation of this book, the structural consequences of social insurance are emphasized. In the main, they emanate from moral hazard effects. The other major problem of asymmetric information, adverse selection, is of no relevance to SI, which is typically organized as a mandatory monopolistic scheme. To recall, moral hazard refers to an unobservable change in individual behavior induced by insurance coverage. Specifically, individuals are predicted to reduce their preventive effort (see Sect. 8.2). This holds regardless of whether insurance is contracted with an IC or provided by SI. For this reason, moral hazard effects of a similar type are to be expected from both private and social insurance.

Nevertheless, the debate in the literature revolves primarily around changes in behavior induced by SI rather than PI. There are four main reasons for such a focus.

1. *Amount of funds involved.* As documented in Sect. 10.1, the contributions to SI exceed the premium volume of PI by far in the relevant domain of personal insurance. The likelihood of SI having important macroeconomic effects is, therefore, greater.
2. *Number of individuals affected.* Being mandatory, SI affects all citizens of a country. In comparison, even a large IC enrolls only a small part of the population. Therefore, any incentive effects of SI have a larger scope than of PI.
3. *Uniformity of incentives.* The uniformity of benefits in SI creates uniform incentives for all insured. As long as preferences are not very heterogeneous, this fact also contributes to the likelihood that modifications of individual behavior have measurable consequences at the macroeconomic level. By way of contrast, competing ICs need to tailor contracts (and hence incentives) to the preferences of their clientele.

4. *External effects.* Moral hazard effects cause the frequency and amount of losses to increase, and with them, administrative expense. Due to the lack of observability, the cost of insurance coverage increases also for consumers who are less prone to moral hazard than others. Moral hazard, therefore, constitutes a negative external effect within a risk pool. Accordingly, ICs seek to mitigate or internalize this external effect, using a progressive increase in premium for additional coverage, copayment in the case of loss, and experience rating. By way of contrast, SI does not dispose of most of these instruments, resulting in more marked moral hazard effects than in PI (see also Sect. 10.2.1.5).

When discussing the macroeconomic impacts of SI and attempts to estimate their magnitude, one has to bear in mind that the crucial quantity is the external cost that consumers under the influence of moral hazard cause to the rest of the population. These external costs may not be that high. For instance, moral hazard may induce some people to opt for early retirement. Only to the extent that the associated reduction in income falls short of its actuarial value, there is an external effect.

An analysis of all branches of SI would go beyond the limits of this textbook. Rather, it focuses on those branches that have especially important or multi-faceted impacts, namely provision for old age, health insurance, and unemployment insurance. They share the property of being financed by a payroll tax in most industrial countries, thus driving a wedge between the gross wage and the net wage earned by workers (augmented by the present value of future claims against SI). The substitution effect of a raised rate of contribution to SI on labor supply is negative *ceteris paribus*.

However, a reduction of the net wage also has an income effect resulting in a reduced demand for the good "leisure", which is equivalent to an increase in labor supply. The total effect, therefore, is theoretically indeterminate, and empirical investigations come to conflicting conclusions [see Blundell 1992 for a survey]. For this reason, the effects caused by the financing of SI are not investigated further. Emphasis will be on the benefits of SI and the incentives they create. More generally, the benefits and contributions of SI are predicted to affect decisions over a person's entire life cycle, from investment in education to efforts to increase longevity. Therefore, SI may affect the entire demographic structure of a country [Zweifel and Eugster 2008].

10.5.1 Impacts of Provision for Old Age

Provision for old age through SI is expected to affect the supply of labor. On the one hand, contributions reduce the net wage rate, being similar to a payroll tax; on the other hand, benefits may influence the timing of retirement and hence lifetime labor supply. In addition, contributions amount to forced savings, with the likely consequence of crowding out private saving at least in part. Finally, emphasizing once more moral hazard effects, it may be argued that payment of a SI pension

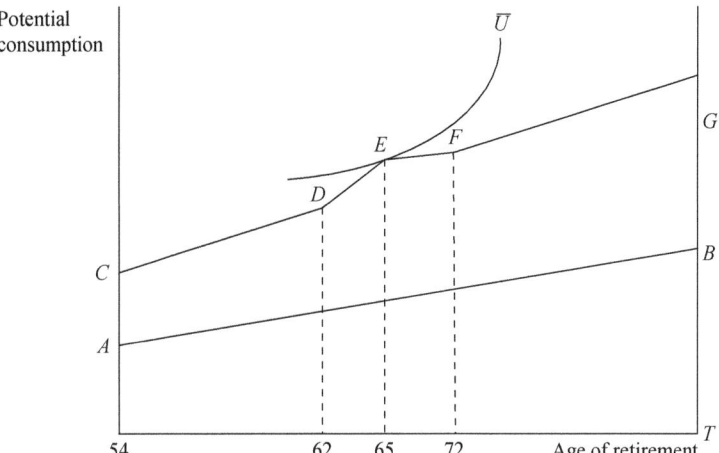

Fig. 10.5 Incentives governing the choice of time of retirement created by U.S. Social Security

T: Length of life, assumed to be certain
AB: Budget constraint without Social Security
CDEFG: Budget constraint with Social Security

(rather than a capital at retirement time, as is often the case with PI) serves to reduce the cost of longevity, thus contributing to the aging of the population.

10.5.1.1 Effects of Old Age Provision on the Labor Market

From a theoretical perspective, there is good reason to expect provision for old age through SI (often called Social Security in the United States) to modify the time of retirement through its benefits, potentially reducing lifetime labor supply. A study examining the financial incentives created by U.S. Social Security in a particularly clear way is by Burtless (1986).

In Fig. 10.5, potential consumption (which coincides with income if there are no savings) is related to age at retirement. Note that this is nothing but a modified two-goods diagram, with leisure on the horizontal axis replaced by its complement, amount of time worked up to retirement. In the absence of social security, possibilities for consumption would increase along the straight line AB, reflecting the annual wage rate. If parts of labor income are saved, the slope of AB would increase due to accrued interest; however, this is neglected for simplicity. In the presence of Social Security, consumption possibilities rise faster along CD than along AB because typically additional contribution years cause claims to benefits to increase rapidly. This effect is especially marked between ages 62 and 65 (segment DE) to the extent that SI penalizes early retirement (or honors continued work until 65, respectively). By way of contrast, income earned past official retirement age of 65 was partially offset by reduced benefits at the time, resulting in the reduced slope of segment EF (up to age 72 but not beyond).

With potential consumption a good and a longer lifetime work a bad, preferences of workers are depicted in Fig. 10.5 by indifference curves having a positive slope. An accumulation of optima at point E is predicted because indifference curves with somewhat different slopes still lead to the same optimum at E. This implies that under the influence of SI, workers in the United States choose to retire at age 65 with increased probability. In the absence of SI, the optimum would be anywhere along a straight line AB, reflecting individual preferences.

Indeed, in a sample of more than 4,000 men that were surveyed between 1969 and 1979, Burtless (1986) finds an accumulation of retirement at age 65. This result is not very surprising since 65 was the official retirement age. However, there were modifications in the time profile of claims to benefits (the line $CDEFG$ of Fig. 10.5) permitting the author to test whether retirement age responds to these changes. The estimated responses found are as predicted and statistically significant but turn out to be rather small. Specifically, they are not large enough to fully explain the observed fall in labor participation rates of old Americans since World War II.

This type of model is subject to restrictive assumptions, however. In particular, it views individuals as deciding their age of retirement once and for all in full knowledge of their future labor incomes, Social Security benefits, and especially their health status. In contrast, the option model tracks their decision process over time.

Indeed, by not taking retirement at time t, a worker retains the option of retiring at time $t+1, t+2, \ldots$. If there is a time of retirement $t+n$ that entails a higher expected utility than retirement at point t, then the worker is predicted to defer retirement.

This option model was tested by Stock and Wise (1990) using the personnel files of a large company. The data include not only the benefits of Social Security but also of employment-related schemes. In a sample comprising some 1,500 employees, the authors found properties of the SI component to have less influence on the decision to retire than those of the employment-related component. In Germany, the option model was used by Börsch-Supan (1992), based on information on 479 retirees in the Socioeconomic Panel of 1984 who had taken retirement between ages 60 and 70. There is a set of dependent variables taking on the value of one if retirement occurs in one of these eleven years and zero otherwise. They thus serve as indicators of the unobservable probability of going into retirement at the age considered. Since participants in the panel are observed only once, the dynamics of the option model are lost. Nevertheless, the option value of deferred retirement turns out to be a highly significant determinant of the transition into retirement. The higher this option value, the lower *ceteris paribus* is the probability of retirement.

▶ **Conclusion 10.11** Research, e.g., from the United States and Germany finds some microeconomic evidence suggesting that more generous provision for old age through SI induces an earlier age at retirement.

However, microeconomic evidence does not suffice to substantiate the claim that provision for old age through SI has an impact at the macroeconomic level. In addition, the reduction of labor supply around retirement age could be balanced by more

10.5 Impacts of Social Insurance Beyond Insurance Markets

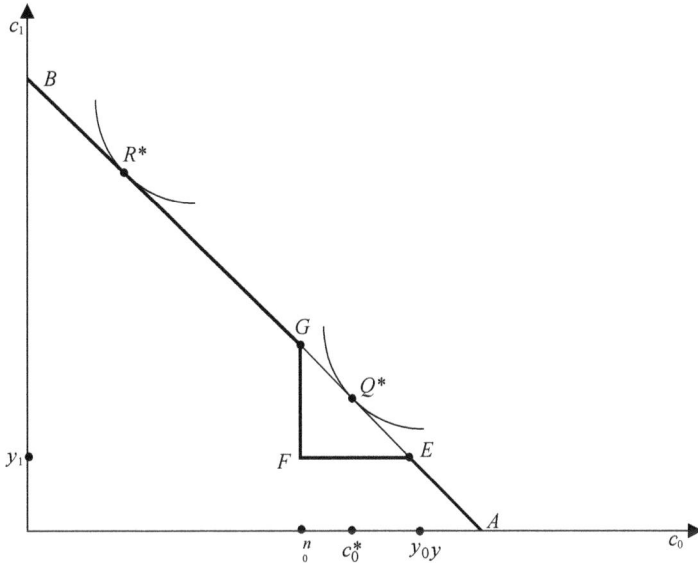

Fig. 10.6 Provision for old age through SI and private saving

work earlier in life. An early study by Burkhauser and Turner (1978) addresses this issue. Using aggregate data, the authors found a positive relationship between the weekly worktime of men aged 25 to 64 and the benefits of U.S. Social Security they were looking forward to. They interpret this result as reflecting workers' response to a reduction in the net wage caused by the creation of Social Security in 1938 and its extension after World War II, with the income effect favoring leisure dominating the substitution effect favoring work. This response may even explain why the negative trend in the length of the workweek came to a halt in the United States after World War II. Note that the accrual of credits through additional work earlier in life results in higher claims against SI. However, additional benefits often exceed their actuarial value, especially for the first generation of beneficiaries. For this reason, the expectation remains that any tendency in favor of early retirement induced by SI burdens society with a negative external effect.

10.5.1.2 Impacts on the Capital Market

Provision for old age through SI could crowd out private savings, thus impinging on the supply of capital. In principle, SI benefits constitute a substitute for income generated by accumulated savings.

To illustrate this substitution, Fig. 10.6 depicts a simple two-goods model, with $c_0 :=$ consumption during the active period and $c_1 :=$ consumption during the retirement period. Both periods have length one for simplicity. Without provision for old age by SI, the budget constraint is represented by the straight line AB with slope $(1 + r)$, with r symbolizing the real rate of interest. This reflects the fact that savings invested in the capital market during the active phase yield a certain rate of return.

The endowment point E compounds to a combination of incomes $\{y_0, y_1\}$, with y_1 symbolizing the minimum income during retirement that is guaranteed by public welfare. Due to provision for old age through SI, the budget constraint changes. The individual is mandated to contribute to SI by an amount that is symbolized by FE. In return, SI pays benefits amounting to FG, which reflects the assumption that it achieves the same rate of return as private investors (if it uses the capital market) or that this rate of return equals the "biological rate of return" of the pay-as-you-go alternative (see Sect. 10.4.1).

When considering the impact of SI, two cases must be distinguished.

1. *High planned savings:* Let the individual considered have a strong preference for income during retirement, resulting in R^* as the optimum. Evidently, there is no change in optimal consumption; however, effective private savings are reduced by the amount of FE going to SI.
2. *Low planned savings:* Let the individual opt for point Q^* in principle. The new budget constraint $AEFGB$ imposed by SI makes the immediate attainment of Q^* impossible. There are three possible responses:

   - The individual seeks to retain Q^* as the optimum. This can be achieved by paying the contribution FE and taking up a credit to bridge the gap between G and Q^* (at a rate of interest equal to the rate of return r on investments). Private savings during the active period decrease accordingly.
   - The individual settles for vertex point G by reducing consumption during the active period in order to finance the contribution to SI. Planned private savings fall from $(y_0 - c_0^*)$ to zero, and there is a utility loss.
   - The individual tries to settle for vertex point E, with consumption rising to equality with the income during the active period. However, contributions to SI take precedence over consumption. Thus, the individual must pay FE, making point G the final optimum. Again, private savings fall from $(y_0 - c_0^*)$ to zero, combined with a loss in utility.

This simple model, therefore, predicts a crowding out of private savings by contributions to SI in several circumstances.

In the pay-as-you-go alternative, total savings decrease in step because contributions are used to finance the benefits for the retired generation. In the funded alternative, contributions become savings of SI. In that case, total savings increase whenever effective consumption is below the planned value. This obtains if individuals with low planned savings are forced to settle for point G in Fig. 10.6. Studies at the microeconomic level indeed find that claims against SI substitute for other forms of wealth. In particular, Kotlikoff (1979) distinguishes between the actuarially fair component of these claims and a possible excess (caused by a high "biological rate of return" at the time). The author estimates that one US\$ of extra actuarial component goes along with a reduction of other forms of wealth of 0.6 US\$. The excess component, by way of contrast, does not seem to reduce private assets but rather to increase them by US$ 0.24. However, this effect is not statistically significant.

Again, microeconomic evidence does not suffice to demonstrate an effect at the macroeconomic level. The famous study by Feldstein (1974) tried to fill this gap. The author constructed a variable "claims against U.S. Social Security" and inserted it as an extra explanatory variable in the consumption function. This can be justified by noting that these claims are part of wealth, which amounts to a stream of future incomes. The estimated propensity of consumption with respect to this "Social Security wealth" turned out to be about double the value with regard to private wealth (0.025 rather than 0.012). By implication, private savings amount to only about one-half of what they might be without provision for old age through SI. However, Leimer and Lesnoy (1982) found these estimates to be mainly caused by an error in the calculation of the "Social Security wealth" variable. In addition, the authors point out that the aggregate data contain young individuals who were unlikely to decide their consumption in view of conditions of SI that were to prevail up to 40 years later. In this situation, the choice of hypothesis concerning the formation of expectations becomes decisive, and indeed results differ strongly depending on this choice.

The notion of whether SI constitutes a substitute of private saving is referred to as "crowding out". A famous study by Gruber and Yelowitz (1999) examines the expansion of U.S. Medicaid (which provides health insurance to the poor) by 500% between 1984 and 1993. Since this expansion occurred at different times in the member states, the authors can test the effect on the net wealth (and hence savings) of low-income households; they find a 8.2% reduction, suggesting substantial crowding out.

Since evidence from outside the United States is rare, the introduction of SI in Germany (the so-called Bismarck reforms of the early 1880s) constitutes an interesting test case. By applying the difference-in-difference method, i.e., correcting for the positive general trend in the development of deposits with savings banks, Streb (2018) identifies a clear crowding out effect. The sum of forced SI and voluntary private savings did increase, at the price of a utility loss (see case no. 2 in the discussion of Fig. 10.6).

▶ **Conclusion 10.12** Provision for old age through SI likely reduces private savings, with the funded alternative possibly increasing total aggregate savings. These impacts cannot be easily quantified at the macroeconomic level, however.

10.5.1.3 Other Impacts of Provision for Old Age Through SI: Number of Children

In addition to impacts on the markets for labor and capital, provision for old age through SI may influence the number of children. In the context of developing countries in particular, the argument that children serve as a substitute for the non-existing provision for old age through SI is widely accepted. Conversely, one may expect that an expansion of this branch of SI causes the number of children to fall. This argument has been formalized by Felderer (1992).

Let a household be interested in consumption during the active period c_t, consumption during retirement c_{t+1}, and the number of offspring e_t. Therefore, the utility function to be maximized reads

$$\max U = U(c_t, c_{t+1}, e_t). \tag{10.10}$$

However, each child gives rise to cost amounting to q_t; moreover, a contribution τ_t is deducted from labor income for SI. Labor supply is exogenous and normalized to one; therefore, net labor income amounts to $w_t(1 - \tau_t)$, with w_t symbolizing the wage rate. Finally, the household is also expected to support parents to the tune of B_t. Its budget constraint, therefore, is given by

$$c_t + q_t \cdot e_t = w_t(1 - \tau_t) - B_t. \tag{10.11}$$

During retirement, each child contributes to the support of parents, at the tune of B_{t+1}. This is augmented by the benefits of SI, z_{t+1}. The budget constraint applying to the retirement period thus becomes

$$c_{t+1} = B_{t+1} \cdot e_t + z_{t+1}. \tag{10.12}$$

A full comparative static analysis is not necessary if one is willing to assume that an increase of the contribution rate τ_t (without an adjustment of benefits) leads to a reduction of consumption in both periods. Therefore, one has $\partial c_{t+1}/\partial \tau_t < 0$. This can be used in the differentiated form of the budget constraint (10.12), yielding

$$\frac{\partial c_{t+1}}{\partial \tau_t} = B_{t+1} \cdot \frac{\partial e_t}{\partial \tau_t} < 0. \tag{10.13}$$

From this, one obtains immediately

$$\frac{\partial e_t}{\partial \tau_t} < 0. \tag{10.14}$$

This result can be interpreted as follows. Any increase in contributions to SI forces households to save. One possible outcome, then, is to reduce the number of children e_t [see Eq. (10.14)], another, to decrease consumption in both periods. However, such a decrease is facilitated by fewer children according to Eq. (10.13).

Whether the result (10.14) still holds in industrial countries with a high level of provision for old age through SI is questionable, however. Quite likely, benefits of SI (denoted by z_{t+1}) have replaced the contributions from children B_{t+1} in their entirety. With $B_{t+1} = 0$, however, the assumption $\partial c_{t+1}/\partial \tau < 0$ does not determine the sign of $\partial e_t/\partial \tau_t$ anymore in (10.14).

10.5.1.4 Other Impacts of Provision for Old Age Through SI: Life Expectancy

Provision for old age through SI nowadays has the form of a pension benefit that is paid until the end of life. Moreover, contributions are not scaled according to the life expectancy of the insured. This encourages moral hazard since consumers have a weakened incentive to avoid a "loss" or to limit it. However, to the insurance scheme, "loss" in this context means the survival of the insured. Only by surviving can they claim the benefits of SI. The prediction, therefore, is that the existence and expansion of provision for old age through SI induces a higher life expectancy.

Philipson and Becker (1998) juxtapose these modalities to those characterizing PI. In the case of an annuity benefit, the IC checks consumers for signs pointing to high life expectancy. By charging consumers with a higher life expectancy a higher premium, it causes them to purchase less insurance coverage, which in turn serves to reduce moral hazard effects.[5] In the case of a capital benefit, the IC faces similar problems of moral hazard and possibilities to deal with them.

Things are different for consumers, who are now exposed to a trade-off (see Fig. 10.7). With a fixed amount of financial resources, they cannot sustain a high rate of consumption for very long. On the one hand, they may use the capital obtained (and their time) for consumption. On the other hand, they can invest the capital in life-prolonging measures (in particular, medical services). In this case, they would have to limit consumption from the beginning to ensure that they have something to live on up to the end of a long life. In Fig. 10.7, the trade-off associated with a capital benefit is shown as the transformation curve KK'. It emphasizes the fact that additional consumption per unit of time can only be had by opting for less effort to increase life expectancy and hence lower life expectancy.

There is a second transformation curve RR' that applies to both an annuity product in PI and the pension benefit that is typical of SI. It represents a different trade-off between quality of life (importantly determined by consumption per unit time) and length of life. The reason is that living on for another year is honored by receipt of the annuity or pension. Therefore, consumers who seek to prolong their life (by maximizing their probability of survival using medical services) need to sacrifice comparably little in terms of consumption and hence the quality of life. Hence, the transformation curve RR' slopes down less steeply than the one associated with the capital benefit KK'.

Let the point of reference be Q^* lying on both KK' and RR'. It, therefore, represents a potential optimum for two types of individuals (symbolized by two sets of indifference curves). One type (I) prefers K^* to Q^* and would, therefore, opt for the capital benefit. In the context of old age provision through SI, however, this

---

[5] In Europe, universal life insurance is prevalent. It offers a capital paid in the case of premature death but also in the case of surviving to a certain age (62, say). However, benefits for premature death have to be paid earlier on average, causing them to have a higher present value than the capital benefit. Therefore, depending on the rate of interest, the IC may benefit from a higher life expectancy for this type of contract since it is associated with lower loss payments in present value terms.

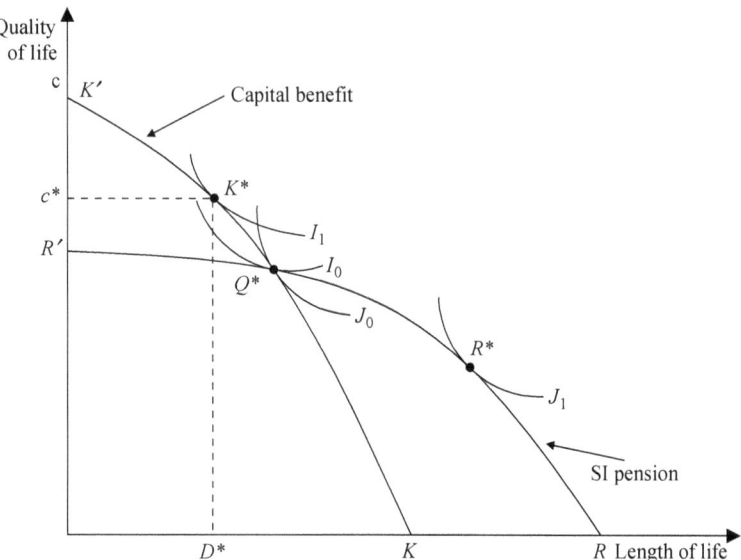

Fig. 10.7 Quality of life and length of life depending on capital versus pension benefit

alternative usually does not exist, constraining type I to remain at Q^*. The other type (J) is represented by indifference curves with steeper slopes, indicating a stronger preference for long life. With provision for old age through SI, this type can move to R^* on RR' where less consumption goes along with an increased length of life. In conclusion, this branch of SI is predicted to induce an increase in life expectancy.

Of course, testing this prediction by comparing countries with and without old age provision through SI is hardly possible. However, a higher SI pension within the same country should still make the optimum R^* more attractive, resulting in a moral hazard effect working in favor of longevity. Indeed, Philipson and Becker (1998) present data on employees of the U.S. Federal Administration showing that out of 100 males surveyed at age 57, who were looking forward to a low pension amounting to US$ 1,000 or less per month, only 42 lived to age 80. However, about 60% of those who were looking forward to a high pension amounting to US$ 3,000 or more lived to that age. Admittedly, this discrepancy can be due to differences in life expectancy at birth, genetics, education, professional activity, and lifestyle. Nevertheless, it constitutes preliminary empirical evidence suggesting that provision for old age through SI may have the predicted moral hazard effect of increasing life expectancy.

Indeed, over the life cycle individual decisions, ranging from education to marriage, divorce, retirement, and desired length of life as discussed above can be shown to be influenced by SI. Whereas much of the debate focuses on the impacts of demographic change on SI, the causality thus is seen to run the other way, from SI to demographic change. The pertinent hypotheses and some empirical evidence are presented in Zweifel and Eugster (2008).

10.5 Impacts of Social Insurance Beyond Insurance Markets

The analysis of this section also suggests that provision for old age through SI creates a tendency for medical services to be used for prolonging life. Such a tendency could be reinforced by social health insurance to be analyzed in the next section.

10.5.2 Impacts of Social Health Insurance

Health insurance is also expected to induce moral hazard effects. They may even be particularly strong because here the insureds are able to influence the amount of loss after its occurrence. As patients, they can opt for more intensive treatment or for the newer therapy, both associated with higher healthcare expenditure [see cases (b2) and (b3) of Sect. 8.2.1]. For simplicity, moral hazard on the part of providers of healthcare services is neglected, although their behavior is affected as well by health insurance. Moral hazard on the part of patients can be analyzed in the following way [adapted from Zweifel et al. 2009, Sect. 7.4.2].

Let there be a state of health and a state of sickness. However, for a patient, only the second state is relevant. Utility then depends on medical services denoted by M (valued as inputs to attain an improved health status) and disposable income y. For simplicity, medical services are assumed to have unit price, causing them to coincide with expenditure for treatment. The patient pays a contribution P and is reimbursed by an amount I that depends on M. One, therefore, has

$$U = U(M, y) = U(M, Y_0 - P - M + I(M)), \text{ with } y := Y_0 - P - M + I(M). \tag{10.15}$$

Gross income Y_0 and the contribution P are assumed to be predetermined; in particular, P does not respond to M, precluding an experience rating of contributions. Asymmetry of information is reflected by the fact that reimbursement I depends on expenditure M rather than health status, which is assumed unobservable. Typically, social health insurance does not reimburse the total amount of M but only a share α (with α denoting the rate of coverage and $1 - \alpha$, the rate of copayment, also called rate of coinsurance). Therefore, the relationship between reimbursement I and healthcare expenditure M becomes

$$I(M) = \alpha \cdot M, \text{ with } I'(M) = \alpha \quad (0 < \alpha < 1), \text{ and therefore,} \tag{10.16}$$
$$y = Y_0 - P - (1 - \alpha) \cdot M. \tag{10.17}$$

In view of (10.15) and (10.17), the first-order condition concerning the use of medical services reads

$$\frac{dU}{dM} = \frac{\partial U}{\partial M}\{M, Y - P - (1 - \alpha)M\} - (1 - \alpha) \cdot \frac{\partial U}{\partial y}\{M, Y - P - (1 - \alpha) \cdot M\} = 0. \tag{10.18}$$

The notation within braces is to remind the reader that the partial derivatives continue to be functions of M and especially α.

In the present context, *ex-post* moral hazard means that an increase in the rate of coverage ($d\alpha > 0$) causes the utilization of medical services to increase ($dM > 0$). Testing for moral hazard, therefore, amounts to establishing the sign of $dM/d\alpha$ in the neighborhood of the optimum defined by Eq. (10.18). This calls for a comparative static analysis (an application of the implicit function theorem). To this end, let the first-order condition (10.18) be disturbed by an impulse $d\alpha > 0$. Since condition (10.18) must again be satisfied after the shock, the right-hand side of the equation below is zero,

$$\frac{\partial^2 U}{\partial M^2} \cdot dM + \frac{\partial^2 U}{\partial M \partial \alpha} \cdot d\alpha = 0. \tag{10.19}$$

The left-hand side of (10.19) states that the marginal utility of medical care is affected by two changes. One is the impulse $d\alpha > 0$ itself, having an effect given by the mixed second-order derivative of utility. The other is the adjustment of M, the only decision variable considered here. Solving (10.19) for $dM/d\alpha$, one obtains

$$\frac{dM}{d\alpha} = -\frac{\partial^2 U / \partial M \partial \alpha}{\partial^2 U / \partial M^2}. \tag{10.20}$$

To simplify matters, the sufficient condition $\partial^2 U / \partial M^2 < 0$ for a maximum (stating that the marginal utility of additional medical services decreases) is assumed to be satisfied. Therefore, the sign of the numerator in (10.20) determines the sign of $dM/d\alpha$. Partial differentiation of Eq. (10.18) results in

$$\frac{\partial^2 U}{\partial M \partial \alpha} = \frac{\partial^2 U}{\partial M \partial y} \cdot M - (1-\alpha) \cdot \frac{\partial^2 U}{\partial y^2} \cdot M + \frac{\partial U}{\partial y}. \tag{10.21}$$

The last two terms are positive given decreasing marginal utility of income; the first one, however, only if $\partial^2 U / \partial M \partial y > 0$.[6] The marginal utility of medical services, therefore, would have to increase with rising income, or conversely, the marginal utility of additional income and consumption would have to increase with more medical care (and hence better health). This is a plausible assumption considering that most consumption possibilities yield full utility only when the individual is restored to good health. One, therefore, has

$$\frac{dM}{d\alpha} > 0. \tag{10.22}$$

An increase in the rate of coverage (or conversely, a decrease in the rate of coinsurance), therefore, is predicted to cause more demand for medical services. This constitutes an *ex-post* moral hazard effect.

---

[6] Note that risk aversion is not relevant here because there is no risk involved (the state of sickness obtains with certainty). Therefore, the argument in favor of $\partial^2 U / \partial M \partial y > 0$ is not in contradiction to the analysis of Sect. 3.2.2, which is in terms of risky wealth.

10.5 Impacts of Social Insurance Beyond Insurance Markets

Table 10.5 Utilization of medical services per capita and year in the Health Insurance Experiment

Type of contract	Likelihood of utilization (%)	Ambulatory care expenditure[a]	Number of physician contacts	Likelihood of hospitalization (%)	Hospital expenditure[a]	Total expenditure[a]
No copayment	86.8	446	4.55	10.3	536	982
25% copayment	78.7	341	3.33	8.4	489	831
50% copayment	77.2	294	3.03	7.2	590	884
95% copayment	67.7	266	2.73	7.9	413	679
Individual deductible	72.3	308	3.02	9.6	489	797

[a] In US$ of 1991
Note All differences between the contract types are statistically significant (at a significance level of 0.02 or better), with the exception of hospital expenditure.
Source Newhouse et al. (1993), p. 41 (Table 3.2)

The most convincing empirical evidence continues to come from the so-called Health Insurance Experiment of the RAND Corporation [see Newhouse et al. 1993]. During the second half of the 1970s, the authors assigned about 2,000 families to different types of health insurance. They purchased the right of contractual choice from the families in order to prevent the ones in good health from selecting contracts with low rates of coinsurance (an adverse selection effect, see Sect. 8.3). Otherwise, contracts with low rates of coverage could be associated with low medical expenses because of attracting favorable risks rather than limiting moral hazards. Contracts differed in terms of their rate of copayment $(1 - \alpha)$ with rates of 0%, 25%, 50%, and 95%, each with a cap of US$ 1,000 per year and family. Finally, a variant called "individual deductible" imposed 95% cost sharing for ambulatory care, no copayment for hospital care, and a limit of US$ 150 per family member.

The evidence is displayed in Table 10.5. With an increasing rate of copayment, the likelihood of utilizing medical services decreases consistently, as is true of ambulatory care expenditure and the number of physician contacts. With regard to the likelihood of one or several hospitalizations, the pattern is again unambiguous up to a copayment of 50% but not beyond. As to hospital expenditure, the contractual variants fail to exhibit a statistically significant difference. However, total expenditure does decrease with an increasing rate of copayment. The price elasticity of the demand for medical services with regard to the rate of copayment, therefore, is not zero but is estimated at -0.2 by Newhouse et al. (1993). In other words, an increasing rate of reimbursement α induces a higher demand for medical care, as predicted by Eq. (10.22).

Non-experimental evidence is subject to the weakness that selection effects can be controlled statistically at best. However, most systems of social health insurance disallow contractual variation in terms of rates of copayment. For this reason, the experience of private health insurers with their varied contracts may still be of interest. A study by Zweifel and Waser (1992) uses individual records provided by three German ICs. The loss distribution for ambulatory care expenditure was analyzed only for values where the insured had a financial incentive to submit billings (which is the case if the billing exceeds the deductible). For a sequence of increasing threshold values, different rates of copayment are related to the likelihood of expenditure to exceed the threshold. Significant reduction effects are found up to a threshold value of 1,000 Deutsche Mark (approximately 500 Euros, in 1985 prices).

However, there is evidence of still stronger reduction effects in contracts featuring premium rebates for no claims (so-called bonus options). This reinforcement effect of experience rating can be integrated in the model of Eq. (10.15) in the following way. For simplicity, the fact that the contribution P reacts to utilization M with a lag of one year or more is neglected. Therefore, predetermined P is replaced by the function $P(M)$, with $P' := dP/dM > 0$. The first-order condition then becomes in analogy to (10.18)

$$\left.\frac{dU}{dM}\right|_{P'>0} = \frac{\partial U}{\partial M}\{M, Y - P(M) - (1-\alpha)M\}$$

$$- (1 - \alpha + P') \cdot \frac{\partial U}{\partial y}\{M, Y - P(M) - (1-\alpha)M\} = 0. \quad (10.23)$$

Recall that a positive value of the mixed derivative in (10.20) indicates a moral hazard effect. Replacing now $(1 - \alpha)$ by $(1 - \alpha + P')$ from (10.23) to obtain the modified version of (10.21), one obtains

$$\left.\frac{\partial^2 U}{\partial M \partial \alpha}\right|_{P'>0} = \frac{\partial^2 U}{\partial M \partial y} \cdot M - (1 - \alpha + P') \underbrace{\frac{\partial^2 U}{\partial y^2}}_{<0} \cdot M + \frac{\partial U}{\partial y} > \left.\frac{\partial^2 U}{\partial M \partial \alpha}\right|_{P'=0}$$

thus,

$$\left.\frac{dM}{d\alpha}\right|_{P'>0} > \left.\frac{dM}{d\alpha}\right|_{P'=0}. \quad (10.24)$$

The added term $P' > 0$ serves to increase the positive second term. Therefore, experience rating ($P' > 0$) reinforces the moral hazard effect $dM/d\alpha > 0$. Conversely, an increase in the rate of copayment $(1 - \alpha)$ has a magnified dampening impact on the demand for medical care and healthcare expenditure if combined with experience rating. This reinforcement effect is indeed found in the study by Zweifel and Waser (1992).

At the macroeconomic level, one can conclude that eliminating copayments in the health branch of SI has the effect of increasing healthcare expenditure. This amounts to a negative externality to the extent that the insured who causes this increase burdens the rest of the insurance pool. Being uniform, the higher rates of contribution have to be paid by everyone rather than those characterized by particularly marked *ex-post* moral hazard.

▶ **Conclusion 10.13** The existence of *ex-post* moral hazard effects in social health insurance cannot be proven due to uniform contractual provisions including copayment. However, they are strongly suspected on the basis of the Health Insurance Experiment and experiences made by private health insurers. The consequences at the macroeconomic level can in part be interpreted as externalities.

10.5.3 Impacts of Unemployment Insurance

Unemployment can be viewed as the net outcome of an inflow into and an outflow from a pool of unemployed individuals, giving rise to a certain stock. Unemployment insurance (UI) influences these two flows in several ways.

- *Inflow into unemployment.* Unemployment insurance may make the transition from the state "employed" to the state "unemployed" more likely in two ways.

  1. On the one hand, there is a moral hazard on the part of the employed. Since they do not have to bear the full financial consequences of unemployment thanks to UI, they may reduce their (unobserved) efforts in terms of punctuality, diligence, and even servility in dealing with superiors.
  2. On the other hand, there is also a moral hazard on the part of employers. Especially large layoffs are not costless to employers, who must come up with a severance package. The benefits of UI provide a costless substitute. Therefore, the existence of UI is predicted to undermine efforts designed at preventing layoffs. This moral hazard effect could again be mitigated by experience rating (as in health insurance, see Sect. 10.5.2), i.e., making employer contributions to UI increase with past layoffs.

- *Outflow from unemployment.* Here, moral hazard is associated with the workers. The more generous UI, the longer can be the search for a new job, implying a lower likelihood of transition into employment during a given period. However, prolonged job search could also result in an improved match between the worker's skills and those required for the new job, resulting in increased productivity and wages. Therefore, a slower outflow due to UI need not cause an efficiency loss, provided the intensity and quality of search effort are unaffected by UI.
- *Stock of unemployed.* Since UI encourages the inflow into unemployment while slowing the outflow, it may well lead to an increased duration and hence stock of unemployment.

In their survey, Atkinson and Micklewright (1991) call attention to the fact that the hypothesized impacts of UI on inflows and outflows neglect important facts.

Most importantly, UI benefits are subject to conditions. As a consequence, more than one-third of the unemployed in the United States and the United Kingdom did not receive benefits as of 1988. Conditions also govern the transition to the public welfare that supports the long-term unemployed after the expiry of UI benefits. These conditions differ between countries, making international comparison difficult.

Also, limiting the analysis to the two flows neglects the influence of UI on employment, which should be the relationship of primary concern. Indeed, there are inflows and outflows to and from non-employment as well, in particular of housewives. Here, UI may encourage employment because entry may occur with a view to UI benefits later in life. In the case of Germany, e.g., these flows are of a comparable size as those into and from unemployment [Burda and Wyplosz 1994].

Empirical research has revolved around the relationship between UI and the outflow from unemployment, with evidence from individual records suggesting rather small effects. For example, Meyer (1990) estimates the impact of an increase in the rate of income replacement provided by UI in the United States (for instance from 50 to 60% of the insured wage). This increase is associated with 1.5 weeks longer duration of unemployment. For the United Kingdom, Nickell (1990) presents similar results.

As mentioned above, the interplay between UI and public welfare is an important consideration when it comes to the flow out of unemployment. Therefore, the study by Hujer and Schneider (1989) for Germany based on data from the Socioeconomic Panel is of particular interest. They relate the outflow from unemployment to the transition from UI to the more stingy public welfare alternative. Surprisingly, they find that increasing proximity to transition time seems to cause the likelihood of re-employment to decrease rather than increase as expected. The authors interpret this as a selection effect. As unemployment continues, only individuals who can be employed with difficulty remain in the pool.

▶ **Conclusion 10.14** There are theoretical reasons for unemployment insurance to cause an increase in the duration of unemployment and hence in the stock of unemployed. However, empirical studies at the micro economic level find small or even unexpected effects.

Even after the refinements cited above, these approaches to UI can be criticized as being too narrow. In particular, Landais et al. (2018) argue that UI impacts not only labor supply but also the demand for labor by employers. First, an increase in UI benefits improves the so-called outside option of the unemployed (i.e., to simply rely on UI), resulting in a tightening of the labor market which causes employers to post fewer job openings. Second, it discourages job search by workers with low qualifications while permitting those with high qualifications to hold out longer, which increases employers' chances of filling their job openings. While the first effect lowers the rate of outflow from unemployment, the second increases it, albeit usually with a lag. According to the authors' calibrated simulations, this renders the net effect of UI on unemployment indeterminate, raising the question of what is the optimal amount of social insurance. This issue is addressed in the next section, again for the case of UI.

10.5.4 Optimal Amount of Social Insurance

When discussing the impacts of SI, moral hazard is emphasized as a negative side effect in this chapter. As shown for the case of health insurance in Sect. 10.5.2, moral hazard can be reined in by copayment, which amounts to a limitation of insurance coverage. This raises the suspicion that the amount of coverage provided by SI could be excessive in view of moral hazard effects, causing inefficiency. However, the efficiency test of Sect. 10.4.2.2 cannot be applied because the issue there is the interplay of different branches of SI in the interest of minimizing volatility rather than the degree of coverage provided. Therefore, the contribution by Anderson (1994) is discussed in this section because it seeks to establish the optimum amount of coverage in one branch of SI, namely UI in the case of the United States. Optimality is defined from the point of view of a worker acting as an IB (insurance buyer).

The point of departure is the slope of the insurance line [see Eq. (3.15) of Sect. 3.2.1],

$$\frac{dW_1}{dW_2} = -\frac{1-\pi}{\pi} \left[\text{and hence } \frac{dW_2}{dW_1} = -\frac{\pi}{1-\pi}, \text{ see below}\right]. \quad (10.25)$$

Recall that W_1 denotes wealth in the loss state (which is equated with UI benefits here) and W_2, wealth in the no-loss state (labor income when employed). The right-hand side of the equation shows the conditions on which an insurer can transform a premium paid in the no-loss state (dW_2) into a net payment in the loss state (dW_1). The lower the probability of loss π, the larger can be payment in the loss state. Note that both π and size of loss L are predetermined at this stage, precluding all moral hazard effects. At the optimum of the IB with full insurance coverage, the left-hand side of (10.25) is equal to the marginal rate of substitution between the wealth levels in the two states [see Eq. (3.11)].

If it were possible to somehow determine this marginal rate of substitution, one could pit the left-hand side of (10.25) against its right-hand side. For instance, let the absolute value of dW_1/dW_2 exceed that of the right-hand side. This would indicate that the IB calls for more net benefit in the loss state for the premium paid than offered by an actuarially fair insurance scheme, implying that the amount of UI should be reduced. If conversely the left-hand side is smaller than the right-hand side, this would be an indication that the amount of coverage should be expanded.

The exposition below focuses on the worker's marginal willingness to pay for UI coverage. It is given by dW_2/dW_1, i.e., the reciprocal of (10.25), indicating how much income in the no-loss state of employment the worker is prepared to give up in return to more income in the loss state of unemployment. When applying this to UI, four adjustments are necessary before arriving at crucial Eq. (10.26) below.

1. One must take into account that given a marginal tax rate on income of τ^e, 1 MU (monetary unit) of extra contribution to UI (which is not subject to income tax) has the same effect on disposable wealth (or income, respectively) as $1/(1-\tau^e)$ MU. Therefore, dW_2 in the second part of Eq. (10.25) is to be replaced by $dW_2/(1-\tau^e)$.

2. Benefits of UI are also subject to income taxation in the United States, albeit at a lower rate $\tau^u < \tau^e$. This means that it takes $1/(1-\tau^u)$ MU to buy 1 MU of net benefit dW_1.
3. UI cannot operate without a loading for administrative expense amounting to λ. For this reason, additional UI coverage costs $\pi(1+\lambda)$ rather than π per unit (see Table 3.2 of Sect. 3.3).
4. An insurer who seeks to ensure its solvency must account for moral hazard by a surcharge to the premium. This surcharge is symbolized by m, reflecting the increase in the probability of unemployment as well as its duration. Using estimates for the United States, Anderson (1994) sets each component equal to 0.26. The surcharge for moral hazard effects, therefore, amounts to $m = 0.52$ US\$ per dollar to be paid as UI benefits in the event of unemployment; this value will be used below. After these adjustments, Eq. (10.25) becomes

$$\frac{dW_2 \cdot (1-\tau^u)}{dW_1 \cdot (1-\tau^e)} = -\frac{\pi \cdot (1+\lambda+m)}{(1-\pi)}. \qquad (10.26)$$

The right-hand side of this equation indicates how much premium an UI scheme seeking to secure its economic viability must charge in order to be able to transfer one MU net from the state "employed" to the state "unemployed". From the point of the individual, one obtains

$$\frac{dW_2}{dW_1} = -\frac{\pi(1+\lambda+m)(1-\tau^e)}{(1-\pi)(1-\tau^u)}. \qquad (10.27)$$

The left-hand side of (10.27) reflects the marginal willingness to pay for a secure labor income. Note that this willingness to pay can be inferred from employment choices in the labor market. There, workers accept a lower wage *ceteris paribus* in return for increased job security. Conversely, in industries with higher employment risks, higher wages must be paid for otherwise comparable work.

This willingness to pay for job security applies also to UI because UI confronts workers with a trade-off between reduced income and increased income security as well. This similarity can be used to infer marginal willingness to pay for income security provided by UI from wage rates observed in industries of different riskiness. Employment risk (symbolized by $RISK$ below) is defined as the share of days without work in a year normalized to 250 workdays. It is a sensible measure since UI benefits are also paid for a certain amount of time. Employment risk defined in this way varies importantly between U.S. industries during the observation period 1980–1987. In agriculture, it amounted to almost 16% (40 out of 250 days), whereas in financial services, it was minimal with some 3%. The average value of $RISK$ is 0.048 or 4.8%.

10.5 Impacts of Social Insurance Beyond Insurance Markets

The author used approximately 25,000 observations of employed workers over the years 1984–1986. The preferred OLS regression reads

$$\begin{aligned} lnWAGE = {}& constant + 0.74^{+}RISK - 2.04^{*}(RISK \cdot REPL.RATE) \\ & + 0.00027^{***}LOSTWORKDAYS - 0.29^{***}FEMALE \\ & - 0.05^{***}NON\text{-}WHITE \\ & + 5 \text{ variables for experience and education;} \\ & + 6 \text{ variables for urban/rural communities, regions; and} \\ & + 5 \text{ variables for employment categories.} \end{aligned} \quad (10.28)$$

[+(*,***): Statistically significant at the 0.1 (0.05, 0.001) level]

lnWAGE: Logarithm of the hourly wage rate at 1981 prices, after tax;
RISK: Average industry-specific share of days without work in a year normalized to 250 workdays;
REPL.RATE: Replacement rate, ratio of UI benefits to the wage rate, after tax;
LOSTWORKDAYS: Workdays lost per 100 workdays due to illness or accident;
FEMALE: = 1 if individual is female, = 0 otherwise;
NON-WHITE: = 1 if individual is not white, = 0 otherwise.

This regression result can be interpreted as follows.

- *RISK:* A higher risk of unemployment must be compensated by a higher wage rate. However, the pertinent coefficient barely reaches the conventional level of significance.
- *RISK · REPL.RATE:* The product of the two variables represents the risk-reducing effect of UI. A given job risk is mitigated by UI according to the level of the replacement rate. The coefficient pertaining to this product is significantly negative. Its effect in the sample is so marked that a higher value of $RISK$ on the whole is associated not with a higher but a lower value of the wage rate.
- *LOSTWORKDAYS:* The share of days lost from illness and accident per 100 workdays constitutes a good indicator of industry-specific health risks on the job. As expected, a higher risk must be compensated by a higher wage rate, as indicated by the positive regression coefficient. However, it is conceivable that this effect would also be mitigated or even reversed by insurance (workplace accident insurance in this context).
- *FEMALE:* Women have a lower wage rate than men after controlling for experience and education. The gap amounts to some 30%, judging from the regression

coefficient of –0.29 indicating the relative change associated with a change of the explanatory variable from 0 to 1.[7]
- *NON-WHITE:* This racial effect on the wage is statistically significant and represents an approximate 5% reduction in wages.

How workers trade off between the wage rate earned (*WAGE*) and the security of income provided by UI can be derived from the regression result (10.28). One has to calculate the number of MU the accepted wage rate decreases (dW_2) if UI benefits increase by 1 MU ($dI = dW_1$). In the absence of market imperfections, this reduction of the wage rate is equal to the subjective willingness to pay workers for increased income security.

This relationship calls for a series of implicit derivatives. For ease of notation, $A := RISK \cdot REPL.RATE$, i.e., A symbolizes the influence of UI,

$$\frac{dW_2}{dW_1} = \frac{40 \cdot dWAGE}{dI} = \frac{40 \cdot \partial WAGE}{\partial \ln WAGE} \cdot \frac{\partial \ln WAGE}{\partial A} \cdot \frac{\partial A}{\partial (I/40WAGE)} \cdot \frac{\partial (I/40WAGE)}{\partial I}$$

$$= 40WAGE \cdot (-2.04) \cdot 0.048 \cdot 1 \cdot \frac{1}{40WAGE} = -0.098.$$
(10.29)

The necessary steps can be explained as follows.

1. The quantity dW_2 is equated to the additional weekly labor income $40 \cdot dWAGE$ in the employed state that results from a variation of the wage rate $dWAGE$ and a 40 hours week.
2. The quantity dW_1 is equated to an additional MU of benefit dI provided by UI.
3. Since the wage rate appears in logarithmic form in Eq. (10.28), *WAGE* in (10.29) must first be differentiated w.r.t. *lnWAGE*. Using the known rule $\partial \ln x / \partial x = 1/x$, one obtains $\partial WAGE / \partial \ln WAGE = WAGE$.
4. The change $d\ln WAGE$ is caused by a change of A which in turn can be traced to a change of the replacement rate *REPL.RATE* in the product $RISK \cdot REPL.RATE$. For this reason, the partial derivative of *lnWAGE* w.r.t. A [which amounts to -2.04 according to (10.28)] needs to be multiplied by *RISK*, using the average value of 0.048.
5. The influence of the UI benefit is the benefit I measured in MU in relation to the weekly wage income amounting to $40 \cdot WAGE$, implying that the derivative of A with respect to $(I/40 \cdot WAGE)$ yields exactly one.
6. The change of the replacement rate $(I/40 \cdot WAGE)$ must be related to the change in the UI benefit I expressed in MU. The differentiation results in $1/(40 \cdot WAGE)$.

---

[7] Basing this estimate directly on the regression coefficient neglects the fact that the log transformation (resulting in $\ln WAGE$) constitutes a non-linear transformation causing a problem of retransformation [for details, see Kennedy 1986].

10.5 Impacts of Social Insurance Beyond Insurance Markets

Plugging in these values step by step results in the value shown in (10.29). It states that workers were prepared to sacrifice 9 to 10 cents of their wage for one US$ additional UI benefit. Therefore, one has an estimate of the left-hand side of Eq. (10.27). This value can now be compared with an estimate of the right-hand side, which derives from known values for π, λ, m, τ^e, and τ^u. The result of this calculation is that the two sides of (10.27) match very precisely, implying that the marginal willingness to pay revealed in the labor market is equal to the trade-off offered by UI. One can, therefore, conclude that the amount of UI coverage is optimal on average. In addition, deviations from the optimum never exceed 3% in subgroups (for instance men versus women, whites versus non-whites). It reaches a maximum of 8% in the state of West Virginia; employed people of that state have a marginal willingness to pay for income security falling 8% short of what an efficient insurer would have to charge.

▶ **Conclusion 10.15** The coverage offered by unemployment insurance in the United States is likely to correspond to the optimum from the point of view of insured workers. In the labor market, they are prepared to pay the premium in the guise of a reduced wage rate that has to be charged by an insurer using an actuarial premium calculation.

Since the actuarial premium calculation mentioned in this conclusion takes into account moral hazard, the conjecture that the rate of coverage of UI is excessive in view of its moral hazard effects is refuted at least for the United States. This finding cannot be simply transferred to other countries, however. The replacement rate was about 43% in the United States of the 1980s [see Anderson 1994], compared to 65%, e.g., in Germany (with children present in the family: 75%); [Hujer and Schneider 1989]. By 2009, this rate increased to 47% (single earners in the United States) and to 72% in Germany, only exceeded by Luxembourg (90%) and Switzerland (83%) [Van Vliet and Caminada 2012]. Whether marginal willingness to pay for income security is so much higher in Germany than in the United States for such a high replacement rate to be optimal is an open question.

However, since unemployment is a process lasting several weeks and up to several months, the optimal UI should be related to this process. Therefore, one can ask whether it is optimal to start with a low-income replacement rate (indemnity) and then to increase it as the unemployment period continues, or to start with a high rate and to lower it over time so as to increase the "pressure" on the unemployed to find a job [see, e.g., Kolsrud et al. 2018].

Exercises

10.1

(a) For explaining the existence of social insurance (SI), reference was made to efficiency arguments. However, there are also arguments relating to political economy. Please explain the two alternative explanations in no more than three sentences each.
(b) What are the implications of the two explanations concerning the size of SI? Which one is more compatible with observations from your country?
(c) What are the implications of the two explanations concerning the development of SI over time? Which one is more compatible with observations from your country?
(d) Is it possible to derive predictions also relating to the structure of SI from the two explanations? Are these predictions compatible with observations from your country?
(e) Which of the two explanations of SI do you prefer based on your preceding considerations?

10.2

(a) When considering the impacts of provision for old age through SI on the capital market, the pay-as-you-go variant and private savings were assumed to achieve the same rate of return. Please explain the condition under which this assumption holds true. Check whether they hold true at present.
(b) Change this assumption to the effect that private savings

  (b1) achieve a higher rate of return than SI;
  (b2) achieve a lower return than SI.

  Derive predictions concerning the amount of private savings using the two-goods model of Sect. 10.5.1.2.
(c) Does the assumption of equal rates of return introduced in (a) turn out to be essential or not?

References

Aaron, H. (1966). The social insurance paradox. *Canadian Journal of Economics and Political Science, 32*, 371–374.

Anderson, D.A. (1994). Compensating wage differentials and the optimal provision of unemployment insurance. *Southern Economic Journal, 60*(3), 644–656.

Atkinson, A.B., & Micklewright, J. (1991). Unemployment compensation and labor market transitions: a critical review. *Journal of Economic Literature, 29*, 1679–1727.

Blanchard, O., & Perotti, R. (2002). An empirical characterization of the dynamic effects of changes in government spending and taxes on output. *Quarterly Journal of Economics, 117*(4): 1329–1368.

Blundell, R. (1992). Labour supply and taxation: a survey. *Fiscal Studies, 13*, 15–40.

Börsch-Supan, A. (1992). Population aging, social security design, and early retirement. *Journal of Institutional and Theoretical Economics/Zeitschrift für die gesamte Staatswissenschaft, 148*(4), 533–557.

Bove, C., Efthyvoulou, G., & Navas, A. (2017). "Political cycles in public expenditure: butter vs. guns." *Journal of Comparative Economics* 45(3), 528–604.

Breyer, F. (1990). *Ökonomische Theorie der Alterssicherung (Economic Theory of Provision for Old Age)*. München: Vahlen.

Buchanan, J.M., & Tullock, G. (1976). Polluters' profits and political response: direct control versus taxes: reply. *American Economic Review, 66*(5), 983–984.

Burda, M.C., & Wyplosz, C. (1994). Gross labor market flows in Europe. *European Economic Review, 38*, 1287–1915.

Burkhauser, R.V., & Turner, J.A. (1978). A time-series analysis on social security and its effect of the market work of men at younger ages. *Journal of Political Economy, 86*(4), 704–715.

Burtless, G. (1986). Social security, unanticipated benefit increases, and the timing of retirement. *Review of Economic Studies, 53*, 781–805.

Crain, M. (1979). Cost and output in the legislative firm. *Journal of Legal Studies, 8*(3), 607–621.

Culyer, A.J. (1980). *The Political Economy of Social Policy*. Oxford: Martin Robertson.

Dahlby, B.G. (1981). Adverse selection and pareto improvements through compulsory insurance. *Public Choice, 37*, 547–558.

Downs, A. (1957). *An Economic Theory of Democracy*. New York: Harper & Row.

Eisen, R. (2006). Adverse selection in the health insurance market after genetic tests. In P.A. Chiappori, & C. Gollier (Eds.), *Competitive Failures in Insurance Markets. Theory and Policy Implications* (pp. 34–54). Cambridge MA: MIT Press.

Eisen, R., & Zweifel, P. (1997). Überlegungen zur optimalen Kombination unterschiedlicher Versicherungsprodukte über die Zeit (On the optimal combination of different insurance products over time). In L. Männer (Ed.), *Langfristige Versicherungsverhältnisse. Ökonomie, Technik, Institutionen* (pp. 355–378). Karlsruhe: Verlag Versicherungswirtschaft.

Felderer, B. (1992). Does a public pension system reduce savings rates and birth rates? *Journal of Institutional and Theoretical Economics/Zeitschrift für die Gesamte Staatswissenschaft, 148*, 312–325.

Feldstein, M.S. (1974). Social security, induced retirement, and aggregate capital accumulation. *Journal of Political Economy, 82*(5), 906–926.

Gruber, J. & Yelowitz, A.S. (1999). Public health insurance and private savings. *Journal of Political Economy, 107*(2), 1249–1274.

Hujer, R., & Schneider, H. (1989). The analysis of labor market mobility using panel data. *European Economic Review, 33*(2/3), 530–536.

Kennedy, P.E. (1986). Interpreting dummy variables. *Review of Economics and Statistics, LXVIII*, 174–175.

Kolsrud, J., Landais, C., Nilsson, P., & Spinnewijn, J. (2018). The optimal timing of unemployment benefits: Theory and evidence from Sweden. *American Economic Review, 108* (4–5), 985–1033.

Kotlikoff, L.J. (1979). Testing the theory of social security and life-cycle accumulation. *American Economic Review, 69*, 396–411.

Ladaique, M. (2011). *The OECD Social Expenditure Database (SOCX)*. Beirut: Expert Group Meeting on Social Security in Western Asia, 8-9 Sept. 2011.

Landais, C., & Michaillat, P., Saez, E. (2018). A macroeconomic approach to optimal unemployment insurance: Applications. *American Journal of Economic Policy, 10*(2), 182–216.

Leimer, D.R., & Lesnoy, S. (1982). Social security and private saving: new time-series evidence. *Journal of Political Economy, 90*, 606–642.

Meyer, B.D. (1990). Unemployment insurance and unemployment spells. *Econometrica, 58*(4), 7575–7782.

Mueller, D.C. (1989). *Public Choice II*. Cambridge MA: Cambridge University Press.
Newhouse, J.P. (1996). Reimbursing health plans and health providers: efficiency in production versus selection. *Journal of Economic Literature, 34*(3), 1236–1263.
Newhouse, J.P., et al. (1993). *Free for All? Lessons from the RAND Health Insurance Experiment*. Cambridge MA: Harvard University Press.
Nickell, S. (1990). Inflation and the UK labour market. *Oxford Review of Economic Policy, 6*(4), 26–35.
Niskanen, W.A. (1971). *Bureaucracy and Representative Government*. Chicago: Aldine.
Olson, M. (1965). *The Logic of Collective Action: Public Goods and the Theory of Groups*. Boston: Harvard University Press.
Philipson, T.J., & Becker, G.S. (1998). Old-age longevity and mortality-contingent claims. *Journal of Political Economy, 106*(3), 551–573.
Samuelson, P.A. (1958). An exact consumption-loan model of interest with or without the contrivance of money. *Journal of Political Economy, 66*, 467–482.
Schlesinger, H. (1997). Mikro-Korrelationen versus Makro-Korrelationen und die zukünftige Versicherungsdeckung (Micro vs. macro correlations and future insurance coverage). In L. Männer (Ed.), *Langfristige Versicherungsverhältnisse. Ökonomie, Technik, Institutionen* (pp. 379–386). Karlsruhe: Verlag Versicherungswirtschaft.
Schneider, F. (1986). The influence of political institutions on social security policies: A public choice view. In J.M. Schulenburg (Ed.), *Essays in Social Security Economics* (pp. 13–31). Berlin: Springer.
Schoder, J., Zweifel, P. & Eugster, P. (2013). Insurers, consumers, and correlated risks. *Journal of Insurance Issues*, 36 (2), 1–29.
Schulenburg, J.M. (1986). Optimal insurance purchasing in the presence of compulsory insurance and uninsurable risks. *Geneva Papers on Risk and Insurance, 11*(38), 5–16.
Sinn, H.W. (1996). Social insurance, incentives and risk taking. *International Tax and Public Finance, 3*, 259–280.
Stock, J.A., & Wise, D.A. (1990). Pensions, the option value of work, and retirement. *Econometrica, 58*(5), 1151–1180.
Streb, J. (2018). Does social security crowd out private savings? The case of Bismarck's system of social insurance. *European Review of Economic History, 22*(3), 298–321.
Van Dalen, H.P., & Swank, O.A. (1996). Government spending cycles: ideological or opportunistic? *Public Choice, 89*, 183–200.
Van Vliet, O. & Carminada, K. (2012). Unemployment replacement rates dataset among 34 welfare states, 1971-2009: an update, extension and modification of the Scruggs' Welfare State Entitlements Data Set. *NEUJOBS Special Report No. 2*. Leiden: Leiden University.
Von Camphausen, S., Bornschein, B. & Wick, R. et al. (2005). Prevalence and incidence of Parkinson's disease in Europe, *European Neuropsychopharmacy* 15, 465–490.
Zweifel, P. (2013). The division of labor between private and social insurance. In G. Dionne (Ed.), *Handbook of Insurance* (pp. 1097–1118). Boston: Kluwer.
Zweifel, P., & Eugster, P. (2008). Life-cycle effects of social security in an open economy: a theoretical and empirical survey. *Zeitschrift für die gesamte Versicherungswissenschaft (German Journal of Risk and Insurance), 97*(1), 61–77.
Zweifel, P., & Frech, H.E. (2016). Why 'optimal' payment for healthcare providers can never be optimal under community rating. *Applied Health Economics and Health Policy, 14*(1), 9–20.
Zweifel, P., & Waser, O. (1992). *Bonus Options in Health Insurance*. Dordrecht: Kluwer.
Zweifel, P., Breyer, F., & Kifmann, M. (2009). *Health Economics*, 2nd ed. New York: Springer.

Challenges Confronting Insurance 11

This chapter is devoted to foreseeable future developments that will pose challenges to the insurance systems of industrial countries. These challenges will call for adjustments by both private insurance (PI) and social insurance (SI), likely also affecting their division of labor. As a matter of principle, the need for adjustment will confront both PI and SI. The reason is that the changes in the economic environment discussed below result in modifications of risk behavior on the part of insurance buyers (IBs).

At least to the extent that insurance companies (ICs) are exposed to competitive pressure, they must adjust to these changes lest they forgo profit opportunities. Changed risk preferences and possibilities of risk management call for the development of new products and adjustments in pricing. Failure to act results in reduced premium volume in risk underwriting that translates into less capital investments and smaller subsequent returns.

Public regulation and cartel-like behavior of PI slow down these adjustments. However, lags in adjustment and the concomitant reform backlog tend to be more marked for SI. Reforms of SI are decided by parliament in representative democracies (and often voters themselves in direct democracies). The prevalent uniformity of SI benefits causes every conceivable solution to run against the interests of some population subgroups. This makes reform of SI a time-consuming parliamentary process. Meanwhile, tensions increase, amounting to challenges of growing acuteness.

Four of these challenges are discussed in this chapter. They confront both PI and SI; however, one of the two may have scope for adjustment that the other is lacking, resulting in a changed division of labor between the two. The exposition starts with the globalization of economic relations, which is of primary importance for both the underwriting of risks and investment of capital by an IC (Sect. 11.1). At first sight, SI appears to be sheltered from this shock; however, it may run into problems of financing as an indirect consequence of globalization.

A second challenge confronting all countries is new developments in science and technology. Section 11.2 deals with the fact that genetic information becomes increasingly available, confronting IBs with the question of whether they want to have it and possibly share it with their IC. In addition, it addresses the consequences of advances in information technology for an IC.

Next, the waves of deregulation and re-regulation that increasingly are becoming of international scope are the topic of Sect. 11.3. These waves raise the question of what constitutes the principal elements of an insurance contract.

Finally, Sect. 11.4 revolves around demographic change. Although often viewed as a domestic challenge, it has similar properties across industrial countries. Demographic change mainly affects the lines of personal insurance and has the potential of importantly modifying the division of labor between PI and SI.

11.1 Globalization of International Economic Relations

Economic integration pursued by institutions such as the North American Free Trade Association (NAFTA), the European Union (EU), and the World Trade Organization (WTO) have not only served to lower barriers to trade but led to increasingly equal treatment of domestic and foreign producers, at least in industrial countries. This process calls for adjustment on the part of insurers both for their commercial and personal lines [see Bernheim 1998].

11.1.1 Globalization and Corporate Insurance

Globalization is to a considerable degree the result of the fact that enterprises can open up sites for production, distribution, and administration in a foreign country at a much lower cost than in the past. This multitude of locations creates possibilities for internal risk diversification (see Sect. 5.2). An event triggering a loss in country A usually affects only the assets invested there but not those invested in countries B, C, etc. In addition, business cycles in North America, the European Union, and the Asian Pacific continue to differ to a sufficient degree for assets and liabilities of geographically diversified enterprises to be less than perfectly correlated.

However, increased possibilities for internal diversification imply that firms considering buying coverage will offer risks to the IC that

- are highly correlated;
- can be mitigated by the enterprise at a high cost only;
- have a loss distribution that cannot be estimated easily;
- can be transferred to the capital market only at a high cost.

The appropriate measures to be taken by IC management can be deduced from Sects. 6.4 and 7.2, assuming that management is closely tied to the interests of shareholders as the owners of the IC.

1. *Increased effort at risk selection.* The properties cited above imply that risks offered by globally active enterprises likely are heterogeneous between countries. For the IC, this means large differences in terms of contribution to expected profit between country-specific risks. As shown in Sect. 6.2.2, optimal risk selection effort is high when contributions to expected profit differ drastically.
2. *Use of innovative insurance products.* ICs offering business interruption coverage can also provide coverage to minimize supply chain disruption worldwide. Other coverages offered by a global network of ICs (perhaps through a broker) can mitigate risks encountered when broadening business services.
3. *Increase of the asset–liability ratio.* Positively correlated risks cause the variance of the surplus, $Var(A - L)$ to increase, *ceteris paribus* (A: assets; L: liabilities). Option pricing theory in general and Sect. 7.2.3 in particular state that this serves to increase the value of the put option held by the owners of the IC while diminishing the effective value of claims held by the IB. This causes a negative demand response by both current and future purchasers of corporate insurance coverage. In an attempt to neutralize this effect, management of the IC would want to adjust upward the asset–liability ratio.
4. *Restructuring of the risk portfolio.* While the risks offered for underwriting by globally active enterprise X may be positively correlated across countries, they may still be negatively correlated with those of another enterprise Y. The management of the IC can benefit from this fact in order to reduce $Var(A - L)$ by having both X and Y in its portfolio, permitting it to lower the asset–liability ratio in the shareholders' interest (see Sect. 7.2.4.2).
5. *Restructuring of capital investment.* An increase in $Var(A - L)$ due to risk underwriting can also be counteracted by investing in assets whose returns tend to increase in line with loss payments.

Note that corporate insurance falls outside the domain of social insurance (SI), where individuals are covered. Therefore, SI is not directly affected by globalization (see Sect. 11.1.2 below, however).

11.1.2 Globalization and Individual Insurance

Individual insurance is indirectly affected by globalization since international business activity requires international mobility of workers [see Van Den Berghe 1998]. With the portability of SI benefits restricted for some time to come, this mobility gives rise to a demand for supplementary insurance coverage tailored to individual needs. Offering a uniform product, SI cannot satisfy this demand. Meeting it constitutes a challenge to PI as well, however, because of the need to develop a supplementary product for each market in principle.[1]

At the same time, globalization aggravates problems encountered by national SI schemes. For instance, facilitation of migration causes wage rates in the target country to fall, and with them the payroll tax revenue per worker of SI [see Wildasin 1991]. This effect is especially marked for lower skilled workers, while the highly skilled tend to benefit from globalization, putting PI at an advantage. However, the decision to migrate at all is influenced by SI, with globalization permitting individuals to choose the SI scheme offering them a favorable benefit–cost ratio. Other decisions along the life cycle are generally influenced by SI as well, and ease of migration arguably exacerbates their negative feedback on SI [see Zweifel and Eugster 2008].

In addition, multinational enterprises have an impact on employment through their choice of location by inducing migration of jobs rather than of workers. Contributions to SI crucially depend on domestic employment, while PI can more easily follow these migrants. On the whole, there is reason to expect globalization to modify the division between PI and SI in favor of PI [see Zweifel 2013].

▶ **Conclusion 11.1** The globalization of economic relationships changes the properties of corporate risks offered to insurers for underwriting; however, PI has instruments of insurance technology to deal with this change. In personal lines, globalization favors PI to the detriment of SI.

11.2 Changes in Science and Technology

Scientific discoveries and technological change can be of great relevance for insurance because they typically modify the exposure to the risk of the IB as well as the IC. This section focuses on two examples, the availability of genetic information and advances in information technology (IT). The first concerns the individual IB, who must decide whether or not to obtain genetic information and to share it (with the health insurer in particular). Second, new IT modifies corporate demand for insurance. At the same time, however, it creates scope for more accurate pricing of insurance products, ultimately contributing to an expansion of their supply. Another instance of change in science and technology creating a challenge to insurance is new medical technology; however, this is discussed as "dynamic" moral hazard in Sect. 8.2.1.

---

[1] Within the European Union, for instance, claims against the SI of the home country can increasingly be presented also while residing in another member country (see Sect. 9.2.3). However, bringing these claims to bear still is associated with considerable cost.

11.2.1 Genetic Information

Future availability of genetic information will enable agents to predict individuals' illnesses and life expectancies with greater precision. In the extreme, the problems of asymmetric information treated in Sect. 8.3 would vanish at least in life and health insurance businesses because the IC knows the risk type. In principle, the IC could force the IB to acquire this information and make it available as a precondition for obtaining insurance coverage. However, if acquiring genetic information should not be in consumers' interest, such a move would push them toward SI which has no use of genetic information since it covers all risk types with uniform conditions.

This issue has been examined by Doherty and Posey (1998). They distinguish between low, high, and uninformed risks. Uninformed (U) risks can have access to a test free of charge which makes them either a low (L) or a high (H) risk type. In addition, the high risks have the possibility of prevention (costing V) that serves to decrease the probability of loss π^H (to be interpreted as a probability of death or illness). Equations (11.1a)–(11.1c) provide more detail:

$$\pi^{H\prime}(V) < 0, \quad \pi^{H\prime\prime}(V) > 0; \tag{11.1a}$$

$$\pi^H[\infty] > \pi^U > \pi^L > 0; \tag{11.1b}$$

$$\pi^U = \rho \pi^L + (1-\rho)\pi^H[0], \quad \text{with} \tag{11.1c}$$

π^i: probability of loss according to risk type i, $i = H, U, L$ (high, uninformed, low);
V: preventive effort (in utility terms, see below);
ρ: share of low risks in the insured population.

Expression (11.1a) states that the possibility of prevention is available only to IBs who know they are high risk. There are decreasing marginal returns to prevention; the decrease is so marked that a high-risk type remains in that category even with infinitely high preventive effort V [assumption (11.1b)]. The uninformed risks are characterized by Eq. (11.1c). They constitute a category in between, representative of the entire population insured. Their probability of loss equals the weighted mean of the probability of loss for the low risks (π^L) and the corresponding baseline probability (i.e., assuming zero preventative effort since the uninformed have no reason to spend on prevention) for the high risks ($\pi^H[0]$).

Figure 11.1 illustrates the decision-making problem of the three risk types, who are represented by three indifference curves with their pertinent slopes. The indifference curve of the uninformed risk has a slope that according to Eq. (11.1c) is between those of the low- and the high-risk types (for the determination of these slopes, see Sect. 8.3 again). Acting on the information provided by the genetic test, the high-risk type is assumed to already have spent preventive effort (to a degree which is still to be determined). Therefore, the indifference curve $\overline{EU}^H[V^*]$ lies between

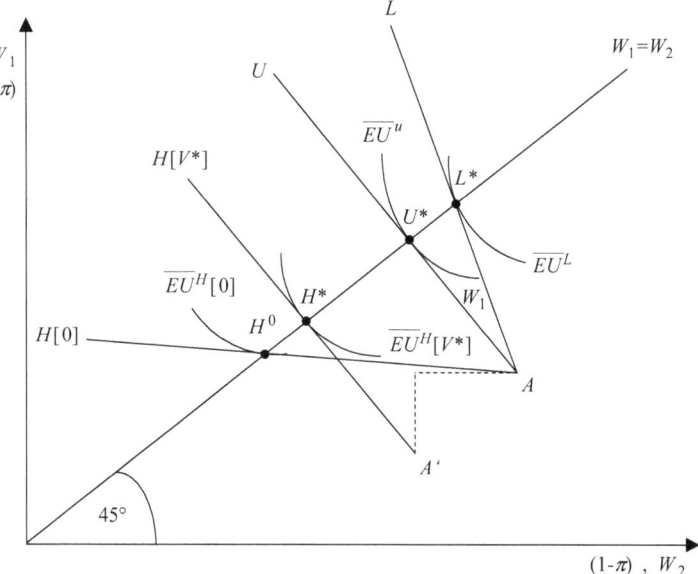

Fig. 11.1 Optima of high, uninformed, and low risks

those of the other two risk types, without however reaching the indifference curve of the uninformed type, in keeping with assumption (11.1b). This also means that the indifference curves pertaining to the risk types can intersect only once pairwise (single crossing property). Finally, those who have been found to be high-risk types will have spent on prevention both in the loss state (W_1) as in the no loss state (wealth W_2). Therefore, their endowment point moves from A to A'.

The genetic information is assumed to be available to the IC, who can thus calculate premiums according to true risk, causing each risk type to choose full coverage (optimum points H^*, U^*, and L^*, respectively). The question to be answered now is whether uninformed individuals have an advantage from taking the test and making its result available. In fact, this means moving away from U^* to participating in a lottery composed of the points $\{H^*, L^*\}$ representing the optima in case the test renders the outcomes "high risk" and "low risk", respectively. The element L^* of this lottery is easy to describe because the individual purchases full coverage at the low premium calculated for the low-risk types. The decisive component is H^*, the properties of which depend on the amount of prevention individuals undertaken at the time they find out to be high risks.

The point H^* of Fig. 11.1 reflects the pair of values $\{I_n^H, V^*\}$ that indicates the optimal amount of coverage as well as the optimal amount of prevention. Here, I_n^i symbolizes the *net insurance benefit* in the case of loss, amounting to $I_n^i = I - P^i$ for the high, uninformed, and low ($i = H, U, L$) risks, respectively. The slope of the budget line for the three risk types is therefore given by

11.2 Changes in Science and Technology

$$\left(\frac{dW_1}{dW_2}\right)^i = -\left(\frac{I-P}{P}\right)^i = -\left(\frac{I_n}{P}\right)^i = -\left(\frac{1-\pi}{\pi}\right)^i, \quad i = H, U, L. \quad (11.2)$$

This equation permits to relate the premiums to the net benefits purchased,

$$P^i = \left(\frac{\pi}{1-\pi}\right) \cdot I_n^i. \quad (11.3)$$

Note that this is nothing but the inverse of the slope of the insurance line in (W_1, W_2)-space of the basic model presented in Sect. 3.2. The optimization for someone who turns out to be a high risk ($i = H$) can now be formulated as follows:

$$\max_{I_n^H, V} EU = \pi^H(V) \cdot v[W_0 - L + I_n^H] + (1 - \pi^H(V)) \cdot v\left[W_0 - \frac{\pi^H(V)}{1 - \pi^H(V)} \cdot I_n^H\right] - V, \quad (11.4)$$

EU: expected utility;
π^H: probability of loss of a high risk, depending on prevention V;
L: loss, predetermined;
I_n^H: net benefit (after payment of premium) in the case of a high-risk type;
V: preventive effort (valued in units of utility).

Prevention V must be undertaken prior to the occurrence of loss and is therefore state-independent. For simplicity, it does not enter the risk utility function but is deducted as an expense (in utility units). The necessary condition of an interior optimum with regard to the amount of *insurance coverage* ($I_n^{H*} \geq 0$) then reads

$$\frac{\partial EU^H}{\partial I_n^H} = \pi^H(V) \cdot v'[W_0 - L + I_n^H] + (1 - \pi^H(V)) \cdot \frac{\pi^H(V)}{1 - \pi^H(V)}$$
$$\cdot (-1) \cdot v'[W_0 - P^H] = 0. \quad (11.5)$$

In this equation, $v'[W_0 - P_H]$ is the marginal utility of wealth in the no-loss state according to (11.4). Cancelation of $(1 - \pi^H(V))$ immediately yields

$$v'[W_0 - L + I_n^H] = v'[W_0 - P^H], \quad (11.6)$$

i.e., the marginal utilities of wealth in the loss and no-loss state must be equal in the optimum. Since the risk utility function is defined only in terms of wealth, this condition implies equal wealth in the two states and therefore full insurance coverage, i.e., the usual first-best solution in the absence of information asymmetry. Accordingly, point H^* of Fig. 11.1 lies on the certainty line ($W_1 = W_2$).

The necessary condition for an interior optimum with regard to *prevention* is obtained by applying the rule for the derivative of a ratio,

$$\frac{\partial EU^H}{\partial V} = \pi^{H'}[V^*] \cdot v[W_0 - L + I_n^H] - \pi^{H'}[V^*] \cdot v[W_0 - P^H]$$

$$+ (1 - \pi^H) v'[W_0 - P^H] \left\{ \frac{-\pi^{H'}[V^*](1 - \pi^H) + \pi^H \{-\pi^{H'}[V^*]\}}{(1 - \pi^H)^2} \right\}$$

$$\cdot I_n^H - 1 = 0. \qquad (11.7)$$

After factoring out $\pi^{H'}[V^*]$, one obtains

$$\pi^{H'}[V^*]\{\underbrace{v[W_0 - L + I_n^H] - v[W_0 - P^H]}_{(0)} - \frac{1}{1 - \pi^H} \underbrace{I_n^H \cdot v'[W_0 - P^H]}_{(+)}\} = 1.$$
$$\phantom{\pi^{H'}[V^*]}{\scriptstyle(-)}$$

$$(11.8)$$

On the right-hand side, one has the marginal cost of prevention that is equal to one unit of utility. The expected marginal return on the left-hand side is the sum of two components. On the one hand, this is the marginal effectiveness of prevention in terms of a reduced probability of loss ($\pi^{H'}[V^*] < 0$) multiplied by the value of this change in terms of utility units. This is the (negative) utility difference between the loss and the no-loss states, which however is reduced to zero in view of full insurance coverage. On the other hand, there is the reduction of premium, to be valued by the marginal utility of wealth. This reduction is the consequence of the fact that due to genetic information, the IC can establish the relationship between π^H and V, enabling it to honor preventive effort (contrary to standard assumptions, e.g., in Sect. 8.2).

Condition (11.8) implicitly determines the optimal amount of prevention through its optimal value of marginal effectiveness, $\pi^{H'}[V^*]$. However, in case a genetic deficiency was found for which there is no successful treatment, $\pi^{H'}[V^*] = 0$ obtains, and the condition (11.8) *cannot be satisfied*. Since the marginal utility of prevention has a value of zero, it falls short of its marginal cost of one, making it optimal to minimize prevention, i.e., to reduce it to zero. Conversely, one can say that the marginal effectiveness of prevention must be sufficiently high for all (in particular, also for small) values of V for condition (11.8) to be satisfied and hence to obtain an optimal solution $V^* > 0$.

After having described the optimal point H^* of Fig. 11.1 through the pair of values $\{I_n^{H*}, V^*\}$, one can return to the crucial concern regarding the interest of a uninformed risk in obtaining genetic information to share with the IC. The *value of information* for the uninformed risk is given by the following comparison of utility values:

$$N = \rho \cdot v[L^*, \pi^L] + (1 - \rho)\{v[H^*, \pi^H[V^*]] - V^*\} - v[U^*, \pi^U] \qquad (11.9)$$

- N: value of genetic information;
- L^*: optimum of the low risk;
- H^*: optimum of the high risk after spending on prevention to the degree of V^*;
- U^*: optimum of the uninformed risk;
- ρ: share of the low-risk types in the population at risk.

The first two terms of Eq. (11.9) reflect the expected value of the lottery referred to in the context of Fig. 11.1. Since the share of low-risk types in the population is ρ, a (fully reliable) genetic test transforms the uninformed risk into a low one with probability ρ and into a high one with probability $(1 - \rho)$. In this latter event, it has to be taken into account that there will be prevention Effort not only resulting in the improved solution H^* but also costing V^*. In Eq. (11.9), this lottery is compared with the certain utility associated with remaining in the state of being uninformed.

This comparison becomes even more intuitive if one splits up the transition from U^* to H^* in Fig. 11.1 into two steps. First, there is the movement to intermediate point H^0, where the individual already knows to be a high-risk type but has not invested in prevention yet. The risky utility associated with that situation is denoted by $v[H^0, \pi^H[0]]$. Second, preventive effort moves the individual to the final optimum H^*. By adding and subtracting $(1 - \rho) \cdot v[H^0, \pi^H[0]]$, one has

$$N = (1 - \rho)\{\underbrace{v[H^*, \pi^H[V^*]] - V^* - v[H^0, \pi^H[0]]}_{(+)}\}$$
$$+ \underbrace{(1 - \rho) \cdot v[H^0, \pi^H[0]] + \rho \cdot v[L^*, \pi^L] - v[U^*, \pi^U]}_{(-)}. \quad (11.10)$$

The value of genetic information N now can be seen to consist of two components.

1. *Value of the prevention option.* The uninformed individual is recognized as a high-risk type with probability $(1 - \rho)$, causing a move to the risky utility $v[H^0, \pi^H[0]]$. However, this move opens up the possibility of reducing the probability of loss through prevention, from $\pi^H[0]$ to $\pi^H[V^*]$. Provided V^* is positive, the whole first component of (11.10) must be positive as well. Now for $V^* > 0$ to obtain, the marginal value of prevention starting from $V = 0$ must have been positive, satisfying condition (11.8). This indicates a possibility of successful preventive treatment for the genetically caused disease in question.
2. *Value of participating in an information lottery.* This is the second main term of (11.10). Its last component $v[U^*, \pi^U]$ reflects the certain utility associated with the state of being uninformed. The other two components describe the expected utility resulting from the uninformed individual achieving lower utility with probability $(1 - \rho)$ prior to investing in prevention and of being recognized as a low-risk type paying a low premium with probability ρ. However, for a risk-averse individual, the sum of these three terms is always negative.

▶ **Conclusion 11.2** The value of obtaining genetic information and sharing it with the IC is ambiguous. Being exposed to uncertainty with regard to their risk status is valued negatively by risk-averse uninformed individuals. Only if prevention is sufficiently effective in lowering the probability of the onset of the genetically caused disease, and hence the premium, can genetic information have a positive private value.

This conclusion is confirmed by a survey in the United States [see Singer 1991]. While almost two-thirds of participants were in favor of prenatal genetic testing in general, 92% opposed it when the purpose was to establish the sex of the baby. This difference could be related to the fact that there is no treatment alternative for the "genetic deficiency" of having the wrong sex.

Therefore, genetic information may not have value for the IB privately. However, it still has value for society because it facilitates self-selection by risk types, enabling the IC to improve the welfare of the low-risk types by writing separating contracts (see Sect. 8.3). Also, prevention induced by genetic testing makes the high-risk types better risk types who then migrate away from their contract with full coverage only when the contract designed for the low-risk types attains more complete coverage (see Sect. 8.3). Finally, Eisen (2006) shows that the welfare loss caused by asymmetric information can be reduced by compulsory SI providing partial coverage, in analogy to the model by Dahlby (1981) discussed in Sects. 8.3 and 10.2.

Doherty and Posey (1998) also extend their analysis to the case of private information where consumers perform the test but keep the result secret from the insurer. Of the two variants of their model, the first one with public information presumably is the more relevant one because the IB expects that the IC will attempt to obtain genetic information as part of their risk-selection effort. As stated in Conclusion 11.2, information has only a positive value for consumers if a sufficiently effective possibility of prevention exists. The challenge for the IC thus consists in selecting those genetic tests that are associated with promising ways of preventing (or deferring) the onset of the disease. In addition, the cost of decision-making, which may be quite important, has been neglected up to this point. For instance, consider the information, "You will fall ill with cancer with probability π at the age of x years." In this situation, one would have to review intentions regarding the number of children and accumulation of assets. Is there a sufficient return to investing in health, acquiring additional skills, and saving for financial assets? By offering counsel, the IC can contribute to lowering this cost of decision-making, inducing uninformed risk types to acquire information.

The availability of genetic information may affect the division of labor between PI and SI to the detriment of PI. Especially, life insurance with its long contract duration and health insurance with its guaranteed renewability implicitly provide cover against deterioration of risk status. Genetic information permits a more accurate prediction of this process. If accuracy should turn into practical certainty in the future, this component of the insurance contract would turn into a subsidy of high risks. Under the pressure of competition, such a cross-subsidy is not possible. In this case, monopolistic SI with uniform contributions could offer a product that cannot be offered by PI (see also the argument in Sect. 10.2.1.1).

11.2.2 Advances in Information Technology

Both the cost of producing information through research and of transmitting and processing it have been lowered by new information technology (IT). However, it is

11.2 Changes in Science and Technology

especially the second type of cost that has been decreasing rapidly. This development has consequences both for the demand and supply of insurance.

11.2.2.1 New IT and Corporate Demand for Insurance
On the demand side, new information technology concerns (1) corporate risk management of "information" as an asset and (2) risk management of information-driven processes.

(1) *Risk management of information.* Information constitutes an asset firms want to protect. Lowered cost of access to and of transfer of information implies that theft of information has become easier. This increased risk can be met by enterprises only to a limited degree by transferring it, i.e., by purchasing insurance coverage. For obtaining coverage, the value of the information would have to be known to both contractual partners. This means that the potential IB would have to disclose the information to the IC. This problem is mitigated if coverage is only partial, because this is attractive for low-risk types while also limiting moral hazard effects (see Sects. 8.3.1.2 and 8.4). For instance, while insurance against hacking attacks has become available, it provides partial coverage only, being limited to observable consequences of information theft such as the interruption of business while excluding the loss of information per se [see Mehl 1998].

(2) *Risk management of information-driven production processes.* Innovation in IT results in a faster succession of production processes. Since identification and assessment of risks require repeated observation under unchanged conditions, enterprises have reduced possibilities of assessing and reducing risks. This makes the transfer of risk more attractive. For the IC, this amounts to underwriting business with a probability of loss π that is especially difficult to assess. Therefore, it will be able to satisfy this additional demand for coverage only at a higher loading and hence price (see Sect. 7.1.3).

11.2.2.2 New IT and the Supply of Insurance
On the supply side, the IC benefits from the new possibilities of IT in at least two important ways, (1) through a lowered cost of distribution, and (2) through an improved structuring and controlling of its underwriting portfolio.

(1) *Reduction of sales expense.* The proverb "All business is local" reflects the fact that the conclusion of an insurance contract is preceded by an exchange of information that traditionally was effected best through personal contact between the IB and a direct writer or broker. This exchange can now be performed electronically, and the initial prediction was that an increasing share of contracts can be concluded through the Internet [Bernheim 1998]. In some cases, this obviates the creation of a local agency, resulting in cost savings especially in international insurance business. Additionally, peer-to-peer insurance platforms have come into existence further improving the supply of insurance. In this way, information technology enhances the globalization of international trade generally and

insurance products in particular (see Sect. 11.1 again).

The Internet also reduces the search cost (an expense) for the IB. Brown and Goolsbee (2002) utilize the model of Stahl (1989) as a theoretical justification of lower search costs due to the Internet. The authors find increased competition (and thus lower premiums) in the term life insurance market during the growth of the Internet in the mid-to-late 1990s.

(2) *Structuring and controlling of the underwriting portfolio.* Risk classification has the objective of determining the expected value of loss (and hence the fair premium) as accurately as possible. Risk classification reflects systematic differences in expected loss, which need to be distinguished from random influences. Specifically, let there be k factors $F_1, \ldots F_k$ such as the age of the driver, make of the car, and local weather determining the amount of loss L in the case of auto insurance. In addition, there is a purely random influence ε that by assumption is not correlated with these factors. As a linear approximation, one can therefore posit

$$L = a_0 + a_1 F_1 + a_2 F_2 + \ldots + a_k F_k + \epsilon. \tag{11.11}$$

This amounts to a linear regression equation with $a_0, a_1, \ldots, a_k$ denoting coefficients to be estimated. Since the expected value ε is assumed to be zero, expected loss is given by

$$EL = a_0 + a_1 F_1 + a_2 F_2 + \ldots + a_k F_k. \tag{11.12}$$

Comparison between Eqs. (11.11) and (11.12) shows that the deviation $(L - EL)$ is given by ϵ. By relating EL to the determining factors $F_1, \ldots F_k$ as precisely as possible, one can minimize the importance of these deviations $(L - EL)$ in expected value. For an IC, this constitutes a sensible goal because it can keep the safety loading in its premiums low, enhancing its competitiveness. Since the expected value of $\epsilon = L - EL$ is zero by assumption, minimizing $E(L - EL)^2 = Var(\epsilon)$ becomes the objective.

Traditionally, ICs have been content to use only a few attributes for risk classification, not least because available IT did not permit analyzing their loss experience in great detail. Today's IT enables them to filter out all relevant determinants from their loss data. Applying regression analysis, they can predict losses more accurately, permitting them to reduce their safety loading to their competitive advantage. As an illustration, let the last factor F_k be the one excluded from the analysis. However, this means that the random term ε in Eq. (11.11) is in fact replaced by a modified $\tilde{\varepsilon}$ that also contains the influence of X_k, resulting in

$$\begin{aligned} L &= a_0 + a_1 F_1 + a_2 F_2 + \ldots + a_{k-1} F_{k-1} + \tilde{\varepsilon}, \\ &\text{with } \tilde{\varepsilon} = a_k F_k + \epsilon. \end{aligned} \tag{11.13}$$

As for the variance of $\tilde{\varepsilon}$, one has

$$\begin{aligned} Var(\tilde{\varepsilon}) &= Var(\epsilon) + 2a_k Cov(\epsilon, F_k) + a_k^2 Var(F_k) \\ &> Var(\epsilon), \quad \text{since } Cov(\epsilon, F_k) = 0 \text{ by assumption.} \end{aligned} \tag{11.14}$$

Conversely, by taking into account all factors that influence the amount of loss in its pricing, the IC can bring down unexplained variance and hence its safety loading to a minimum. This is facilitated by new IT.

In addition, IT has made it easier to discover uncorrelated or even negatively correlated components of a risk portfolio, again contributing to a reduced need for costly reserves. The starting point is the efficient frontier of insurers, defined in terms of expected return on surplus and its volatility, as defined in Sect. 6.8. For efficiency, this volatility needs to be minimized for a given rate of return. Underwriting policy can contribute to this objective by keeping the variance in the change of losses $(L_t - L_{t-1})/L_t$ small. As an approximation, this is achieved by minimizing $Var(lnL_t - lnL_{t-1})$, which in turn calls for a low value of $Var(L_t)$. Now let Eq. (11.11) hold for two lines of business, giving rise to losses L_1 and L_2. From $Var(L_1 + L_2) = Var(L_1) + Var(L_2) + 2Cov(L_1, L_2)$, it becomes evident that a negative correlation between the two lines would constitute an instance of internal risk diversification, in analogy to Sect. 5.2. However, to the extent that premiums correctly reflect expected losses EL_1 and EL_2, it is the correlation between ε_1 and ε_2 that determines the amount of reserves needed to keep the probability of insolvency at a predetermined level. Through statistical inference, two density functions $f(\hat{\varepsilon}_1)$ and $f(\hat{\varepsilon}_2)$ of the estimated residuals can be constructed. The aim is to calculate the convolution of these two density functions, in analogy to the simple example of Sect. 7.1.1.2. With the help of modern IT, this can be done for several lines of underwriting business. If the variance of the convoluted density function is found to decrease, reserves can be freed for additional underwriting.

▶ **Conclusion 11.3** New information technology facilitates the theft of information, which is not fully insurable; it also shortens information-driven production processes, which increases the cost of coverage. Its use by the IC to reduce acquisition expense and safety loadings can counterbalance these effects.

11.3 Changes in Legal Norms

One of the major challenges facing insurers is a change of legal norms. One instance of this type of change is exemplified by the decision of the West Virginia Supreme Court in 1993 admitting claims from gradual environmental pollution although insurance policies contained the qualification "sudden and accidental". Another example is the decision of the European High Court in 2011 to ban premium differentiation according to gender in life and pension insurance in spite of differences in longevity. Other changes in legal norms occur at the level of public ordinances but may be just as important. They are subsumed by the generic term "insurance regulation" for simplicity.

In accordance with arguments given in Sect. 9.2 (see Conclusion 9.6), there is a market for insurance regulation. As a consequence, changes in the supply or demand for regulation again and again result in the changed intensity of regulation, creating waves of regulation and deregulation. These waves give rise to the question of whether some elements of insurance contracts are likely to be unaffected by them. They can be found in a set of principal elements of insurance contract law which are discussed first before turning to the consequences of deregulation and re-regulation.

11.3.1 Principal Elements of Insurance Contract Law

Following Rea (1993), there are at least *four dimensions* of the insurance contract that are likely to be permanent since they help to avoid negative external effects, making them efficiency-enhancing.

1. *Existence of an insurable interest.* As a rule, the IB must be the owner of the asset to be insured. To see the importance of this, let someone enter into a contract involving the property of a third party Z. In that case, the IB would have an interest in causing a loss event affecting the property of Z, giving rise to a negative external effect on Z. This external effect is internalized, however, if there is a contractual agreement between the IB and Z regarding the purchase of insurance.
2. *Prohibition of excess coverage (indemnification).* This norm is designed to avoid the extreme form of moral hazard. An IB with coverage beyond the value of the insured asset has a definitive incentive to cause a loss (see also the transition from moral hazard to insurance fraud discussed in Sect. 8.2.1). A negative externality exists as soon as the IC lacks the information to identify the IB prone to moral hazard because premiums rise generally rather than in response to individual moral hazard effects.
3. *The good faith principle.* The IB is required to make available all information necessary for calculating the premium. Since this permits predicting losses with greater precision, the safety loading can be reduced, creating a positive externality. Therefore, not only the low but also the high risks benefit, resulting in a Pareto improvement.
4. *Exclusion of intentional action.* In the event of a loss, there is a probability of the property of a third party Z being affected. Even if Z should be insured, he or she has to bear some of the cost (e.g., finding a repair service and cost of information associated with looking for replacement). Therefore, the IB who intentionally causes a loss may burden Z with an avoidable negative externality.

However, these four characteristics fail to define the insurance contract in many other dimensions, leaving a great deal of room for modifications of legal norms in the wake of deregulation and re-regulation. In particular, the extent of product and especially premium regulation is left undefined, giving rise to the expectation that there will be variation in the intensity of regulation in the future, as predicted by Conclusion 9.5 of Sect. 9.2. Experience with banking regulation especially after the subprime mortgage crisis of the years 2007–2009 indicates that a phase of re-regulation is likely to begin for insurance as well. The implication of the most recent tendency toward re-regulation cannot be fully assessed at this time in view of the many new regulatory initiatives proposed and supervisory agencies created (see Sect. 9.4.4). Therefore, the analysis below is confined to the phase of deregulation ushered in by the Uruguay Round of the World Trade Organization (WTO) concluded in 1997.

11.3.2 Consequences of Deregulation

The Uruguay Round of the World Trade Organization stipulates non-discriminatory access of foreign competitors to national insurance markets. This amounts to a reduction of regulatory intensity. The predicted consequence is a reduction of premiums for a given level of loss payments and hence the pressure on profit margins. This was indeed observed in the United States; however, the decrease in regulatory intensity was of a domestic nature in that favorable tax treatment of pensions was extended to products that are offered by non-insurance companies [see Santomero 1977].

The challenge of deregulation in the main affects the supply side of the market. The ICs need to adjust their use of the insurance technology to reduced profit margins. One important way is to benefit from economies of scope and scale, frequently through mergers and acquisitions.

- *Economies of scope:* One issue here is whether an IC that operates in several lines of underwriting can transact a given volume of business at lower acquisition and administrative expense than a competitor with just one line of business (scope effects in the narrow sense). The empirical evidence cited in Sect. 7.2.2 concerning expense ratios leads to the conclusion that economies of scope in the narrow sense may not be very marked except for the very large IC. However, there can be scope effects in a wider sense in that a multi-line IC offers a more attractive investment alternative in the capital market due to diversification effects. According to the CAPM, this would be the case if several of the betas linking line-specific rates of return to the rate of return on the capital market were particularly low or even negative. However, the empirical evidence cited in Sect. 7.3.3 fails to support this view. Moreover, one would have to know whether economies of scope are the result of mergers or growth of an existing IC (*organic growth*). Since building a new line of business requires a considerable amount of time, IC management often prefers mergers because they facilitate quick expansion, which is of special value when insurance markets integrate (e.g., in the European Union after 1992). In addition, as long as legal norms continue to differ between countries, a merger allows the conglomerate to come up with tailor-made products satisfying these norms right away.
- *Economies of scale:* Whenever the extension of existing lines of business is associated with less than proportionally increasing cost, the quest for size is an appropriate response to decreasing profit margins in underwriting activity. The evidence presented in Sects. 7.4.2 and 7.4.3 leads to the expectation that limited scale economies exist both in life and non-life insurance. Again, there is the open question of whether the path leading to large size should involve mergers or

organic growth. The balance is tipped in favor of mergers when the speed of the process is of particular advantage.

▶ **Conclusion 11.4** The quest for size as a response to deregulation of insurance markets can be interpreted as a way to reap economies of scope and scale, with the latter somewhat better supported by empirical evidence. In the choice between mergers and acquisitions and organic growth, the first alternative prevails when deregulation goes along with the integration of national markets.

11.4 Demographic Change

A broad definition of demographic change will be adopted in this section comprising not only modifications in the age and sex composition of population but also in structural characteristics such as civilian status and education. However, since the aging of population is the most salient aspect of demographic change, it is addressed first.

11.4.1 Aging of Population

At the microeconomic level of the individual, aging of population means an increase in life expectancy. Traditionally, life expectancy is measured at birth. However, since in industrial countries almost 90% of a cohort reach retirement age by now, remaining life expectancy after retirement increasingly becomes an issue. With the remaining life expectancy increasing, the livelihood of retired persons must be financed during a longer period of time. Publications by the Geneva Association propose novel answers to this challenge, in particular, the creation of a "fourth pillar" through part-time work during retirement (http://genevaassociation.org/PDF/4Pillars).

As Table 11.1 reveals, the remaining life expectancy, e.g., of a 60-year-old woman in Japan at present amounts to an estimated 28.9 years, an increase of 11.1 years since 1960. Increases of similar magnitude have occurred in all the other countries sampled. However, they need not result in more demand for insurance. Individuals could also generate more savings (which would affect the demand for banking services) or defer retirement. The comparative advantage of insurance (of the annuity type) is that it relieves individuals of the risk of "excessive" longevity. However, likely due to improved control over health status, the standard error of life expectancy in industrialized countries has decreased on average, from 19 years in 1960 to 15 years in 2005 [see Schoder and Zweifel 2011]. This reduction of uncertainty taken by itself serves to limit the surge in demand for insurance.

Yet there is another development that may enhance demand again. While a fair share of the life years gained is spent in good health, some of them call for long-term care. With potential caregivers either beyond age 60 themselves or unwilling to give up lucrative market work, costly formal care (frequently in nursing homes) is becoming increasingly prominent [for a survey of long-term care issues and solutions, see

Eisen 1997]. Again, this makes insurance (of the annuity type) attractive compared to the accumulation of savings.

Given there is an increased demand for insurance coverage, it can be met by private (PI) or social (SI) insurance. The challenge confronting PI is to offer annuities rather than paying out a lump sum when the policy expires. It thus bears a longevity risk, which however should be manageable in view of the decrease in the variance of life expectancy mentioned above. The decisive issues have to do with asymmetric information. For one, individuals who expect to live long are particularly likely to buy an annuity contract, giving rise to an adverse selection effect. In addition, an annuity may induce individuals to undertake even more effort to prolong life (see Sect. 10.5.1.3).

Counteracting these effects is very difficult for PI:

- Limiting moral hazard by checking for and sanctioning of efforts to prolong life (in analogy to the model of Sect. 6.6) is out of question because this would be considered an attack on the right to (long) life;
- Loading the premium to account for moral hazard (see Sects. 6.6 and 8.2.2.1) would affect IBs during their active lives but would have little impact on their behavior in retirement (when efforts to prolong life mostly occur);
- Any increase in copayment in health insurance (see Sect. 10.5.2) would have to be targeted at medical interventions that are (solely) designed to prolong life.

Since moral hazard effects are largely beyond control for the IC in the case of an annuity, they may accentuate genetic differences between consumers and hence their contributions to expected profit. According to the theory laid out in Sect. 6.5.2, this may cause the IC to strongly rely on risk selection in an attempt to eschew high-risk types with high longevity. However, since this "deficiency" cannot be prevented, consumers are unlikely to be interested in this type of genetic information (see Sect. 11.2.2).

In sum, increasing life expectancy in retirement arguably exacerbates problems of asymmetric information, which in turn triggers additional risk selection effort. This seems to put PI at a disadvantage compared to SI.

Yet, SI also is confronted with problems when the remaining life expectancy increases. This holds true although the comparison of rates of return between the pay-as-you go system (which is typical of SI) and the capital-based system (typical of PI) is not directly affected. Regardless of the system, "biological" returns and capital market returns must be sufficient to finance a prolonged retirement phase. However, contributions to SI are not easily adjusted to this fact because current beneficiaries may be tempted to bring about a majority decision in parliament calling for a shifting of the burden to future generations.

Failing such a decision, benefits must be cut. But this makes SI an insurance that honors its commitment with a probability of less than one. As argued above, willingness to pay for such risky insurance drops considerably compared to coverage devoid of risk. Therefore, SI may well become less attractive to citizens in the course of an aging population.

▶ **Conclusion 11.5** The increase in remaining life expectancy strengthens the demand for insurance coverage in general. However, private insurance is confronted with exacerbated problems of asymmetric information, while social insurance may be unable to honor its commitments. This makes it impossible to predict the effect on the future division of labor between private and social insurance.

11.4.2 Increasing Share of One-Person Households

Demographic change not only affects the age and gender composition of a population. Indeed, it can be interpreted as the aggregate reflection of individual decisions ranging from investment into education, marriage, number of children, time of retirement to efforts at prolonging life [see Zweifel and Eugster 2008]. When it comes to the demand for insurance, however, one of the most important choices is whether or not to take a partner.

Specifically, persons living together in a household provide each other with a degree of mutual insurance. For instance, let A and B live together, each having a 20% probability of having an accident. They are prepared to support each other financially, in this case, a commitment that fails to be honored only if A and B simultaneously have an accident. Assuming independence, the probability of this combined event is a mere 0.04 or 4% ($0.2 \cdot 0.2 = 0.04$). A household consisting of A and B therefore has less demand for accident insurance than do A and B living separately.

Now the share of persons living in a single-person household has been increasing significantly in industrialized countries. This can be inferred from Table 11.1 by noting that remaining life expectancy (mainly in retirement) differs markedly between the two sexes. Women presumably outlive men by up to 6 years while in 1960, the differential was between 3 and 4 years. This means that women must increasingly

Table 11.1 Remaining life expectancy at age 60 (1960 and 2005)

Country	Men				Women			
	1960	2005	2016[a]	Δ	1960	2005	2016[a]	Δ
China	n.a.	18.0	18.7			20.2	21.0	n.a.
France	15.6	21.5	23.8	8.2	19.5	26.3	27.9	8.4
Germany	15.4	20.7	21.8	6.4	18.1	24.4	25.3	7.2
Italy	16.7	21.4	23.2	6.5	19.3	25.7	26.6	7.3
Japan	14.8	22.1	23.7	8.9	17.8	27.7	28.9	11.1
United States	15.8	20.8	21.8	6.0	19.5	24.0	24.7	5.2
United Kingdom	15.0	20.9	23.0	8.0	18.9	23.8	25.5	6.6

[a] World Health Organization: Global Health Observatory data repository
Δ: change in expectancy between 1960 and 2016
Source Eurostat (1999, 2008)

be prepared to live alone, with the implication of an increased demand for insurance protection.[2] In some countries, insurance contracts tailored to the special needs of women have been in existence for several years (Wall Street Journal Europe, 13 Nov. 1998, 12).

In principle, private insurers have both the incentive and the capability of adjusting to changes in demand of this type. For instance, a life insurance policy could stagger benefits according to age and number of children. A surviving head of a household who has to provide for children for another 15 years faces a very different situation from another whose children have already finished their education. Calculating the net premium for a conditional benefit, however, requires knowledge of mortality rates not only according to age and gender but also according to age and number of children. It is clear that the development of such differentiated insurance products entails considerable cost in terms of information gathering and research effort.

Turning to SI, notice that its benefits may be differentiated as well, taking into account the specific circumstances of the insured. For instance, unemployment insurance in several countries conditions benefits on the number of children in the household. However, SI quite generally cannot respect individual preferences to a great degree because the choice would permit net contributors to eschew the redistribution of income that typically occurs through SI.

▶ **Conclusion 11.6** Adaptation to demographic change generally is the hallmark of private insurance. However, the development of suitably differentiated products may require considerable effort and expense.

11.5 Final Remarks

The challenges discussed in Sects. 11.1–11.4 lead to the expectation that the demand for security will tend to increase in the future. However, one has to verify that this will result in an enhanced demand for insurance. In a first step, this is done without distinguishing between PI and SI. The following arguments are of relevance here:

1. *Insurance as an efficiency-enhancing social invention.* An IB who has been paying insurance premiums for years without ever presenting a claim may have doubts concerning the contribution of insurance to efficiency. However, maximization of utility or profit is not possible in the presence of risk. The objective can at best be the maximization of expected utility and expected profit, which is necessarily associated with deviations between the realized outcome and the aimed-at optimum. For evaluation, the relevant criterion is ex ante, i.e., at the point in time when the decision in the face of risk is made. Ex ante, insurance certainly extends the set of instruments for risk management available to both households and busi-

---

[2] For same-sex households, the analysis would change with respect to mortality concerns.

nesses. Still, there are other social inventions besides insurance that can serve as well.

2. *Insurance as one of several possibilities of risk diversification.* For households, possibilities of risk diversification are usually quite limited. Within today's small families, mutual insurance involves too few individuals to let the law of large numbers become operational (see Sect. 7.1.2). Also, households as well as small businesses lack the opportunity of securitizing claims to their assets by floating them on the capital market. Transaction costs associated with the writing of terms of sale, placement on the stock exchange, and advertisement would be excessive. This means that the only intermediary remaining for risk diversification is the insurer.

By way of contrast, shareholders as owners of large corporations have ample possibilities for risk diversification. In particular, they can limit their engagement to a few shares of a single enterprise as part of their market portfolio. One remaining downside could be a strong positive correlation between the rate of return of this firm with that of the market portfolio (a high positive beta). In this event, the purchase of insurance by the enterprise can contribute to risk diversification. Still, this demand could conceivably be met by SI rather than PI. For instance, social accident insurance prevents liability claims that otherwise could be presented by employees or third parties.

Whether the increase in the future demand for insurance will benefit PI rather than SI therefore depends on the price–performance ratio of the two alternatives. However, compared to SI, private insurers have a wide array of tools to render this ratio favorable.

3. *Using insurance technology for enhancing competitiveness.* The instruments of insurance technology (from the choice of distribution system up to capital investment) can be applied to meet the challenges described in Sects. 11.1–11.4. Ultimately, the IC must be able to compete for funding in the capital market in order to survive. This means that the use of insurance technology needs to be in accordance with the interests of shareholders. This applies in particular to decisions that affect the insurer's risk of insolvency. Improved economic literacy will cause consumers to eschew an IC with a high solvency risk. Their willingness to pay for "risky insurance" is considerably lower than for "certain insurance" which honors its commitments with probability one. The reason is that "risky insurance" exposes the IB to a possible combination of two losses, one being the original loss, and the other, failure to receive the indemnity. Risk-averse individuals seek to avoid an accumulation of losses. Considerations of insolvency risk immediately raise issues concerning the role of insurance regulation because regulators are crucially concerned about insolvency. The question then arises on how insurance regulation can enhance the ability of the IC to confront the challenges discussed in Sects. 11.1–11.4. Two recommendations can be deduced from the analysis performed.

A. *Lifting the separation of insurance lines.* The separation of lines in insurance regulation emanates from the desire to prevent reserves accumulated in life insurance business from being used up in non-life business. This amounts to a cross-

11.5 Final Remarks

subsidization which is a sign of imperfect competition. Increased transparency with regard to premiums (see below) would at least mitigate this imperfection. In addition, since large, diversified ICs are often listed on stock exchanges, the capital market serves as a monitoring device. A shift of reserves with the intention of cross-subsidizing non-life business would likely be detected because it means a dilution of claims against the IC held by an IB to the advantage of shareholders. The share price of the IC would therefore have to increase, *ceteris paribus*. On the other hand, consumers could benefit considerably from a combined regulation of insurance lines. Under the current separation of lines, they face the risk of being confronted with an accumulation of copayments. Consider a liability case (with insurance governed by non-life regulation) and an illness episode (with health insurance governed by life regulation) occurring simultaneously. Given the separation of lines, the two contracts have their own cost-sharing provisions, causing the combined copayment to potentially reach an unaffordable level. With the separation of lines lifted, consumers could purchase a combined policy with a cap on total copayment. Beyond the cap, they could enjoy full coverage (unless moral hazard effects loom very large). This would serve to reduce the variability of consumers' final wealth, enhancing the contribution of PI to efficiency and welfare.

B. *Safeguarding of price competition.* The maxim of price competition is less accepted in the insurance industry than in the remainder of the economy. In risk underwriting, the scepter of ruinous competition is frequently evoked (i.e., the price falling below the fair premium level, jeopardizing the solvency of the IC). However, premium regulation induces product regulation, as shown in Sect. 9.2.1. This is true regardless of whether the regulation is imposed by a public agency or a cartel that can mete out sanctions for noncompliance. The ensuing standardization of products deprives ICs of their capability to respond to the challenges discussed above in due time. The transition of solvency regulation toward a capital-based approach (Risk-based Capital in the United States, Solvency II in the European Union) already avoids the downsides of premium and product regulation. It could be complemented by two efficiency-enhancing measures.

- *Pooling of loss data.* Small ICs do not have a sufficiently large insured population to relate the net premium to all the relevant influences in the way outlined in Sect. 11.2.2. Many combinations of characteristics occur too infrequently to permit an accurate estimation of their impact. For this reason, it is appropriate to permit ICs to create joint databases for the calculation of the net (fair) premium.
- *Prohibition of collusion with regard to the gross premium.* Even though ICs may calculate a joint net premium, this does not entail a uniform gross premium since loadings for acquisition effort, administrative expense, and risk bearing may well differ between them. Price competition can therefore be safeguarded by prohibiting collusion with regard to gross premiums.

Admittedly, preventing collusion between ICs who cooperate in the calculation of net premiums poses a formidable challenge to competition authorities. However, finding the dividing line between permitting limited cooperation between competitors which benefits the economy and eliminating collusion which burdens the economy with inefficiency should be worth the effort!

References

Bernheim, A. (1998). Challenges in insurance markets. *Geneva Papers on Risk and Insurance, 23*(89), 479–489.

Brown, J.R. & Goolsbee, A. (2002). "Does the internet make markets more competitive? Evidence from the life insurance industry." *Journal of Political Economy* 110(3): 481–507.

Dahlby, B.G. (1981). Adverse selection and pareto improvements through compulsory insurance. *Public Choice, 37*, 547–558.

Doherty, N.A., & Posey, L.L. (1998). On the value of a checkup: adverse selection, moral hazard and the value of information. *Journal of Risk and Insurance, 65*(2), 189–211.

Eisen, R. (1997). Long-term care systems in Europe – results of an international comparison. In Institute of Actuaries of Australia (Ed.) *Financing of Long Term Health and Community Care*, Seminar Proceedings, Sidney, 1–8.

Eisen, R. (2006). Adverse selection in the health insurance market after genetic tests. In P.A. Chiappori, & C. Gollier (Eds.), *Competitive Failures in Insurance Markets. Theory and Policy Implications* (pp. 34–54). Cambridge MA: MIT Press.

Eurostat database, European Commission. Accessed 1999.

Eurostat database, European Commission. Accessed 2008.

Mehl, C. (1998). Insurability of risks on the information highway. *Geneva Papers on Risk and Insurance, 23*(89), 103–111.

Rea, S.A. (1993). The economics of insurance law. *International Review of Law and Economics, 13*, 145–162.

Santomero, A.M. (1977). Insurers in a changing competitive financial structure. *Journal of Risk and Insurance, 64*(4), 727–732.

Schoder, J., & Zweifel, P. (2011). Flat-of-the-curve medicine: a new perpective on the production of health. *Health Economics Review* 1(2), https://doi.org/10.1186/2191-1-2.

Singer, E. (1991). Public attitudes toward genetic testing. *Population Research and Policy Review, 10*, 235–255.

Stahl, D.O. (1989). Oligopolistic Pricing with sequential consumer search. *American Economic Review, 79*, 700–712.

Van Den Berghe, L. (1998). Shaping the future of the insurance sector. *Geneva Papers on Risk and Insurance, 23*(89), 506–518.

Wildasin, D.E. (1991). Income redistribution in a common labor market. *American Economic Review, 81*, 757–774.

Zweifel, P. (2013). The division of labor between private and social insurance. In G. Dionne (Ed.), *Handbook of Insurance* (pp. 1097–1118). Boston: Kluwer.

Zweifel, P., & Eugster, P. (2008). Life-cycle effects of social security in an open economy: a theoretical and empirical survey. *Zeitschrift für die gesamte Versicherungswissenschaft (German Journal of Risk and Insurance), 97*(1), 61–77.

References

Aaron, H. (1966). The social insurance paradox. *Canadian Journal of Economics and Political Science, 32*, 371–374.

Abdellaoui, M. (2000). Parameter-free elicitation of utility and probability weighting functions. *Management Science, 46*(11), 1497–1512.

Abraham, K.S. (1995). *Insurance Law and Regulation: Cases and Materials*, 2nd ed. Westbury NY: The Foundation Press.

ADAC (2009) (German Automobile Club, 2009). Versicherungsunfälle in Deutschland (http://www.adac.de/verkerhr/statistiken/unfalldaten, visited 5 Jan. 2009).

Adams, M.B., & Tower, G.D. (1994). Theories of regulation: some reflections on the statutory supervision of insurance companies in Anglo-American countries. *Geneva Papers on Risk and Insurance Issues and Practice, 71*, 156–177.

Aldy, L.E., & Viscusi, W.K. (2008). Adjusting the value of a statistical life for age and cohort effects. *Review of Economics and Statistics, 90*, 573–581.

Allais, M. (1953). Le Comportement de l'homme rational devant le risque: Critique des postulats et axiomes de l'école américaine. *Econometrica, 21*, 503–546.

Anderson, D.A. (1994). Compensating wage differentials and the optimal provision of unemployment insurance. *Southern Economic Journal, 60*(3), 644–656.

Arnott, R., & Stiglitz, J.E. (1990). The welfare economics of moral hazard. In H. Loubergé, (Ed.), *Risk, information and insurance* (pp. 91–122). Boston, Kluwer.

Arrow, K.J. (1974). Optimal insurance and generalized deductibles. *Scandinavian Actuarial Journal, 57*, 1–42.

Arrow, K.J. (1965). *Aspects of the Theory of Risk-Bearing*, Yrjö Johnsson Lectures, Säätiö, Helsinki Y.J.

Arrow, K.J. (1963). Uncertainty and the welfare economics of medical care. *American Economic Review, 53*, 941–973.

Arrow, K.J. (1951). Alternative approaches to the theory of choice in risk-taking situations. *Econometrica, 19*, 404–437; In K.J. Arrow (Ed.), *Essays in the Theory of Risk Bearing* (pp. 1–43). Amsterdam: North-Holland.

Association of German Insurers GDV (1997). *Jahrbuch 1997 (Yearbook 1997)*. Karlsruhe: Verlag Versicherungswirtschaft.

Association of German Insurers GDV (1996). *Wettbewerbsfaktoren der Kompositversicherungsunternehmen in Deutschland (Determinants of competitiveness of multiline insurers in Germany)*. Karlsruhe: Verlag für Versicherungswirtschaft.

Atkinson, S.E., & Halvorsen, R. (1990). The valuation of risks to life: evidence from the market for automobiles. *The Review of Economics and Statistics, 72*(1), 133–136.

Atkinson, A.B., & Micklewright, J. (1991). Unemployment compensation and labor market transitions: a critical review. *Journal of Economic Literature, 29*, 1679–1727.

Bakx, P., de Meijer, C., Schut, F., & van Doorslaer, E. (2014). Going formal or information, who cares? The influence of public long-term care insurance. *Health Economics, 24*: 631–643.

Baltensperger, E., Buomberger, P., Iuppa, A., Wicki, A., & Keller, B. (2008). Regulation and intervention in the insurance industry – fundamental issues. *The Geneva Reports, Risk and Insurance Research* 1.

Barberis, N.C. (2012). Thirty years of prospect theory in economics: A review and assessment. *NBER Working Paper 18621*.

Barrese, J., & Nelson, J.M. (1992). Independent and exclusive agency insurers: a reexamination of the cost differential. *Journal of Risk and Insurance, 59*(3), 375–397.

Barsky, R.B., Juster, T.F., Kimball, M.S., & Shapiro, M.D. (1997). Preference parameters and behavioral heterogeneity: an experimental approach in the health and retirement study. *Quarterly Journal of Economics, 112*(2), 537–579.

Basel Committee on Banking Supervision (2019). Basel framework: Scope and definitions. Basel: Bank for International Settlements.

Baumol, H.J., Panzar, J.C., & Willig, R.D. (1982). *Contestable Markets and the Theory of Industry Structure*. New York: Harcourt Brace Jovanowitsch.

Beard, R.E., Pentikäinen, T., & Pesonen, E. (1984). *Risk Theory. The Stochastic Basis of Insurance*. London: Chapman and Hall.

Becker, S.W., & Brownson, F.O. (1964). What price ambiguity? Or the role of ambiguity in decision-making. *Journal of Political Economy 72*, 62–73.

Bell, D.E. (1982). Regret in decision making under uncertainty. *Operations Research, 30*(5), 961–981.

Ben Ammar, S., Braun, A. & Eling, M. (2015). Alternative risk transfer and insurance-linked securities: Trends, challenges, and new market opportunities. University of St. Gallen: Institute of Insurance Economics.

Berger, L.A., Cummins, J.D., & Tennyson, S. (1992). Reinsurance and the liability insurance crisis. *Journal of Risk and Uncertainty, 5*, 253–272.

Bernasconi, M., & Loomes, G. (1992). Failures of the reduction principle in an Ellsberg-type problem. *Theory and Decision 32*, 77–100.

Bernheim, A. (1998). Challenges in insurance markets. *Geneva Papers on Risk and Insurance, 23*(89), 479–489.

Bernoulli, D. (1730/31). Specimen theoriae novae de mensura sortis. *Comentarii Academiae Petropolis, Petersburg, 5,* 175–192; English translation (Exposition of a new theory on the measurement of risk), *Econometrica, 22*(1), 23–46.

Black, F. (1972). Capital market equilibrium with restricted borrowing. *Journal of Business, 45,* 444–455.

Black, F., & Scholes, M. (1973). The pricing of options and corporate liabilities. *Journal of Political Economy, 81,* 637–659.

Blake, D. (1996). Efficiency, risk aversion and portfolio insurance: an analysis of financial asset portfolios held by investors in the United Kingdom. *Economic Journal, 106*(438), 1175–1192.

Blanchard, O., & Perotti, R. (2002). An empirical characterization of the dynamic effects of changes in government spending and taxes on output. *Quarterly Journal of Economics, 117*(4): 1329–1368.

Blomquist, G. (1979). Value of life saving: implications of consumption activity. *Journal of Political Economy, 87,* 540–558.

Blundell, R. (1992). Labour supply and taxation: a survey. *Fiscal Studies, 13,* 15–40.

Bonato, D., & Zweifel, P. (2002). Information about multiple risks: the case of building and content insurance. *Journal of Risk and Insurance, 69*(4), 469–487.

Borch, K.H. (1990). The price of moral hazard. *Scandinavian Actuarial Journal,* 173–176 (1980); *reprinted.* In K.H. Borch (Ed.), *Economics of Insurance* (pp. 346–362). Amsterdam: North-Holland.

Borch, K.H. (1962). Equilibrium in a reinsurance market. *Econometrica,* 30, 424–444.

Borch, K.H. (1961). Some elements of a theory of reinsurance. *Journal of Insurance,* 28(3), 35–43.

Borch, K.H. (1960a). An attempt to determine the optimum amount of stop loss reinsurance, in XVIth International Congress of Actuaries, Brussels 1960, Vol. 1, 597–610.

Borch, K.H. (1960b). The safety loading of reinsurance premiums. *Scandinavian Actuarial Journal,* 3–4, 163–184.

Borch, K.H. (1960c). Reciprocal reinsurance treaties seen as a two-person cooperative game. *Scandinavian Actuarial Journal,* 1–2, 29–58.

Börsch-Supan, A. (1992). Population aging, social security design, and early retirement. *Journal of Institutional and Theoretical Economics/Zeitschrift für die gesamte Staatswissenschaft, 148*(4), 533–557.

Bove, C., Efthyvoulou, G., & Navas, A. (2017). "Political cycles in public expenditure: butter vs. guns." *Journal of Comparative Economics* 45(3), 528–604.

Breyer, F. (1990). *Ökonomische Theorie der Alterssicherung (Economic Theory of Provision for Old Age).* München: Vahlen.

Brigo, D., Pallavicini, A. & Torresetti, R. (2010). *Credit Models and the Crisis: A Journey into CDOs, Copulas, Correlations and Dynamic Models.* New York: Wiley.

Brown, J.R. & Finkelstein, A. (2008). "The interaction of public and private insurance: Medicaid and the long-term care insurance market." *American Economic Review* 98(3): 1083–1102.

Brown, J.R. & Goolsbee, A. (2002). "Does the internet make markets more competitive? Evidence from the life insurance industry." *Journal of Political Economy* 110(3): 481–507.

Bruner, D.M. (2007). Risk Aversion and stochastic dominance. *Working Paper*, Calgary: University of Calgary.

Buchanan, J.M., & Tullock, G. (1976). Polluters' profits and political response: direct control versus taxes: reply. *American Economic Review, 66*(5), 983–984.

Buchholz, W., & Wiegard, W. (1992). Allokative Überlegungen zur Reform der Pflegevorsorge (Allocative considerations concerning long-term care provision). *Jahrbücher für Nationalökonomie und Statistik, 209*, 441–457.

Buckley, A., Ross, S.T., Westerfield, R., & Jaffe, J. (1998). *Corporate Finance Europe*. London: McGraw-Hill.

Burda, M.C., & Wyplosz, C. (1994). Gross labor market flows in Europe. *European Economic Review, 38*, 1287–1915.

Burkhauser, R.V., & Turner, J.A. (1978). A time-series analysis on social security and its effect of the market work of men at younger ages. *Journal of Political Economy, 86*(4), 704–715.

Burtless, G. (1986). Social security, unanticipated benefit increases, and the timing of retirement. *Review of Economic Studies, 53*, 781–805.

Bühlmann, H. (1970). *Mathematical Methods in Risk Theory*. Heidelberg: Springer.

Bühlmann, H., & Gisler, A. (2005). *A Course in Credibility Theory and its Applications*. New York: Springer.

Calford, E.M. (2017). Uncertainty aversion in game theory: Experimental evidence. *Purdue University Economics Department Working Paper No. 1291*.

Camerer, C., & Karjalainen, R. (1994). Ambiguity-aversion and non-additive beliefs in non-cooperative games: Experimental evidence. In B. Munier and M.J. Machina (eds.), *Models and Experiments in Risk and Rationality*, Springer Netherlands, 325–358.

Camerer, C., & Weber, M. (1992). Recent developments in modeling preferences: Uncertainty and ambiguity. *Journal of Risk and Uncertainty 5,* 325–370.

Carlin, P.S., & Sandy, R. (1991). Estimating the implicit value of a young child's life. *Southern Economic Journal, 58*(1), 79–105.

Carlton, D.W., & Perloff, J.M. (1999). *Modern Industrial Organization*, 3rd ed. Reading MA: Addison-Wesley.

Carnap, R. (1950). *Logical Foundations of Probability*. Chicago: Chicago University Press.

Cawley, J., & Philipson, T.J. (1999). An empirical examination of information barriers to trade in insurance. *American Economic Review, 89*(4), 827–846.

Chen, H., Cummins, J.D., Sun, T., & Weiss, M.A. (2020). Property-casualty insurers: Microstructure, insolvency risk, and contagion. *Journal of Risk and Insurance.* 87(2), 253–284.

Chiappori, P.A., Jullien, B., Salanié, B., & Salanié, F. (2006). Asymmetric information in insurance: general testable implications. *RAND Journal of Economics, 37*(4), 738–789.

Chipman, J.S. (1960). Stochastic choice and subjective probability. In D. Willner (ed.), *Decisions, Values and Groups*. Volume I. Oxford, England: Pergamon Press, 70–95.

Christiansen-Szalanski, J. & Bushyhead, J.B. (1981). Physicians' use of probabilistic information in a real clinical setting. *Journal of Experimental Psychology: Human Perception and Performance* (7), 928–935.

Cohen, M., & Jaffray, J-Y., & Said, T. (1985). Individual behavior uncertainty: an experimental study. *Theory and Decision 18*, 203–228.

Conte, A., & Hey, J.D. (2013). Assessing multiple prior models of behavior under ambiguity. *Journal of Risk and Uncertainty, 46*(2), 113–137.

Cook, P.J., & Graham, D.A. (1977). The demand for insurance and protection: the case of irreplaceable commodities. *Quarterly Journal of Economics, 91*(1), 143–156.

Copeland, T.E., Weston, J.F. & Shastri, K. (2004). *Financial Theory and Corporate Policy*, 4th ed. Reading MA: Pearson Addison Wesley.

Crain, M. (1979). Cost and output in the legislative firm. *Journal of Legal Studies, 8*(3), 607–621.

Culyer, A.J. (1980). *The Political Economy of Social Policy*. Oxford: Martin Robertson.

Cummins, J.D. (1991a). Statistical and financial models of insurance pricing and the insurance firm. *Journal of Risk and Insurance, 85*(2), 261–302.

Cummins, J.D. (1991b). Capital structure and fair profits in property-liability insurance. In J.D. Cummins, & R.A. Derrig, (Eds.), *Managing the Insolvency Risk of Insurance Companies*, 295–308. Dordrecht: Kluwer.

Cummins, J.D., & Harrington, S.E. (1985). Property-liability insurance rate regulation: estimation of underwriting betas using quarterly profit data. *Journal of Risk and Insurance, 52*, 18–43.

Cummins, J.D., & Phillips, R.D. (2005). Estimating the cost of equity capital for property-liability insurers. *Journal of Risk and Insurance, 72*, 441–478.

Cummins, J.D., & Sommer, D.W. (1996). Capital and risk in property-liability insurance markets. *Journal of Banking and Finance, 20*, 1069–1092.

Cummins, J.D., & Van Derhei, J. (1979). A note on the relative efficiency of property-liability distribution systems. *Bell Journal of Economics and Management Science, 10*, 709–719.

Cummins, J.D., & Weiss, M.A. (2014). Systemic risk and the U.S. insurance sector. *Journal of Risk and Insurance*, 81(3), 489–528.

Cummins, J.D., & Weiss, M.A. (2009). Convergence of insurance and financial markets: hybrid and securitized risk-transfer solutions. *Journal of Risk and Insurance, 76*(3), 493–545.

Cummins, J.D., & Weiss, M.A. (1993). Measuring cost efficiency in the property-liability insurance industry. *Journal of Banking and Finance, 17*(2–3), 463–481.

Cummins, J.D., Weiss, M.A., Xiaoying, X., & Hongmin, Z. (2010). Economies of scope in financial services: a DEA efficiency analysis of the U.S. insurance industry. *Journal of Banking & Finance, 34*, 1525–1539.

Cummins, J.D., & Zi, H. (1998). Comparison of frontier efficiency methods: an application to the US life insurance industry. *Journal of Productivity Analysis, 10*(2), 131–152.

Curley, S.P., & Yates, J.F. (1989). An empirical evaluation of descriptive models of ambiguity reactions in choice situations. *Journal of Mathematical Psychology 33*, 397–427.

Curley, S.P., & Yates, J.F. (1985). The center and range of the probability affecting ambiguity preferences. *Organizational Behavior and Human Decision Processes 36*, 272–287.

Curley, S.P., & Yates, J.F., & Abrams, R.A. (1986). Psychological sources of ambiguity avoidance. *Organizational Behavior and Human Decision Processes 38*, 230–256.

Cutler, D.M., & Reber, S.J. (1998). Paying for health insurance: the trade-off between competition and adverse selection. *Quarterly Journal of Economics, 113*, 433–466.

Dahlby, B.G. (1981). Adverse selection and pareto improvements through compulsory insurance. *Public Choice, 37*, 547–558.

Dardis, R. (1980). Economic analysis of current issues in consumer product safety: fabric flammability. *Journal of Consumer Affairs, 14*(1), 109–123.

Davidson, W.N.III, Cross, M.L., & Thornton, J.H. (1992). Corporate demand for insurance: some empirical and theoretical results. *Journal of Financial Services Research, 6*, 61–72.

Daykin, C.D., Pentikäinen, T., & Pesonen, M. (1994). *Practical Risk Theory for Actuaries*. London: Chapman & Hall.

De Alessi, L. (1987). Why corporations insure. *Economic Inquiry, XXV*, 429–438.

DiMasi, J.A., Hansen, R.W., & Grabowski, H.G. (2003). The price of innovation: new estimates of drug development costs. *Journal of Health Econonomics, 22*(2), 151–185.

Dionne, G., & Doherty, N.A. (1994). Adverse selection, commitment, and renegotiation: extension and evidence from insurance markets. *Journal of Political Economy, 102*(2), 209–235.

Dionne, G., & Lasserre, P. (1985). Adverse selection, repeated insurance contracts and announcement strategy. *Review of Economic Studies, 50*(7), 719–723.

Dionne, G., & Michaud, P-C. & Dahchour, M. (2013). Separating moral hazard from adverse selection and learning in automobile insurance: longitudinal evidence from France. *Journal of the European Economic Association, 11*(4), 897–917.

Dionne, G., & St-Michel, P. (1991). Workers' compensation and moral hazard. *Review of Economics and Statistics, LXXXIII*(2), 236–244.

Doherty, N.A. (1997). Innovations in managing catastrophic risks. *Journal of Risk and Insurance, 64*(4), 713–718.

Doherty, N.A. (1985). *Corporate Risk Management, a Financial Exposition.* New York: McGraw-Hill.

Doherty, N.A. (1981). The measurement of output and economies of scale in property-liability insurance. *Journal of Risk and Insurance, 48*(3), 391–402.

Doherty, N.A., & Garven, J.R. (1986). Price regulation in property-liability insurance: a contingent claims approach. *Journal of Finance, 41*, 1031–1050.

Doherty, N.A., & Posey, L.L. (1998). On the value of a checkup: adverse selection, moral hazard and the value of information. *Journal of Risk and Insurance, 65*(2), 189–211.

Doherty, N.A., & Schlesinger, H. (1983). Optimal insurance in incomplete markets. *Journal of Political Economy, 91*, 1045–1054.

Donni, O., & Fecher, F. (1997). Efficiency and productivity of the insurance industry in the OECD countries. *Geneva Papers on Risk and Insurance, 22*, 523–535.

Dowd, K. (1998). *Beyond Value at Risk. The New Science of Risk Management.* New York: Wiley.

Downs, A. (1957). *An Economic Theory of Democracy.* New York: Harper & Row.

Dreyfuss, M.K., & Viscusi, W.K. (1995). Rates for time preference and consumer valuations of automobile safety and fuel efficiency. *Journal of Law & Economics, 38*(1), 79–105.

Dumm, R.E., Eckles, D.L., Nyce, C., & Volkman-Wise, J. (2020). The representative heuristic and catastrophe-related risk behaviors. *Journal of Risk and Uncertainty, 60, 2*, 157–185.

Dumm, R.E., Eckles, D.L., Nyce, C., & Volkman-Wise, J. (2017). Demand for windstorm insurance coverage and the representative heuristic. *Geneva Risk and Insurance Review, 42*, 117–139.

Eeckhoudt, L., Gollier, C., & Schlesinger, H. (2005). *Economic and Financial Decisions under Risk.* Princeton: Princeton University Press.

Eeckhoudt, L., Meyer, J., & Ormiston, M.B. (1997). The interaction between the demand for insurance and insurable assets. *Journal of Risk and Uncertainty, 14*, 25–39.

Eeckhoudt, L., & Schlesinger, H. (2006). Putting risk in its proper place. *American Economic Review, 96*(1), 280–289.

Ebner, M. & Neumann, T. (2005). Time-varying betas of German stock market returns. *Financial Markets and Portfolio Management, 19* (1), 29–46.

Eckles, D.L. & Hilliard, J.I. (2015). Government intervention through an implicit federal backstop: is there a link to market power? *The Geneva Papers on Risk and Insurance–Issues and Practice, 40*, 538–555.

Eckles, D.L., Hoyt, R.E., & Miller, S.M. (2014). The impact of enterprise risk management on the marginal cost of reducing risk: Evidence from the insurance industry. *Journal of Banking and Finance, 43*, 247–261.

Ehrlich, J., & Becker, G.S. (1972). Market insurance, self-insurance and self-protection. *Journal of Political Economy, 80*, 623–648.

Eichberger, J., Kelsey, D., & Schipper, B.C. (2008). Granny versus game theorist: Ambiguity in experimental games. *Theory and Decision, 64*, 333–362.

Einav, L., Finkelstein, A., & Cullen, M.R. (2010). Estimating welfare in insurance markets using variation in prices. *The Quarterly Journal of Economics, 125* (3), 877–921.

Einhorn, H.J., & Hogarth, R.M. (1986). Decision making under ambiguity. *Journal of Business 59*, S225–S250.

Einhorn, H.J., & Hogarth. R.M. (1985). Ambiguity and uncertainty in probabilistic inference. *Psychology Review 92*, 433–461.

Eisen, R. (2006). Adverse selection in the health insurance market after genetic tests. In P.A. Chiappori, & C. Gollier (Eds.), *Competitive Failures in Insurance Markets. Theory and Policy Implications* (pp. 34–54). Cambridge MA: MIT Press.

Eisen, R. (1997). Long-term care systems in Europe – results of an international comparison. In Institute of Actuaries of Australia (Ed.) *Financing of Long Term Health and Community Care*, Seminar Proceedings, Sidney, 1–8.

Eisen, R. (1991). Market size and concentration: insurance and the European market 1992. *Geneva Papers on Risk and Insurance Issues and Practice, 60*, 263–281.

Eisen, R. (1990). Problems of equilibria in insurance markets with asymmetric information. In H. Loubergé (Ed.), *Risk, Information and Insurance* (pp. 123–141). Boston: Kluwer.

Eisen, R. (1989). Regulierung und Deregulierung in der Deutschen Versicherungswirtschaft (Regulation and deregulation in the German insurance industry). *Zeitschrift für die gesamte Versicherungswirtschaft, 78*(2), 157–175.

Eisen, R. (1982). Competition and regulation in insurance – informational equilibria and rate regulation. *Discussion Paper* IIM/IP 82-3, International Institute of Management, Berlin.

Eisen, R. (1979a). Equilibrium in risk bearing: The principle of equivalence– different implications of alternative interpretations. *Geneva Papers*, No. 11 (January), 14–33.

Eisen, R. (1979b). *Theorie des Versicherungsgleichgewichts (Theory of Equilibrium on Insurance Markets)*. Berlin: Duncker & Humblot.

Eisen, R., Müller, W., & Zweifel, P. (1993). Entrepreneurial insurance: a new paradigm for deregulated markets. *Geneva Papers on Risk and Insurance, 66*, 3–56.

Eisen, R. & Nell, M. (1994). Die Wirkung von Versicherungsschutz auf Drittmärkte: Das externe moralische Risiko, in: Hesberg, D. et al. (Hrsg.): Risiko, Versicherung, Markt: Festschrift für Walter Karten zur Vollendung des 60. Lebensjahres, pp. 221–241, Karlsruhe: Verlag Versicherungswirtschaft.

Eisen, R. & Sloan, F.A. (1996). *Long-Term Care: Economic Issues and Policy Solutions*. Boston MA: Kluwer.

Eisen, R., & Zweifel, P. (1997). Überlegungen zur optimalen Kombination unterschiedlicher Versicherungsprodukte über die Zeit (On the optimal combination of different insurance products over time). In L. Männer (Ed.), *Langfristige Ver-

sicherungsverhältnisse. Ökonomie, Technik, Institutionen (pp. 355–378). Karlsruhe: Verlag Versicherungswirtschaft.

Ellsberg, D. (1961). Risk, ambiguity, and the Savage axioms. *Quarterly Journal of Economics, 75*, 643–669.

Embrechts, P., Lindskog, F., & McNeil, A. (2001). Modelling dependence with copulas and applications to risk management. In S. Rachev (Ed.), *Handbook of Heavy Tailed Distributions in Finance* (pp. 329–384). Dordrecht: Elsevier.

European Commission (2010). *White Paper on Insurance Guarantee Schemes.*

Evans, D. (2012). Risk intelligence. In S. Roeser et al. (eds.), *Handbook of Risk Theory*, Vol. 2, 604–620.

Evans, M.F., & Schaur, G. (2010). A quantile estimation approach to identify income and age variation in the value of statistical life. *Journal of Environmental Economics and Management, 59*, 260–270.

Ezra, D. (1991). Asset allocation by surplus optimization. *Financial Analysts Journal* Jan./Feb.: 51–57.

Fama, E., & French, K.R. (2015). A five-factor asset pricing model. *Journal of Financial Economics, 116*(1), 1–22.

Fama, E., & French, K.R. (1992). The cross-section of expected stock returns. *Journal of Finance, 47*(2), 427–465.

Fama, E., & Jensen, M.C. (1983). Separation of ownership and control. *Journal of Law and Economics, 26*, 301–325.

Fecher, E., Perelman, S.D., & Pestieau, P. (1991). Scale economies and performance in the French insurance industry. *Geneva Papers on Risk and Insurance Issues and Practice, 60*, 315–326.

Federal Drug Administration (2006). *Guidance for Industry. Q9 Quality Risk Management.* Rockville MD.

Felderer, B. (1992). Does a public pension system reduce savings rates and birth rates? *Journal of Institutional and Theoretical Economics/Zeitschrift für die Gesamte Staatswissenschaft, 148*, 312–325.

Feldstein, M.S. (1974). Social security, induced retirement, and aggregate capital accumulation. *Journal of Political Economy, 82*(5), 906–926.

Fellner, W. (1961). Distortion of subjective probabilities as a reaction to uncertainty. *The Quarterly Journal of Economics, 75*(4), 670–689.

Finsinger, J. (1983a). *Versicherungsmärkte (Insurance Markets).* Frankfurt: Campus Verlag.

Finsinger, J. (1983b). The performance of property line insurance firms under the German regulatory system. *Journal of Theoretical and Institutional Economics/Zeitschrift für die gesamte Staatswissenschaft, 139*, 473–489.

Finsinger, J., Hammond, E., & Tapp, J. (1985). *Insurance: Competition or Regulation?*, Reprint No. 19. London: Institute for Fiscal Studies.

Finsinger, J., & Schmidt, F.A. (1994). Prices, distribution channels, and regulatory intervention in European insurance markets. *Geneva Papers on Risk and Insurance Issues and Practice, 70*, 22–36.

Frech, H.E. III, & Samprone, J.C. (1980). The welfare loss of excess nonprice competition: the case of property-liability insurance regulation. *Journal of Law and Economics, XXI*, 429–440.

Frech, H.E. III, & Smith, M.P. (2015). Anatomy of a slow-motion health insurance death spiral. *North American Actuarial Journal, 19*, 60–72.

Frech, H.E. III, & Zweifel, P. (2017). Market Socialism and community rating in health insurance. *Comparative Economic Studies, 59*, 3, 405–427.

Friend, I., & Blume, M.E. (1975). The demand for risky assets. *American Economic Review, 65*(5), 900–922.

Froot, K.A. (1999). Introduction. In K.A. Froot (Ed.), *The Financing of Catastrophe Risk* (pp. 1–22). Chicago: University of Chicago Press.

Garbacz, C. (1989). Smoke detector effectiveness and the value of saving a life. *Economic Letters, 31*(3), 281–286.

Garven, J.R., & D'Arcy, S. (1991). A synthesis of property-liability insurance pricing techniques. In J.D. Cummins, & R.A. Derrig (Eds.), *Managing the Insolvency Risk of Insurance Companies*, Chap. 8. Boston: Kluwer.

Garven, J.R., & Lamm-Tennant, J. (2003). The demand for reinsurance: theory and empirical tests. *Insurance and Risk Management, 7*(3), 217–237.

Gayor, T., Hamilton, J.T., & Viscusi, W.K. (2000). Private values of risk tradeoffs at superfund sites: housing market evidence on learning about risk. *Review of Economics and Statistics, 82*(3), 439–451.

Giarini, O., & Stahel, W. (2000). *Die Performance-Gesellschaft (The Performance Society)*. Marburg: Metropolis Verlag.

Gigliotti, G., & Sopher, B. (1990). The testing principle: a resolution of the Ellsberg paradox, working paper, Department of Economics, Rutgers University.

Gilboa, I., & Schmeidler, D. (1989). Maxmin expected utility with non-unique prior. *Journal of Mathematical Economics, 18*, 141–153.

Goldsmith, R.W., & Sahlin, N-E. (1983). The role of second-order probabilities in decision making. In. P. Humphreys, O. Svenson, and A. Vari (eds.), *Analysing and Aiding Decision Processes*. Amsterdam: North-Holland, 455–467.

Gonzalez, R., & Wu, G. (1999). On the shape of the probability weighting function. *Cognitive Psychology, 38*(1), 129–166.

Goavaerts, M.J., & Laeven, J.A. (2008). Actuarial risk measures for financial derivative pricing. *Insurance: Mathematics and Economics, 42*(2), 540–547.

Grabowski, H.G., Viscusi, W.K., & Evans, W.N. (1989). Price and availability tradeoffs of automobile insurance regulation. *Journal of Risk and Insurance, 56*, 275–299.

Grace, M.F. & Timme, S.G. (1992). An examination of cost economies in the United States life insurance industry. *Journal of Risk and Insurance, 59*, 72–103.

Greene, W.H. (1997). *Econometric Analysis*, 5th ed. Upper Saddle River NY: Prentice Hall.

Greenwald, B.C., & Stiglitz, J.E. (1990). Asymmetric information and the new theory of the firm. *American Economic Review Papers and Proceedings, 80*(2), 160–165.

Grossman, S.J. (1976). On the efficiency of competitive stock markets when traders have diverse information. *Journal of Finance, 31*, 573–585.

Grossman, S.J., & Hart, O.D. (1986). The costs and benefits of ownership: theory of vertical and lateral integration. *Journal of Political Economy, 94*, 691–719.

Gruber, J. & Yelowitz, A.S. (1999). Public health insurance and private savings. *Journal of Political Economy, 107*(2), 1249–1274.

Gurley, J.G., & Shaw, E.S. (1960). *Money in a Theory of Finance*. Washington DC: The Brookings Institution.

Hackmann, M.B. & Pohl, R. (2018). Patient vs. provider incentives in long-term care. *Working Paper*, Loss Angeles: UCLA Dept. of Economics.

Hadar, J., & Russell, W. (1969). Rules for ordering uncertain prospects. *American Economic Review, 59*, 25–34.

Halek, M., & Eisenhauer, G.M. (2001). Demography of risk aversion. *Journal of Risk and Insurance, 68*(1), 1–24.

Hardwick, P. (1997). Measuring cost inefficiency in the UK life insurance industry. *Applied Financial Economics, 7*(1), 37–44.

Harrington, S.E. (1987). A note on the impact of auto insurance rate regulations. *Review of Economics and Statistics, 69*, 166–170.

Hax, K. (1964). *Grundlagen des Versicherungswesens* (Basics of Insurance), Wiesbaden: Gabler.

Heath, C., & Tversky, A. (1991). Preference and belief: ambiguity and competence in choice under uncertainty. *Journal of Risk and Uncertainty 4*, 5–28.

Heilmann, W.R. (1989). Decision theoretic foundations of credibility theory. *Insurance: Mathematics and Economics, 8*(1), 77–95.

Heilmann, W.R. (1988). *Fundamentals of Risk Theory*. Karlsruhe: Verlag für Versicherungswissenschaft.

Hempel, C.G. (1965). *Aspects of Scientific Explanation, and Other Essays in the Philosophy of Science*. New York: Free Press.

Hersch, J., & Viscusi, W.K. (2010). Immigrant status and the value of statistical life. *Journal of Human Resources, 45*, 749–771.

Hey, J.D., & Orme, C. (1994). Investigating generalizations of expected utility theory using experimental data. *Econometrica, 62*(6), 1291–1326.

Hirao, Y., & Inoue, T. (2004). On the cost structure of the Japanese property-casualty insurance industry. *Journal of Risk and Insurance, 71*(3), 501–530.

Hirshleifer, J. (1965/6). Investment decisions under uncertainty. *Quarterly Journal of Economics, 79*, 509–536 and *Quarterly Journal of Economics, 80*, 252–277.

Hirshleifer, J., Glazer, A., & Hirshleifer, D. (2005). *Price Theory and Applications*. Cambridge University Press: Cambridge Books.

Hoevenaars, R.P., et al. (2008). Strategic asset allocation with liabilities: beyond stocks and bonds. *Journal of Economic Dynamics & Control, 32*, 2939–2970.

Hogarth, R.M., & Einhorn, H.J. (1990). Venture theory: a model of decision weights. *Management Science 36*, 780–803.

Holmström, B. (1979). Moral hazard and observability. *Bell Journal of Economics, 10*(1), 74–91.

Houser, A. (2007). Women and long-term care. *Fact Sheet* 77R, Washington DC: AARP Public Policy Institute.

Houston, D., & Simon, R. (1970). Economies of scale in financial institutions: a study in life insurance. *Econometrica, 38*, 856–864.

Hujer, R., & Schneider, H. (1989). The analysis of labor market mobility using panel data. *European Economic Review, 33*(2/3), 530–536.

Hull, J. (2018). *Options, Futures, and Other Derivative Securities*, 10th ed. New York NY: Pearson.

Human Mortality Database (2010). http://www.mortality.org/cgi-bin/hmd/country.php?cntr=USA&level=1.

International Association of Insurance Supervisors (IAIS) (2010). Macroprudential surveillance and (re)insurance, *Global Reinsurance Market Report*, Mid-year ed.

Intriligator, M.D. (1971). *Mathematical Optimization and Economic Theory*, Englewood Cliffs, New Jersey: Prentice-Hall.

Ippolito, P.M., & Ippolito, R.A. (1984). Measuring the value of life saving from consumer reactions to new information. *Journal of Public Economics, 25*(1–2), 53–81.

Jackwerth, J.C. (2000), Recovering risk aversion from option prices and realized returns. *Review of Financial Studies* 13(2), 433–451.

Jenkins, R.R., Owens, N., & Wiggins, L.B. (2001). Valuing reduced risks to children: the case of bicycle safety helmets. *Contemporary Economic Policy, 19*(4), 3978–408.

Jensen, M.C., & Meckling, W.H. (1976). Theory of the firm: managerial behavior, agency costs, and ownership structure. *Journal of Financial Economics, 3*, 306–360.

Joskow, P.L. (1973). Cartels, competition and regulation in the property-liability insurance industry. *The Bell Journal of Economics and Management Science, 4*(2), 375–427.

Kahn, B.E., & Sarin, R.K. (1988). Modeling ambiguity in decisions under uncertainty. *Journal of Consumer Research 15*, 265–272.

Kahneman, D., & Tversky, A. (1979). Prospect theory: An analysis of decision under risk. *Econometrica, 47*(2), 263–292.

Kellner, S., & Matthewson, G.F. (1983). Entry, size, distribution, scale and scope economies in the life insurance industry. *Journal of Business, 56*(1), 25–44.

Kelsey, D. & le Roux, S. (2015). An experimental study on the effect of ambiguity in a coordination game. *Theory and Decision, 79*, 667–688.

Kennedy, P.E. (1986). Interpreting dummy variables. *Review of Economics and Statistics, LXVIII*, 174–175.

Keppe, H.J., & Weber. M. (1991). Judged knowledge and ambiguity aversion, working paper no. 277, Christian-Albrechts-Universität, Kiel, Germany.

Kimball, M.S. (1990). Precautionary saving in the small and in the large. *Econometrica, 58*(1), 53–73.

Kniesner, T.J., & Viscusi, W.K. (2005). Value of a statistical life: relative position vs. relative age. *American Economic Review, 95*, 142–146.

Kniesner, T.J., Viscusi, W.K., Woock, C., & Ziliak, J.P. (2012). The value of statistical life: evidence from panel data. *Review of Economics and Statistics, 94*, 19–44.

Kniesner, T.J., Viscusi, W.K., & Ziliak, J.P. (2014). Willingness to accept equals willingness to pay for labor market estimates of statistical life. *Journal of Risk and Uncertainty, 48*, 187–205.

Kniesner, T.J., Viscusi, W.K., & Ziliak, J.P. (2010). Policy relevant heterogeneity in the value of statistical life: new evidence from panel data quantile regressions. *Journal of Risk and Uncertainty, 40*, 15–31.

Kniesner, T.J., Viscusi, W.K., & Ziliak, J.P. (2006). Life-cycle consumption and the age-adjusted value of life. *Contributions to Economic Analysis and Policy, 5*, 1–34.

Knight, F.H. (1921). *Risk, Uncertainty and Profit*. Chicago: Chicago University Press.

Kochi, I., & Taylor, L.O. (2011). Risk heterogeneity and the value of reducing fatal risks: further market-based evidence. *Journal of Benefit-Cost Analysis, 2*, 1–26.

Kofler, E., & Menges, G. (1976). *Entscheidungen bei unvollständiger Information* (Decisions Based on Incomplete Information). Lecture Notes in Economics and Mathematical Systems 136. Berlin: Springer.

Kofler, E., & Zweifel, P. (1988). Exploiting linear partial information for optimal use of forecasts: with an application to U.S. economic policy. *International Journal of Forecasting, 4*(1), 15–32.

Kolsrud, J., Landais, C., Nilsson, P., & Spinnewijn, J. (2018). The optimal timing of unemployment benefits: Theory and evidence from Sweden. *American Economic Review, 108* (4–5), 985–1033.

Konetzka, R.T., He, D., Dong, J., & Nyman, J.A. (2019). Moral hazard and long-term care insurance. *Geneva Papers on Risk and Insurance Issues and Practice, 44*, 231–251.

Kotlikoff, L.J. (1979). Testing the theory of social security and life-cycle accumulation. *American Economic Review, 69*, 396–411.

Kreier, R., & Zweifel, P. (2010). Health insurance in Switzerland: a closer look at a system often offered as a model for the United States. *Hofstra Law Review, 39*(1), 89–111.

Krelle, W. (1959). A theory of rational behavior under uncertainty. *Metroeconomica, 11*(1–2), 51–63.

Kreps, D.M. (1990). *A Course in Microeconomic Theory*. Princeton: Princeton University Press.

Kumbhakar, S., & Lovell, C.A. (2000). *Stochastic Frontier Analysis*. Cambridge: Cambridge University Press.

Kunreuther, H., & Pauly, M.V. (1985). Market equilibrium with private knowledge: an insurance example. *Journal of Public Economics, 26*(3), 269–288.

Kunreuther, H., & Pauly, M.V. & McMorrow, S. (2013). *Insurance and Behavioral Economics*. Cambridge: Cambridge University Press.

Ladaique, M. (2011). *The OECD Social Expenditure Database (SOCX)*. Beirut: Expert Group Meeting on Social Security in Western Asia, 8-9 Sept. 2011.

Laffont, J.J. (1990). *The Economics of Uncertainty and Information*. Cambridge MA: Cambridge University Press.

Landais, C., & Michaillat, P., Saez, E. (2018). A macroeconomic approach to optimal unemployment insurance: Applications. *American Journal of Economic Policy, 10*(2), 182–216.

Lange, O. (1943). A note on innovations. *Review of Economics and Statistics, 25*, 19–25.

Larson Jr., J.R. (1980). Exploring the external validity of a subjectively weighted utility model of decision making. *Organizational Behavior and Human Performance 26*, 293–304.

Leeth, J.D., & Ruser, J. (2003). Compensating wage differentials for fatal and nonfatal injury risk by gender and race. *Journal of Risk and Uncertainty, 27*, 257–277.

Leimer, D.R., & Lesnoy, S. (1982). Social security and private saving: new time-series evidence. *Journal of Political Economy, 90*, 606–642.

Leland, H.E. (1968). Saving and uncertainty: the precautionary demand for saving. *Quarterly Journal of Economics, 82*(3), 465–473.

Levinthal, D. (1988). A survey of agency models of organizations. *Journal of Economic Behavior and Organization, 9*, 153–185.

Lichtenstein, S., et al. (1978). Judged frequency of lethal events. *Journal of Experimental Psychology, 4*, 551–578.

Lintner, J. (1965). The valuation of risky assets and the selection of risky investments in stock portfolios and capital budgets. *Review of Economic Studies, 47*, 13–37.

Loomes, G., & Sugden, R. (1987). Testing for regret and disappointment in choice under uncertainty. *Economic Journal, 97*(388a), 118–129.

Loomes, G., & Sugden, R. (1983). Regret theory and measurable utility. *Economics Letters, 12*(1), 19–21.

Loomes, G., & Sugden, R. (1982). Regret theory: an alternative theory of rational choice under uncertainty. *The Economic Journal, 92*(368), 805–824.

MacCrimmon, K.R. (1968). Descriptive and normative implications of the decision-theory postulates. In K. Borch, & J. Mossin (Eds.), *Risk and Uncertainty* (pp. 3–23). London: MacMillan.

MacCrimmon, K.R., & Larsson, S. (1979). Utility theory: axioms versus 'paradoxes.' In M. Allais and O. Hagen (eds.), *Expected Utility and the Allais Paradox*. Dordrecht, Holland: D. Reidel, pp. 333–409.

Machina, M.J. (1995). Non-expected utility and the robustness of the classical insurance paradigm. *Geneva Papers on Risk and Insurance Theory, 20*, 9–50.

Machina, M.J. (1987). Choice under uncertainty: problems solved and unsolved. *Journal of Economic Perspectives, 1*, 121–154.

Machina, M.J., & Viscusi, W.K. (2014). *Handbook of the Economics of Risk and Uncertainty*. Amsterdam: Elsevier.

Mahlberg, B., & Url, T. (1993). Effects of the single market on the Austrian insurance industry. *Empirical Economics, 28*, 813–838.

Markowitz, H.M. (1959). *Portfolio Selection - Efficient Diversification of Investments*. New Haven and London: Yale University Press.

Markowitz, H.M. (1952). Portfolio selection. *Journal of Finance, 7*, 77–91.

Mayers, D., & Smith, C.W. (1988). Ownership structure across lines of property-liability insurance. *Journal of Law and Economics, 31*, 351–378.

Mehl, C. (1998). Insurability of risks on the information highway. *Geneva Papers on Risk and Insurance, 23*(89), 103–111.

Menezes, C.F., & Hanson, D.L. (1970). On the theory of risk aversion. *International Economic Review, 11*, 481–487.

Merkin, R., & Rodger, A. (1997). *EC Insurance Law*. London: Longman.

Merton, R.C. (1972). An analytical derivation of the efficient portfolio frontier. *Journal of Financial and Quantitative Analysis, 7*(4), 1851–1872.

Meyer, B.D. (1990). Unemployment insurance and unemployment spells. *Econometrica, 58*(4), 7575–7782.

Meyer, R. & Kunreuther, H.C. (2017). *The ostrich paradox: why we underprepare for disasters*. Wharton Digital Press.

Mooney, S.F., & Salvatore, J.M. (1990). Insurance fraud project: report on research. In G. Dionne, A. Gibbens, & P. Saint-Michel (Eds.) (1993), *Analyse Économique de la Fraude. Risques, 16*, 9–34.

Morall, J.F. (1986). A review of the record. *Regulation, 10*, 25–34.

Morin, R.A., & Suarez, A. (1983). Risk aversion revisited. *Journal of Finance, 38*(4), 1201–1216.

Mork, K.A., & Hall, R.E. (1979). Energy prices, inflation, and recession, 1974-75. NBER Working Paper w0369. New York: National Bureau of Economic Research.

Mossin, J. (1968). Aspects of rational insurance purchasing. *Journal of Political Economy*, 553–568.

Mossin, J. (1966). Equilibrium in a capital asset market. *Econometrica, 34*, 768–783.

Murphy, A.H. & Winkler, R.I. (1977). Reliability of subjective probability forecasts of precipitation and temperature. *Journal of the Royal Statistical Society C, Applied Statistics* 26 (1), 41–47.

Murray, J.D., & White, R.W. (1983). Economics of scale and economics of scope in multiproduct financial institutions: a study of British Columbia credit unions. *Journal of Finance, 38*, 887–901.

Mueller, D.C. (1989). *Public Choice II*. Cambridge MA: Cambridge University Press.

Müller, W. (1995). Informationsprodukte (Informational products). *Zeitschrift für Betriebswirtschaft, 65*, 1017–1044.

Müller, W. (1981). Theoretical concepts of insurance production. *Geneva Papers on Risk and Insurance, 6*(21), 63–83.

National Association of Insurance Commissioners (NAIC) Capital Adequacy Task Force (2009). *Risk-Based Capital. General Overview*, http://www.naic.org/documents/committees_e_capad_RBC.

Negoita, C.V. (1981). The current interest in fuzzy optimization. *Fuzzy Sets and Systems, 6*(3), 261–269.

Nektarios, M. (2010). Deregulation, insurance supervision and guaranty funds. *The Geneva Papers on Risk and Insurance-Issues and Practice, 35*, 452–468.

Newhouse, J.P. (1996). Reimbursing health plans and health providers: efficiency in production versus selection. *Journal of Economic Literature, 34*(3), 1236–1263.

Newhouse, J.P., et al. (1993). *Free for All? Lessons from the RAND Health Insurance Experiment*. Cambridge MA: Harvard University Press.

Nickell, S. (1990). Inflation and the UK labour market. *Oxford Review of Economic Policy, 6*(4), 26–35.

Niehans, J. (1948). Zur Preisbildung bei ungewissen Erwartungen (On the formation of prices under uncertain expectations). *Schweizerische Zeitschrift für Volkswirtschaft und Statistik (Swiss Journal of Economics and Statistics), 84*, 433–456.

Nieto, B., Orbe, S. & Zaraya, A. (2011). Time-varying beta estimators in the Mexican emerging market. Working Paper, Bilbao: Universidad del País Vasco.

Niskanen, W.A. (1971). *Bureaucracy and Representative Government*. Chicago: Aldine.

Norton, E.C. (2000). Long-term care. In Culyer, A.S. & Newhouse, J.P. (Eds.), *Handbook of Health Economics* (pp. 955–994), New York: North Holland.

OECD (2020). *OECD Insurance Statistics 2019*. Paris: OECD Publishing.

OECD (2019). *OECD Insurance Statistics 2018*. Paris: OECD Publishing.

OECD (2017). *OECD Insurance Statistics 2016*. Paris: OECD Publishing.

OECD (2010a). *The Impact of the Financial Crises on the Insurance Sector and Policy Respones*. Paris: OECD Publishing.

OECD (2010b). *Insurance Statistics Yearbook 2010*. Paris: OECD Publishing.

Olson, M. (1965). *The Logic of Collective Action: Public Goods and the Theory of Groups*. Boston: Harvard University Press.

Outreville, J.F. (2002). Introduction to insurance and reinsurance coverage. In D.M. Dror, & A.S. Preker (Eds.), *Social Reinsurance. A new Approach to Sustainable Community Health Financing*, (pp. 59–74). Washington: The World Bank.

Outreville, J.F. (1998). *Theory and Practice of Insurance*. Dordrecht: Kluwer.

Panzar, J.C., & Willig, R.D. (1977). Economics of scale in multi-output production. *Quarterly Journal of Economics, 81*, 481–493.

Pauly, M.V. (1990). The rational non-purchase of long-term care insurance. *Journal of Political Economy* 98(1): 153–168.

Pauly, M.V. (1974). Overinsurance and public provision of insurance: the role of moral hazard and adverse selection. *Quarterly Journal of Economics, 88*(1), 44–62.

Pauly, M.V., & Held, P.J. (1990). Benign moral hazard and the cost-effectiveness analysis of insurance coverage. *Journal of Health Economics, 9*(4), 447–461.

Pauly, M.V., Kunreuther, H., & Kleindorfer, P. (1986). Regulation and quality competition in the U.S. insurance industry. In J. Finsinger, & M.V. Pauly (Eds.), *The Economics of Insurance Regulation* (pp. 65–107). London: Macmillan.

Peltzman, S. (1976). Towards a more general theory of regulation. *Journal of Law and Economics, 19*, 211–240.

Philipson, T.J., & Becker, G.S. (1998). Old-age longevity and mortality-contingent claims. *Journal of Political Economy, 106*(3), 551–573.

Plantin, G., & Rochet, J. (2007). *When Insurers Go Bust: an Economic Analysis of the Role and Design of Prudential Regulation.* Princeton: Princeton University Press.

Popper, K. (1945). *The Open Society and Its Enemies.* London: Oxford University Press.

Portney, P.R. (1981). Housing prices, health effects and valuing reductions in risk of death. *Journal of Environmental Economics and Management, 8*(1), 72–78.

Posner, R.A. (1974). Theories of economic regulation. *Bell Journal of Economics and Management Science, 5*(2), 335–358.

Praetz, P.D. (1980). Returns to scale in the U.S. life insurance industry. *Journal of Risk and Insurance, 47*(4), 525–532.

Pratt, J. (1964). Risk aversion in the small and in the large. *Econometrica, 32*, 122–136.

Pricewaterhouse Coopers (2008–2011). *Countdown to Solvency II*, several editions, https://www.pwc.com/gx/en/insurance/solvency-ii/.

Pritchett, S.T. (1973). Operating expenses of life insurers, 1961–1970: implications for economies of size. *Journal of Risk and Insurance, 40*(2), 157–165.

Pyle, D. (1971). On the theory of financial intermediation. *Journal of Finance, 26*, 737–747.

Quiggin, J. (1982). A theory of anticipated utility. *Journal of Economic Behavior and Organization, 3*, 323–343.

Ramaswami, M. (1997). Value at risk and asset-liability based asset allocation for a pension fund, a foundation, and an insurance company, *Working Paper*, Bankers Trust, New York.

Raviv, A. (1979). The design of an optimal insurance policy. *American Economic Review, 69*, 84–96.

Rea, S.A. (1993). The economics of insurance law. *International Review of Law and Economics, 13*, 145–162.

Rees, R., Gravelle, H., & Wambach, A. (1999). Regulation of insurance markets. *Geneva Papers on Risk and Insurance Theory, 24*(1), 55–68.

Riley, J.G. (2001). Silver signals: twenty-five years of screening and signaling. *Journal of Economic Literature, 39*, 432–478.

Riley, J.G. (1979). Informational equilibrium. *Econometrica, 47*(2), 331–359.

Riley, W.B., & Chow, K.V. (1992). Asset allocation and individual risk aversion. *Financial Analysis Journal* Nov./Dec.: 32–37.

Roll, R. (1977). A critique of the asset pricing model theory's tests: part I. On past and potential testability of the theory. *Journal of Financial Economics, 4*, 129–176.

Rothschild, M., & Stiglitz, J.E. (1976). Equilibrium in competitive insurance markets: an essay on the economics of imperfect information. *Quarterly Journal of Economics, 90*(4), 629–650.

Rothschild, M., & Stiglitz, J.E. (1970). Increasing risk: I. A definition. *Journal of Economic Theory, 2,* 225–243.

Samuelson, P.A. (1958). An exact consumption-loan model of interest with or without the contrivance of money. *Journal of Political Economy, 66,* 467–482.

Sandmo, A. (1970). The effect of uncertainty on saving decisions. *Review of Economic Studies, 37*(3), 353–360.

Santomero, A.M. (1977). Insurers in a changing competitive financial structure. *Journal of Risk and Insurance, 64*(4), 727–732.

Savage, L.J. (1951). The theory of statistical decision. *Journal of the American Statistical Association,* 45, 57–67.

Savage, L.J. (1954). *The Foundations of Statistics.* New York.

Schlesinger, H. (1997). Mikro-Korrelationen versus Makro-Korrelationen und die zukünftige Versicherungsdeckung (Micro vs. macro correlations and future insurance coverage). In L. Männer (Ed.), *Langfristige Versicherungsverhältnisse. Ökonomie, Technik, Institutionen* (pp. 379–386). Karlsruhe: Verlag Versicherungswirtschaft.

Schneider, F. (1986). The influence of political institutions on social security policies: A public choice view. In J.M. Schulenburg (Ed.), *Essays in Social Security Economics* (pp. 13–31). Berlin: Springer.

Schoder, J., & Zweifel, P. (2011). Flat-of-the-curve medicine: a new perspective on the production of health. *Health Economics Review* **1**(2), doi: 10.1186/2191-1-2.

Schoder, J., Sennhauser, M., & Zweifel, P. (2010). Fine-tuning of health insurance regulation – unhealthy consequences for an individual insurer. *International Journal of the Economics of Business, 17*(3), 313–327.

Schoder, J., Zweifel, P. & Eugster, P. (2013). Insurers, consumers, and correlated risks. *Journal of Insurance Issues,* 36 (2), 1–29.

Schölzel, C., & Friederichs, P. (2008). Multivariate non-normally distributed random variables in climate research – introduction to the copula approach. *Nonlinear Processes in Geophysics, 15,* 761–772.

Schradin, H.R. (1994). Kritische Erfolgsfaktoren in der Versicherung. Untersuchungsansätze und Methodische Grundlagen für die Analyse organisatorischer Teileinheiten (Critical determinants of profitability in insurance: approaches and methods for the analysis of organizational units). *Zeitschrift für die gesamte Versicherungswissenschaft, 83*(4), 531–561.

Schubert, R., Brown, M., Gysler, M., & Brachinger, H.W. (1999). Financial decisionmaking: are women really more risk-averse? *American Economic Review, 89,* 381–385.

Schulenburg, J.M. (1986). Optimal insurance purchasing in the presence of compulsory insurance and uninsurable risks. *Geneva Papers on Risk and Insurance, 11*(38), 5–16.

Scotton, C.R., & Taylor, L.O. (2011). Valuing risk reductions: incorporating risk heterogeneity into a revealed preference framework. *Resource and Energy Economics, 33,* 381–397.

Seiford, L.M., & Thrall, R.M. (1990). Recent developments in DEA. *Journal of Econometrics, 46*, 7–38.

Setbon, M., Raude, J., Fischler, C., & Flauhault, A. (2005). Risk perception of the "Mad Cow Disease" in France: determinants and consequences. *Risk Analysis, 25*(2), 813–826.

Shackle, G.L. (1949). *Expectations in Economics*. Cambridge MA: Cambridge University Press.

Sharpe, W.F. (1964). Capital asset prices: a theory of market equilibrium under risk. *Journal of Finance, 19*, 425–442.

Shrieves, R.E., & Dahl, D. (1992). The relationship between risk and capital in commercial banks. *Journal of Banking & Finance, 16*, 439–457.

Siegel, F.W., & Hoban, J.P. (1982). Relative risk aversion revisited. *Review of Economics and Statistics, 64*(3), 481–487.

Singer, E. (1991). Public attitudes toward genetic testing. *Population Research and Policy Review, 10*, 235–255.

Sinn, H.W. (1996). Social insurance, incentives and risk taking. *International Tax and Public Finance, 3*, 259–280.

Sinn, H.W. (1988). Gedanken zur volkswirtschaftlichen Bedeutung des Versicherungswesens (On the importance of insurance for the economy). *Zeitschrift für die gesamte Versicherungswissenschaft, 77*, 1–27.

Sinn, H.W., & Weichenrieder, A.J. (1993). The biological selection of risk preferences. In Bayerische Rückversicherung (Ed.), *Risk as a Construct*. Munich: Knesebeck, 67–73.

Skipper, H.W., & Kwon, W.J. (2007). Risk Management and Insurance: Perspectives in a Global Economy. Malden, MA: Blackwell Publishing.

Sklar, A. (1973). Random variables, joint distribution functions and copulas. *Kybernetika, 9*, 449–460.

Slovic, P., Fischhoff, B., & Lichtenstein, S. (1980). Facts and fears – understanding risks. In Schwing, R.C., & Albers, W.A. (Eds.), *Societal Risk Assessment*. New York: Plenum Press.

Slovic, P., & Tversky, A. (1974). Who accepts Savage's axiom? *Behavioral Science 19*, 368–373.

Smith, V. (1968). Optimal insurance coverage. *Journal of Political Economy, 76*, 68–77.

Stahl, D.O (1989). Oligopolistic Pricing with sequential consumer search. *American Economic Review, 79*, 700–712.

Statistisches Bundesamt (2018). *Statistisches Jahrbuch für das Ausland (Statistical Yearbook for Foreign Countries)* Wiesbaden.

Statistisches Bundesamt (1999). *Statistisches Jahrbuch für das Ausland (Statistical Yearbook for Foreign Countries)* Wiesbaden.

Stewart, J. (1994). The welfare implications of moral hazard and adverse selection in competitive insurance markets. *Economic Inquiry, 32*, 193–208.

Stock, J.A., & Wise, D.A. (1990). Pensions, the option value of work, and retirement. *Econometrica, 58*(5), 1151–1180.

Strain, R.W. (1989). *Reinsurance*. New York: The College of Insurance.

Streb, J. (2018). Does social security crowd out private savings? The case of Bismarck's system of social insurance. *European Review of Economic History*, 22(3), 298–321.

Suret, M. (1991). Scale and scope economies in the Canadian property and casualty insurance industry. *Geneva Papers on Risk and Insurance Issues and Practice*, 59, 236–256.

Swiss Re (2019). World insurance: the great pivot east continues. *sigma* 3/2019.

Swiss Re (2018a). Natural Catastrophes and man-made disasters in 2017: A year of record-breaking losses. *sigma* 1/2018.

Swiss Re (2018b). World Insurance in 2017: Solid, but mature life markets weigh on growth. *sigma* 3/2018.

Swiss Re (2010a). Nature and man-made disasters in year 2009: Catastrophes claim fewer victims, insured losses fall. *sigma* 1/2010.

Swiss Re (2010b). World insurance in 2009. *sigma* 2/2010.

Swiss Re (2003). The Picture of ART. *sigma* 1/2003.

Swiss Re (1999). Assekuranz Global 1997: Stark expandierendes Lebensgeschäft. Stagnierendes Nicht-Lebengeschäft (Global Insurance 1997: Strongly expanding life business. Stagnating non-life business). *sigma* 3/1997.

Swiss Re (1998). Der globale Rückversicherungsmarkt im Zeichen der Konsolidierung (The global reinsurance market in consolidation). *sigma* 9/1998.

Swiss Re (1996). Risikotransfer über Finanzmärkte: Neue Perspektiven für die Absicherung von Katastrophenrisiken in den USA (Risk transfer through capital markets: New perspectives for hedging catastrophic risks in the USA). *sigma* 5/1996.

Swiss Re (1993). Ökonomische Analyse der Versicherungsnachfrage (Economic analysis of insurance demand). *sigma* 5/1993.

Szpiro, G.G. (1997). The emergence of risk aversion. *Complexity*, 2(4), 31–39.

Szpiro, G.G. (1986a). Measuring risk aversion: an alternative approach. *Review of Economics and Statistics*, 68, 156–159.

Szpiro, G.G. (1986b). Über das Risikoverhalten in der Schweiz (On risk behavior in Switzerland). *Swiss Journal of Economics and Statistics*, 122(III), 463–470.

Taylor, K.A. (1991). Testing credit and blame attributions as explanations for choices under ambiguity, working paper, Department of Decision Sciences, University of Pennsylvania.

Taylor, P.R. (2012). The mismeasure of risk. In S. Boeser, R. Hillebrand, P. Sandin, M. Peterson (eds.), *Handbook of Risk Theory*. Vol. 1. New York: Springer, 441–475.

Toivanen, A.M. (1993). Economies of scale and scope in the Finnish non-life insurance industry. *Journal of Banking and Finance*, 21(6), 759–779.

Tversky, A. & Kahneman, D. (1992). Advances in prospect theory: Cumulative representation of uncertainty. *Journal of Risk and Uncertainty*, 5, 297–323.

Tversky, A. & Kahneman, D. (1973). Availability: a heuristic for judging frequency and probability. *Cognitive Psychology*, 5, 207–232.

Tversky, A. & Kahneman, D. (1971). Belief in the law of small numbers. *Psychological Bulletin*, 76, 2, 105–110.

Urquhart, J., & Heilmann, K. (1983). *Riskwatch: The Odds of Life*. New York: Facts on File.
US Department of Transportation, Bureau of Transportation Statistics (2020). Traffic Safety Facts: Research Note. October.
US Department of Transportation, Bureau of Transportation Statistics (2016). Traffic Safety Facts: Research Note. March.
US Department of Transportation, Bureau of Transportation Statistics (2011). http://www.bts.gov/data_and_statistics.
Van Dalen, H.P., & Swank, O.A. (1996). Government spending cycles: ideological or opportunistic? *Public Choice, 89*, 183–200.
Van de Ven, W.P., & Ellis, R.P. (2000). Risk adjustment in competitive health plan markets. In A.J. Culyer, & J.P. Newhouse (Eds.), *Handbook of Health Economics* (pp. 155–845), Chap. 14. London: Elsevier.
Van Den Berghe, L. (1998). Shaping the future of the insurance sector. *Geneva Papers on Risk and Insurance, 23*(89), 506–518.
Van Vliet, O. & Carminada, K. (2012). Unemployment replacement rates dataset among 34 welfare states, 1971-2009: an update, extension and modification of the Scruggs' Welfare State Entitlements Data Set. *NEUJOBS Special Report No. 2*. Leiden: Leiden University.
Viscusi, W.K. (2014). The value of individual and societal risks to life and health. In Mark J. Machina & W. Kip Viscusi (Eds.), *Handbook of the Economics of Risk and Uncertainty*, Chap. 7. Amsterdam: North-Holland.
Viscusi, W.K. (2004). The value of life: estimates with risks by occupation and industry. *Economic Inquiry, 42*, 29–48.
Viscusi, W.K. (2003). Racial differences in labor market values of a statistical life. *Journal of Risk and Uncertainty, 27*, 239–256.
Viscusi, W.K. (1993). The value of risks to life and health. *Journal of Economic Literature, 31*, 1912–1946.
Viscusi, W.K., & Aldy, L.E. (2007). Labor market estimates of the senior discount for the value of statistical life. *Journal of Environmental Economics and Management, 53*, 377–392.
Viscusi, W.K., & Aldy, L.E. (2003). The value of a statistical life: a critical review of market estimates around the world. *Journal of Risk and Uncertainty, 27*(1), 5–76.
Volkman-Wise, J. (2015). Representativeness and managing catastrophe risk, *Journal of Risk and Uncertainty, 51*, 267–290.
Von Bomhard, N. (2010). The advantages of a global solvency standard. *The Geneva Papers, 35*, 79–91.
Von Camphausen, S., Bornschein, B. & Wick, R. et al. (2005). Prevalence and incidence of Parkinson's disease in Europe, *European Neuropsychopharmacy* 15, 465–490.
Von Eije, J.H. (1989). *Reinsurance Management. A Financial Exposition*, Dissertation, Erasmus University (Foundation for Insurance Science), Rotterdam.
Von Neumann, J., & Morgenstern, O. (1944). *Theory of Games and Economic Behavior*. Princeton: Princeton University Press.

Vroomen, J., & Zweifel, P. (2011). Preferences for health insurance and health status: does it matter whether you are Dutch or German? *European Journal of Health Economics, 12*(1), 87–95.

Wald, A. (1945). Statistical decision functions which maximize the maximum risk. *Annuals of Mathematics, 46,* 265–280.

Walden, M.L. (1985). The whole life insurance policy as an options package: an empirical investigation. *Journal of Risk and Insurance, 52*(4), 44–58.

Weiss, M.A. (1995). A multivariate analysis of loss reserving estimates in property-liability insurers. *Journal of Risk and Insurance, 52*(2), 199–221.

White, H. (1980). A heteroscedasticity-consistent covariance estimator and a direct test for heteroscedasticity. *Econometrica, 48,* 817–838.

Wildasin, D.E. (1991). Income redistribution in a common labor market. *American Economic Review, 81,* 757–774.

Williams, J.T. (1977). Capital asset prices with heterogeneous beliefs. *Journal of Financial Economics, 5,* 219–239.

Wilson, C.A. (1977). A model of insurance markets with incomplete information. *Journal of Economic Theory, 16*(2), 167–207.

Winter, R.A. (1992). Moral hazard and insurance contracts. In G. Dionne (Ed.) *Contributions to Insurance Economics* (pp. 61–96). Dordrecht: Kluwer.

Xu, X. & Zweifel, P. (2014). Bilateral intergenerational moral hazard: empirical evidence from China. *Geneva Papers on Risk and Insurance: Issues and Practice, 39*(4), 651–667.

Yates, J.F., & Zukowski, L.G. (1976). Characterization of ambiguity in decision making. *Behavioral Science 21,* 19–25.

Zimmermann, A., Zweifel, P., & Kofler, E. (1985). Application of the linear partial information model to forecasting the Swiss timber market. *Journal of Forecasting, 4,* 387–338.

Zweifel, P. (2020). Innovation in long-term care insurance: Joint contracts for mitigating relational moral hazard. Working paper.

Zweifel, P. (2019). Planned Solvency III regulation: Should it be adopted outside the European Union? *Asia-Pacific Journal of Risk and Insurance*, 13(1), 1–12.

Zweifel, P. (2015). Solvency regulation of banks and insurers: A two-pronged critique. *International Journal of Financial Research*, 6(3), 86–105.

Zweifel, P. (2013). The division of labor between private and social insurance. In G. Dionne (Ed.), *Handbook of Insurance* (pp. 1097–1118). Boston: Kluwer.

Zweifel, P. (2009). Technological change and health insurance. In J. Costa-Font, C. Courbage, & A. McGuire (Eds.), *The Economics of New Health Technologies. Incentives, Organization and Financing* (pp. 93–107). Oxford: Oxford University Press.

Zweifel, P. (2007). The theory of social health insurance. *Foundations and Trends in Microeconomics, 3*(3), 183–273.

Zweifel, P. (2001). Improved risk information, the demand for cigarettes, and anti-tobacco policy. *Journal of Risk and Uncertainty, 23*(3), 299–303.

Zweifel, P. (1994). EEA (European Economic Area) and insurance in the EFTA countries. In Sveriges Riksbank (Ed.), *Financial Integration in Western Europe, Occasional Paper, 10*, 93–99.

Zweifel, P., & Auckenthaler, C. (2008). On the feasibility of insurers' investment policies. *Journal of Risk and Insurance, 75*(1), 193–206.

Zweifel, P., & Breuer, M. (2006). The case for risk-based premiums in public health insurance. *Health Economics, Policy and Law, 1*(2), 171–188.

Zweifel, P., Breyer, F., & Kifmann, M. (2009). *Health Economics*, 2nd ed. New York: Springer.

Zweifel, P. & Courbage, C. (2016). Long-term care: is there crowding-out of informal car, private insurance as well as saving? *Asia Pacific Journal of Risk and Insurance* 10(1), 107–132.

Zweifel, P., & Eugster, P. (2008). Life-cycle effects of social security in an open economy: a theoretical and empirical survey. *Zeitschrift für die gesamte Versicherungswissenschaft (German Journal of Risk and Insurance), 97*(1), 61–77.

Zweifel, P., Felder, S., & Meier, P. (1999). Ageing of population and healthcare expenditure: A red herring? *Health Economics, 8*(6), 485–496.

Zweifel, P., & Frech, H.E. (2016). Why 'optimal' payment for healthcare providers can never be optimal under community rating. *Applied Health Economics and Health Policy, 14*(1), 9–20.

Zweifel, P., & Ghermi, P. (1990). Exclusive vs. independent agencies: A comparison of performance. *The Geneva Papers on Risk and Insurance Theory, 15*(2), 171–192.

Zweifel, P., & Strüwe. W. (1998). Long-term care insurance in a two-generation model. *Journal of Risk and Insurance* 65(1): 33–56.

Zweifel, P., & Strüwe. W. (1996). Long-term care insurance and bequests as instruments for shaping intergenerational relationships. *Journal of Risk and Uncertainty* 12(1): 65–76.

Zweifel, P., & Telser, H. (2009). Cost-benefit analysis for health. In R.J. Brent (Ed.), *Handbook of Research on Cost-Benefit Analysis* (pp. 31–54). Northhampton MA: Edward Elgar.

Zweifel, P., Telser, H., & Vaterlaus, S. (2006). Consumer resistance against regulation: the case of health care. *Journal of Regulatory Economics, 29*(3), 319–332.

Zweifel, P., & Waser, O. (1992). *Bonus Options in Health Insurance*. Dordrecht: Kluwer.

Author Index

A
Aaron, H., 437
Abdellaoui, M, 134
Abraham, K.S., 390
Abrams, R.A., 135, 136
ADAC, 32
Adams, M.B., 390, 393, 395
Aldy, L.E., 66, 68
Allais, M., 118
Anderson, D.A., 459, 460, 463
Arnott, R., 330
Arrow, K.J., 3, 40, 53, 54, 58, 75, 95, 103, 317
Association of German Insurers, 200, 319
Atkinson, A.B., 457
Atkinson, S.E., 67
Auckenthaler, C., 241, 388

B
Bakx, P., 351
Baltensperger, E., 411, 415
Barberis, N.C., 134
Barrese, J., 210, 211, 214
Barsky, R.B., 58
Basel Committee on Banking Supervision, 415
Baumol, H.J., 290
Beard, R.E., 224, 261, 262
Becker, G.S., 109, 110, 327, 451, 452
Becker, S.W., 134, 135
Bell, D.E., 116, 130
Ben Ammar, S., 247
Berger, L.A., 234

Bernasconi, M., 135
Bernheim, A., 468, 477
Bernoulli, D., 39, 81
Black, F., 276, 280
Blake, D., 57
Blanchard, O., 443
Blomquist, G., 66, 67
Blume, M.E., 57
Blundell, R., 444
Bonato, D., 331, 332
Borch, K.H., 218, 223, 224, 317
Bornschein, B., 433
Börsch-Supan, A., 446
Bove, V., 430, 431
Brachinger, H.W., 58
Braun, A., 247
Breuer, M., 387
Breyer, F., 30, 64, 66, 86, 317, 320, 435, 453
Brigo, D., 120
Brown, J.R., 348, 478
Brown, M., 58
Brownson, F.O., 134, 135
Bruner, D.M., 74
Buchanan, J.M., 422
Buckley, A., 169
Bühlmann, H., 215, 269
Buomberger, P., 411, 415
Burda, M.C., 458
Burkhauser, R.V., 447
Burtless, G., 445, 446
Bushyhead, J.B., 36

© The Editor(s) (if applicable) and The Author(s), under exclusive license to Springer Nature Switzerland AG 2021
P. Zweifel et al., *Insurance Economics*, Classroom Companion: Economics,
https://doi.org/10.1007/978-3-030-80390-2

C

Calford, E.M., 136
Camerer, C., 134–136
Carlin, P.S., 67
Carlton, D.W., 208
Carminada, K., 463
Carnap, R., 128
Cawley, J., 316
Chen, H., 412
Chiappori, P.A., 316
Chipman, J.S., 135
Chow, K.V., 57, 58
Christensen-Szlanski, J., 36
Cohen, M., 135, 136
Conte, A., 116
Cook, P.J., 90
Copeland, T.E., 159, 164, 169
Courbage, C., 342, 346, 352
Crain, M., 432
Cross, M.L., 180, 181
Cullen, M.R., 365
Culyer A.J., 423
Cummins, J.D., 179, 202, 205, 210–212, 234, 264, 275, 276, 283, 284, 286, 289, 293, 305, 306, 411–413
Curley, S.P., 135, 136
Cutler, D.M., 361–365

D

Dahchour, M., 375, 377, 402
Dahlby, B.G., 425, 476
Dahl, D., 285
D'Arcy, S., 282, 283
Dardis, R., 67
Davidson, W.N.III, 180, 181
Daykin, C.D., 224
De Alessi, L., 177, 178
De Meijer, C., 351
DiMasi, J.A., 197
Dionne, G., 334, 337, 339, 366, 371, 372, 375, 377, 378, 402
Doherty, N.A., 95, 104, 142, 158, 176, 204, 211, 223, 245, 278, 297, 366, 371, 372, 471, 476
Dong, J., 350, 351
Donni, O., 305
Dowd, K., 243
Downs, A., 429
Dreyfuss, M.K., 67
Dumm, R.E., 132, 136

E

Ebner, M., 248
Eckles, D.L., 132, 136, 146, 390
Eeckhoudt, L., xi, 55, 56, 92, 102, 345
Efthyvoulou, G., 430, 431
Ehrlich, J., 109, 110, 327
Eichberger, J., 136
Einav, L., 365
Einhorn, H.J., 135, 136
Eisen, R., 86, 97, 109, 224, 307, 317, 325, 327, 330, 342, 358, 373, 384, 425, 441, 476, 483
Eisenhauer, G.M., 58
Eling, M., 247
Ellis, R.P., 386
Ellsberg, D., 118
Embrechts, P., 245
Eugster, P., 441–444, 452, 470, 484
European Commission, 414
Evans, D, 36
Evans, M.F., 68
Evans, W.N., 400, 402
Ezra, D., 235

F

Fama, E.F., 162, 286
Fecher, E., 300–303, 305
Federal Drug Administration, 121
Felderer, B., 450
Felder, S., 73
Feldstein, M.S., 449
Fellner, W., 124
Finkelstein, A., 348, 365
Finsinger, J., 209, 404, 406, 408, 409, 416
Fischhoff, B., 35
Fischler, C., 38
Flauhault, A., 38
Frech, H.E. III, 365, 386, 397, 398, 400, 402, 409, 427
French, K.R., 162
Friederichs, P., 245
Friend, I., 57
Froot, K.A., 245

G

Garbacz, C., 67
Garven, J.R., 223, 226, 227, 230, 232, 234, 278, 282, 283
Gayor, T., 67
Ghermi, P., 209, 212
Giarini, O., 5
Gigliotti, G., 135
Gilboa, I., 116, 127
Gisler, A., 215
Glazer, A., 370
Goavaerts, M.J., 267, 268
Goldsmith, R.W., 135

Author Index

Gonzalez, R., 134
Gollier, C., xi
Goolsbee, A., 478
Grabowski, H.G., 197, 400, 402
Grace, M.F., 205
Graham, D.A., 90
Gravelle, H., 385
Greene, W.H., 287, 331
Greenwald, B.C., 197
Grossman, S.J., 208, 276
Gruber, J., 449
Gurley, J.G., 277
Gysler, M., 58

H

Hackmann, M.B., 352
Hadar, J., 71
Halek, M., 58
Hall, R.E., 130
Halvorsen, R., 67
Hamilton, J.T., 67
Hansen, R.W., 197
Hanson, D.L., 53
Hardwick, P., 305
Harrington, S.E., 179, 400
Hart, O.D., 208
Hax, K., 3
Heath, C., 136
He, D., 350, 351
Heilmann, K., 33, 34
Heilmann, W.R., 267, 268, 270
Held, P.J., 319
Hempel, C.G., 3
Hersch, J., 66, 68
Hey, J.D., 116
Hilliard, J.I., 390
Hirao, Y., 305
Hirshleifer, D., 370
Hirshleifer, J., 86, 370
Hoban, J.P., 57
Hoevenaars, R.P., 237, 239
Hogarth, R.M., 135, 136
Holmström, B., 208
Houser, A, 342
Houston, D., 298
Hoyt, R.E., 146
Hujer, R., 458, 463
Hull, J., 171, 172
Human Mortality Database, 30

I

IAIS, 411, 414
IMF, 19
Inoue, T., 305

Intriligator M.D., 129
Ippolito, P.M., 67
Ippolito, R.A., 67
Iuppa, A., 411, 415

J

Jackwerth, J.C., 173
Jaffe, J., 169
Jaffray, J-Y., 135, 136
Jenkins, R.R., 67
Jensen, M.C., 208, 286
Joskow, P.L., 209
Jullien, B., 316
Juster, T.F., 58

K

Kahn, B.E., 135, 136
Kahneman, D., 115, 121, 122, 124, 126, 131–134
Karjalainen, R., 136
Keller, B., 411, 415
Kellner, S., 300
Kelsey, D., 136
Kennedy, P.E., 212, 462
Keppe, H.J., 136
Kifmann, M., 30, 64, 66, 86, 317, 320, 453
Kimball, M.S., 54, 56, 58
Kleindorfer, P., 400
Kniesner, T.J., 68
Knight, F.H., 2, 31
Kochi, I., 68
Kofler, E., 2, 116, 128–130
Kolsrud, J., 463
Konetzka, R.T., 350, 351
Kotlikoff, L.J., 448
Kreier, R., 386
Krelle, W., 81
Kreps, D.M., 354
Kumbhakar, S., 305
Kunreuther, H.C., 132, 133, 371, 400
Kwon, J., xi

L

Ladaique, M., 420
Laeven, J.A., 267, 268
Laffont, J.J., 72
Lamm-Tennant, J., 226, 227, 230, 232, 234
Landais, C., 458, 463
Lange, O., 81
Larson Jr., J.R., 135
Larsson, S., 135
Lasserre, P., 366
Leimer, D.R., 449

Leland, H.E., 54
Le Roux, S., 136
Lesnoy, S., 449
Levinthal, D., 177, 208, 318
Lichtenstein, S., 35
Lindskog, F., 245
Lintner, J., 160
Loomes, G., 116, 130, 135
Lovell, C.A., 305

M
MacCrimmon, K.R., 117, 134, 135
Machina, M.J., 81
Mahlberg, B., 308
Markowitz, H.M., 124, 154, 160, 240, 271
Matthewson, G.F., 300
Mayers, D., 175, 179, 285
McMorrow, S., 132
McNeil, A., 245
Meckling, W.H., 208
Mehl, C., 477
Meier, P., 73
Menezes, C.F., 53
Menges, G., 116, 128, 129
Merkin, R., 391
Merton. R.C., 154
Meyer, B.D., 458
Meyer, J., 102
Meyer, R., 132, 133
Michaillat, P., 458
Michaud, P-C., 375, 377, 402
Micklewright, J., 457
Miller, S.M., 146
Molenaar, R.D.J., 239
Mooney, S.F., 319
Morall, J.F., 64
Morgenstern, O., 82
Morin, R.A., 57
Mork, K.A., 130
Mossin, J., 95, 98, 160
Mueller, D.C., 429
Müller, W., 3, 204, 384
Murphy, A.H., 36
Murray, J.D., 292

N
NAIC, 404
Navas, A., 430, 431
Negoita, C.V., 2
Nektarios, M., 389
Nell, M., 317
Nelson, J.M., 210, 211, 214
Neumann, T., 248
Newhouse, J.P., 427, 455

Nickell, S., 458
Niehans, J., 81, 116, 130
Nieto, B., 248
Nilsson, P., 463
Niskanen, W.A., 393, 422, 433
Norton, E.C., 342
Nyce, C., 132, 136
Nyman, J.A., 350, 351

O
OECD, 7, 8, 15, 16, 409, 410, 420
Olson, M., 392, 429
Orbe, S., 248
Orme, C., 116
Ormiston, M.B., 102
Outreville, J.F., 222, 224
Owens, N., 67

P
Pallavicini, A., 120
Panzar, J.C., 290
Pauly, M.V., 132, 317, 319, 325, 342, 371, 400
Peltzman, S., 391, 392
Pentikäinen, T., 224, 261, 262
Perelman, S.D., 300–303
Perloff, J.M., 208
Perotti, R., 443
Pesonen, E., 224, 261, 262
Pesonen, M., 224
Pestieau, P., 300–303
Philipson, T.J., 316, 451, 452
Phillips, R.D., 275
Plantin, G., 385, 412, 414
Pohl, R., 352
Popper, K., 61
Portney, P.R., 67
Posey, L.L., 471, 476
Posner, R.A., 392
Praetz, P.D., 300
Pratt, J., 49, 50, 53, 54, 58, 75
Pricewaterhouse Coopers, 413
Pritchett, S.T., 300
Pyle, D., 277

Q
Quiggin, J., 115, 125

R
Ramaswami, M., 240
Raude, J., 38
Raviv, A., 103
Rea, S.A., 480

Reber, S.J., 361–365
Rees, R., 385
Riley, J.G., 358, 360
Riley, W.B., 57, 58
Rochet, J., 385, 412, 414
Rodger, A., 391
Roll, R., 160
Ross, S.T., 169
Rothschild, M., 51, 69, 317, 352, 357
Ruser, J., 68
Russell, W., 71

S
Saez, E., 458
Sahlin, N-E., 135
Said, T., 135, 136
Salanié, B., 316
Salanié, F., 316
Salvatore, J.M., 319
Samprone, J.C., 397, 398, 400, 402, 409
Samuelson, P.A., 435
Sandmo, A., 54
Sandy, R., 67
Santomero, A.M., 481
Sarin, R.K., 135, 136
Savage, L.J., 116, 119, 130
Schaur, G., 68
Schipper, B.C., 136
Schlesinger, H., xi, 55, 56, 92, 104, 345, 443
Schmeidler, D., 116, 127
Schmidt, F.A., 209, 404, 416
Schneider, F., 430
Schneider, H., 458, 463
Schoder, J., 74, 217, 482
Scholes, M., 280
Schölzel, C., 245
Schotman, P.C., 239
Schradin, H.R., 203
Schubert, R., 58
Schulenburg, J.M., 421
Schut, F., 351
Scotton, C.R., 68
Seiford, L.M., 305
Sennhauser, M., 217
Setbon, M., 38
Shackle, G.L., 81
Shapiro, M.D., 58
Sharpe, W.F., 160
Shaw, E.S., 277
Shrieves, R.E., 285
Siegel, F.W., 57
Simon, R., 298
Singer, E., 476
Sinn, H.W., 13, 38, 423

Skipper, H.W., xi
Sklar, A., 244
Sloan, F.A., 342
Slovic, P., 35, 135
Smith, C.W., 175, 179, 285
Smith, M.P., 365
Smith, V., 95
Sommer, D.W., 202, 283, 284, 286, 289
Sopher, B., 135
Spinnewijn, J., 463
Stahel, W., 5
Stahl, D.O., 478
Statistisches Bundesamt, 17, 18, 32
Steenkamp, T.B.M., 239
Stewart, J., 373
Stiglitz, J.E., 51, 69, 197, 317, 330, 352, 357
St-Michel, P., 334, 337, 339, 378
Stock, J.A., 446
Strain, R.W., 224
Streb, J., 449
Strüwe, W., 342
Suarez, A., 57
Sugden, R., 116, 130
Sun, T., 412
Suret, M., 291, 293
Swank, O.A., 430, 432
Swiss Re, 5, 6, 10, 17–21, 246–248
Szpiro, G.G., 38, 58–60

T
Taylor, K.A., 136
Taylor, L.O., 68
Taylor, P.R., 120
Telser, H., 69, 297
Tennyson, S., 234
Thornton, J.H., 180, 181
Thrall, R.M., 305
Timme, S.G., 205
Toivanen, A.M., 305
Torresetti, R., 120
Tower, G.D., 390, 393, 395
Tullock, G., 422
Turner, J.A., 447
Tversky, A., 115, 121, 122, 124, 126, 131–136

U
Url, T., 308
Urquhart, J., 33, 34
U.S. Department of Transportation, 32

V
Van Dalen, H.P., 430, 432

Van Den Berghe, L., 469
Van Derhei, J., 210–212
Van de Ven, W.P., 386
Van Doorslaer, E., 351
Van Vliet, O, 463
Vaterlaus, S., 297
Viscusi, W.K., 66–68, 400, 402
Volkman-Wise, J., 131, 132, 136, 137
Von Bomhard, N., 415
Von Camphausen, S., 433
Von Eije, J.H., 224
Von Neumann, J., 82
Vroomen, J., 297

W
Wald, A., 81
Walden, M.L., 297
Wambach, A., 385
Waser, O., 455, 456
Weber, M., 134–136
Weichenrieder, A.J., 38
Weiss, M.A., 195, 293, 305, 411–413
Westerfield, R., 169
Weston, J.F., 159, 164, 169
White, H., 287
White, R.W., 292
Wicki, A., 411, 415
Wick, R., 433
Wiggins, L.B., 67
Wildasin, D.E., 470
Williams, J.T., 276

Willig, R.D., 290
Wilson, C.A., 357, 358, 361
Winkler, R.I., 36
Winter, R.A., 218
Wise, D.A., 446
Woock, C., 68
Wu, G., 134
Wyplosz, C., 458

X
Xiaoying, X., 293, 305
Xu, X., 349, 350

Y
Yates, J.F., 135, 136
Yelowitz, A.S., 449

Z
Zaray, A., 248
Zi, H., 205, 293, 305, 306
Ziliak, J.P., 68
Zimmermann, A., 129
Zukowski, L.G., 135
Zweifel, P., 13, 30, 37, 64, 66, 69, 73, 74, 86, 129, 130, 205, 209, 212, 217, 241, 296, 297, 317, 320, 331, 332, 341, 342, 346, 349, 350, 352, 365, 384, 386–388, 390, 395, 411, 415, 416, 427, 441–444, 452, 453, 455, 456, 470, 482, 484

Subject Index

A
Administrative expense, 87, 93
Adverse selection, 352, 455
  and market equilibrium, 353
  and moral hazard combined, 373
  and risk type, 353
  and social insurance, 425
  consequences of, 318
  death spiral property, 318
  definition of, 318
  efficiency loss subadditive, 375
  empirical evidence, 361, 362
  estimate of welfare loss, 364
  in a multi-period context, 366, 367
  in a single-period context, 352, 353, 357
  in health insurance, 361
  triggered by change, 362
  unsustainable pooling contract, 355
  unsustainable pooling equilibrium, 353, 357
Altruism, 346
Ambiguity aversion, 116, 126
  empirical evidence, 134
Amnesia bias, 132
Arbitrage, 165
Arbitrage pricing theory, 162–164
Asset, 4, 22, 28, 80, 141, 148, 167, 278
  and private insurance, 433
  and social insurance, 433
  claims against, 279
  classes in capital investment, 237
  present value of, 278
  redistribution of claims against, 285
  stochastic disturbance, 279
  systematic change, 279
Asset and liability
  discrepancy in maturity, 412
  maturity matching, 412
Asset–liability ratio
  and globalization, 469
Asymmetric information
  and adverse selection, 315, 352, 483
  and adverse selection and moral hazard combined, 373
  and cost of acquiring information, 315
  and observability of behavior, 337
  and rationing of coverage, 356, 366, 426
  basic assumption, 316
  concerning preventive behavior, 316
  concerning risk type, 316
  moral hazard, 315, 483

B
Banks, 7, 8
Barriers to entry, 88
Behavioral risk audit, 133
Bequest, 342, 346
Bernoulli principle, 40, 41, 83
Beta
  of a security, 158
  of firm's assets, 174
  of insurance company, 274
  of security, 157, 160, 272
  of underwriting, 174, 179–181, 274, 275

Binary prospect, 44, 48
Black–Scholes formula, 280
Broker, 207, 210, 212

C

Call option, 166
Capital asset pricing model, 156, 160, 271
  applied to insurance, 274, 276
  critique of, 276
Capital investment policy, 235, 236
  and funding rate, 237
  and optimal liability hedging, 239
  asset-only optimization, 239
  efficiency frontier of, 237
Capital market, 142, 272–274
  and return on investment, 410
  and risk diversification, 148
  efficient, 281
  equilibrium on, 160
  regulation of, 411
Capital market line, 155, 156, 159
Capital user cost
  and capital asset pricing model, 296
Cash-flow underwriting, 202, 272
Catastrophic event
  and copulas, 243
Certainty equivalent, 42, 45, 46
  of option value, 228
Certainty line, 86, 361
Certainty preference, 39
Chebyshev's inequality, 263
Claim
  aggregate, 259, 260
Claim process, 255
  and loss distribution, 254
  uncertain amount of a claim, 254
  uncertain number of claims, 254
Coefficient of variation, 265
Combined ratio, 180
  in life insurance, 192
  in non-life insurance, 192
Comparative-static analysis, 99, 101, 230, 323, 326, 335, 344, 454
Competition, 215
  and market equilibrium, 328
  and no information exchange, 328
  and product innovation, 487
  prohibition of collusion, 487
Consequence, 39, 82, 83
  death, 31–33
  severity of, 31
  subjective valuation of, 37
  valuation of, 40, 82
Consumer's optimum, 88

Consumer surplus, 8, 364
  affected by regulation, 397
Contingent claim, 82
Contract
  pooling, 354, 356
  pooling, first period, 369
  renegotiation-proof, 366, 367, 369
  separating, 358, 359, 367, 425, 476
  separating, implementation of, 359
  separating, open to challenge, 360
  separating, second period, 370
  type of, 95
Convolution, 479
Copula
  and catastrophic events, 243
  and lower tail dependence, 245
  and upper tail dependence, 245
  and "value at risk", 245
  for combining distribution functions, 245
Corporate demand for insurance, 173
Correlation
  between catastrophic events, 243
  between deviations of benefits, 441, 442
  between returns, 150, 151, 153
  between returns on assets and liabilities, 284
  between risks in the tails, 243
  between risk units, 143
  between securities, 153
  between total and retained loss, 224
  coefficient, 151
  degree of coverage and losses paid, 316
  instantaneous, 280
  lack of, 152
  of catastrophic risks, 242
  of losses, 222
  with market portfolio, 156
Cost–benefit analysis, 64, 65
  human capital approach, 66
  willingness-to-pay approach, 66
Cost-effectiveness analysis, 65
Cost of capital, 161, 174, 175, 236
Cost–utility analysis, 65
Credibility, 130
Cumulative distribution function
  joint, 244
  partial, 244

D

Data envelopment analysis, 293, 305
  and economies of scale, 305
  and economies of scope, 293
  and efficiency measurement, 305
  applied to life insurance, 293

Subject Index 521

applied to non-life insurance, 293
compared to stochastic frontier analysis, 305
Deadweight loss
due to adverse selection, 365
Decision rule under uncertainty, 81
expected utility criterion, 82
Decision under uncertainty, 39, 40
Deductible, 95, 103
Demand for insurance, 15, 84, 90
and capital asset pricing model, 179–181
and degree of coverage, 98, 99
and genetic information, 473
and imperfect agency, 177
and one-person households, 484
and premium rate, 98
and prevention, 109
and quasi rents, 178
and risk aversion, 97
and sunk costs, 178
corporate, 175
degree of coverage, 105
dependent on price, 18
dependent on wealth, 16
excessive coverage, 107
full coverage, 96, 105
household, 80
income elasticity of, 20, 21
irreplaceable asset, 89–91
multiple risks, 104
no coverage, 94, 96, 108
partial coverage, 96
price elasticity, 20, 21, 255
Demand for reinsurance
and correlation between investments and claims, 231
and cost of insolvency, 228
and leverage, 230
and "longer-tail" lines, 231
and surplus, 229
and taxation, 229, 231
based on option pricing theory, 226, 227, 232, 233
empirical evidence, 232–234
funds generating factor, 228
Demographic change
aging of population, 482
and demand for insurance, 484
and life cycle decisions, 484
and life insurance, 485
and one-person households, 484
and social insurance, 485
challenge to insurance, 482
life expectancy, 482

Dependent agent, 213
Deregulation, 270
and capital asset pricing model, 481
and economies of scope, 481
challenge to insurance, 481
in the European Union, 391
in the United States, 400
Derivative, 164
Direct selling, 210
Direct writer, 208
insurance technology, 207
Disaster, 4
man-made, 4
natural, 4
Diversification, 141, 439

E
Economies of scale
and competition, 296
and definition of output, 297
and deregulation, 481
and elasticity of cost with respect to output, 296, 301
and measurement of cost, 296
and the law of large numbers, 295
and translog cost function, 300
critique of empirical evidence, 300, 303
defined, 295
empirical evidence, 298, 302
in life insurance, 298
in non-life insurance, 302
Economies of scope
and cost function, 289–291
and deregulation, 481
and subadditivity of cost, 290
empirical evidence, 292, 293
sources of, 290
stochastic, 293, 294
Efficiency frontier, 153, 237
and data envelopment analysis, 305
in (μ, σ)-space, 271, 272, 294, 388, 439, 441
in (μ_S, σ_S)-space, 306
insurance-specific, 439
with regard to cost, 304
Efficient frontier, 154, 159
Elasticity of cost with respect to output
and economies of scale, 301, 302
Elections, 430, 432
Enterprise Risk Management, 146
Equity capital, 279
Equity capital-to-asset ratio, 288
Exclusive agent, 209, 210
insurance technology, 207

Expected profit, 86
Expected rate of return, 163
Expected utility, 39, 40, 46, 82, 103
 decision criterion, 71
 difference from utility, 89
Expected utility criterion, 82
Expected utility theorem, 83
Expected utility theory
 Allais paradox, 118
 alternatives, 115
 Ellsberg paradox, 118
 independence axiom, 117
 weaknesses, 120
Expected value of return, 160
 and volatility, 149
Expense ratio, 192, 213, 299
 and premium growth, 201
 and regulation, 407
 and type of company, 407
Experience rating, 215, 366, 367
 and commitment, 366, 369
 and guaranteed renewability, 371, 373
 description of contracts, 370
 empirical evidence, 371
Externality, 62

F
Forward contract, 164, 166
Funding rate, 236, 238
Funds generating factor, 228, 273
Future contract, 166

G
Game situation, 1, 218
Genetic information, 217
 and cost of decision-making, 476
 and demand for insurance, 473
 and information lottery, 475
 and outcome of test, 475
 and risk type, 472
 and uninformed risk, 472
 as a challenge confronting insurance, 471
 private value of, 475
 public value of, 476
 value of, 475
Genetic risk type
 and outcome of test, 472
Globalization
 as a challenge confronting insurance, 468
Governance problems
 absence of powerful creditor, 412
Guaranteed renewability, 366
Guarantee fund, individual
 as an internalization measure, 388

Guarantee fund, joint
 and adverse selection, 390
 and moral hazard, 390
 as an internalization measure, 390
Guaranteeing claims with tax money
 and moral hazard, 390
 no internalization of externality, 390

H
Hedging, 165
 through stock option, 167
Herding, 132
Heuristics, 131
 availability, 132
 representative, 131
  empirical evidence, 136

I
Implicit Function Theorem, 345
Incentive compatibility
 and separating equilibrium, 358
Indemnity, 103
 optimal, 103
Independent agent, 207, 209, 210, 212, 288
Indifference curve, 86, 94, 98, 328, 446
 according to consumer type, 451
 according to risk type, 353, 354, 362, 471
 equation of, 85
 single crossing property of, 354, 357
 slope of, 85, 86, 96
Inheritance taxation, 348
Innovation, 163
 on capital market, 163
 specific, 163
 systematic, 163
Insurance, 2, 3
 and changes in science and technology, 470
 and cost of bankruptcy, 176
 and crowding out of owners, 176
 and economies of scope, 293
 and financial development, 17
 and genetic information, 471–476
 and globalization, 468–470
 and long-term care, 483
 and moral hazard, 319
 and new information technology, 477
 and protection of creditors, 176
 and risk diversification, 486
 and risk management, 486
 and risk selection, 468
 and saving, 54, 55
 and systematic risk, 174
 and tax burden, 176

Subject Index 523

and the capital market, 165, 166
as a Giffen good, 102
as a luxury good, 102
capital asset pricing model of, 272, 274, 275
challenges confronting, 467, 468, 479, 480
changes in legal norms, 479, 480
commercial, 180
contract law, 479, 480
contract law, permanent elements, 480
contribution to welfare, 89
corporate demand for, 468
corporate demand for and new information technology, 477
cost of, 296
definition of, 3
definition of output, 204
division of labor between private and social, 467–470, 476, 482, 483, 485
efficiency-enhancing, 486
employment, 7
fraud, 319
full coverage, 88
function of, 11, 14
household demand for, 470
importance of, 4, 7
individual demand for, 470
in life insurance, 7, 13, 189, 195
in non-life insurance, 7, 13, 189, 194, 275, 293
optimal coverage, 88
penetration, 18
portfolio, 266
premium income, 7
price of, 258
private vs social, 467, 469, 470, 476, 482, 483, 485
supply of, 253
technology, 63
transaction cost of, 175
world market, 15
Insurance benefit
uniform in pooling contract, 354
Insurance company, 7
and public authorities, 199
and risk aversion, 197, 199
and solvency, 198
and stakeholder approach, 198
as financial intermediary, 277
asset–liability ratio, 283–285
assets and liabilities, 278, 279
balance sheet, 190
deciding amount of coverage, 327
equity capital-to-asset ratio, 286

financial statement, 186
importance of objectives, 200–203
income statement, 192, 194–196
insurance reserves, 190
insurance technology, 203
investments, 14
losses paid as output, 297, 302
material, 407
mutual, 234, 300, 301
objectives of, 196, 198
offering pooling contract, 354
operational statement, 192, 194–196
output, 211
output defined, 203–205, 297
premium income as output, 297
premiums ceded to reinsurance, 193
price setting, 282, 283
property-liability, 282, 283
public, 300, 301, 407
risk management, 283
stock, 212, 234, 300, 301, 407
viewed as an option, 278
viewed as a security, 278
Insurance contract
pareto-optimal, 103
Insurance fraud
control of, 320
Insurance industry, 4, 8, 9
contribution to welfare, 8–10
importance of, 8
Insurance line, 87, 93, 110, 112, 356, 360, 363, 425, 427, 459, 473
according to risk type, 355
for a pooling contract, 355
slope of, 96
Insurance market, 7, 62
incompleteness of, 104
Insurance pricing
according to capital asset pricing model, 272, 282, 283
according to option pricing theory, 277, 282, 283
actual compared to rules, 282
financial, 270, 272
monopolistic, 402
recommended by insurance commissioners, 282
Insurance technology, 185, 201, 202
and competitiveness, 486
and risk selection, 215, 216
and "value at risk", 241
capital investment policy, 235, 236
control of moral hazard, 218, 219
cost of distribution, 209, 210, 212, 213

demand for reinsurance
  empirical evidence on reinsurance, 233
  distribution and regulation, 209, 213
  distribution channel, 206, 207
  empirical evidence on distribution, 210, 212
  empirical evidence on reinsurance, 232, 234
  experience rating of, 215
  incentives in distribution, 212, 213
  instruments, 205, 206
  new elements of, 241
  purchase of reinsurance coverage, 221, 222, 224
  scale economies, 210
  underwriting policy, 214
Insurer's risk, 264, 265, 267
  absolute, 265
  definition of, 267
  relative, 264, 265
Intergenerational moral hazard, 342

K
Kuhn-Tucker condition
  and boundary solution, 321
  and interior solution, 321

L
Lapse rate, 299, 300
Law of large numbers, 264, 266
Leverage, 198, 273
Liability, 62, 278
  stochastic disturbance, 280
  systematic change, 280
Life expectancy, 30
Linear partial information, 116, 128
Loading, 180, 267
  and probability of loss, 477
  due to moral hazard, 221
  fixed, 88, 93
  for administrative expense, 460
  for *ex-ante* moral hazard, 324
  for moral hazard, 460
  for safety, 93, 254, 256–258, 264, 267, 406
  for safety and new information technology, 478
  for safety in social insurance, 427
  price of insurance, 399
  proportional, 88, 93, 95, 97, 112
Long-term care insurance, 341, 345
  expenditures, 341
  private, 350
Loss, 4, 12, 100, 101
  accumulation of, 92

coefficient of correlation, 144
expected value of, 38, 79, 143
incurred but not reported (IBNR), 13, 266
mean, 264
negatively correlated, 107
of asset, 2
of utility, 92
paid, 180, 210
payment, 254, 279
positively correlated, 105, 107
present value of, 281
probability of, 100
standard deviation, 143
standard deviation of portfolio, 144
total, 143, 145, 256, 265, 273
uninsurable, 105
Loss distribution, 28, 256, 267
  amount of claim fixed, 256
  amount of claim uncertain, 259
  convolution of, 259, 260
  expected value of, 260
  irrelevance of, 281
  normal power approximation, 260, 262
  number of claims uncertain, 256
  Poisson distribution, 257
  skewness, 260, 262
  standard deviation, 262
  variance, 260
Loss ratio, 191
  and experience rating, 371
  and premium growth, 201, 371
  and U.S. regulation, 400
Lottery, 42

M
Market equilibrium
  and absence of Pareto-optimality, 329
  and degree of concentration, 357
  and experience rating, 330
  and planning horizon, 360
  Pareto-optimality of, 355
  pooling, 357, 362
  possibility of separating, 359
  reaction type, 358
  separating, 358, 426
  separating, conditions for, 358
  separating, high risks, 359
  separating, low risks, 359
  with ex-ante moral hazard, 327–329
  with information exchange, 330
  with public information, 355
Market failure, alleged
  due to insolvency, 386
  due to lack of co-operation, 384

due to lack of transparency, 384
due to ruinous competition, 384
due to uncertain premium calculation, 384
Market failure, possible
  due to adverse selection, 386
  due to moral hazard, 386
  in health insurance, 386
Market rate of return
  with correlation, 157
maxEmin criterion, 129
Maximization
  of equity, 146
  of expected profit, 196, 199
  of expected utility, 82, 197
  of growth, 197
  of profit, 196
Mean preserving spread, 70
Mean square error, 283
Minimum efficient scale, 302
  and definition of cost, 296
  and economic interpretation, 308
  and economies of scale, 296
  and size of market, 307
  in life insurance, 307
  in non-life insurance, 307
Monopoly, 10
Moral hazard, 13, 90, 217, 219, 220
  and adverse selection combined, 373
  and criminal activity, 319
  and duration of payment, 338
  and insurance fraud, 319
  and law of demand, 319
  and observability of behavior, 318
  and partial coverage, 374
  and principal-agent relationship, 318, 334
  and progressively increasing premium, 325, 329
  and rate of income replacement, 338
  and risk type, 374
  consequences of, 318
  definition and importance, 318
  definition of, 318
  demand-increasing effect, 319
  efficiency gains, 375
  ex-ante, 319
  ex-ante, affecting amount of loss, 325, 326
  ex-ante, affecting probability of loss, 320
  ex-ante and absence of Pareto-optimality, 330
  ex-ante and existence of equilibrium, 330
  ex-ante and individual behavior, 324
  ex-ante and insurance coverage, 323
  ex-ante and linear premium function, 327

ex-ante and maintenance of equilibrium, 328
ex-ante and market equilibrium, 327, 328
ex-ante and partial coverage, 325
ex-ante and probability of loss, 324
ex-ante and risk aversion, 324
ex-ante, empirical evidence, 331, 332
ex-post, 334
ex-post and observability, 337, 338
ex-post and search for physician, 335
ex-post, dynamic, 319, 453
ex-post, empirical evidence, 337, 338
ex-post of disability insurance, 334
ex-post, predicted, 336
ex-post, static, 319, 453, 456
loading for, 318
spillover from public insurance, 332
Myopia, 133

N

Nash equilibrium, 347
Nature, 129
New information technology
  and corporate demand for insurance, 477
Normal distribution, 260, 262
Normal power approximation, 260

O

One-person household
  and demand for insurance, 485
  and social insurance, 485
Optimism, 133
Optimization, 272
Option, 164
  Black-Scholes formula, 171
  call, 279–281, 284, 285
  call and put combined, 168
  call held by shareholders, 229
  European, 169
  insolvency put, 281
  intrinsic value of, 166
  premium, 168
  price, 277
  put, 168, 279–281, 284, 285
  put-call parity, 169, 172
  strike price, 166, 167
  time value, 166
  underlying assumptions, 172
  value of, 170
Option pricing theory, 236, 271, 284
  and total risk exposure, 281
  criticism of, 282
  crucial assumptions, 282

P

Perfect competition, 9
Perils, 28
   multitude of, 33
Poisson distribution, 257
Pooling
   in social insurance, 441
   of loss data, 487
   of risks, 441
Portfolio, 296
   composed of portfolios, 153, 439
   critique of, 240
   market, 159
   minimum variance, 152, 154, 238, 239, 440
   of claims against insurance, 439
   of equities, 148, 150
   of losses, 144
   of risks, 267, 290
   of securities, 152, 153, 158
   optimal composition, 238, 239
   optimization, 240
   planing horizon, 240
Portfolio optimization, 271
Portfolio theory, 271
Premium, 9, 86
   according to option price theory, 280
   and guaranteed renewability, 371
   and Herfindahl measure of concentration, 405
   and solvency level, 281
   ceded to reinsurance, 301
   experience rating of, 366, 455
   fair, 79, 88, 254, 268, 355, 358
   function, 95
   growth and experience rating, 372
   highballing of, 371
   income, 280, 299, 301
   linear given competition, 328
   lowballing of, 371
   marginally fair, 87, 88, 90, 97, 106, 324
   not fair, 93
   pooling, 354
   pure, 268
   rate, 19, 21, 100, 102, 223, 297
   rebate for no claims, 367, 369, 455
   regulated, 387, 405, 406
   transfer to low risks, 367
   volume, 9, 19, 21
   with ex-ante moral hazard, 324
   without exchange of information, 327
Premium calculation
   traditional, 254, 257
Premium function, 221
Premium growth, 213
Premium income, 29
Premium principle, 267, 268
   alternatives to, 271
   arbitrariness of, 269
   based on the net premium, 268
   critique of, 270
   defined implicitly, 269
   equivalence principle, 268
   Esscher principle, 270
   expected value of, 268
   exponential, 269
   loss function, 269
   of zero utility, 269
   standard deviation, 269
   variance, 268
Prevention, 3, 219, 220
   affecting amount of loss, 326
   and ex-ante moral hazard, 320
   and full coverage, 326
   and full insurance coverage, 322
   and genetic information, 471
   and probability of loss, 320, 471–474
   and risk type, 471–474
   and risk utility function, 322
   decreasing marginal effectiveness of, 320
   marginal benefit, 373, 474
   marginal cost, 322, 373, 474
   marginal cost according to risk type, 373
   marginal effectiveness, 373
   marginal return, 322
   observability, 218
   zero in market equilibrium, 328
Preventive effort
   and insurance, 109, 110
   in response to insurance coverage, 323
   loss-reducing, 110
   observability of, 111
   optimal, 110
Price
   of risk bearing, 268, 275
Price elasticity
   of demand for insurance, 365, 399
   of demand for medical care, 455
"price equal to marginal cost" condition, 355
Price of insurance, 9
   and moral hazard, 401
   and no-fault regulation, 401
   and regulation, 401
Principal-agent relationship, 285
   and moral hazard, 218
   applied to distribution, 208
   applied to distribution channels, 208
Private insurance

claims as components of portfolio, 438
funded scheme for old age provision, 436
Probability, 2
  conditional, 104
  density function, 81–83
  estimation error, 36
  estimation of, 36
  of death, 66, 257, 471
  of illness, 471
  of insolvency, 263, 284, 414
  of loss, 81, 86, 101, 331, 425, 459
  of loss, according to risk type, 354, 360
  of loss and ex-ante moral hazard, 332
  of loss and moral hazard, 320
  of loss, average, 354
  of occurrence, 29, 31, 39, 80, 84, 113
  of retirement at 65, 446
  of retirement at a given age, 446
  of ruin, 256, 257, 266, 268
  of solvency, 258, 261, 262
  of successful search, 335, 337
  relative frequency, 30
  subjective, 2, 31
  unknown, 2
Probability of occurrence, 43, 48
  perception of, 34
Probability theory, 263
Probit estimation, 331
Propensity to save, 346
Prospect theory, 115, 121
  cumulative, 125
  decision weights, 124, 125
  editing phase, 122
  empirical evidence, 133
  evaluation phase, 123
  loss aversion, 124
  reference point, 124
  value function, 124
Prudence, 54
Public goods, 69

Q
Quality-adjusted life year, 65

R
Random variable, 172, 263, 267
  combination of, 244
Rank Dependent Utility, 115, 125
  empirical evidence, 134
  weighting function, 126
Rate of coinsurance, 95
Rate of return
  "biological" of social insurance, 437, 440, 441, 448

equilibrium, 272
expected, 273
  of claims against private insurance, 438
  of claims against social insurance, 438
  of funded scheme, 437
  of insurance company, 273, 276
  of pay-as-you-go scheme, 435
  of private insurance, 441
  on capital investment, 273
  on equity capital, 273
  on savings, 448
  on underwriting, 273, 274, 282, 283
  risk-adjusted, 272, 273, 275, 296
  risk-free, 155
  volatility of, claims against insurance, 438
Reaction curve, 348
Regret theory, 116, 130
Regulation, 10, 20, 177
  and assignment pool, 402
  and bureaucracy, 396
  and competition, 487
  and financial crisis 2007–2009, 409, 410
  and lobbying expenditure, 396
  and loss ratio, 400
  and non-price competition, 397
  and product improvement, 397
  and re-regulation, 481
  and risk-adjustment, 386
  and separation of insurance lines, 486
  and small insurance companies, 396
  as substitute of powerful creditor, 386
  Basel II agreement, 412
  capture theory of, 392
  capture theory of, hypotheses, 395
  capture theory of, weaknesses, 392
  carry-over from banks, 414
  changes in demand for, 393
  changes in supply of, 393, 412
  common standard, 415
  company action level, 404
  competing theories of, 391
  conflict of interest, 406
  conflict of objectives, 391, 403
  convergence of banking and insurance, 411
  demand by consumers and industry, 393
  demand due to crises, 395
  demand for, 410
  designed to avoid insolvency, 384, 406
  designed to mitigate external costs, 384, 388
  difference between banks and insurers, 386
  differences between countries, 390

dilemma of, 387
double trigger approach, 414
effect on consumer surplus, 397
effect on efficiency, 409
effect on expense ratio, 407
effect on life insurance, 404
effect on marketing expenses, 407
effect on premiums, 404
effect on purchase of reinsurance, 407, 408
effects in United States, 397
effects of, 396
empirical evidence, Europe, 404
empirical evidence, Germany, 406, 407
empirical evidence, United States, 399, 401–403
evaluation of past performance, 414
formal type, countries, 390
formal type of, 384, 388
guarantee fund, 414
implications of convergence, 411
influence of small interest groups, 395
insolvency and external effects, 384
instruments of, 387
intensity, exogenous, 396
intensity in equilibrium, 393
intermediate, United States, 390
justification of, 383, 386
market for, 393
market for, applied to insurance, 393
market for regulation model, 392, 413
market theory, hypotheses, 395
material type, 384, 388, 406, 411
material type and consumer protection, 387
material type, countries, 390
material type, justification, 384
minimum required capital, 413
new authorities, 415
objectives of, 384
of capital investment, 388
of capital market products, 411
of insolvency level, 414
of insolvency risk, 486
of life insurance, 487
of non-life insurance, 487
of premiums, 215, 276, 387, 403
of premium-to-surplus ratio, 390
of products, 387
of rate of return, 197
of solvency, 198
of solvency level, 285, 288, 289, 386, 404
public interest theory, 391, 395, 413
recent initiatives, 413, 414
regulatory arbitrage, 415

solvency capital, layers, 413
solvency II, 412, 413
stringency of, 400–402
supply by government, 393
systematic risk as justification, 412
theories of, 391
theories of, using risk-based capital, 388, 395, 403, 404
Reinsurance, 7
aggregate
  excess contract, 225
and surplus, 241
and "value at risk", 241
as an internalization measure, 388
automatic, 224
facultative, 224
function of, 222
mandatory, 224
per-occurrence excess contract, 226
per-risk excess contract, 225
proportional, 224, 227
retention limit, 225
Relational moral hazard, 345–348, 352
empirical evidence, 349
long-term care insurance, 341
Return
value of, 149
Return on security
non-systematic component, 157
systematic component, 157
Return on surplus
expected, 294
volatility, 294
Risk, 1, 2, 4, 13, 27
allocation of, 12
and chance, 28
as a random variable, 28
catastrophic, 4
definition of, 28
diversification of, 272
downside, 55
idiosyncratic, 158
importance of, 33
insolvency, 255
in unemployment, 460
market, 158
measurement of, 29
minimization of, 146
non-systematic, 158
of a portfolio, 149
of future inflation, 439
of insolvency, 256, 486
of insolvency valued, 281, 284
perception of, 34

political, 439
pooling of, 267, 272
related to underwriting, 222
systematic, 158, 159, 174
systematic, and regulation, 412
systematic, banks vs insurers, 412
transfer and globalization, 468
transfer to reinsurance, 223
underwriting and new information technology, 478
Risk aversion, 34, 39, 42, 46, 81, 439
and age, 57
and ex-ante moral hazard, 324, 332
and gender, 58
and initial wealth, 57
and prevention, 111
and "value at risk", 243
coefficient of, 49
coefficient of absolute risk aversion, 50, 99, 269, 336
coefficient of partial risk aversion, 53
coefficient of relative risk aversion, 51, 52, 59, 61
constant absolute risk aversion, 59, 100, 101
constant relative risk aversion, 56, 57, 59
decreasing absolute risk aversion, 57, 59, 100, 101
decreasing relative risk aversion, 57, 59
increasing absolute risk aversion, 101, 102
increasing relative risk aversion, 57, 59
of insurance company, 286
of investors, 155
origin of, 38
Risk-based capital, 272
and "value at risk", 243
correlation between risk categories, 404
diversification effects, 404
risk categories, 404
Risk diversification
and insurance, 486
internal and new information technology, 479
Risk-free alternative, 42
Risk management, 7, 141, 161
etiological measure, 61
financial, 148
instruments of, 63, 79
internal diversification, 142
of information, 477
of information-driven production processes, 477
of the insurance company, 283
palliative measure, 61

risk prevention, 61
risk reduction, 61
risk transfer, 62
Risk neutrality, 86, 103, 111
and ex-post moral hazard, 336
Risk portfolio
and globalization, 469
Risk premium, 47–50
relative, 52
Risk selection, 215, 216
and genetic information, 217
and product innovation, 217
Risk type, 198, 215, 216
Risk utility function, 39, 41, 42, 55, 59, 83, 119, 269, 321, 342, 368
and prevention, 322
concavity of, 39, 41
construction of, 43, 45
curvature of, 48, 56
illustrated, 41
state-dependent, 89
Risky prospect, 80, 82, 84

S

Safety scale, 33
Saint Petersburg game, 39
Saint Petersburg paradox, 39
Scale economies
and size of market, 307
Securitization
and financial crisis 2007–09, 409
and subprime mortgages, 409
Security, 62
equilibrium portfolio of, 156
risk-free, 155
Security market line, 160, 272
Self-insurance, 111
Shareholders, 148
Share price, 148
Simplification, 133
Single crossing property, 472
Social insurance
absence of currency diversification, 440
absence of experience rating, 428
alleviating transaction costs, 426
and adverse selection, 375
and age at retirement, 445
and decisions over the life cycle, 452
and delegates to parliament, 430
and life cycle decisions, 470
and life expectancy, 451
and life expectancy, empirical evidence, 452
and moral hazard, 375, 443, 455, 463

and moral hazard as external effect, 444, 456
and popularity of government, 430
and private saving, empirical evidence, 449
and public administration, 433
and reelection, empirical evidence, 430
and retirement as an option, 446
and social policy, 421
and systematic redistribution, 429
as an efficiency-enhancing institution, 422
as an extreme outcome of regulation, 430
as an instrument in the hands of politicians and bureaucrats, 422, 429
as a public good, 423, 424
"biological" rate of interest, 437
"biological" rate of return, 449
claims as components of portfolio, 438
competitive advantage, 427
contributions, 450
correcting adverse selection, 424
correcting excessive time preference, 422
correcting free riding on altruism, 423, 424
differences between countries, 421
different from private insurance, 443, 444
effect on healthcare expenditure, 453–455
effect on labor supply, 445, 447
effect on number of children, 449, 450
effect on private saving, 447, 448
effects of financing, 444
efficiency comparison, qualifications, 437
efficiency-enhancing institution correcting market failure, 422
evidence from U.S. health insurance experiment, 455
existence of, 422, 424, 428, 430, 433
experience rating of, 455
favoring high risks, 427
growth of, 422, 424, 428, 430, 433
health insurance, 348, 453–455
importance of, 443
interplay with private insurance, 442
limited by moral hazard, 428
loading for administrative expense, 460
loading for moral hazard, 460
macroeconomic impact of, 443
mandatory for poor, 424
neglecting risk preferences, 427
optimal amount, empirical evidence, 461–463
optimal amount of, 459–463
paradox of social security, 437
partial mandatory coverage, 425
pay-as-you-go scheme, 441

pension vs. capital benefit, 451
permitting fair premium, 427
potential for Pareto improvement, 426
private vs. social, 419
provision for old age, 444, 445, 448
provision for old age, efficiency, 435
rate of coinsurance, 453–455
rate of return, funded scheme, 437
rate of return, pay-as-you-go scheme, 435, 436
requirements for efficiency, 435, 437, 441
substitution with private insurance, 421
taxation of benefits, 460
testing for efficiency, 441, 442
uniformity of incentives, 443
used to secure reelection, 430
Social spending
  inertia, 431
Social vs private insurance, 421, 486
Social welfare
  and defense spending, 432
Solvency II, 272
  and "value at risk", 243
Solvency level, 257
Speculation, 165
Statistical human life, 66, 69
Status quo bias, 132
Stochastic dominance
  first-degree, 69, 70
  second-degree, 71, 72
Stochastic frontier analysis
  and economics of scale, 304
  and economics of scope, 305
Stochastic process, 278
Subcertainty, 125
Supply of insurance and new information technology, 477–479
Surplus, 235, 254, 266, 279, 284
  initial, 256–258
  rate of return on, 235, 236
  returns, 289
  standard deviation of, 289
  standard deviation of returns, 287, 288
  variance of returns, 284, 287
  volatility of, 284
Survival curve, 30
Systematic risk
  banking and insurance, 412
  contributing factors, 412
  criteria for, 412
  difference between banks and insurers, 412

Subject Index

T

Temperance, 56
Transfer of policyholders to a competing insurer
    as an internalization measure, 390
    effect on surplus, 390
Translog cost function, 292, 300, 302

U

Uncertainty, 1, 2
    degrees of, 1
Unemployment
    impact of social insurance, empirical evidence, 458
    impact of social insurance on inflow, 457, 458
    impact of social insurance on outflow, 457, 458
    impact of social insurance on stock, 457, 458
    moral hazard effect of social insurance, 459, 463
Unemployment benefits, 457
Unemployment insurance, 457, 458
    optimal amount, 459
Utility function, 80

V

"Value at risk"
    and catastrophic events, 241
    and copula, 245
    and density function, 242
    and normal power approximation, 242
    and reinsurance, 241
    and risk aversion, 243
    and skewness, 242
    and stochastic dominance, 243
    of surplus, 241
Vector autoregression, 238
Volatility, 149
Vulnerability, 5, 7

W

Wealth
    changes in, 81
    expected value of, 80
    final, 81, 98
    insurable, 98, 100, 101
    marginal utility of, 86, 89–92
    optimal level, 90
    risky vs. certain, 92
Wiener process, 280
Willingness to pay
    for certainty, 43, 46, 49, 269
    for certainty, labor market, 460
    for income security, 462, 463
    for insurance coverage, 48, 459

GPSR Compliance

The European Union's (EU) General Product Safety Regulation (GPSR) is a set of rules that requires consumer products to be safe and our obligations to ensure this.

If you have any concerns about our products, you can contact us on

ProductSafety@springernature.com

In case Publisher is established outside the EU, the EU authorized representative is:

Springer Nature Customer Service Center GmbH
Europaplatz 3
69115 Heidelberg, Germany

www.ingramcontent.com/pod-product-compliance
Ingram Content Group UK Ltd.
Pitfield, Milton Keynes, MK11 3LW, UK
UKHW022120230426

12048UKWH00010BA/633